Handbook of Global Navigation Satellite Systems

Handbook of Global Navigation Satellite Systems

Edited by **Margaret Ziegler**

CLANRYE INTERNATIONAL

New Jersey

Published by Clanrye International,
55 Van Reypen Street,
Jersey City, NJ 07306, USA
www.clanryeinternational.com

Handbook of Global Navigation Satellite Systems
Edited by Margaret Ziegler

International Standard Book Number: 978-1-63240-272-1 (Hardback)

Printed in China.

Contents

Preface

It is often said that books are a boon to mankind. They document every progress and pass on the knowledge from one generation to the other. They play a crucial role in our lives. Thus I was both excited and nervous while editing this book. I was pleased by the thought of being able to make a mark but I was also nervous to do it right because the future of students depends upon it. Hence, I took a few months to research further into the discipline, revise my knowledge and also explore some more aspects. Post this process, I begun with the editing of this book.

Lucid descriptions of Global Navigation Satellite Systems (GNSS) are provided in this thorough and reader-friendly book. GNSS plays a significant role in high accuracy navigation, positioning, and technical questions connected to accurate positioning. It is an extremely accurate, constant, all-weather and real-time method. This book presents current outcomes and innovations in GNSS technology, organization, signal, receiver, technique and sources of error. Its various functions in hybrid positioning, multi-sensor incorporation, height system, NRTK, etc. have also been evaluated and analyzed. This book intends to help experts and students in knowing the topic in a better way.

I thank my publisher with all my heart for considering me worthy of this unparalleled opportunity and for showing unwavering faith in my skills. I would also like to thank the editorial team who worked closely with me at every step and contributed immensely towards the successful completion of this book. Last but not the least, I wish to thank my friends and colleagues for their support.

Editor

Part 1

GNSS Signals and System

1

High Sensitivity Techniques for GNSS Signal Acquisition

Fabio Dovis[1] and Tung Hai Ta[2]
[1]*Politecnico di Torino*
[2]*Hanoi University of Science and Technology*
[1]*Italy*
[2]*Vietnam*

1. Introduction

The requirements of location based and emergency caller localization services spurred by the E-911 mandate (USA) and the E-112 initiative (EU) have generated the demand for the availability of Global Navigation Satellite Systems (GNSS) in harsh environments like indoors, urban canyons or forests where low power signals dominate. This fact has pushed the development of High Sensitivity (HS) receivers

To produce positioning and timing information, a conventional GNSS receiver must go through three main stages: code synchronization; navigation data demodulation; and Position, Velocity and Time (PVT) computation. Code synchronization is in charge of determining the satellites in view, estimating the transmission code epoch and Doppler shift. This stage is usually divided into code acquisition and tracking. The former reduces the code epoch and Doppler shift uncertainties to limited intervals while the latter performs continuous fine delay estimation. In particular, code acquisition can be very critical because it is the first operation performed by the receiver. This is the reason for lots of endeavors having been invested to improve the robustness of the acquisition process toward the HS objective.

Basically, the extension of the coherent integration time is the optimal strategy for improving the acquisition sensitivity in a processing gain sense. However, there are several limitations to the extension of the coherent integration time T_{int}. The presence of data-bit transitions, as the 50bps in the present GPS Coarse-Acquisition (C/A) service, modulating the ranging code is the most impacting. In fact, each transition introduces a sign reversal in successive correlation blocks, such that their coherent accumulation leads to the potential loss of the correlation peak. Therefore, the availability of an external-aiding source is crucial to extend T_{int} to be larger than the data bit duration T_b (e.g. for GPS L1 C/A, $T_b = 20$ ms). This approach is referred as the aided (or assisted) signal acquisition, and it is a part of the Assisted GNSS (A-GNSS) positioning method defined by different standardization bodies (3GPP, 2008a;b; OMA, 2007).

However, without any external-aiding source, the acquisition stage can use the techniques so-called post-correlation combination to improve its sensitivity. In general, there are 3 post-correlation combination techniques, namely: coherent, non-coherent and differential combination. In fact, the coherent combination technique is equivalent to the T_{int} extension with the advantage that in this stand-alone scenario $T_{int} \leq T_b$. The squaring loss (Choi et al., 2002) caused by the non-coherent combination makes this technique less competitive than the others. However, its simplicity and moderate complexity make it suitable for conventional GNSS receivers. Among the three techniques, the differential combination can be considered as a solution trading-off sensitivity and complexity of an acquisition stage (Schmid & Neubauer, 2004; Zarrabizadeh & Sousa, 1997). As an expanded view of the conventional differential combination technique, generalized differential combination is introduced for further sensitivity improvement (Corazza & Pedone, 2007; Shanmugam et al., 2007; Ta et al., 2012).

In addition, modern GNSSes broadcast new civil signals on different frequency bands. Moreover, these new signals are composed of two channels, namely data and pilot (data-less) channels (e.g. Galileo E1 OS, E5, E6; GPS L5, L2C, L1C). These facts yield another approach, usually named *channel combining acquisition* (Gernot et al., 2008; Mattos, 2005; Ta et al., 2010) able to fully exploit the potential of modern navigation signals for sake of sensitivity improvement.

This book chapter strives to identify the issues related to HS signal acquisition and also to introduce in details possible approaches to solve such problems. The remainder of the chapter is organized as follows. Section 2 presents fundamentals of signal acquisition including the common representation of the received signal, the conventional acquisition process. Furthermore, definition of the the performance parameters, in terms of detection probabilities and mean acquisition time are provided. HS acquisition issues and general solutions, namely stand-alone, external-aiding and channel combining approaches, are introduced in Section 3. In Section 4, the stand-alone generalized differential combination technique is presented together with its application to GPS L2C signal in order to show the advantages of such a technique. Section 5 focuses on introducing a test-bed architecture as an example of the external-aiding signal acquisition. The channel combining approach via joint data/pilot signal acquisition strategies for Galileo E1 OS signal is introduced in Section 6. Eventually, some concluding remarks are drawn.

2. Fundamentals of signal acquisition

2.1 Received signal representation

The received signal after the Analog to Digital Converter in a Direct Sequence Code Division Multiple Access (DS-CDMA) GNSS system can be represented as

$$r[n] = \sqrt{2C}d[n]c[n+\tau]\cos(2\pi(f_{IF}+f_D)nT_S+\varphi)+n_W[n] \tag{1}$$

where C is the carrier power (W); $d[n]$ is the navigation data; $c[n]$ is the spreading code, f_{IF}, f_D denote the Intermediate Frequency (IF) and Doppler shift (Hz) respectively; $T_S = 1/F_S$ stands for the sampling period (s) (F_S is the sampling frequency (Hz)); φ is the initial carrier phase (rad); τ is the initial code delay (samples) ; and n_W is the Additive White Gaussian Noise (AWGN) with zero mean ($\mu = 0$) and variance σ_n^2 ($n_W \sim \mathcal{N}(0, \sigma_n^2)$).

In fact, most of the current and foreseen signals of GNSSes use either BPSK or BOC modulations (Ta, 2010). For these modulations, $c[n]$ has the representation as follows:

- BPSK(f_c):

$$c(t) = \sum_{k=-\infty}^{+\infty} q_k \Pi(t - kT_c) \tag{2}$$

where Π is the rectangular function; q_k is the PRN code. Because of the properties of the PRN code, q_k is a periodic sequence with the period N chips, q_k can be rewriten as $q_k = q_{mod(k,N)}$, then the digital version of (2) is

$$c[n] = c(nT_S) = \sum_{k=-\infty}^{+\infty} q_{mod(k,N)} \Pi\left(nT_S - kT_c\right) \tag{3}$$

being T_c, and $f_c = 1/T_c$ the chip duration (s) and chipping rate (chip per second - cps) respectively.

- BOC(f_s, f_c): Similarly,

$$c[n] = \sum_{k=-\infty}^{\infty} q_{mod(k,N)} s_{mod(k,a/2)} \Pi\left(n - kT_c\right) \tag{4}$$

with $s_{mod(k,a/2)} \in \{-1, 1\}$is the sub-carrier with the frequency f_s and $a = 2\frac{f_s}{f_c}$. Usually in GNSS f_s is a multiple of f_c (i.e. $a/2$ is an integer value) and both the values of f_c and f_s are normalized by 1.023 MHz; for instance BPSK(5) and BOC(10,5) mean $f_c = 5 \times 1.023$ MHz and $f_s = 10 \times 1.023$ MHz. The subcarrier $s[n]$ can be sine-phased, $s[n] = \text{sgn}[\sin(2\pi f_s n T_S)]$; or cosine-phased, $s[n] = \text{sgn}[\cos(2\pi f_s n T_S)]$ with sgn(x) being the signum function of x.

2.2 Conventional acquisition process

As introduced in (Kaplan, 2005), the conventional acquisition process (see Fig. 1) strives to determine the presence of a desired signal defined by PRN code (c), code delay (τ) and Doppler offset (f_D) in the incoming signal. The uncertainty regions of (c, τ, f_D) form a signal search-space, each cell ($\hat{c}, \hat{\tau}, \hat{f}_D$) of which is used to locally generate an equivalent tentative signal, see Fig. 2(a). The acquisition process correlates the incoming signal ($r[n]$) with the tentative signal ($\hat{r}[n]$) to measure the similarity between the two signals.

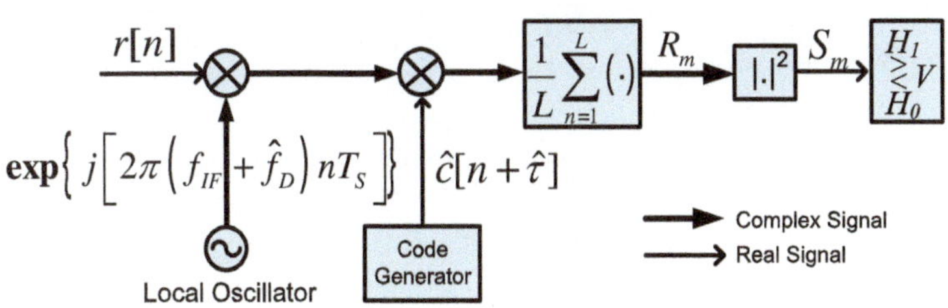

Fig. 1. Conventional signal acquisition architecture

It is well known that there are several general approaches to code acquisition of a GNSS signals. The basic functional operation is a correlation between a local replica of the code and the incoming signal as depicted in Fig. 1, where a serial approach scheme is reported. Time (or frequency) parallel acquisition approaches, are often efficiently implemented by using Fast Fourier Transform algorithms (Tsui, 2005).

In general, the complex-valued correlation R, which is also referred as Cross Ambiguity Function (CAF), between the incoming and the local generated signals is:

$$R_m = \frac{1}{L} \sum_{n=(m-1)L}^{mL} \{r[n]\hat{c}[n+\hat{\tau}]e^{j(2\pi(f_{IF}+\hat{f}_{D_m}))nT_S}\} \triangleq s_m + w_m \tag{5}$$

where m stands for the index of the coherent integration interval $[(m-1)L, mL]$, $\lfloor L = T_{int}F_S \rfloor$ denotes the coherent integration time T_{int} (s) in samples; s_m, w_m are the signal and the noise components respectively, and (Holmes, 2007)

$$\begin{cases} s_m = \sqrt{2C}\mathcal{R}[\theta]\mathrm{sinc}(\triangle\overline{f}_{d_m}T_{int})e^{j(\pi\triangle\overline{f}_{d_k}T_{int}+\phi_m)} \triangleq G_m e^{j\Phi_m} \\ w_m = \frac{1}{L}\sum_{n=(m-1)L}^{mL} n_W[n]\hat{c}[n+\hat{\tau}]e^{j[2\pi(f_{IF}+\hat{f}_{D_m})nT_S]} \end{cases} \tag{6}$$

where $\theta = \tau - \hat{\tau}$ is the difference between actual and estimated code delays and $\triangle\overline{f}_{d_m} = f_D - \hat{f}_{D_m}$ is the difference between Doppler shifts during the interval m, as depicted in Fig. 2(a). $(\phi_m = 2\pi\triangle\overline{f}_{d_{m-1}}T_{int} + \phi_{m-1})$ is the phase mismatch at the end of the m-th interval, and $\mathcal{R}[\theta]$ is the cross-correlation function between the incoming signal and the local PRN codes. In an ideal, noiseless case, such cross-corelation would results to be the autocorrelation function of the two PRNs that can be written for a BPSK signal as

$$\mathcal{R}[\theta] = -\frac{1}{L} + \frac{L+1}{L}\Lambda_0\left(\frac{\theta}{\lambda}\right) \otimes \sum_{m=-\infty}^{\infty} \delta[\theta + mL] \tag{7}$$

and for a BOC signal as (Betz, 2001):

$$\mathcal{R}[\theta] = \left[\Lambda_0\left(\frac{\theta}{\frac{\lambda}{a}}\right) + \sum_{l=1}^{a-1}(-1)^{|l|}\frac{a-|l|}{a}\Lambda_{l\frac{\lambda}{a}}\left(\frac{\theta}{\frac{\lambda}{2}}\right) + \sum_{l=-(a-1)}^{-1}(-1)^{|l|}\frac{a-|l|}{a}\Lambda_{l\frac{\lambda}{a}}\left(\frac{\theta}{\frac{\lambda}{2}}\right)\right] \otimes \sum_{m=-\infty}^{\infty} \delta[\theta + mL] \tag{8}$$

where λ is the samples per chip, and Λ is the triangle function of x, centered at z, with a base width of y

$$\Lambda_z\left(\frac{x}{y}\right) = \begin{cases} \left(1 - \frac{|x|}{y}\right) & z \leq |x| \leq z + y - 1 \\ 0 & \text{elsewhere} \end{cases} \tag{9}$$

From (7) and (8), it can be noted that, when observed over the interval $[-T_c, T_c]$ around the main peak, the autocorrelation function of BPSK signal has the main peak only, whilst the BOC has $(2a-1)$ peaks. Fig. 2(b) shows the theoretical autocorrelation functions of a BPSK(1) and a BOC(1,1). As seen from Fig. 2(b) and Fig. 2(c), the estimation residuals $(\theta, \triangle f_d)$ cause correlation loss on both dimensions. To limit this loss, the cell size $(\triangle\tau, \triangle f_D)$ must be chosen carefully taking into account also the pull-in range of the tracking stage. In general, for BPSK

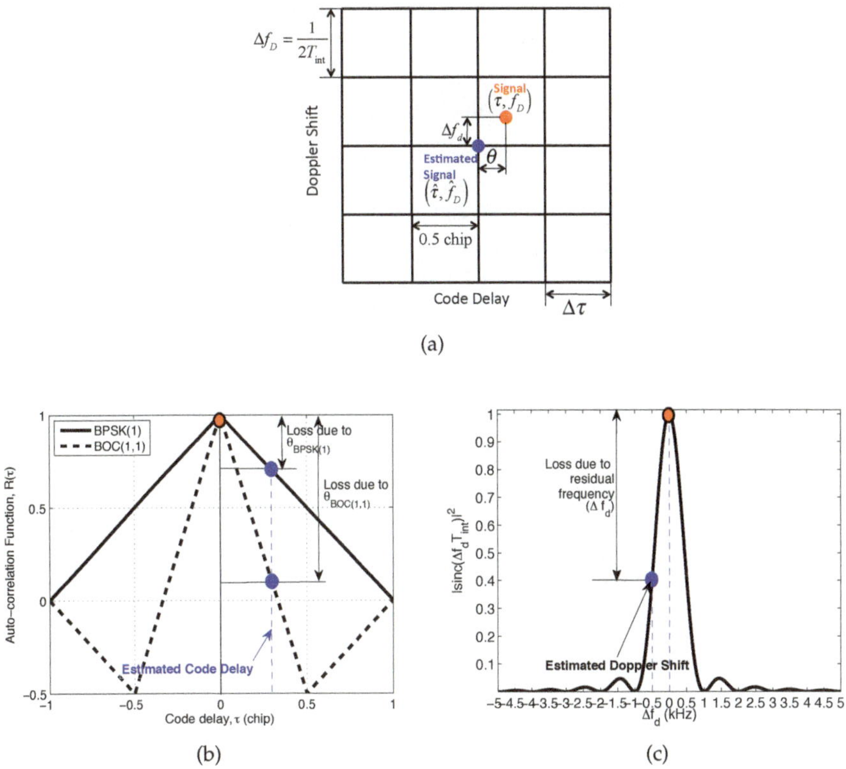

Fig. 2. (a) Acquisition search-space; (b) Auto-correlation functions of BPSK(1) and BOC(1,1); (c) Sinc function

signal $\triangle\tau_{BPSK} = 0.5$ chip. However, for BOC signal, due to the appearance of side-peaks, $\triangle\tau$ is chosen so that the tracking stage can avoid to lock to the side-peaks. For BOC(1,1), in order to achieve the same average correlation loss as for a BPSK signal, $\triangle\tau_{BOC(1,1)} = 0.16$ chip (Wilde et al., 2006). As for Doppler shift dimension, $\triangle f_D = \frac{2}{3T_{int}}$ as in (Kaplan, 2005) or $\triangle f_D = \frac{1}{2T_{int}}$ as in (Misra & Enge, 2006) are often chosen concerning the trade-off between complexity and sensitivity.

2.3 Acquisition performance parameters

When dealing with real signals, the incoming code is affected by several factors such as propagation distortion and noise, thus resulting in a distorted correlation function. In order to achieve an optimal detection process, the Neyman-Pearson likelihood criterion is used. In fact, the magnitude $S_m = |R_m|^2$ of each complex correlator output can be modeled as a random variable with statistical features. Thus, S_m is compared with a predetermined threshold (V) in order to decide which hypothesis between H_0 ($S_m < V$) and H_1 ($S_m > V$) is true, where H_0 and H_1 respectively represent the absence or presence of the desired peak. Once the decision is taken, the parameters $\hat{f}_D, \hat{\tau}$ are taken. Such values must belong to the pull-in range of the tracking stage of the receiver.

2.3.1 Statistical characterization of the detection process

As previously remarked, the signal acquisition can be seen as a statistical process, and the value taken by the correlator output for each bin of the search space can be modeled as a random variable both when the peak is absent (i.e. H_0) or present (i.e. H_1). In each case the random variable is characterized by a probability density function (pdf). Fig. 3(a) shows the signal trial hypothesis test decision when both pdfs are drawn. The threshold

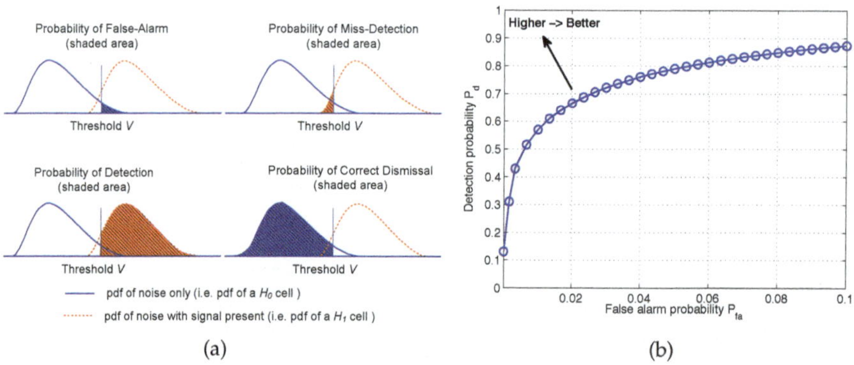

Fig. 3. (a) Possible pdfs of a hypothesis test; (b) Receive Operating Characteristic (ROC) curve

V is pre-determined based on the requirements of: (i) false-alarm probability (P_{fa}), e.g. $P_{fa} = 10^{-3}$, or (ii) mean acquisition time ($\overline{T}_A$), e.g. $\overline{T}_A$ is minimum.

For a specific value of V, there are four possible outcomes as shown in Fig. 3(a). Each outcome is associated with a probability which can be computed by an appropriate integration as (Kaplan, 2005):

- Probability of false-alarm (P_{fa}):

$$P_{fa} = \int_V^{+\infty} f(s|H_0)ds \tag{10}$$

- Probability of correct dismissal (P_{cd}):

$$P_{cd} = 1 - P_{fa} \tag{11}$$

- Probability of detection (P_d):

$$P_d = \int_V^{+\infty} f(s|H_1)ds \tag{12}$$

- Probability of miss-detection (P_{md}):

$$P_{md} = 1 - P_d \tag{13}$$

As described, once P_{fa} and P_d are known, the others can be easily computed. These two probabilities are also used to plot the Receiver Operational Characteristic (ROC) curve (see Fig. 3(b)) depicting the behaviors of the P_{fa} versus P_d for different values of V. This curve is useful for performance comparison among different acquisition strategies.

2.3.2 Peak-to-floor ratios

Theoretical assessment of acquisition performance is not always possible, since it requires also the knowledge of the pdf of the decision variables. For such a reason Monte-Carlo simulation are often employed. In such a case, in order to have suitable confidence in the results, each simulated value of the ROC curve (as in in Fig. 3(b)) has to be the result of the average of million of simulated cases. Therefore, if the sensitivity of a single acquisition scheme in different conditions has to be assessed, it is also useful to consider easy-to-compute parameters, named peak-to-floor ratios, $(\alpha_{max}, \alpha_{mean})$. They are defined as:

$$\alpha_{max} = \frac{\left|S_{peak}\right|^2}{\max\left|S_{floor}\right|^2}; \; \alpha_{mean} = \frac{\left|S_{peak}\right|^2}{E\left[\left|S_{floor}\right|^2\right]} \tag{14}$$

where S_{peak} is the maximum of the CAF magnitude and S_{floor} is the floor of the CAF magnitude (i.e outside the main correlation peak which is 2 chips wide). These metrics highlight the overall trend of post-correlation Signal-to-Noise Ratio (SNR), avoiding time-consuming calculations or simulations. Anyhow, it is important to point out that the comparison of different acquisition schemes based on the peak-to-floor ratios may be not fair if their decision variables show different statistical properties (Ta et al., 2008).

2.3.3 Mean acquisition time

Let us consider a search-space with N_c columns and N_f rows as in Fig. 2(a), and denote A as a successful detection of a serial acquisition engine (Fig. 1) after some miss-detections and false-alarms. The mean duration from the beginning of the process to the instant when A happens is named mean acquisition time, and can be written as (Park et al., 2002)

$$\overline{T}_A = (N_c N_f - 1)(T_d + T_{fa}P_{fa})\frac{2 - P_d}{2P_d} + \frac{T_d}{P_d} \tag{15}$$

with T_d and T_{fa} being the dwell time and the penalty time respectively.

Equation (15) shows that $\overline{T}_A$ depends on the values of :

- The false-alarm (P_{fa}) and detection (P_d) probabilities at a single cell.
- The search space size $N_c \times N_f$
- The penalty time T_{fa} and the dwell time T_d. In fact, T_{fa} is represented through T_d and the penalty coefficient k_p, $T_{fa} = k_p T_d$. Obviously, T_d depends on each strategy.

Therefore, $\overline{T}_A$ can be seen as the performance parameter taking into account both the computational complexity and the sensitivity of a strategy.

3. High sensitivity acquisition problems

3.1 Acquisition in harsh environments

The conventional acquisition stage in Fig. 1 is designed to work in open-sky conditions. However, in harsh environments, high sensitivity (HS) acquisition strategies are required. In principle, as a nature of DS-CDMA, the longer the coherent integration time (T_{int}) between the local and the received signals is, the better the de-spreading gain (i.e. signal-to-noise ratio improvement) that can be obtained after the correlation process. However, the presence of unknown data bit transitions limits the value of $T_{int} \leq T_b$ (e.g. $T_{int} \leq 20$ ms as for GPS L1 C/A signal) to avoid the correlation loss. This limitation is only neglected if there is an external-aiding source, which provides the data transition information.

The sensitivity improvement obtained by increasing T_{int} is traded-off with an increased computational complexity. As pointed out in Section 2.2, the size of the Doppler step ($\triangle f_D$) reduces as T_{int} becomes larger and this fact increases the search-space size. Furthermore, the instability of the receiver clock causes difficulties for the acquisition stage, especially if T_{int} is large, because of the carrier and code Doppler effects. Therefore, one should consider the trade-off between the sensitivity improvement and the complexity increase when changing the value of T_{int}.

Considering the availability of external-aiding sources and the trade-off between the sensitivity and the complexity, the HS strategies can be divided into:

- Stand-alone approach (to deal with light harsh environments, e.g. light indoor)
- External-aiding approach (to deal with harsh environments, e.g. indoor).

Modern GNSSes broadcast new civil signals on different frequency bands and the new GNSS signals embed the combination of the data channel and a pilot (data-less) channel, per carrier frequency. Examples are E1 OS, E5, E6 signals of Galileo and L5, L2C, L1C signals of GPS. All these facts make possible another approach designed to provide improved acquisition sensitivity:

- Channel combining acquisition approach.

These three approaches are presented in details in the following.

3.2 Stand-alone approach for light harsh enviroments

Without the availability of external aiding sources, the strategies of this approach use $T_{int} \leq T_b$. The sensitivity obtained at a specific value of T_{int} is improved by combining the correlator outputs in different ways: coherent, non-coherent and differential combining. These techniques are referred as post-correlation combination techniques.

3.2.1 Coherent combination

For each cell $(\hat{c}, \hat{\theta}, \hat{f}_D)$ of the search-space, M correlator outputs $\{R_1, R_2, ..., R_m, ..., R_M\}$ obtained by correlating the incoming and the local signals at length T_{int}, see (5), are considered. As for the coherent technique, these M samples are combined as

$$S_C = \left| \sum_{m=1}^{M} R_m \right|^2 \tag{16}$$

However, (16) can be rewritten to

$$S_C = \left| \frac{1}{N} \sum_{n=0}^{MN} \{ r[n] \widehat{c}[n + \widehat{\tau}] e^{j(2\pi(f_{IF} + \widehat{f}_{D_M}))nT_S} \} \right|^2 \tag{17}$$

As seen in (17), the true value of the coherent integration time is no longer T_{int} but increases to MT_{int}. Hence, it is fair to state that the coherent combination of $\{R_1, ..., R_M\}$ is equivalent to increase T_{int} to MT_{int}, at the cost of an increased complexity.

3.2.2 Non-coherent combination

Unlike the coherent combination, the non-coherent technique combines the squared-envelops of the correlation values $\{R_1, ..., R_M\}$. The mathematical representation of the decsion variable is then

$$S_N = \sum_{m=1}^{M} |R_m|^2 . \tag{18}$$

By using this technique, the main correlation peak also tends to emerge from the noise floor. However, the noise floor is averaged towards a non-zero value. This value is referred as the squaring loss (Choi et al., 2002) and makes the non-coherent combination less effective than the coherent one. However, the effect is not equivalent to an increasing of T_{int}.

3.2.3 Differential combination

This technique was first introduced in the communication field by (Zarrabizadeh & Sousa, 1997). As far as the satellite navigation field is concerned , (Elders-Boll & Dettmar, 2004; Schmid & Neubauer, 2004) are among the first works using this technique and its variants. The mathematical representation of the conventional differential combination is

$$S_D = \left| \sum_{m=2}^{M} R_m R_{m-1}^* \right|^2 \tag{19}$$

As presented in (19), the complex correlator output R_m is multiplied by the conjugate of the one obtained at the previous integration interval R_{m-1}. Then the obtained function is accumulated and its envelope becomes the ultimate decision variable. The fact that the signal component remains highly correlated between consecutive correlation intervals, while the noise tends to be de-correlated, results in the improvement of the technique with respect to the non-coherent one. In comparison with the coherent combination, this technique obtains less de-spreading gain, but also requires less computational resources because the search-space size is unchanged (Yu et al., 2007). Therefore, this technique can be seen as a trade-off solution concerning the pros and cons of the coherent and the non-coherent combination techniques.

However, this technique might suffer from the combination loss due to the unknown data transitions. Assuming that the chance of changing data bit sign after each data bit period is 50%, then if full code correlation (i.e. $T_{int} = 1$ ms) is used, the average degradation due to data overlay is $20\log(18/19) \approx 0.47$ dB. However, in the Galileo case, this loss is scaled to $20\log(1/2) \approx 6$ dB, because the data bit duration is equal to the code length of 4 ms. This fact causes difficulties in applying the differential technique for Galileo E1 OS receivers.

As an expanded view of the conventional differential combination technique, generalized differential combination techniques are introduced to further improve the sensitivity of the acquisition process. These advanced differential techniques will be discussed in details in Section 4.

3.3 External aiding approach for harsh environments

For this approach, basically, the availability of external aiding sources makes the value of T_{int} able to be larger than T_b (i.e. $T_{int} > T_b$). Therefore, in this scenario, increasing T_{int} (or coherent combination of the correlator outputs) is the most suitable solution to give the best sensitivity improvement to the acquisition stage operating in harsh environments. In literature, this approach is also referred as assistance or assisted approach.

As pointed out in (Djuknic & Richton, 2001), the assisted technique enables HS acquisition, since it provides the signal processing chain with preliminary (but approximate) code-phase / Doppler frequency estimates along with fragments of the navigation message. This allows for wiping off data-bit transitions and for extending the coherent integration time. The concept of data-bit assistance has been also introduced by the 3rd Generation Partnership Project (3GPP) in its technical specifications of the Assisted GNSS (A-GNSS) for UMTS (3GPP, 2008a) and GSM/EDGE (3GPP, 2008b) networks.

In general, with all post correlation processing techniques presented in Section 3.2, sensitivity losses are experienced due to

- the residual Doppler error (including the finite search resolution in frequency and the contribution of the user dynamics)
- the uncertainty on the Local Oscillator (LO) frequency.

These effects impact the observed Radio Frequency (RF) carrier frequency and can be more relevant with long coherent integrations (Chansarkar & Garin, 2000) as the case of the coherent combination in this external aiding approach.

Finally, a trade-off between sensitivity and complexity is always necessary, particularly for mass-market receivers (e.g. embedded in cellular phones) which require real-time processing but low power consumption. Despite the recent improvements in chip-set sizes and speeds, a real-time indoor-grade high-sensitivity receiver for cellular phones does not exist yet. Reduced sampling rates are mandatory to minimize the computational load of the baseband processing as well as the optimization of the assistance information exchange is fundamental in order to minimize the communication load which is likely to be paid by the user, according to the latest trends, such as the Secure User Plane for Location (SUPL) defined by Open Mobile Alliance (OMA), (Mulassano & Dovis, 2010; OMA, 2007).

The mentioned A-GNSS specifications, basically define the procedures for requesting and sending information on user position and assistance data. These are typically of two paradigms:

- Mobile-based: assistance data are provided to the User Equipment (UE), which measures the pseudo-ranges and provides the position estimation to the proper network service.
- Mobile-assisted: the UE measures the pseudoranges and sends them to a location server which performs the positioning and service-related tasks.

In both these two modes, the position estimation may benefit of the knowledge of additional information available to the location server gathered from one or more reference receivers (e.g. differential corrections, precise timing and ephemeris, etc.). In the followings, the challenges of the external aiding approach are discussed. It should be noted that the chosen signal for analyses is GPS L1 C/A.

3.3.1 Navigation data wipe-off

The typical effects of both the data wipe-off and non-removed bit transitions are in Fig. 4(a) and Fig. 4(b) respectively. In the first case, the main correlation peak is easily identified whilst in the other one no peak can be distinguished over the floor. Under the AWGN assumption, in

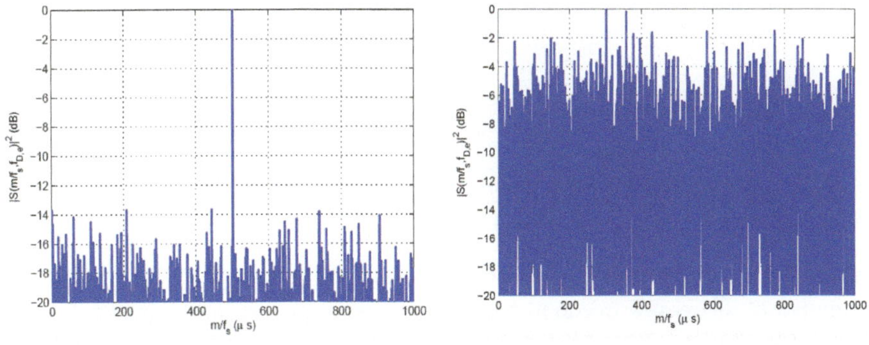

Fig. 4. CAF along code-phase - simulated GPS L1 C/A signal with code-phase of 500 μs, C/N_0 = 24 dB-Hz and T_{int} = 1 s: (a) with data wipe-off; (b) without data wipe-off

the correct Doppler and code-phase bins α_{mean} is theoretically proportional to post-correlation Signal-to-Noise Ratio (SNR) and it is expected to increase by 3dB when T_{int} doubles. This can be seen in Fig. 5, where we show the effect on α_{mean} and α_{max} of a coherent correlation with $C/N_0 = 24$ dB-Hz and $T_{int} = \{100, 500, 1000\}$ ms performed on simulated GPS signals, both with and without data wipe-off. In Fig. 5(a), we observe that at the highest values of C/N_0 the peak-to-floor ratios change linearly, i.e. α_{mean} increases by 3 dB when T_{int} doubles (e.g. from 500 ms to 1000 ms). In this case, R_{peak} is the correct correlation peak. At the lowest C/N_0, α_{max} is practically 0 dB and the detected peak is likely a noise peak, thus $|R_{peak}|^2 \approx \max|R_{floor}|^2$

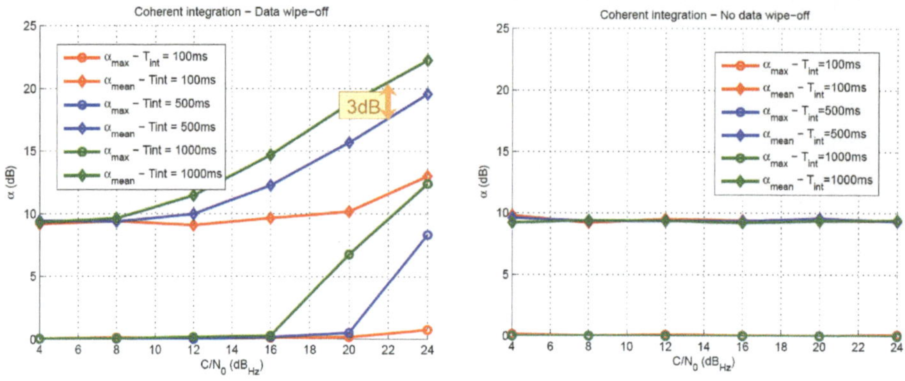

(a) (b)

Fig. 5. α_{mean} and α_{max} vs. C/N_0 and different integration windows: (a) with data wipe-off; (b) without data wipe-off

. α_{mean} is constant for low C/N_0 values because at such a noise level, R_{floor} is a zero-mean Gaussian random variable and for most of $R[k,m]$ samples:

$$|R_{floor}|^2 \leq E\{|R_{floor}|^2\} + \eta\sqrt{Var\{|R_{floor}|^2\}} \tag{20}$$

where η is an arbitrary constant. Then:

$$\alpha_{mean} = \frac{\max\{|R_{floor}|^2\}}{E\{|R_{floor}|^2\}} = 1 + \eta\frac{\sqrt{Var\{|R_{floor}|^2\}}}{E\{|R_{floor}|^2\}} \tag{21}$$

Since R_{floor} is complex and Gaussian distributed, then $|R_{floor}|^2 = \mathcal{R}\{R_{floor}\} + \mathcal{I}\{R_{floor}\}$ is χ^2 distributed (2 degrees of freedom) and thus the ratio of mean and variance is constant (Kreiszig, 1999). In Fig. 5(b), it can be seen that without data wipe-off the CAF envelope behaves as if it is made of noise only, even at the highest values of C/N_0.

3.3.2 Doppler effects on carrier and code

The Doppler effect observed at the receiver location is caused by the time-variant propagation delay of the transmitted signal along its path toward the receiver. This delay changes over time even in case of a low-dynamics user (e.g. pedestrians, etc.), as at least the SV is moving along its own orbit. Even if the rate of change is relatively slow, when long coherent integration windows are used, it can be shown that it impacts on the acquisition sensitivity. Let (22) be the general expression of the received RF signal (noiseless for simplicity):

$$s_{RX}(t) = \sqrt{2C}c[t-\tau(t)]\cos\{2\pi f_{RF}[t-\tau(t)]\} \tag{22}$$

where $\tau(t)$ is the time-variant propagation delay. With a first-order expansion of the time-variant delay, i.e. $\tau(t) = \tau_0 + a \cdot t + ...$, the carrier phase becomse:

$$\begin{aligned} 2\pi f_{RF}[t - \tau(t)] &= 2\pi f_{RF}(t - \tau_0 - a \cdot t) = \\ &= 2\pi f_{RF} t - 2\pi f_{RF}\tau_0 - 2\pi f_{RF} a \cdot t = \\ &= [2\pi(f_{RF} + f_D)t + \varphi_0] = \end{aligned} \tag{23}$$

Let us denote $\varphi_0 = -2\pi f_{RF}\tau_0$ and $f_D = -2\pi f_{RF} a$ then

$$2\pi f_{RF}[t - \tau(t)] = \left[2\pi f_{RF}\left(1 + \frac{f_D}{f_{RF}}\right)t + \varphi_0\right] \tag{24}$$

where $f_D = -af_{RF} = -f_{RF}\dfrac{d\tau(t)}{dt}$ is the usual Doppler frequency shift. Due to the Doppler effect, the observed carrier frequency is different from the nominal RF carrier frequency. With a second-order expansion for $\tau(t)$, we could see that also f_D changes in time and we could take into account a Doppler-rate term r_D. For a ground GPS receiver in low-dynamics conditions, the typical intervals are $f_D = -5$ kHz $\div 5$ kHz and $r_D = -1$ Hz/s $\div 0$ Hz/s.

The IF down-conversion leaves unmodified the Doppler frequency, as the IF carrier results:

$$BPF\{\cos[2\pi(f_{RF} + f_D)t + \phi_0] \cdot 2\cos[2\pi(f_{RF} - f_{IF})t]\} = \cos[2\pi(f_{IF} + f_D)t + \phi_0 + \phi_{RX}] \tag{25}$$

where BPF {} refers to the front-end filtering operation performed by the down-conversion stage and φ_{RX} is the related additional phase contribution.

The code component is theoretically periodic with fundamental frequency equal to the inverse of the code period. When propagating from the satellite to the receiver, the same time-variant delay impacts on all the harmonic components:

$$\begin{aligned} c[t - \tau(t)] &= \sum_{h=0}^{\infty} \mu_h e^{j2h\pi f_0[t - \tau(t)]} \\ &= \sum_{h=0}^{\infty} \mu_h e^{j2h\pi f_0\left(1 + \frac{f_D}{f_{RF}}\right)t + \vartheta_0} \end{aligned} \tag{26}$$

Due to the Doppler effect, each harmonic is shifted of the same relative frequency offset $\left(1 + \dfrac{f_D}{f_{RF}}\right)$. Thus the fundamental frequency of the delayed code is now $f_0\left(1 + \dfrac{f_D}{f_{RF}}\right)$ and its period duration is:

$$T_{code} = \frac{\tilde{T}_{code}}{\left(1 + \frac{f_D}{f_{RF}}\right)} \tag{27}$$

where $\tilde{T}_{code}$ is the nominal one. Consequently, the true chip rate is

$$R_c = \tilde{R}_c\left(1 + \frac{f_D}{f_{RF}}\right) \tag{28}$$

f_D (kHz)	α_{max} (dB)	α_{mean} (dB)	Doppler-induced code-phase estimation error (chips)
-5	0.01	14.18	1.683
-2.5	2.66	18.88	0.830
0	12.77	22.48	0
2.5	2.22	18.93	-0.839
5	0.19	14.47	-1.838

Table 1. Peak-to-floor ratios and Doppler effect on estimated code phase, $C/N_0 = 24$ dB-Hz and $T_{int} = 1$ s.

During the acquisition phase, if the local code is generated at the nominal chip rate R_c, the correlation between local and received codes suffers a loss due to the difference with the true received chip rate R_c. Furthermore, such a loss increases with the integration times. A loss of about 8 dB in α_{mean} can be estimated at $C/N_0 = 24$ dB-Hz ($T_{int} = 1$ s). Table 1 shows the degradation of the correlation peak and the code-phase estimation error.

3.3.3 Local oscillator stability

The uncertainty on the nominal value f_{LO} of the LO frequency is usually expressed as fractional frequency deviation (Audoin & Guinot, 2001):

$$y_{LO}(t) = \frac{\triangle f(t)}{f_{LO}} = \frac{f(t) - f_{LO}}{f_{LO}} = \frac{f(t)}{f_{LO}} - 1 \tag{29}$$

where $f(t)$ is the true instant frequency. y_{LO} is affected by environmental conditions (e.g. temperature, pressure), dynamic stress (e.g. acceleration, jerk, etc.), circuital tolerances, etc. The time deviation (i.e. the time difference between the clock with the true oscillator and an ideal clock), is given by:

$$x_{LO}(t) = \int_{-\infty}^{t} y_{LO}(u)du \tag{30}$$

With the zero-th order expansion $y_{LO}(t) = y_0 + ...$, (y_0 is a constant frequency offset), the time deviation results:

$$x_{LO}(t) = x_0 + y_0 \cdot t \tag{31}$$

where x_0 is an initial synchronization error between real and ideal clocks and t is the time elapsed since the initial synchronization epoch. This model can be used to evaluate the effect of the local oscillator accuracy on both the down-conversion and the sampling stages.

During the down-conversion the true mixing signal (used in (25)) is:

$$2\cos[2\pi(f_{RF} - f_{IF})(1 + y_0 t)] \tag{32}$$

The true IF carrier is actually affected by an additional unpredictable shift, that prevents the exact carrier frequency estimation, even with very accurate Doppler aiding information. By means of (31) we can evaluate the impact of the LO on the sampling process. With the true sampling clock, the sampling timescale can be defined as:

$$t_S(s)\Big|_{t=\frac{n}{f_S}} = \frac{n}{f_S} + x_{LO}\left(\frac{n}{f_S}\right) = \frac{n}{f_S} + x_0 + y_0\frac{n}{f_S} \qquad (33)$$

$$n = 0, 1, 2, ...$$

where n/f_S is the ideal sampling instant and f_S is the sampling frequency. The sampled version of the IF signal is:

$$r[n] = r\left(\frac{n}{f_S} + x_0 + y_0\frac{n}{f_S}\right) \qquad (34)$$

and it is affected by a time-variant delay with respect to the ideal case. This gives rise to an equivalent Doppler effect, as previously discussed, and hence to an additional correlation loss. Oscillators typically used in GNSS receivers are mostly Crystal Oscillators (XOs) with some degree of frequency stabilization, e.g. Thermally-Compensated Crystal Oscillator (TCXO), with typical accuracy $y_{LO} \sim 10^{-6}$ and Oven-Controlled Crystal Oscillator (OCXO), with typical accuracy $y_{LO} \sim 10^{-8}$ (Vig, 2005). Table 2 shows how a constant offset on the LO frequency may impact both on α_{mean}, α_{max} and on the accuracy of the code-delay estimation in case of a 1 s coherent integration.

f_D/f_{LO}	$\alpha_{max}(dB)$	α_{mean} (dB)	Code-phase error (chips)
0	22	31	0
$0.5 \cdot 10^{-6}$	18	31	0.75
$1.5 \cdot 10^{-6}$	0	27	1.5

Table 2. Constant offset on LO frequency. $T_{int} = 1$ s, $C/N_0 = +\infty$

3.4 Channel combining approach:

- Channel Combining on Different Carrier Frequencies

In a new or upgraded GNSS, there are several civil signals broadcast in different frequencies. This fact assures a future for civil GNSS dual-frequency receivers, which are now used only in high-value professional or commercial applications such as survey, machine control and guidance, etc. Beside the predictable advantages, such as ionosphere error elimination and carrier phase measurement improvement, civil dual-frequency receivers also offer sensitivity improvement by making possible combined acquisition strategies. The combined acquisition on different carrier frequencies is guaranteed by the fact that the signal channels belonging to a common GNSS are time synchronized, and the Doppler shifts of these channels are related by the ratio among the carrier frequencies. In literature, (Gernot et al., 2008) uses this approach for combined acquisition of GPS L1 C/A and L2C signals.

- Channel Combining on a Common Frequency:

New GNSS signals are composed of data and pilot (data-less) channels. These two channels can be multiplexed by Coherent Adaptive Subcarrier Modulation (e.g. Galileo E1 OS), Time Division Multiplexing (GPS L2C) and Quadrature Phase-Shift Keying (Galileo E5; GPS L5,

L1C). The transmitted power is shared between two channels. Therefore, if the acquisition is performed on both channels, then the better sensitivity improvement can be obtained. In literature, (Mattos, 2005; Ta et al., 2010) use this approach for Galileo E1 OS signal acquisition.

Essentially, for the channel combining acquisition approach (common or different frequencies), in each involved channel, an acquisition strategy belonging to either the stand-alone or the external-aiding approach is performed. Then the acquisition outputs from all the channels are combined in different ways. In Section 6, the joint data/pilot acquisition strategies for Galileo E1 OS signal is introduced as an example for this approach.

4. Stand-alone approach: Generalized differential combination technique

4.1 Technique description

As seen in (19), the decision variable of the Conventional Differential Combination (CDC) technique is an accumulation of the products between two consecutive correlator outputs R_m, R_{m-1}. In a broader manner, the Generalized Differential Combination (GDC) has been introduced (Corazza & Pedone, 2007; Shanmugam et al., 2007). This technique considers the products of two consecutive correlator outputs as in CDC as well as the products of two correlator outputs at all sample distances or referred as all possible spans, see Fig. 6(a). Let us

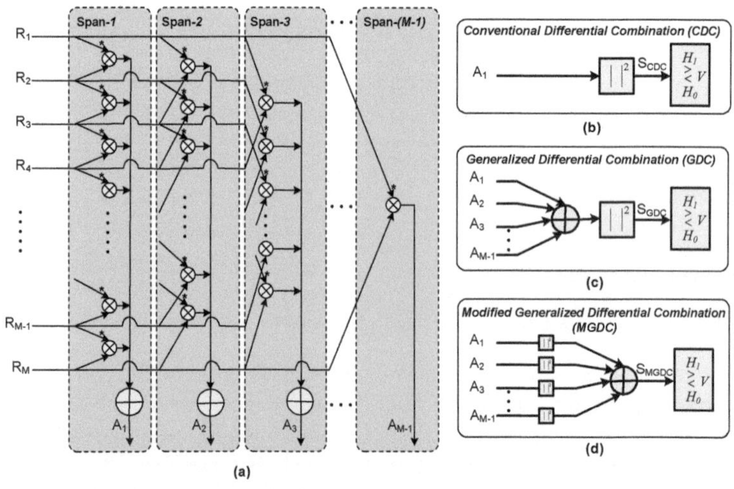

Fig. 6. Differential Post Correlation Processing Architecture: (a) Differential operations; (b) Conventional Differential Combination (CDC); (c) Generalized Differential Combination (GDC); (d) Modified Generalized Differential Combination (MGDC)

define a span-i term as:

$$A_i = \sum_{m=i+1}^{M} R_m R_{m-i}^* \tag{35}$$

Then the decision variable of GDC (Fig. 6(c)) is

$$S_{GDC} \triangleq \left| \sum_{i=1}^{M-1} A_i \right|^2 \tag{36}$$

Note that the CDC technique is in fact the GDC taking into account span-1 A_1 only, see Fig. 6(b). Basically, the GDC technique can be considered as a coherent integration of the differential combinations at different sample distances. Following the analyzes in (Ta et al., 2012), with small M (e.g. $M \leq T_b/T_{int}$), in normal circumstances with normal user dynamic and frequency standards, the average frequency drift is small and tends to zero. Therefore, the values of $G_m, \triangle \bar{f}_{d_m}$ in (6) are constant for all $m \in [0, M-1]$. The signal component A_i^S of an arbitrary span-i (A_i) in (35) can be represented as

$$A^S_{i|\tau,\triangle\bar{f}_d} = \sum_{m=i}^{M} G^2 e^{j2\pi i \triangle \bar{f}_d T_{int}} \tag{37}$$

with

$$\begin{cases} \triangle\bar{f}_d = \triangle\bar{f}_{d_1} = ... = \triangle\bar{f}_{d_M} \\ G = G_1 = ... = G_M = \sqrt{2C}\mathcal{R}[\tau]\text{sinc}(\triangle\bar{f}_d T_{int}) \end{cases} \tag{38}$$

For the GDC technique, substituting (37) into (36), S_{GDC} is computed

$$S_{GDC} = |D|^2 = \left| \sum_{i=1}^{M-1} G^2 e^{j2\pi i \triangle \bar{f}_d T_{int}} \right|^2 \tag{39}$$

Equation (39) shows that the residual carrier phase is still present in the d_{GDC}. This fact causes an unpredictable loss, which depends on the specific value of $\triangle\bar{f}_d$. To eliminate this loss, Modified Generalized Differential Combination (MGDC) technique (Ta et al., 2012) can be used, see Fig. 6(d). Following this technique, the decision variable of the MGDC technique is

$$S_{MGDC} = \sum_{i=1}^{M-1} |A_i|. \tag{40}$$

If the noise is neglected, (40) becomes

$$\begin{aligned} S_{MGDC} = \sum_{i=1}^{M-1} |A_i^S| &= |(M-1)G^2 e^{j2\pi\triangle\bar{f}_d T_{int}}|^2 + |(M-2)G^2 e^{j4\pi\triangle\bar{f}_d T_{int}}|^2 + ... \\ &+ |G^2 e^{j2\pi(M-1)\triangle\bar{f}_d T_{int}}|^2 = (M-1)G^2 + (M-2)G^2 + ... + G^2 = \frac{M(M-1)}{2} G^2 \end{aligned} \tag{41}$$

By forming the decision variable in this way, the unpredictable loss caused by the residual carrier phase is canceled completely. However, the non-coherent integrations between all the spans make the noise averaging process worse than for GDC.

Note: for the GDC and MGDC techniques, the number of spans involved can vary from 1 to $M-1$. By default, all $(M-1)$ possible spans are considered as in (40). If a different number

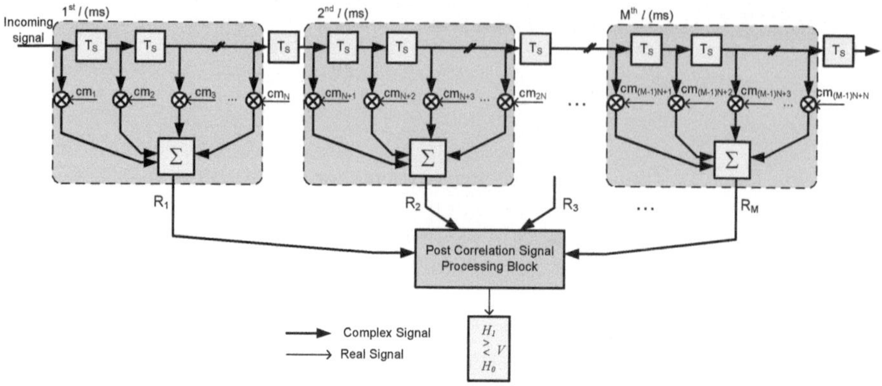

Fig. 7. L2C Partial acquisition using matched filter

of spans i ($1 \leq i \leq M-1$) is used, in the following, the notations for the two techniques will be GDC(i) and MGDC(i).

4.2 Application of technique to L2C signal

In this Section, the MGDC technique is used to acquire GPS L2C signal. This signal is chosen because it employs a long PRN code period, which can be used to generate partial correlator outputs with the same sign. Hence, there is no combination loss due to data bit transitions in differential accumulation (see Section 3.2.3).

4.2.1 L2C signal acquisition

The L2C signal has advantages in interference mitigation due to its advanced PRN code format. This signal is composed of two codes, namely L2 CM and L2 CL. The L2 CM code is 20-ms long containing 10230 chips; while the L2 CL code has a period of 1.5 s with 767250 chips. The CM code is modulo-2 added to data (i.e. it modulates the data) and the resultant sequence of chips is time-multiplexed (TM) with CL code on a chip-by-chip basis. The individual CM and CL codes are clocked at 511.5 kHz while the composite L2C code has a frequency of 1.023 MHz. Code boundaries of CM and CL are aligned and each CL period contains exactly 75 CM periods. This TM L2C sequence modulates the L2 (1227.6 MHz) carrier (GPS-IS, 2006). The original L2C data rate is 25 bps but a half rate convolutional encoder is employed to transmit the data at 50 sps. Consequently, each data symbol matches the CM period of 20 ms.

With these specifications, the common signal representation in (1) is changed to

$$r[n] = \sqrt{2C}\{d[n]c_{\mathrm{m}}[n+\tau] + c_{\mathrm{l}}[n+\tau+kP]\}\cos[2\pi(f_{IF}+f_D)nT_S+\varphi] + n_W[n] \quad (42)$$

where $c_{\mathrm{m}}[n]$ and $c_{\mathrm{l}}[n]$ are the received CM and CL codes respectively (samples); θ is the received signal delay; P refers to the number of samples in a full CM code period (i.e. 20 ms), $0 \leq k \leq 74$ is an integer that gives the CL code delay relative to CM code.

Fig. 7 shows an architecture of the partial acquisition suitable for L2C CM signal. A segmented matched filter (MF) is used as a correlator (Dodds & Moher, 1995; Persson et al., 2001). The

MF is loaded with one full modified CM code. The modified CM code is obtained from the original CM code with every alternative sample being zero padded to account for the TM structure. The MF does not produce the correlation results equivalent to the full code period, i.e. $T_{int} = 20$ ms. Nevertheless, it provides M partial correlation results with $T_{int} = l$ ms as in Fig. 7. It can be thought of as the partial acquisition process using M different local codes of 1-ms length. By setting the local codes in this way, the signal components of all M correlator outputs $R_1, ..., R_M$ have the same sign. Therefore, the differential combination can be used among these M outputs without any loss from the data transition effect. These M correlator outputs are then directed to Post Correlation Signal Processing Block, which contains 3 differential combination solutions, namely CDC, GDC and MGDC, as presented in Section 4. The analytical expressions of the performance parameters of these techniques can be found in (Ta et al., 2012).

4.2.2 Performance analyses

Summarizing the techniques introduced in the previous sections, there are five strategies that have to be investigated: non-coherent, CDC, GDC, MGDC and 20-ms coherent combination (full code acquisition). Fig. 8 shows the behavior of the detection probabilities of all the strategies when $T_{int} = 1$ ms, $P_{fa} = 10^{-3}$ and the signal strength (C/N_0) varies. The 20-ms coherent technique, as expected, has the best performance. Among the others, all the differential post correlation processing techniques, i.e. GDC, MGDC, CDC, are better than the non-coherent one. The CDC technique taking into account only Span-1 provides the lowest improvement of 1 dB with respect to the non-coherent. The performance of MGDC with different numbers of spans involved (i.e. span size) is also shown in Fig. 8(a). It can be observed that as the span size increases, the detection capability also improves. For the highest span size (i.e. 19 in the figure), the MGDC can offer an advantage of more than 1 dB over the CDC as well as more than 2 dB over the non-coherent combination. These improvements are preserved even the worst case is considered as can be seen in Fig 8(b). Among the differential techniques, the GDC has the highest performance. If all the spans are considered, the GDC performance approaches that of the coherent one. However, this performance is only guaranteed when the residual carrier phase is known (i.e. the perfect case). In Fig. 8(b), the detection probability of the GDC technique reduces dramatically due to the residual carrier phase. Table 3 compares the simulation results of $\overline{T}_A$ for the normal

T_{int} ms	$\overline{T}_A$ ($\times 10^5$) ms	Relative Savings
0.5	0.08527	97.15%
1	0.1624	94.5%
2	0.313	89.5%
5	0.769	74.3%
10	1.519	49.3%
20	2.996	0%

Table 3. Reduction of Mean Acquisition Time by using MGDC at different partial coherent integration times with respect to full 20-ms acquisition ($C/N_0 = 23$ dB-Hz)

outdoor operating range of signal power, i.e. above 32 dB-Hz. It can be observed that a significant saving in $\overline{T}_A$ of MGDC (with respect to the full CM period correlation acquisition) can be achieved by shortening T_{int}.

5. External aiding acquisition technique for indoor positioning

In this section, a test-bed architecture, which is proposed by (Dovis et al., 2010), is introduced as an example of the external-aiding acquisition approach.

5.1 Test-bed architecture

The test-bed as seen in Fig. 9 includes two chains:

Test receiver chain: The main task of this chain is to collect a snapshot of the digitized GPS signal and sends it to a location server through a cellular communication channel. The chain consists of a GPS L1 front-end with the antenna at the test location. The RF front-end is connected to a PC which collects digital sample streams into binary files. The local oscillator is a rubidium (Rb) frequency standard (Datum8040, 1998) running the front-end through a waveform synthesizer (HP, 1990).

Reference receiver chain: The main task of this chain is to perform the HS acquisition process taking advantage of the available assistance information. The chain consists of a reference GPS receiver which processes open-sky signals from a fixed (known) location and provides measurements to an assistance server. The latter provides the necessary aiding information to the HS acquisition engine and the GPS Time indication for the synchronization of the sample-stream recorder, performed before starting each signal collection session. The synchronization process introduces an uncertainty on the GPS Time tags, since it is performed by the software running at the PC, which is assumed to be 2s as in this work.

The assistance server is a software tool developed at Telecom Italia Laboratories to support several test activities on Assisted GPS (A-GPS) technologies. It collects data from the reference receiver and generates time-tagged log files with several kind of assistance information to be provided to the HS acquisition engine. Each line of the log file, for each visible SV, contains code-phase, Doppler frequency and Doppler rate estimates.

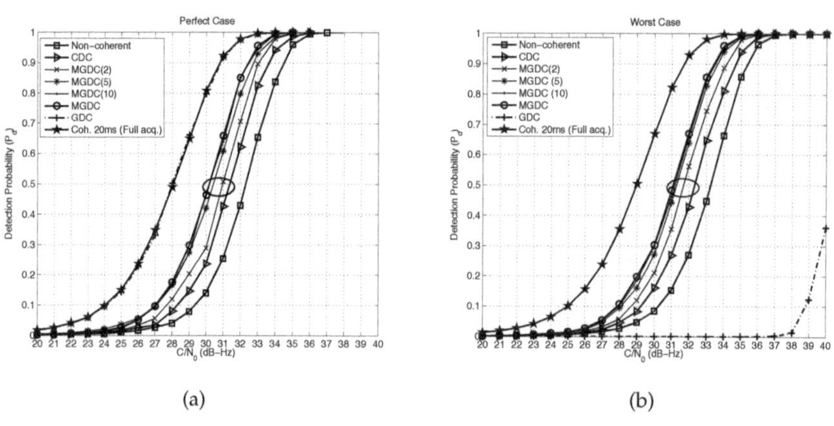

(a) (b)

Fig. 8. Detection probability (P_d) of all post-correlation processing techniques at different signal power levels in (a) perfect case: $\triangle \bar{f}_d = 0$ Hz; (b) worst case: $\triangle \bar{f}_d = 12.5$ Hz for coherent combination (i.e. full 20-ms acquisition) and $\triangle \bar{f}_d = 250$ Hz for the other techniques.

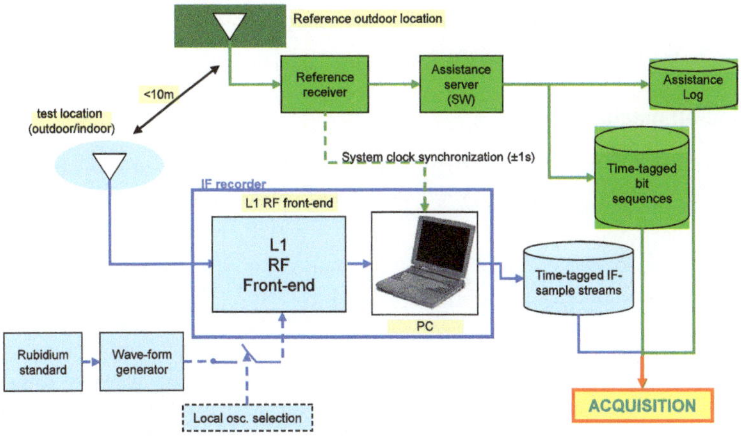

Fig. 9. Test bed architecture: reference chain (green) and test chain (blue)

5.2 Acquisition procedure

- Step 1 - Preliminary fast detection of the strongest PRN

In this step, the FFT-based circular correlation stage is used to quickly detect the best PRN selected on the basis of the predicted C/N_0 and elevation. The value of T_{coh} is 10ms and; the number of M correlator outputs is then non-coherently combined to achieve a sufficient post-correlation SNR; $\triangle f_D$ =100 Hz (trade-off between frequency resolution and search complexity), while the code-phase search space spans over a full code period.

- Step 2 - Determination of the assistance offsets

Code-phase and frequency offsets are caused by: (i) space displacement of test and reference antennas (mostly code-phase offset); (ii) the time offset between the reference receiver and test receiver clocks (code-phase offsets); and (iii) the uncertainty on the test receiver LO frequency (Doppler frequency offset). In this step, these offsets, which are the same for all the PRNs, can be computed by considering the difference of the preliminary estimates (from step 1) with those provided by the assistance data.

- Step 3 - Aided long coherent correlation with data wipe-off on weaker PRNs

The offsets obtained with the strong PRN can be used to correct the assistance predictions and finely determine the code-phase/Doppler frequency of other PRNs at the last step (aided long coherent correlation), ensuring the best achievable post-correlation SNR by means of a low-complexity data wipe-off technique. At this step, the frequency bin size of $\triangle f_{D,3} = 1/T_{coh}$ (e.g. $\triangle f_{D,3} = 1$ Hz if $T_{coh} = 1$ s) is used, over a frequency search space 100 Hz wide (i.e. the residual uncertainty from step 1), and a code-phase search range 6 chips wide. The knowledge of aiding data would allow for a narrower search space, but the acquisition has to account for possible residual errors between the true and predicted code phases.

The code-phase resolution is as low as 1 sample (for both step 1 and 3). The reference signal bandwidth is $B = 2.046$ MHz chosen to match the main lobe (two-sided) of the GPS signal spectrum. Therefore the performed tests have been run with a sampling rate $f_S = 4.092$ MHz designed to meet the Nyquist criterion. Finally the local code rate taking into account the Doppler effect, as presented in (27), is used.

5.3 Data wipe-off mechanism

In order to increase the coherent integration over the data bit duration (i.e. 20 ms), the acquisition stage performs data wipe-off process. Basically, the conventional data wipe-off process is done as follows

$$R = \frac{1}{N}\sum_{n=1}^{MN} \widehat{d}[n] \cdot \{r[n]\widehat{c}[n+\widehat{\tau}]e^{j(2\pi(f_{IF}+\widehat{f_D}))nT_S}\} \tag{43}$$

with $\{\widehat{d}[n]|\ n = 1...MN\}$ being the data sequence provided by the assisted data. However, at the acquisition stage, the signal snap-shot and the assisted data are not synchronized. Therefore, in order to determine the correct bit sequence for the signal snap-shot, the acquisition stage needs to test all possible data sequence in a predetermined uncertainty. Then the maximum likelihood estimator is used for decision. Hence, it can be said that the acquisition stage in this scenario searches for the presence of a desired signal on 4-dimensions, namely: PRN, code-phase, frequency and bit-phase (i.e. 4D search-space).

In fact, this mechanism requires an unacceptable computational effort for a single position fix, because for each bit-phase (i.e. a data bit sequence candidate), the whole search-space must be re-computed. As a result, the number of elementary steps (i.e. multiply&add) is

$$(T_{coh} \cdot f_S) \times (N_{cp} \cdot N_f) \times N_{bit-seq} = 4.092 \cdot 10^8 \cdot N_{cp} \cdot N_f \tag{44}$$

with N_{cp}, $N_f, N_{bit-seq}$ being the numbers of code-phase, Doppler frequency and bit-phase bins respectively; $f_S = 4.092$ MHz and $T_{coh} = 1$ s.

However, (43) can be rewritten as

$$R = \sum_{m=1}^{M} \widehat{d}_m R_m \tag{45}$$

where R_m is partial correlation value with representation in (5). With this approach, the acquisition stage can compute $R_1, R_2, ..., R_M$ then save these values for testing with all possible values of bit-phase. This approach in fact utilizes the coherent combination presented in (16). For this mechanism, the number of elementary steps is

$$[M(f_S \cdot T_{coh_1}) + M \cdot N_{bit-seq}] \cdot N_{cp}N_f = 4.192 \cdot 10^6 \cdot N_{cp} \cdot N_f \tag{46}$$

with M being the number of partial correlations obtained after 1-ms coherent integration time (T_{coh_1}). From (44) and (46), the computational complexity of the partial correlation approach has a reduction of approximately 2 orders of magnitude with respect to the conventional one.

5.4 Performance analyses

This section demonstrates the application of the test-bed for indoor signal acquisition. The required integration time for indoor signals is longer than for outdoor ones. The sky plot, see Fig. 10, has been generated by means of an auxiliary receiver with the antenna placed out of the lab window, so to have and indication of the available GPS constellation. The distance between the antenna of the auxiliary receiver and the test indoor antenna is ≤ 10 m. The sky

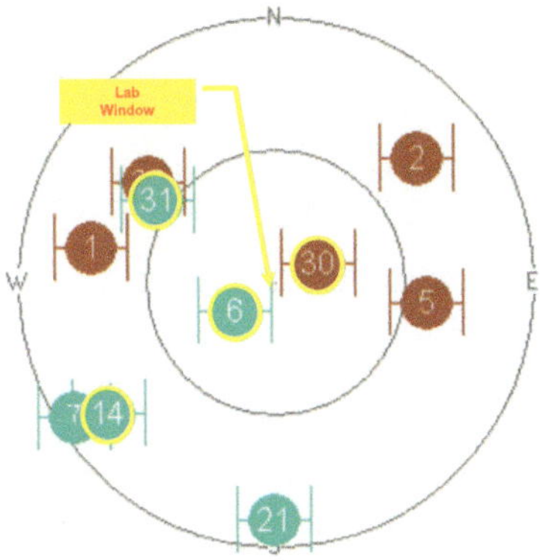

Fig. 10. Skyplot, indoor, Rb

plot relative to this case-study is depicted in Fig. 10. PRN6 and PRN30 are considered in this section. The assistance log is summarized in Table 4. Then the 3-step procedure in Section 5.2 is applied. Firstly, the strongest signal, which is PRN6 as seen in Table 4, is determined. After that, FFT-based acquisition is activated to search for PRN6 in the signal snapshot collected in indoor environment. Then the following procedure has been used to determine the assistance offsets and the corrected aiding data. For PRN6 the code phase from the assistance log is τ_6^a =120 chips. The preliminary fast acquisition on PRN6 estimated a code-phase τ_6^p = 1000.3 chips (Table 6). The code-phase offset is:

$$\delta\tau_6 = \tau_6^p - \tau_6^a = 880.3 \text{ chips} \tag{47}$$

As the signal snapshot is the same for the two PRNs, there are no time drifts to take into account. Therefore:

$$\tau_{30}^p = \delta\tau_{30} + \tau_{30}^a = \delta\tau_6 + \tau_{30}^a = 1060.3 \text{ chips} \tag{48}$$

It should be noted that because the full code length is 1023 chips, therefore, τ_{30}^p can also equal to 37.3 chips.

The same approach is used to compute the aiding value of Doppler frequency, $f_{D,30}^p = -0.6025$ kHz. Finally, after step 2, the aiding parameters are listed in Table 5.

The aiding parameters are used for acquiring the weaker satellite, PRN30, in indoor environment. The correlation results are shown in Fig. 11 and in Table 5, it can be noticed that the 3 dB rule still holds. In fact The signal of PRN 30 pass through the roof and the walls of the laboratory. Thus, it was good realizations of typical indoor signals and and it is detected by assisted coherent correlation.

6. Channel combination approach: Joint data/pilot acquisition strategies

In this section, the channel combination approach to improve the sensitivity of the acquisition is described. The considered signal is Galileo E1 Open Service signal. The current definition

PRN	Elevation (^{o})	C/N_0 (dB-Hz)	Code-phase (chips)	f_D (Hz)	r_D(Hz/s)
30	60	-	180	-635.0	-0.4
6	73	32.7	120	1067.5	-0.5

Table 4. Assistance log, indoor, Rb

PRN	Aiding source	$\tau^{(pred)}$ (chips)	$f_D^{(pred)}$ (kHz)
6	FFT	1000.3	1.1
30	Assistance server	1060.3 or 37.3	-0.6025

Table 5. Aiding data, indoor, Rb

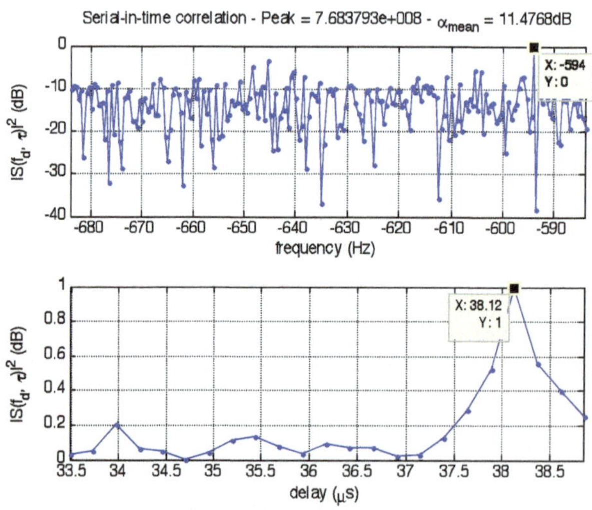

Fig. 11. Long coherent correlation, indoor, Rb, $T_{int} = 2000$ ms, PRN30

of the this signal (GalileoICD, 2008) includes data (B) and pilot (C) channels which are multiplexed by Coherent Adaptive Sub-carrier Modulation (CASM) (Dafesh et al., 1999). Each channels shares 50 % of the total transmitted power. To represent this signal, the common representation in (1) is changed to

$$r[n] = \frac{1}{\sqrt{2}}\sqrt{2C}\left(d[n+\tau]b[n+\tau] - c_{2nd}[n+\tau]c[n+\tau]\right)\cos\left(2\pi(f_{IF}+f_D)nT_S+\phi\right) + n_W[n] \tag{49}$$

$b[n]$, $c[n]$ are respectively the 4-ms primary PRN codes of the data (B) and pilot (C) channels modulated by BOC(1,1) scheme; $d[n]$ is the navigation data in the B channel; $c_{2nd}[n]$ is the secondary code, which together with $c[n]$ form a 100-ms tiered code for the C channel (GalileoICD, 2008). Basically, the conventional acquisition stage in Fig. 1 can perform on either B or C channels. This strategy is referred here as Single Channel (SC). However, SC also implies a waste of half of the real capability. Therefore, joint data/pilot acquisition strategies are introduced to utilize the full potential of the E1 OS signal (Mattos, 2005; Ta et al., 2010). In the followings, these strategies are described together with the performance evaluation.

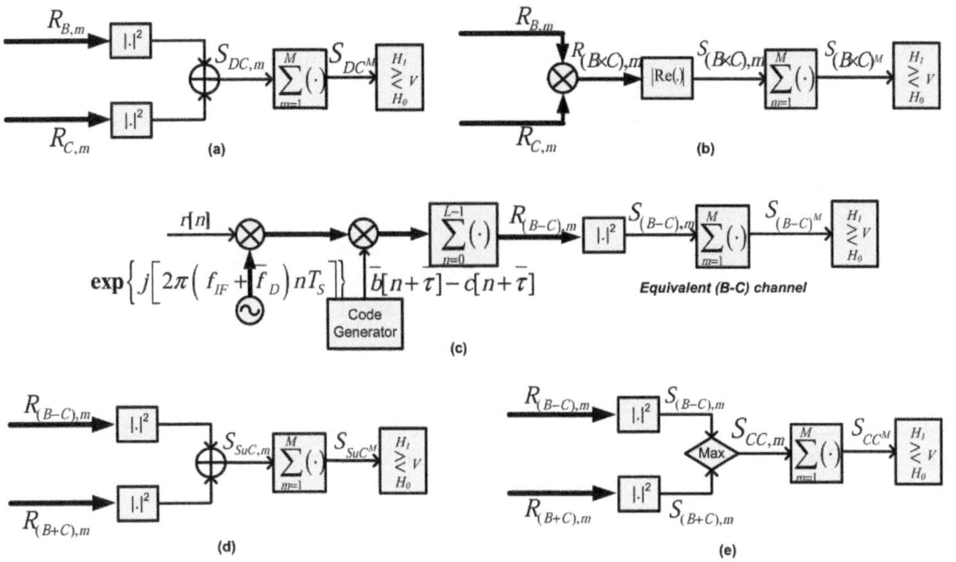

Fig. 12. Joint data/pilot acquisition architectures: (a) Dual Channels - DC; (b) (B×C); (c) Assisted (B-C); (d) Summing Combination - SuC; (e) Comparing Combination - CC

6.1 Joint data/pilot acquisition strategies

The correlation process performs on both channels to produce $R_{B,m}$ and $R_{C,m}$ (5). After that, these correlation values are combined as follows:

- Dual Channel (DC):

This strategy sums the square envelopes from the two channels, see Fig. 12(a). The decision variable is

$$S_{DC^K} = \sum_{m=1}^{M} (|R_B, m|^2 + |R_{C,m}|^2) = \sum_{m=1}^{M} (|S_{B,m}|^2 + |S_{C,m}|^2) \tag{50}$$

- (B×C):

In this strategy [see Fig. 12(b)], the decision variable is

$$S_{(B\times C)^M} = \sum_{i=1}^{M} |\{R_{B,i} \cdot R^*_{C,i}\}|^2 \tag{51}$$

This strategy can be seen as another realization of the conventional differential technique presented in Section 3.2.3. The correlator output in a channel is combined with the one from the other channel instead of the delayed copy of itself as in the conventional differential technique.

- Assisted (B-C):

The baseband E1 OS signal has the form $[d(t)b(t) - c_{2^{nd}}(t)c(t)]$. Due to the bi-polar nature of the data and secondary codes, the digital received baseband signal in each code period is always in one of the two representations

$$|b[n] - c[n]| \text{ or } |b[n] + c[n]| \tag{52}$$

This fact paves the way for a new strategy using one of the two equivalent codes $(\bar{b}[n] - \bar{c}[n])$ or $(\bar{b}[n] + \bar{c}[n])$ as the local code with the decision depending on the signal representation. Consequently, the two new equivalent channels (B-C) and (B+C) are defined. At a time instance, without the availability of an external-aiding source, because of the unknown navigation data bit, the acquisition stage cannot know the correct representation of the received signal, i.e. (B-C) or (B+C). In addition, the two new equivalent codes are orthogonal and still preserve the properties of the PRN codes (Ta et al., 2010). Therefore, if the chosen equivalent local code is incorrect, the correlation value in the equivalent channel might be null although the tentative parameters (i.e. PRN number, Doppler and code delay) are correct, because of the unknown data bit sign. Hence, the availability of an external-aiding source is crucial.

Without loss of generality, let us assume that the external-aiding source assures the signal structure is $(b[n] - c[n])$, therefore, the (B-C) strategy is applied, see Fig. 12(c). The decision variable of the assisted (B-C) is

$$S_{(B-C)^M} \triangleq \left|R_{(B-C)^M}\right|^2 = \left|\sum_{m=1}^{M} R_{(B-C),m}\right|^2 \tag{53}$$

Note that: for this external-aiding scenario, the coherent combination is used.

However, in one full primary code period, the signal can be only in one of the two representations in (52), it is worth to test both the strategies [i.e. (B-C) and (B+C)] and combine their results. This leads to two new strategies so-called Summing Combination and Comparing Combination.

- Summing Combination (SuC):

In this strategy (see Fig. 12(d)), the (B-C) and (B+C) strategies are simultaneously performed. The square envelope outputs are summed up to form the new decision variable

$$S_{SuC} = S_{(B-C)} + S_{(B+C)} = |R_{(B-C)}|^2 + |R_{(B-C)}|^2 = 2(|R_B|^2 + |R_C|^2) \tag{54}$$

In this way, the overall decision variable is no longer affected by the unknown polarity of the data and secondary codes of the received signal. However, multiplying the decision variable by any coefficient does not affect the ultimate performance of a strategy because the signal and the noise powers are increased by the same rate. Therefore, the SuC strategy shares the performance with the DC strategy. For this reason, in the following sections, only the DC strategy is considered.

- Comparing Combination (CC):

This strategy (see Fig. 12(e)) uses a comparator instead of the adder as in the SuC strategy to combine the square envelope outputs of the two equivalent channels. The larger value is chosen to be the decision variable

$$S_{CC^M} = \sum_{m=1}^{M} \max \left\{ S_{(B-C),m}, S_{(B+C),m} \right\} \tag{55}$$

The analytical expressions of the performance parameters of these strategies are presented in (Ta et al., 2010).

6.2 Performance analyses

Fig. 13 clearly shows the improvement of the joint data/pilot strategies over the conventional SC. The benchmark values $P_{fa} = 10^{-3}$ and $P_d = 0.9$ for the hypothesis testing in GNSS receivers are used to quantitatively estimate the improvement. When only one full code period is considered (i.e. $M = 1$), as shown in Fig. 13(a), the joint data/pilot strategies holds the sensitivity enhancement $\sim$ 3 dB over the conventional SC. Among the joint strategies, the assisted (B-C) outperforms the others, because the assistance data always guarantees the local generated signal matching the most to the received one. As for the other stand-alone joint strategies, the difference in P_d is small, but one still can realize that CC is the best one.

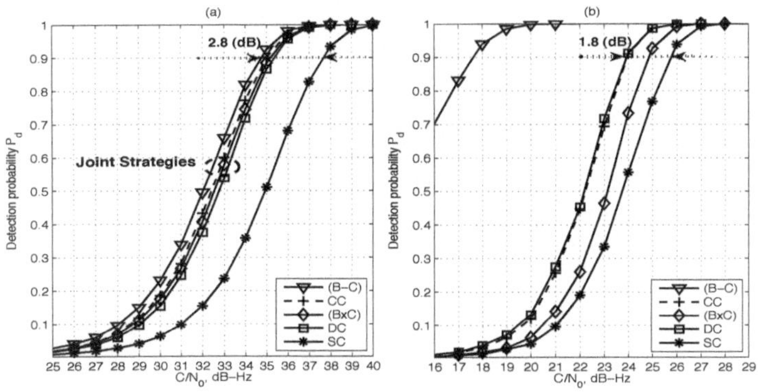

Fig. 13. Detection probability of all the strategies vs. C/N_0 values when $P_{fa} = 10^{-3}$: (a) $M = 1$; (b) $M = 50$

When $K = 50$, the assisted (B-C) is far better than the others, because the coherent combination applied in this strategy brings more performance improvement than the other strategies using the non-coherent technique suffering from the squaring loss phenomenon. This loss also reduces the enhancement (from 2.8 dB to 1.8) dB) of the stand-alone joint strategies with respect to the SC, see Fig. 13(b). Among the stand-alone joint strategies, in this scenario, DC takes the position of CC to be the best. While (B$\times$C) degrades significantly, because unlike $K = 1$, for $K > 1$, to secure the accumulation, the absolute values of the differential operation's outputs are used in the non-coherent combination. This fact makes the averaging not thorough.

Fig. 14 shows the $\overline{T}_A$ values of all the strategies. It should be note that $\overline{T}_A$ simultaneously consider the influences of both the computational complexity and the sensitivity of a strategy.

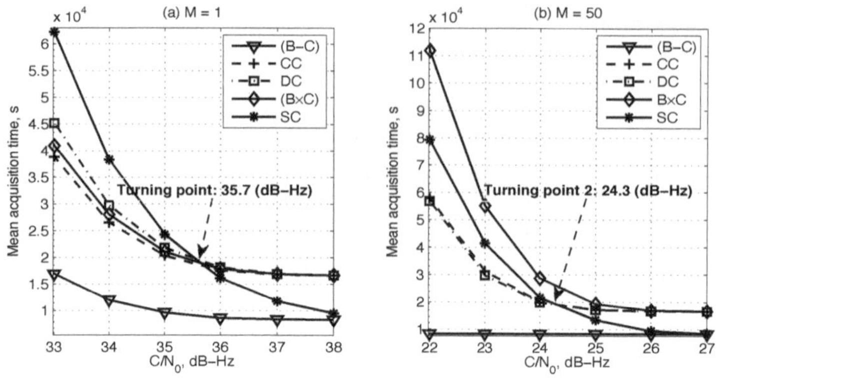

Fig. 14. Mean acquisition time $\overline{T}_A$ vs. C/N_0 values when $P_{fa} = 10^{-3}$: (a) $M = 1$; (b) $M = 50$

For all C/N_0 and K values, (B-C) results in the smallest $\overline{T}_A$, because of its high sensitivity (i.e. detection capability) and also moderate complexity (only one correlator required, but assistance is needed). For $K = 1$ and 33 (dB-Hz) $\leq C/N_0 <$ 36 (dB-Hz), due to the significant sensitivity improvement of the DC, (B$\times$C), and CC strategies with respect to the SC strategy, their $\overline{T}_A$ values are smaller than that of the SC strategy, see Fig. 14(a). However, for $C/N_0 \geq$ 35.7 (dB-Hz), the sensitivity improvement of the SC strategy is sufficient to reduce its $\overline{T}_A$ to be lower than that of the stand-alone joint strategies. For $K = 50$, due to the sensitivity improvement of all the strategies, the turning point appears earlier at $C/N_0 = 24.3$ (dB-Hz). see Fig. 14(b).

7. Conclusions

This Chapter focused on high sensitivity signal acquisition problems. Throughout the chapter, some challenges to HS signal acquisition such as unknown data transitions, Doppler effects on carrier frequency and PRN code rate, local oscillator instability as well as sensitivity-complexity trade-off were discussed in details in order to define the requirements of HS acquisition strategies suitable for different operational scenarios. Then three HS acquisition approaches, namely stand-alone, external-aiding and channel combining, have been introduced. Finally, the applications of these approaches to specific GNSS signals are demonstrated for readers' better understanding.

8. References

3GPP (2008a). Radio resource control (RRC) - release 7, *Organizational Partners* .

3GPP (2008b). Radio resource LCS protocol (RRLP) - release 7, *Organizational Partners* .

Audoin, C. & Guinot, B. (2001). *The Measurements of Time - Time, Frequency and the Atomic Clock*, Cambridge University Press.

Betz, J. W. (2001). Binary Offset Carrier Modulations For Radionavigation, *Journal of Navigation* 48: 227–246.

Chansarkar, M. & Garin, L. (2000). Acquisition of GPS Signals at Very Low Signal to Noise Ratio, *Proceedings of the 2000 National Technical Meeting of the Institute of Navigation (ION NTM 2000), Anaheim, CA, USA*, pp. 731–737.

Choi, I. H., Park, S. H., Cho, D. J., Yun, S. J., Kim, Y. B. & Lee, S. J. (2002). A Novel Weak Signal Acquisition Scheme for Assisted GPS, *Proceedings of the ION GPS 2002, Portland, OR, USA*, pp. 177–183.

Corazza, G. E. & Pedone, R. (2007). Generalized and Average Likelihood Ratio Testing for Post Detection Integration, *IEEE Transactions on Communications* 55: 2159–2171.

Dafesh, P. A., Nguyen, T. M. & Lazar, S. (January 1999). Coherent Adaptive Subcarrier Modulation (CASM) For GPS Modernization, *Proceedings of ION NTM 1999, San Diego, CA, USA*, pp. 649–660.

Park, S. H., Choi, I. H., Lee S. J., & Kim, Y. B. (2002). A Novel GPS Initial Synchronization Scheme using Decomposed Differential Matched Filter, *Proceedings of ION NTM 2002, San Diego, CA, USA*, pp. 246–253.

Datum8040 (1998). DATUM 8040 Rubidium frequency standard - Technical specifications.

Djuknic, G. M. & Richton, R. E. (2001). Geolocation and assisted GPS, *IEEE Computer* 34(2): 123–125.

Dodds, D. & Moher, M. (1995). Spread Spectrum Synchronization for a LEO Personal Communications Satellite System, *Canadian Conference on Electrical and Computer Engineering 1995 (CCECE '95), Montreal, PQ, Canada*, pp. 20–23.

Dovis, F., Lesca, R., Boiero, G. & Ghinamo, G. (2010). A Test-bed Implementation of An Acquisition System for Indoor Positioning, *GPS Solutions* 14(3): 241–253.

Elders-Boll, H. & Dettmar, U. (2004). Efficient Differentially Coherent Code/Doppler Acquisition of Weak GPS Signals, *Proceedings of the IEEE International Symposium on Spread Spectrum Techniques and Applications (ISSSTA 2004), Sydney, Australia*, pp. 731–735.

GalileoICD (2008). Galileo Open Service, Signal In Space Interface Control Document Draft 1, *Technical report*, European GNSS Supervisory Authority / European Space Agency.

Gernot, C., Keefe, K. O. & Lachapelle, G. (2008). Comparison Of L1 C/A L2C Combined Acquisition Techniques, *Proceedings of ENC-GNSS 2008, Toulouse, France*.

GPS-IS (2006). Navstar GPS Interface Specification IS-GPS-200 revision D, *Technical report*, Navstar GPS Joint Program Office.

Holmes, J. K. (ed.) (2007). *Spread Spectrum Systems for GNSS and Wireless Communications*, Artech House.

HP (1990). 325B Synthesizer/Function Generator Service Manuald, *Agilent Technologies* .

Kaplan, E. D. (ed.) (2005). *Understanding GPS: Principles and Applications*, 2nd edn, Artech House.

Kreiszig, E. (1999). *Advanced Engineering Mathematics*, John Wiley and Sons.

Mattos, P. G. (2005). Acquisition of the Galileo OAS L1b/c signal for the mass-market receiver, *Proceedings of ION GNSS 2005, Long Beach, CA, USA*, pp. 1143–1152.

Misra, P. & Enge, P. (2006). *Global Positioning System: Signals, Measurements, and Performance*, 2nd edn, Ganga-Jamuna Press.

Mulassano, P. & Dovis, F. (2010). *Assisted Global Navigation Satellite Systems: An Enabling Technology for High Demanding Location-Based Services*, CRC Press.

OMA (2007). Secure user plane for location (SUPL), *Open Mobile Alliance* .

Persson, B., Dodds, D. & Bolton, R. (2001). A Segmented Matched Filter for CDMA Code Synchronization in Systems with Doppler Frequency Offset, *in* 1 (ed.), *Proceedings of IEEE Globecom '01, San Antonio, Texas, USA*, pp. 648–653.

Schmid, A. & Neubauer, A. (2004). Perfomance Evaluation of Differential Correlation for Single Shot Measurement Positioning, *Proceedings of ION GNSS 2004, Long Beach, CA, USA*, pp. 1998–2009.

Shanmugam, S. K., Nielsen, J. & Lachapelle, G. (2007). Enhanced Differential Detection Scheme for Weak GPS Signal Acquisition, *Proceedings of ION GNSS 2007, Fort Worth, TX, USA*, pp. 2499–2509.

Shanmugam, S. K., Watson, R., Nielsen, J. & Lachapelle, G. (2005). Differential Signal Processing Schemes for Enhanced GPS Acquisition, *Proceedings of ION GNSS 2005, Long Beach, CA, USA*, pp. 212–222.

Ta, T. H. (2010). *"Acquisition Architecture for Modern GNSS Signals"*, PhD thesis, Polytechnique University of Turin, Italy.

Ta, T. H., Dovis, F., Lesca, R. & Margaria, D. (2008). Comparison of Joint Data/Pilot High-Sensitivity Acquisition Strategies for Indoor Galileo E1 Signal, *Proceedings of ENC-GNSS 2008, Toulouse, France.*

Ta, T. H., Dovis, F., Margaria, D. & Presti, L. L. (2010). Comparative Study on Joint Data/Pilot Strategies for High Sensitivity Galileo E1 Open Service Signal Acquisition, *IET Radar, Sonar and Navigation* 4, Issue 6: 764–779.

Ta, T. H., Qaisar, S. U., Dempster, A. & Dovis, F. (2012). Partial Differential Post Correlation Processing for GPS L2C Signal Acquisition, *IEEE Transactions on Aerospace and Electronic Systems* 48, Issue 2.

Tsui, J. B.-Y. (2005). *Fundamentals of Global Positioning System Receivers: a Software Approach*, 2nd edn, Wiley-Interscience.

Vig, J. (2005). Quartz crystal resonators and oscillators for frequency control and timing applications - a tutorial, *IEEE Ultrasonics, Ferroelectrics, and Frequency* .

Wilde, W. D., Sleewaegen, J.-M., Simsky, A., Vandewiele, C., Peeters, E., Grauwen, J. & Boon, F. (2006). New Fast Signal Acquisition Unit for GPS/Galileo Receivers, *Proceedings of European Navigation Conference ENC-GNSS 2006.*

Yu, W., Zheng, B., Watson, R. & Lachapelle, G. (2007). Differential combining for acquiring weak GPS signals, *Signal Processing* 87(5): 824–840.

Zarrabizadeh, M. H. & Sousa, E. S. (1997). A Differentially Coherent PN Code Acquisition Receiver for CDMA Systems, *IEEE Transactions on Communications* 45(11): 1456–1465.

2

Baseband Hardware Designs in Modernised GNSS Receivers

Nagaraj C. Shivaramaiah and Andrew G. Dempster
The University of New South Wales
Australia

1. Introduction

The Global Positioning System (GPS) receiver has come a long way from being a specialised tool to a more general purpose everyday use mainstream gadget. This transformation is not only due to the advancements in semiconductor technology and embedded systems but also due to a highly concentrated research effort in the past decade that targeted a high performance, low power and affordable GPS receiver design. Before the ideas for such an efficient GPS receiver design could attain the saturation stage, the GPS modernisation and the development of several satellite navigation systems under the broader "Global Navigation Satellite Systems" (GNSS) umbrella, have brought a new dimension to the problem of efficient GNSS receiver design. The baseband signal processing engine forms an integral part of any GNSS receiver and is a key contributor to the overall cost and power consumption.

This chapter discusses the challenges involved in designing baseband signal processing algorithms for a modernised GNSS receiver. The modernised GNSS receiver in this context includes processing elements not only for the GPS civilian signals GPS L1C/A, GPS L2C, GPS L5 and GPS L1C, but also for the Open Service (OS) signals from other satellite navigation systems that share the same frequency band as that of GPS. The Galileo satellite navigation system is one such example with its E1 and E5 OS signals sharing the GPS L1 and L5 (partial) frequency bands respectively. Though the underlying concept used in all these signals is "spread spectrum", the structure of these signals differ due to different modulation techniques and signal parameters such as chipping rate, spreading code length, signal bandwidth and navigation data rate. These differences make the efficient baseband hardware design an interesting and useful research topic.

The key objectives of this chapter are:

1. To revisit the existing efficient GPS L1 C/A baseband hardware methodologies and list best practices / learnings,
2. To analyse the complexity of the modernised GNSS baseband hardware and to identify key contributors to this complexity,
3. To explore design alternatives that deal with the key complexity contributors and to analyse the implementation feasibility of these design alternatives,
4. To ascertain the practicality of incorporating the design alternatives by implementing them on a FPGA based hardware platform, and

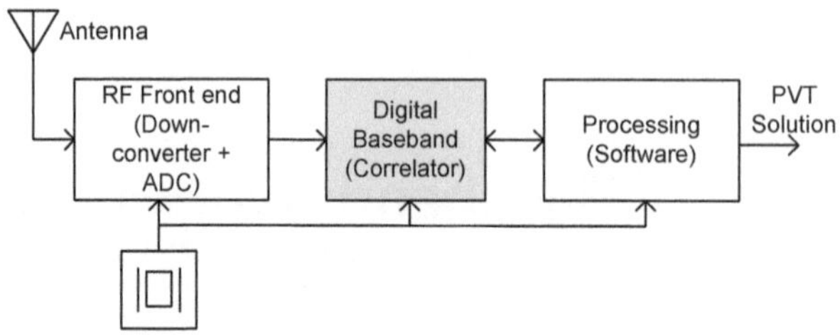

Fig. 1. Typical architecture of a GNSS receiver

5. To provide recommendations and guidelines for the design of a low power, high performance, affordable multi-GNSS baseband hardware.

This chapter substantially draws on one of the authors' conference papers published in ISCAS 2010 (Shivaramaiah & Dempster, 2010).

2. GNSS receiver and baseband hardware

2.1 GNSS receiver architecture

Fig. 1 shows the typical architecture of a GNSS receiver. Each signal from a different frequency band is down-converted and passed through an Analog-to-Digital-Converter (ADC) to obtain the Intermediate Frequency (IF) samples. The baseband signal processing hardware (widely known as the correlator) is usually implemented in hardware. With the help of feedback control algorithms (implemented either as a part of the digital hardware or as a part of the processing in software), the baseband circuit provides accurate estimates of the delay, phase and frequency of the carrier and spreading code in the received signal (tracking). The baseband circuit is also used for the initial coarse estimates of these parameters (acquisition). The processing, usually implemented in software, computes the Position-Velocity-Time (PVT) solution (Braasch & van Dierendonck, 1999; Kaplan & Hegarty, 2006; Parkinson & Spilker, 1995).

2.2 Generic baseband architecture for the tracking process in a GNSS receiver

This section describes a generic architecture for the GNSS baseband that allows the basic functionality of tracking the signal. Though the same architecture can be used for the signal acquisition process, the signal acquisition is not the focus here. The GNSS baseband hardware in its usual definition is comprised of all the signal processing circuits bounded on the input side by the sampled and digitised IF signal, and on the output side by the received signal measurements (carrier phase, code phase, navigation data bits, signal strength, etc.). Fig. 2 shows the functional diagram of generic GNSS baseband hardware for a single signal component. The functionality of each block is described in detail elsewhere in the literature (e.g. Kaplan & Hegarty (2006)) and will be discussed briefly here.

The baseband functionality can be broadly divided into two parts.

1. Core Correlator Hardware

   The core hardware is responsible for correlating the input signal with the local replica and producing the correlation values.

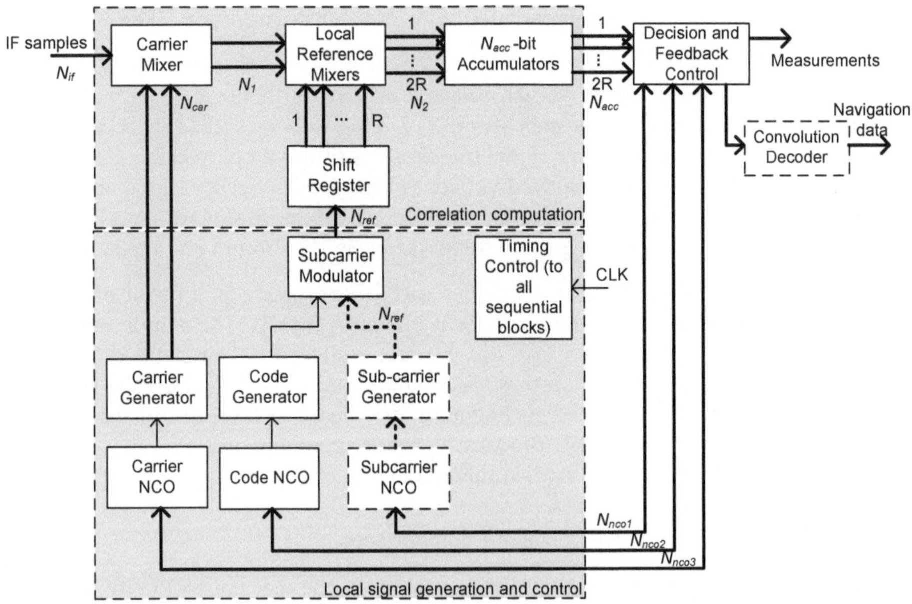

Fig. 2. A functional diagram of the baseband hardware (thick lines carry $N_\bullet$ bits, dashed boxes are optional)

2. Correlator Controller

   The controller processes the correlation values produced by the core hardware and makes decisions based on a set of feedback control algorithms. The threshold detection during the acquisition, the carrier and code locked loops during the tracking, the process of dictating the parameters for the local replica carrier and replica code generation, are all included in this part.

The core correlator hardware functionality can again be divided into two parts.

1. Correlation Computation
2. Local Signal Generation

The necessity of this second level functionality division is due to the new signals that will be dealt with in future sections. This type of segregation of the core correlator hardware functionality helps accommodate new signals both at the same frequency that may belong to a different constellation and the signals at different frequency bands of the same constellation.

The input to the baseband hardware is the sampled and digitized IF signal with N_{if}-bit quantization at a sampling frequency of f_s Hz. In order to demodulate the navigation data bits, the baseband module must first remove the carrier and the spreading code from the signal (Braasch & van Dierendonck, 1999).

The IF samples are mixed with the locally generated carrier in the carrier mixer. The local carrier frequency generator aims to match the frequency of the input signal. Both in phase and quadrature phase signals are generated with N_{car}-bit quantization. The carrier mixer output results in N_1-bit values.

The local replica code+subcarrier signal, referred to here as the local "reference signal" is N_{ref}-bit wide. In the absence of the subcarrier, N_{ref}=1 because the spreading code takes only the values of 1 or 0. Since most of the signal tracking algorithms employed in a GNSS receiver use the delay tracking principle, delayed versions of the local reference signal are generated with the help of shift registers. R is the number of local reference signal "arms" (sometimes referred to as "taps" or "fingers"), typically three: the Early, the Prompt and the Late).

The local reference mixer generates $2R$ values each N_2-bits wide as a result of combining in phase and quadrature values with the local reference signal. These individual sample correlation values are accumulated in a N_{acc}-bit accumulator for a predefined "integration duration". The tracking loops act on these accumulator outputs and adjust the local carrier frequency and the code delay so as to maintain lock (to be at the peak of the correlation function). The tracking loops also produce the measurements and also demodulate the navigation data bits present in the signal (Shivaramaiah, 2004).

2.3 Bit-width requirements of the correlator components

The parameters of interest for the complexity analysis of the core correlator are the number of bits required to represent the intermediate signals, the bit-width of the accumulator and other registers and the minimum frequency of operation required for a particular signal (or any component of a signal thereof). The notations for the number of bits at different stages are shown in Fig. 2, as $N_\bullet$ along with the thick lines. In the following paragraphs a brief description of each of the underlying modules is given and the number of bits required for the accumulator is derived.

2.3.1 ADC/IF (N_{if})

The signal loss due to the quantisation beyond 2-bits is insignificant as long as the sampling thresholds are sensibly set (Hegarty, 2009). However, 3-bits and more have been used to alleviate the problems with the AGC in the presence of RF interference (Kaplan & Hegarty, 2006). Commercial mass-market receivers normally use 2-bit uniform sign-magnitude quantisation with 4 levels $\{\pm 1, \pm 3\}$(Zarlink, 1999, 2001). Therefore for the examples in this chapter it is safe to assume $N_{if} = 2$.

2.3.2 Local carrier generator (N_{car})

The loss due to the local carrier quantisation is studied in Namgoong et al. (2000). Typically, 3-bit uniform NCO phase quantisation and 2-bit amplitude quantisation with 4 levels $\{\pm 1, \pm 2\}$ is used. More bits in the phase and amplitude quantisation increases the Spurious-Free-Dynamic-Range (SFDR) and reduces the quantisation noise. However this has a significant impact on the size of succeeding stages.

2.3.3 Carrier mixer (N_1)

The carrier mixer basically multiplies the input signal with the local carrier bits. Since the resulting values will only have 8 levels $\{\pm1, \pm2, \pm3, \pm6\}$, a 3-bit encoding is sufficient. Observe that with the 3-bit encoding, arithmetic operation cannot be directly performed. Hence if the succeeding stage requires an arithmetic representation then four bits should be used.

2.3.4 Subcarrier generator & subcarrier modulator (N_{ref})

The local code takes on values of either 0 or 1 and hence 1-bit is sufficient for its representation. However, the number of bits required to represent the subcarrier depends on the number of levels in the subcarrier used for the modulation. BOC signals use a 2-level $\{\pm1\}$ subcarrier thus requiring only 1-bit for the representation. AltBOC uses 4-levels (dominant component of the subcarrier) which require more bits for the representation and in such situations approximation needs to be used to use smaller bit-width representations. The local spreading code modifies only the sign of the subcarrier at the output of the subcarrier modulation. Hence, N_{ref} will depend on the number of bits used for the subcarrier representation.

2.3.5 Local reference mixer (N_2)

This can be easily determined from the number of levels of the two inputs. However, the succeeding stage (the accumulator) is an arithmetic operation and requires binary two's complement representation. This leads to an additional bit at the output. For example, with the 8-level N_1 $\{\pm1, \pm2, \pm3, \pm6\}$ and the 4-level $N_{ref}\{\pm1, \pm2\}$, the resultant set will have only 12 levels $\{\pm1, \pm2, \pm3, \pm4, \pm6, \pm12\}$, but due to the later requirement of signed binary representation the output must be 5-bit wide. Let the sample-maximum (magnitude) of the output at this stage be denoted by A_2.

2.3.6 Accumulator (N_{acc})

The interval between two consecutive accumulator resets is generally determined by the coherent integration duration and the coherent integration duration in turn in most cases will be a multiple of the spreading code period. Let N_{acc} denote the number of bits required to represent the worst-case value at the output of the accumulator. Then

$$N'_{acc} = \left\lceil \log_2^{-1}\left(A_2 \left\lfloor \frac{f_s}{f_{co}} M_c L \right\rfloor\right) + C + 1 \right\rceil \tag{1}$$

where $f_s \in R^+$ is the sampling frequency in Hz, $f_{co} \in R^+$ is the chipping rate (with any associated Doppler frequency) in Hz, $L \in \mathbb{N}$ is the primary code length, $M_c \in \mathbb{Q}$ is the number (or fraction) of primary code periods in the coherent integration and C is the complex modulation indicator, $C \in \{0 = Normal, 1 = Complex\}$. (1) clearly satisfies the Design-For-Test (DFT) guidelines, but it is an overkill as all the samples may not end up with a value of A_2. In reality the sample-maximum is controlled by the input signal strength and the local carrier frequency. Hence the required accumulator width $N_{acc} < N'_{acc}$.

An R-arm correlator will have $2R(C+1)$ accumulators (due to the in-phase and quadrature carrier components) and hence accumulator width plays a very important role in correlator complexity. Some correlators use re-sampling prior to the local reference mixer stage (e.g. (Namgoong et al., 2000)), to reduce the number of samples input to the accumulator. However those special techniques are outside the scope of the discussion here.

2.4 Efficient realisation of the correlator core for the GPS L1 C/A signal

As mentioned in the previous section, the input to the correlator is the sampled IF signal. Each sample in the sampled IF signal, when mixed with local carrier and the local reference signal, produces a correlation value ("sample correlation value") which is then fed to the accumulator. Therefore, in the correlator core of Fig. 2, all the blocks do not require sequential logic. The carrier mixer, subcarrier modulation and the local reference mixer are typically implemented as combinational logic. Latching the input signal, carrier NCO, code NCO and the accumulator are implemented as sequential logic. As a result, the combined propagation delay of all the combination logic blocks should be less than the sampling period $t_{pd} + t_{su}^{acc} < 1/f_s$, where t_{pd} is the propagation delay and t_{su}^{acc} is the setup time of the accumulator.

The combinational block has to compute the sample correlation value from the three inputs viz. the incoming signal, the local carrier and the local reference signal. The carrier mixer and the local reference mixer can be realised using Look-Up-Tables (LUTs) separately or together. For the single-bit reference signals, the circuit can be further simplified by feeding the local code to select the add or subtract operation of the accumulator. Fig. 3(a) shows a generic way to realise the correlation computation blocks' combinational logic. The number of instantiations of each block is mentioned above the block. Observe that two carrier mixer blocks are required (I and Q), six reference signal mixer blocks are required (early, prompt and late version of reference signals mixed with I and Q carrier mixer outputs) and six accumulator blocks are required for the complete operation.

Fig. 3(b) shows a realisation of the combinational logic using the LUT method for the GPS L1 C/A signal. In Fig. 3, the input signal and the local carrier are assumed to be 2-bit wide and the local reference signal (in this case only the local code) is 1-bit wide. Observe that the sample correlation output is represented using four bits even though there are only eight possible values. This is because the succeeding stage (which is the signed addition, a part of the accumulation process) is an arithmetic operation and hence the sample correlation values need to be represented in 2's complement format. The local code mixer is eliminated by using the local code output as the Add/Sub selection input of the accumulator. The output therefore consists of six correlation values: inphase-early, inphase-prompt, inphase-late, quadrature-early, quadrature-prompt and quadrature-late. These correlation values are fed to the tracking loops for further processing.

3. Impact of the signal structure on the core correlator architecture

This section analyses the impact of the change in certain parameters of the signal (due to the structure of the new signals) on the architecture of the core correlator.

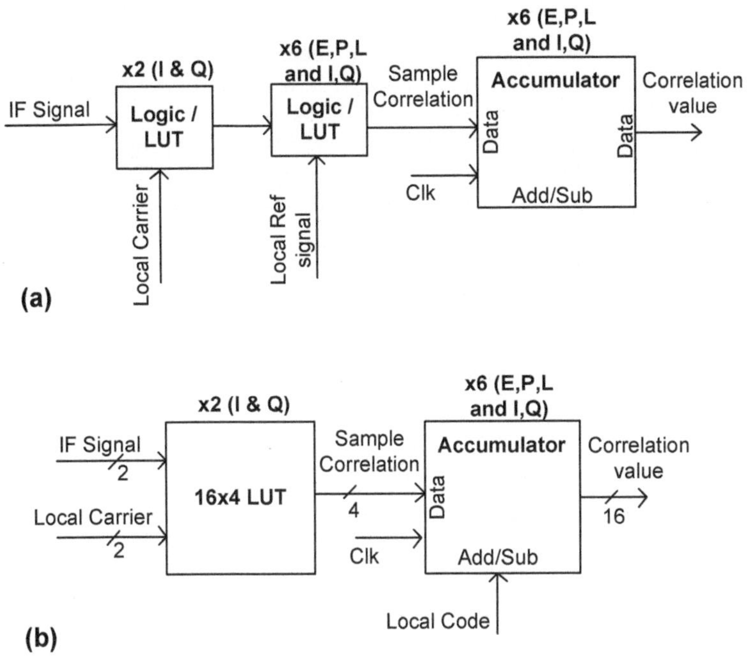

Fig. 3. Realisation of the correlation computation blocks (a) Generic implementation (b) an implementation for the GPS L1 C/A signal

3.1 Longer codes (or longer code period)

3.1.1 Shift register generated codes

Longer codes are usually obtained by wide shift registers or a combination of shift registers. Typically the baseband circuit will have the same number of code generators as the number of channels. If the baseband has to implement multiple tracking channels to simultaneously process multiple signals then the additional number of bits in the shift register brings in additional hardware which is not insignificant.

3.1.2 Memory codes

Memory codes eliminate the need for a code generator (i.e the shift registers and XOR gates used for the code generation). However the codes for all the pseudo random noise (PRN) sequences must be stored in a circular buffer or ROM. The decision on whether to use a circular buffer or ROM depends on the overall architecture of the receiver. For tracking the signal, it is enough to read the buffer sequentially (like in a FIFO) and no address generation is required. However if there is a requirement to read the local code from a particular delay (which could be the case when the receiver wants to reacquire the signal) then it is better to use the ROM which then demands a separate address generator. The read clock to the FIFO or the ROM is nothing but the output of the code NCO.

Another issue with the memory codes is the way the codes are stored. The codes for all the PRNs cannot be stored in a single memory because it will limit the access of the memory from different channels. Hence the code for each PRN should be stored in a separate memory block. Even in this situation, there is a constraint on the architecture. During the signal acquisition or during the tracking if there is a requirement for more than one GNSS channel to use the same PRN, then the memory block will have to have more than one port which is expensive in terms of the resource and power consumption.

3.1.3 Effect on the accumulator bit-width

Another consequence of longer codes is that the number of bits in the accumulator has to be increased, i.e. the N_{acc} requirement increases (assuming that the accumulator is used to integrate the correlation values for the duration of one code period).

3.2 Subcarrier modulation

3.2.1 Two-level (1-bit) subcarriers

With the subcarrier modulation an additional NCO, subcarrier generator and subcarrier modulator may be required depending on the requirement of the tracking loops. If the subcarrier has only two levels then the subcarrier and the replica code bit can be combined with the help of a single XOR gate and the result will also be a 1-bit value. This does not change the other parts of the correlation computation circuit and also the reference signal can still be fed to the Add/Sub input of the accumulator.

3.2.2 Multi-level (> 1-bit) subcarriers

If the subcarrier has multiple levels (i.e. requiring more than 1-bit) then the process of combining the replica code bit and the subcarrier is not a simple XOR operation, but requires a negation operation which results in the same number of bits as the subcarrier (N_{ref}). Secondly, the width of the shift register that generates the early, prompt and late values should be increased to N_{ref}. Since the reference signal is not represented by a single bit it cannot be used directly as an input to the accumulator and therefore there needs to be a dedicated reference signal mixer block. Thirdly, the reference signal mixing operation should accommodate this bit-width increase in one of the inputs. As a result, the number of bits required to represent the sample correlation value will increase, which in turn increases the number of bits in the accumulator.

3.3 Modulation type

The BOC family of signals has a narrow autocorrelation main peak. As a result the spacing between the R delayed versions of the reference signals should be reduced in order to achieve better tracking performance (Shivaramaiah & Dempster, 2009). Reduction in the spacing requires the code and the subcarrier NCO to be operating at a higher clock frequency. This constrains the minimum clock frequency requirement of these NCOs. As a result, the overall operating frequency requirement of the correlator will go up and also an additional clock divider circuit is required.

3.4 Multiple signal components

When a signal has more than one component (say pilot and data components), it is wise to compute the correlation values independently for each signal component, thus allowing the subsequent processing blocks to use efficient tracking techniques (Shivaramaiah, 2011). One can optimise the correlation computation blocks by combining the logic for the signal components but that would give a combined correlation value to the tracking loops. This combined correlation value may suffer from loss due to the data and or code bit inversions between the signal components. Therefore it is wise to isolate the different signal components at the correlation computation stage (and combine in the succeeding stages if required).

3.5 Receiver bandwidth and the operating frequency

Receiver bandwidth has a direct impact on the sampling frequency and hence the operating frequency of the circuit. While some baseband blocks can be fed a slower clock than the sampling frequency (but still derived from the sampling frequency), some other blocks have to operate at the sampling frequency itself. Any bandwidth reduction below the minimum required (which is typically the bandwidth occupied by the main lobe(s)) done before the correlation operation stage, will result in rounded auto-correlation peaks, which in turn result in noisier range measurements. Therefore it is a good practice to keep the operating frequency at least equal to the sampling frequency until the carrier mixing stage and at least equal to four times the subcarrier frequency (or the twice the code frequency in the absence of subcarrier) from the reference signal mixer stage onwards.

3.6 Complex modulation

In the case of AltBOC signals the lines generated within the core correlator portion in Fig. 2 carry complex signals. The local reference mixer LUT must cater for the complex correlation operation.

There are basically two ways to realise this complex reference signal mixer: with the logic or with LUTs. With the logic one would be using adders/subtracters and multipliers of appropriate length to compute the reference signal mixer outputs. With the LUT, there are many ways, each using different sizes of the LUT. In both the cases the resource requirement would significantly increase compared to the GPS L1 C/A correlator (which requires no reference signal mixer).

4. Core correlator architectural modifications for the new signals

4.1 New GNSS signals and general requirements

Table 1 revisits the centre frequency, typical receiver bandwidth and code lengths of some of the new open service signals. These parameters largely determine the architecture and complexity of the baseband signal processing stage in a GNSS receiver.

The following are the important points to note from the table.

Signal name	Centre frequency (Typical receiver bandwidth) in MHz	Modulation type	Code length * (memory code? Y/N)	Chipping rate (MHz)
GPS L1 C/A	1575.42 (2)	BPSK	1023 (N)	1.023
GPS L2C	1227.6 (2)	BPSK	CM-20460 (N), CL-767250 (N)	1.023
GPS L5	1176.45 (20)	BPSK	20460 (N)	10.23
GPS L1C, Galileo E1, Compass B1	1575.42 (14)	MBOC / CBOC	1023 (N), 4096 (Y), **	1.023
Galileo E5, Compass B2	1191.795 (50)	AltBOC	10230 (N), ***	10.23

** Primary code only, *** Yet to be available for the Compass signal

Table 1. Some new GNSS signals and their parameters of interest

1. increased signal bandwidths demand higher sampling frequencies
2. increased spreading code lengths and chipping rates demand higher shift register clock frequencies,
3. use of multi-level subcarriers, as in the case of AltBOC type of modulation, increases the number of bits in the local reference signal,
4. use of memory codes demands additional memory to hold the spreading code for all the satellites, and
5. increased minimum operating frequency of the baseband hardware mainly due to a) and b)

The operating frequency and the circuit complexity determine the energy efficiency of digital logic and therefore the design of an efficient baseband logic circuit becomes extremely important in the context of baseband hardware targeted to process multi-GNSS signals (Shivaramaiah et al., 2009).

This section discusses the major contributors for the resource utilisation of the correlators for the new signals. The parameters of the correlator that processes GPS L1 C/A signal are used as the reference.

4.2 GPS L2C

4.2.1 GPS L2C - CM

The L2C - CM code generation requires a 27-bit shift register instead of the 10-bit code generator shift register that is used to generate the L1 C/A signal. This in turn increases the code generator read /write and control register bit-widths. The operating frequency remains the same and hence any increase in the power consumption is only due to the increase in the number of shift register bits.

4.2.2 GPS L2C - CM and CL)

The additions to the L2C - CM only correlator are : another 27-bit shift register, another set of code mixers and accumulators. Since the CM and CL codes are time-multiplexed, the number of accumulators remains the same if the CM and CL correlation values are combined (the spreading codes can be combined in time similar to what is done at the transmitter). However, the combination will lead to data bit ambiguity problem (Dempster, 2006). When the components are combined, the increase in the power consumption with respect to the L2C- CM only signal case is negligible. If both the CM and the CL signal components are processed independently then the resource utilisation almost doubles compared to the CM only processing.

4.3 Galileo E1

4.3.1 Single signal component (E1b or E1c)

Because of the use of memory codes, the baseband can eliminate the shift register and store the local spreading code in memory. Therefore there is a small saving in terms of the flip-flops/registers compared to the GPS L1 C/A architecture. However, because of the 8 MHz sampling frequency requirement assumption, one expects to see an increase in the power consumption.

4.3.2 Both the signal components (E1b and E1c)

Here two sets of memory codes are used each occupying 4092 bits. In addition the number of local reference mixers and accumulators are not only doubled, but also need to operate at higher frequencies due to the higher sampling frequency. For this reason the expected power consumption is close to twice that of the single signal component (E1b or E1c).

4.4 GPS L5 (pilot and data)

For the GPS L5 signal, the code generator shift register requires 13 bits, which is not a significant increase from the 10-bits of GPS L1 C/A. However, the major difference is the higher chipping rate which demands a higher sampling frequency and in turn a higher correlator operating frequency. Due to the longer code length, the accumulators also have to be wide compared to that of the GPS L1 C/A correlator. As a result of the increased operating frequency, the power consumption requirement is expected to drastically increase though the resource utilisation would go up only slightly more than twice that of the GPS L1 C/A correlator.

4.5 Galileo E5

4.5.1 Galileo E5a or E5b (pilot and data)

For the Galileo E5a and E5b signals, the code generator shift register requires 14-bits. This is only a 1-bit change from the case of GPS L5 correlator and hence all the other circuit parameters (such as bit widths) will be very close to that of the GPS L5 correlator. Hence the expected power consumption for E5a or E5b signal when processed individually, would be close to that of the GPS L5 correlator.

4.5.2 Galileo E5 wideband

In this case both the E5a and E5b signals are processed together as a single wideband signal (with a bandwidth of at least 51.15 MHz). The local code has to be generated individually for all the four components (E5a-pilot, E5a-data, E5b-pilot and E5b-data) of the signal and the generators require 14-bit shift registers. However, because the four signal components and the complex modulation, the local reference mixer is computationally intensive (more LUTs). In addition, a quadruple number of accumulators are required. As mentioned earlier, independent correlation for all the four signals is performed to allow design freedom for the subsequent stages in combining these four components. As a result of a very high operating frequency and drastic increase in the resource requirements compared to GPS L1 C/A correlator, the power consumption is expected to be very high.

4.6 Baseband architecture overview for the GPS and Galileo OS signals

Fig. 4 shows the baseband architecture assuming that the correlator processes GPA L1 C/A, L2C, L5 and Galileo E1, E5 signals. The computation block is marked separately to the local signal (local carrier and replica reference signal) generation block. This sort of grouping the blocks helps the design because different signals (and different channels in some cases) share common parameters and optimising the hardware becomes easier. In this architecture, it is also assumed that the baseband is commanded (assigned PRNs, Doppler and delay parameters etc) to operate from an external processor. The correlation values of different signals are read and processed in the succeeding decision feedback and control stage.

5. Resource requirements for the new signals and recommendations

5.1 Core resource requirements using a straightforward extension of the GPS L1 C/A design

In order to gauge the resource requirements in terms of the number of registers and combinational logic, the core correlators for the GPS and Galileo open service signals have been implemented on the Altera Cyclone-III family device EP3C120F780C8. The FPGA resource utilisation parameters are listed in Table 2.

The resource and the power consumption values closely match the expected outcomes mentioned in the previous section. While the Galileo E1b or E1c core requires almost the same resources as that of GPS L1 C/A, the power consumption is higher. The power consumption for the single component Galileo E1 is 0.8mW more than that of the GPS L1 C/A. This is due to the presence of the memory block, increased accumulator width and increased operating frequency. The Galileo E1 correlator where both the E1b and E1c signals are processed together has a power estimate of 2.24mW, only about 0.6mW more than the single component. This is because some of the blocks such as the carrier NCO and the carrier mixer are common for both the signal components.

The resource for the GPS L5 signal is increased to 701 registers and 204 combinational units which is due to the increase in the accumulator width and also due to the presence of two signal components. The power consumption estimate of the GPS L5 signal is about 11 times that of the GPS L1 C/A signal and is attributed mainly to the operating frequency. The

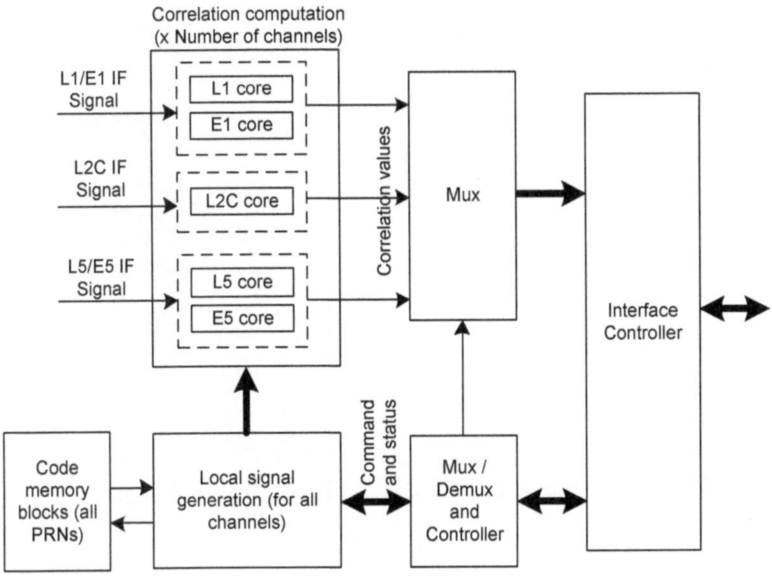

Fig. 4. Components of the baseband module for GPS L1/L2C/L5 and Galileo E1/E5 signals

Signal / Component	Correlator Operating Frequency (MHz)	Resource Utilisation			Power estimate (mW)
		Registers	Combina-tional	Memory (bits)	
GPS L1 C/A	4	446	151	-	1.06
Galileo E1b or E1c	8	436	149	4092	1.84
Galileo E1 (E1b and E1c)	8	631	176	8184	2.24
GPS L2C CM only	4	478	210	-	1.13
GPS L2C (CM and CL)	4	737	245	-	1.61
GPS L5 (Pilot and Data)	40	701	204	-	11.93
Galileo E5a or E5b	40	694	203	-	11.80
Galileo E5	100	1010	253	-	39.28

Table 2. Resource utilisation and power consumption estimates of the core correlator (single channel) for different signals

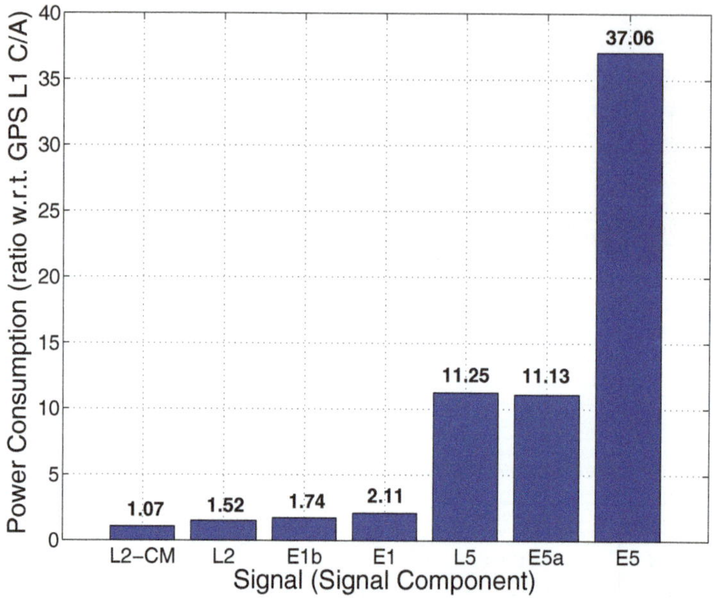

Fig. 5. Ratio of the power estimate for new signals with respect to GPS L1 C/A

resource and power consumption of the Galileo E5a and E5b signals is close to that of the GPS L5 correlator, as expected. As a result of a very high operating frequency the power consumption of the wideband Galileo E5 correlator shoots up to almost 37 times that of the GPS L1 C/A signal. The power consumption for the E5 signal can be reduced a little bit further by focusing more on how the complex mixers are realised as discussed in Shivaramaiah (2011).

The ratio of the power consumption estimate with respect to the GPS L1 C/A is shown in Fig.5. The power consumption was estimated using the PowerPlay Analyzer tool with real IF signal samples provided as an input[1] to the baseband module.

5.2 Complexity comparison results for different baseband configurations

Fig. 6 shows the power consumption of different signals vs. the number of channels. A "channel" comprises the core correlator, timing control, address and data multiplexer/demultiplexer (for a memory mapped interface to the subsequent stage), and some housekeeping operations. Although the resource consumption is not described in detail here, it should be mentioned that the two major memory spreading code sets in the case of the Galileo signal occupy around 410K bits (E1, 4092 bits, 2 signal components, 50 PRNs) of memory and 10K bits (E5 secondary code, 100 bits, 2 components, 50 satellites) which are totally new additions to the GNSS receiver baseband hardware.

[1] The PowerPlay tool estimates the toggle rate of the internal nets and the output pins based on the input signal and the associated clock-frequency.

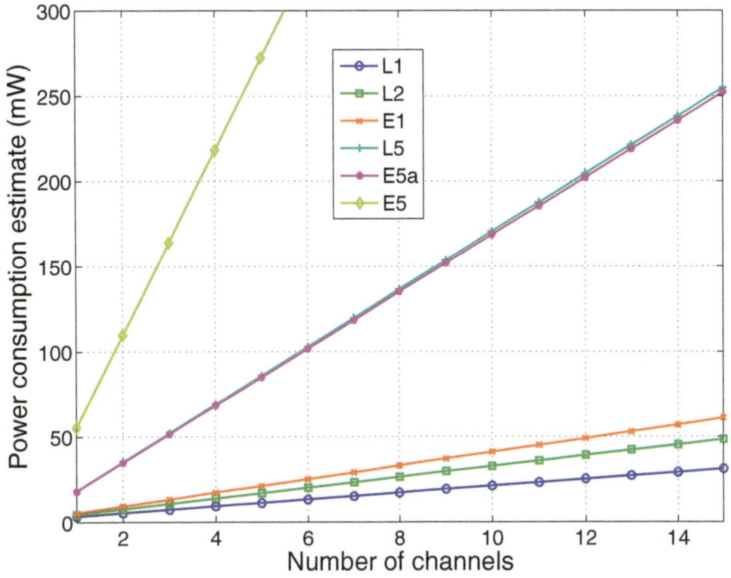

Fig. 6. Power consumption of the entire baseband circuit

Fig. 7 shows the power consumption for different combinations of signals where each signal has been assumed to be using 12 channels. It is interesting to note that a GNSS receiver designed to process all the civilian signals of GPS and Galileo would require slightly short of one watt for the baseband hardware (using the Altera Cyclone-III family device EP3C120F780C8), which is 38 times that of GPS L1 C/A baseband hardware.

5.3 Recommendations for the multi-GNSS baseband design

The challenges that are faced in designing the baseband hardware for a multi-GNSS receiver can be broadly categorized into three groups

- complexity reduction challenges,
- power consumption reduction challenges, and
- resource requirement reduction challenges.

The complexity reduction challenges are not of significant concern because of the availability of design tools that help an engineer to handle the kind of complexity present in this situation. However, it is a good practice to have a modular design keeping in mind the scalability of the architecture to additional signals. The complexity issues are not discussed here.

In most of the situations, the resource and power consumption are highly interrelated. Exceptions to these situations are generally the changes in the operating frequency. Reduction in the operating frequency will basically reduce only the power consumption though it may indirectly reduce the resource requirement to some extent (such as a simplified clock tree or

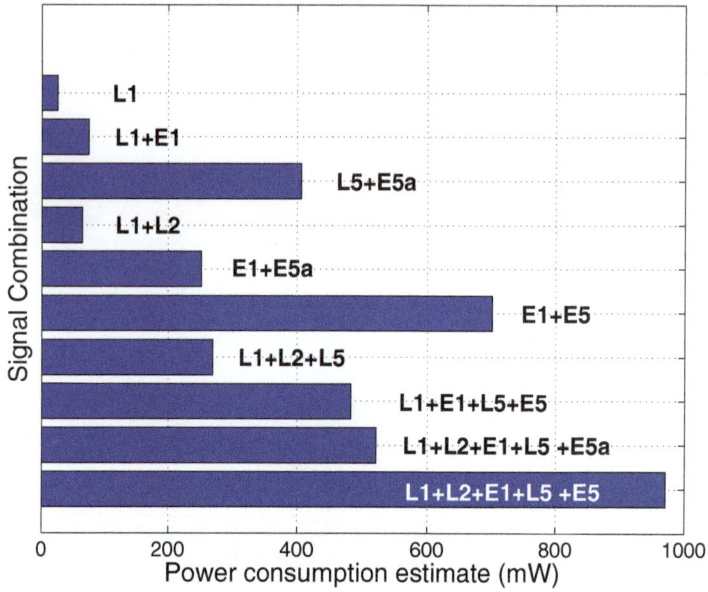

Fig. 7. Power consumption for different multi-signal configurations

reduced fanout requirements due to increased clock period etc.) However, if the reduction in the operating frequency demands a modification in the signal processing chain, then resource requirements may go up. On the other hand, reduction in the resource utilisation will almost always help reduce the power consumption.

The next few paragraphs explore some techniques that enable some progress in overcoming these challenges.

5.4 Resource and power consumption reduction opportunities

5.4.1 Design optimisation of the core correlator blocks

One example of resource reduction is the reference signal mixer for the Galileo E5 signal. The reference signal mixer should be carefully designed to address the complexity vs. propagation delay trade-off.

An architecture for the AltBOC(15,10) modulation (used in Galileo E5 and Compass B3 for example) is shown in Fig. 8. In Fig. 8 it is assumed that the input and the local carrier use two bits and the succeeding stage is not the last arithmetic operation in the chain and hence the carrier mixer output can be encoded with three bits. The local reference signal is assumed to be 2-bit wide which is obtained from a 2-bit subcarrier and 1-bit local code.

The implementation shown in Fig. 8 offers a good trade-off between the complexity and propagation delay requirements compared to both the brute-force logic type of implementation and the brute-force single large-size LUT implementation.

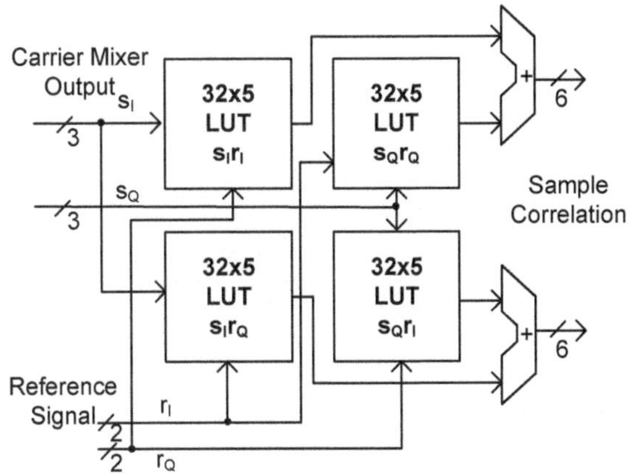

Fig. 8. Local reference mixer for the complex modulation signals

It should be noted that the reference signal mixer example for the complex signal is chosen and dealt in more detail here because it is the correlator block which has a major impact due to the signal structure and is drastically different to the implementation of the reference signal mixer for the GPS L1 C/A signal. It is not difficult to identify other such resource hungry and power hungry blocks and is essentially a part of the baseband hardware design process.

5.4.2 Operating frequency considerations

One of the major contributors for the higher power consumption of the correlators that process new signals is the correlator operating frequency. The operating frequency of the correlator is typically the sampling frequency at which the IF signal samples are received. However in most of the situations, once the signal is brought to the baseband after the carrier mix operation (the signal at this point may still contain residual Doppler) the result can be resampled to a lower sampling frequency. The minimum operating frequency for the stages after the carrier mix operation can then be reduced to twice the spreading code chipping rate (Namgoong & Meng, 2001a,b; Namgoong et al., 2000). Reduction below twice the spreading code chipping rate is possible but care should be taken to trade-off wisely the signal loss vs. correlator power consumption advantage. The carrier mixer output should undergo proper filtering before the sampling frequency reduction which will increase the resource requirement (by the amount of resource consumed by the filter). Initial implementation results show that the resource requirements of the filter are not significant and hence it is not a significant overhead.

5.4.3 Processing signal components separately vs. processing together

The accumulator at the end of the correlator computation chain is a power hungry block. Typically six accumulators are required for the correlator that implements three delayed (early prompt and late) reference signals for the reference signal correlation. The requirement of separate correlation values for the individual signal components increases the requirement of the number of accumulators. For example, the GPS L5 signal requires 12 sets of accumulators

Signal / Component	Correlator Operating Frequency (MHz)	Resource Utilisation			Power estimate (mW)
		Registers	Combina-tional	Memory (bits)	
Galileo E5	100	667	519	-	31.98

Table 3. Resource utilisation and power consumption estimates for the Galileo E5 AltBOC correlator; the reference signal generation is implemented with the help of AltBOC LUT (OSSISICD, 2010)

for each channel and the Galileo E5 requires 24 accumulators per channel. Combining the signal components before the correlation operation is possible but with significant performance degradation. The performance degradation arises mainly due to the data-bit ambiguity. Methods that try to avoid the data-bit ambiguity compromise on the performance parameters of the signal in question. Therefore, again a careful consideration is required to trade-off the performance vs. resource (or power) consumption advantage.

Table 3 shows the power consumption for the Galileo E5 AltBOC correlator if all the four signal components are processed simultaneously. In this case there are only six accumulators required as in the single signal component case. The reference signal in this case is generated according to the AltBOC LUT provided in the Galileo ICD (OSSISICD, 2010). However, the presence of data-bits (assuming that the secondary code phase resolution has already happened) hampers the correlator output and hence the performance. Observe that the power reduction compared to the correlator processing the signal components separately is about 19% which is a significant reduction. In other words it is possible to reduce the correlator power consumption without losing the performance if there is a data aiding mechanism.

5.4.4 Optimising the correlator blocks across signals

Correlator design optimisation is a separate topic of itself as there are several ways to tackle the resource utilisation issue. Moreover the optimisation is often receiver specific. Three examples are given below where the optimisation is possible in specific correlator blocks.

First, the need for subcarrier NCO can be eliminated (even when the multiplication required is not a power of two) by implementing clock multipliers with simple gates. For example, in the case of Galileo E5, the x1.5 clock can be generated by simple gates that implement $\frac{3}{2}$ multiplier.

Second, the carrier and code NCO for different signals from the same satellite can be combined. This is done by programming and generating the required carrier for one of the signals and deriving the difference in the relative Doppler for the second signal.

Third, the operating frequency for the signals can be adjusted such that the operating frequencies can be derived from a single clock with simple dividers. The advantages of such a clock domain construction are simplification of generation of control and timing signals as well as ease of data transfers across different correlation stages of different signals.

6. Summary

This chapter analysed the core correlator complexities of modernised GNSS receiver baseband hardware. A core correlator architecture description has been given and the number of bits for the accumulator has been derived. Power consumption estimates were provided for the new signals at the core correlator level and at the channel level.

It was shown that a GPS and Galileo civil signal receiver baseband would consume approximately 38 times the power of a GPS L1 C/A baseband. The dominant contributor to this increased complexity and power consumption is the Galileo E5 AltBOC signal. In addition, implementation of the core baseband signal processing blocks in FPGA hardware reveals up to eight times the resource requirement compared to the GPS L1 C/A only correlator.

It is possible to optimise the hardware targeting the power consumption with the help of resampling and external aiding. However, the performance trade-off should be carefully looked into. Because the enormous resource and power consumption for the Galileo E5 AltBOC correlator is due to the signal structure itself, it is of interest to explore efficient alternatives to the AltBOC signal and one such attempt is made in Shivaramaiah (2011).

Even if a dedicated Application Specific Integrated Circuit (ASIC) replaces the FPGA baseband hardware, as a rule of thumb, and the authors' own experience with multiple generations of GPS L1 C/A correlator ASIC design, there will be a best case reduction of the FPGA power consumption by a factor of 5. In other words, a baseband ASIC will consume about 100 mW for the L1-L5 and about 200-mW for the all civil GPS+Galileo baseband. This power consumption is very high given that it is only for the baseband hardware and not for the entire receiver. Finally, if other global and regional satellite navigation systems (such as GLONASS, Compass, QZSS, IRNSS) are included, then, the "200 times" estimate of Dempster (2007) would not be far away. Hence it can be concluded that development of a baseband hardware for the commercial general purpose multi-GNSS receiver is still a challenging task. A direction towards a promising solution would be to explore the correlator level reconfigurability across the GNSS signals.

7. References

Braasch, M. & van Dierendonck, A. (1999). GPS receiver architectures and measurements, *Proceedings of the IEEE.*

Dempster, A. G. (2006). Correlators for L2C: Some Considerations, *Inside GNSS* pp. 32–37.

Dempster, A. G. (2007). Satellite navigation: New signals, new challenges, *Circuits and Systems, 2007. ISCAS 2007. IEEE International Symposium on*, pp. 1725 –1728.

Hegarty, C. (2009). Analytical Model for GNSS Receiver Implementation Losses, *U.S. Institute of Navigation International Technical Meeting, ION GNSS.*

Kaplan, E. D. & Hegarty, C. J. (eds) (2006). *Understanding GPS: Principles and Applications*, Artech House.

Namgoong, W. & Meng, T. (2001a). Minimizing power consumption in direct sequence spread spectrum correlators by resampling IF samples-Part I: performance analysis, *Circuits and Systems II: Analog and Digital Signal Processing, IEEE Transactions on* 48(5): 450–459.

Namgoong, W. & Meng, T. (2001b). Minimizing power consumption in direct sequence spread spectrum correlators by resampling IF samples-Part II: implementation issues, *Circuits and Systems II: Analog and Digital Signal Processing, IEEE Transactions on* 48(5): 460–470.

Namgoong, W., Reader, S. & Meng, T. (2000). An all-digital low-power IF GPS synchronizer, *Solid-State Circuits, IEEE Journal of* 35(6): 856–864.

OSSISICD (2010). European gnss (galileo) open service signal in space interface control document.

Parkinson, B. & Spilker, J. (eds) (1995). *Global Positioning System: Theory and Applications*, American Institute of Aeronautics and Astronautics.

Shivaramaiah, N. C. (2004). *A Fast Acquisition Hardware GPS Correlator*, Master's thesis, Center for Electronics Design and Technology, Indian Institute of Science, Bangalore, India.

Shivaramaiah, N. C. (2011). *Enhanced Receiver Techniques for Galileo E5 AltBOC Signal Processing*, PhD thesis, School of Surveying and Spatial Information Systems, University of New South Wales, Sydney, Australia.

Shivaramaiah, N. C. & Dempster, A. G. (2009). Design challenges of a Galileo E1 correlator on the Namuru platform, *IGNSS Symp*, Gold Coast, Australia.

Shivaramaiah, N. C. & Dempster, A. G. (2010). On the baseband hardware complexity of modernized GNSS receivers, *IEEE ISCAS*, pp. 3565 –3568.

Shivaramaiah, N. C., Dempster, A. G. & Rizos, C. (2009). Application of Prime-factor and Mixed-radix FFT Algorithms in Multi-band GNSS Receivers, *Journal of GPS* 8: 174–186.

Zarlink (1999). *GPS Receiver Hardware Design Application Note AN4855*, 2.0 edn, Zarlink Semiconductor.

Zarlink (2001). *GPS 12 channel correlator*, issue 3.2 edn, Zarlink Semiconductor.

3

Evolution of Integrity Concept – From Galileo to Multisystem

Mario Calamia, Giovanni Dore and Alessandro Mori
University of Florence
Italy

1. Introduction

The Galileo navigation system introduced the integrity concept, intended as a continuous control of the information broadcasted by satellites. Although the RAIM technique represents the first example of integrity monitoring, it is able to detect only local errors made at the receiver level. The integrity monitoring applied by EGNOS could instead be seen as the forerunner of the Galileo system. Even if there are many differences in the definition of integrity for the two systems, the aim is the same for both: to protect the user against the failure of the system, warning him in the shortest time and with the greatest precision possible.

The integrity of a navigation system can be defined as follows: "integrity relates to the trust that can be placed in the correctness of information supplied by a navigation system. Specifically, a navigation system is required to deliver an alarm when the error in the derived user position solution exceeds an allowable level (alarm limit). This warning must be issued to the user within a given period of time (time-to-alarm) and with a given probability (integrity risk)" (Oehler et al., 2004).

In the near future a central role will be played by the integrity receiver's capability. This service can be considered essential in the safety critical application domain, particularly in aviation. For these applications, the system's capability of protecting the user against system failure is of primary importance.

Integrity includes the system's ability to supply, at the right time, reliable warnings to the user (alarm). The main problem with this service is how to determine what can be considered safe. This depends on the requirements of the different fields of application. The following parameters are traditionally used to define the safety of the service for a specific application:

- Alarm Limit (AL): the maximum error allowed in the position domain before an alarm is generated.
- Time To Alarm (TTA): the time that elapses between an error's overcoming of the AL and the reception of the alarm by the user's receiver.
- Integrity Risk (IR): the probability that the alarm will not be delivered within the TTA.

Allowable values of AL, TTA and IR depend on the specific application of the navigation system. The Galileo system provides a high level of integrity of the navigation signal. The global integrity concept is the answer to the needs of different types of users who are all looking for different services in terms of signal and performance.

A new concept of Integrity will be introduced in the following paragraphs. In particular, starting from the Galileo Integrity concepts, we will illustrate a few solutions to the integrity problem and describe a new one, in which data of different constellations (GPS/EGNOS and Galileo) are combined in order to improve the accuracy and the availability of the navigation data.

2. Galileo integrity

The integrity concept developed in Galileo has the aim of ensuring the correct computation of the user's position and provide a valid alarm to the user if the error in the position solution has exceeded a fixed threshold - the Alert Limit - relative to the specific application (Martini, 2006). The user can be in one of the following conditions (Table 1):

Case	System Case	System State	System Alert for Satellite	Satellite User Msg.	Comment
1	Fault-Free	Nominal	NO	OK	
2	Fault-Free	Nominal	YES	NOT-OK	False Alert
3	Faulty	Non-nominal	YES	NOT-OK	True Alert
4b	Faulty	Non-nominal	YES	NOT-Monitored	True Alert
4c	Faulty	Non-nominal	NO	OK	Error below Threshold

Table 1. Examples of integrity

In order to estimate all the errors that might occur in different situations, we have adopted a Gaussian model (J. Rife et al., 2004), whose standard deviation derives from the standard deviation of the error distribution and from the accuracy of the system. Moreover, each Gaussian distribution might have a bias, representing the presence of a faulty condition. The following Figure (Figure 1) shows the system's estimate of the error distribution, illustrating the situations displayed in Table 1. The first two cases concern a faulty free condition: the error is modelled with a zero-mean Gaussian distribution. In this case, the system only has an estimation of the error. This estimation could be considered as a sample of the above-mentioned Gaussian distribution, and this sample could be above (1) or below (2) the specific threshold. In case 1, the system is working in nominal condition, whereas case 2 concerns a False Alarm condition. The failure is modelled as the presence of a bias in the error distribution. This bias could be higher than the threshold (case 3), and in that case the system would certainly detect it.

Otherwise, the mentioned bias could be higher than the threshold, but the sample of the distribution could be below this limit (case 4). This case is referred as Missed Detection condition (Martini, 2006).

The Galileo system provides three elements to preserve user integrity:

- Signal-in-Space Accuracy (SISA): this is the expectation of the errors relative to the SW's clock and ephemerides, based on long term observations.

- Integrity Flags (IF): this is a warning relative to a satellite that is transmitting a signal with an excessive error. IF is based on the short term observation of the clock's variations, the ephemerides and the RF signals.
- Signal-in-Space Monitoring Accuracy (SISMA): this is an estimation of the accuracy of the Signal-in-Space Error (SISE).

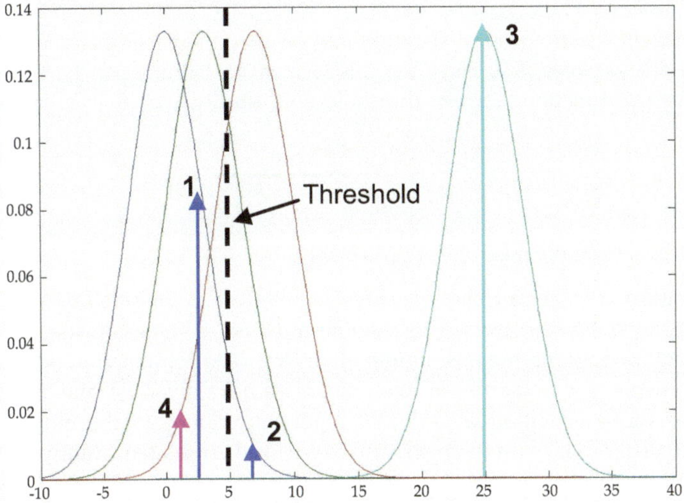

Fig. 1. Integrity events

Using the parameters described above, the user could check the integrity, as follows:

- In a faulty free condition, SISA overbounds the SISE distribution.
- SWs set as NOT OK or NOT MONITORED are discarded from the position computation.
- The user receiver computes the Protection Level using SISA and SISMA parameters.
- PL is compared with the specific AL.

2.1 Faulty free protection level

The Galileo Integrity system is based on the concept of Protection Level. Its main purpose is to calculate the error's bound in the position estimate, in order to be able to control this error with a sufficient level of confidence.

The user receiver judges the accuracy of the computed position solution, typically in term of Horizontal Protection Level (HPL) and Vertical Protection Level (VPL), by means of an estimate of the system errors, an estimate of the local errors and the knowledge of the number and geometry of the SWs used for the positioning algorithm. The computed Protection Level is then compared with a specific Alert Limit, in order to determine the availability of the navigation service.

The original definition of integrity belongs to the position domain, but it can be translated into the Signal-in-Space domain. As a matter of fact, the position error can be replaced by the SISE and the Protection level by the SISA.

2.2 Faulty case

In case of a system failure, the range measurement will be affected by a bias that gets added to the other errors. The aim of the system is to detect this bias. For this reason the Galileo system consists of a Ground Segment (GSS) that is able to monitor range measurements. If the bias exceeds an established integrity threshold, the user will become aware of this via an alarm.

The error detected by the ground segment can be modelled using a zero-mean Gaussian distribution with variance σ^2. Since the false alarm probability can be considered as the area limited by this function between threshold and infinite, we can calculate this threshold as follows:

$$TH = k_{fa} \cdot \sqrt{\sigma_{SISA}^2 + \sigma_{u,L}^2} \tag{1}$$

where k_{fa} derives from the false alarm probability.

The alarm is notified by setting the Integrity Flag relative to the satellite with failure in the information delivered to the users. This satellite must not then be considered by the user in the xPL computation and in the positioning algorithm. The combination between the IF and the PL can ensure the integrity of the information received in the position domain. Moreover, the implementation of a RAIM algorithm comes to the aid of the integrity monitoring, in order to face up to errors caused by local effects (i.e., multipath, interference, jamming and ionospheric effects).

2.3 Evolution of integrity concept

The evolution of the Galileo integrity concept concerns only the verification of system integrity. In particular, based on the above-mentioned definitions, the checking methodology has been modified: the vertical and the horizontal protection levels have been combined in a unique concept, and the user has to compute a probability, named Hazardous Misleading Information Probability (P_{HMI}), which will be compared to the threshold. Once the distribution of the error in the desired reference frame is known (Gaussian overbounding distributions with SISA and SISMA), it will be simple to derive the associated integrity risk both in the faulty and the faulty free conditions appointed to the user equations. Therefore, the error distributions for the vertical (one dimensional Gaussian distribution) and horizontal (Chi Squared distribution with two degrees of freedom) cases need to be derived, and the corresponding integrity risk can be easily computed by analyzing the integral for both distributions with the given alert limits. The integrity risk at the alert limits VAL and HAL are finally computed by adding the vertical and horizontal contributions (Dore & Calamia, 2009).

2.4 Galileo integrity risk

Based on the aforementioned quantities (SISE, SISA, SISMA, IF and TH), the user receiver can derive the integrity risk for the user position solution. This integrity risk is always computed for a given alert limit. Whenever the derived IR at the AL is larger than the allowed IR, the user equipment will raise an alert (Oehler et al., 2004).

The assumptions made for the derivation of the user integrity equation are summarized as follows:

- In a Faulty Free mode, the true SISE for a satellite is zero mean Gaussian distributed with standard deviation SISA.
- In general, a faulty satellite will be flagged as Don't Use.
- For each instance in time, one satellite of those flagged as OK is considered to be faulty but not detected (Faulty Mode). The distribution for the SISE of a faulty satellite is Gaussian with an expectation value TH and a standard deviation SISMA.

Once the distribution of the error in the reference frame is known (Gaussian overbounding distribution with SISA and SISMA respectively), the derivation of the associate integrity risk is straightforward.

Therefore, the error distribution for the vertical (one dimensional Gaussian distribution) and horizontal (Chi Squared distribution with two degree of freedom) cases needs to be derived, and the corresponding integrity risk can be easily computed by analyzing the integral for both distributions with respect to the given alert limit. Finally, the integrity risk at the alert limits HAL (Horizontal) and VAL (vertical) are computed by adding the vertical and horizontal contributions (Oehler et al., 2004).

$$\begin{aligned} P_{HMI}(VAL,HAL) &= P_{IntRisk,V} + P_{IntRisk,H} = \\ &= 1-erf\left(\frac{VAL}{\sqrt{2}\sigma_{u,V,FF}}\right) + e^{-\frac{HAL^2}{2\xi_{FF}^2}} + \\ &+\frac{1}{2}\sum_{j=1}^{N} P_{fail,sat_j}\left(\left(1-erf\left(\frac{VAL+\mu_{u,V}}{\sqrt{2}\sigma_{u,V,FM}}\right)\right)+\left(1-erf\left(\frac{VAL-\mu_{u,V}}{\sqrt{2}\sigma_{u,V,FM}}\right)\right)\right)+ \\ &+\sum_{j=1}^{N} P_{fail,sat}\left(1-\chi^2_{2,\delta_u,H}cdf\left(\frac{HAL^2}{\xi_{FM}^2}\right)\right) \end{aligned} \tag{2}$$

where N is the number of satellites used for the positioning algorithm.

The Integrity Risk computed by the user represents the probability of exceeding the specified alert limits, since the system works according to the hypothesis described above. The Integrity Risk guaranteed by Galileo is partially allocated to user computation and partially to the system itself. This means that a proper design and implementation of the system must guarantee that the system have a sufficiently low probability of being in a condition in which the performance relevant assumption is no longer valid. Only this will ensure that the true overall integrity risk is below the required limit, in accordance with the specified level of service when this service is declared available by the integrity system.

2.5 HMI probability computation algorithm (HPCA)

In order to better understand the P_{HMI} formula (Eq. 2) and all the elements contributing to its design, it is necessary to show the main passages leading to the construction of that equation. These passages could be collected into an algorithm leading to the HPCA algorithm (HMI Probability Computation Algorithm) (Luongo et al., 2004).

The objective of HPCA is to compute the predicted HMI probability in any integrity exposure time interval (150 s) for a given GMS integrity by monitoring state and user geometry.

This algorithm includes the following modules:

- UERE Computation Module
- Position Solution Matrix Computation Module
- Fault Free Position Error Computation Module
- Faulty Position Error Computation Module
- HMI Probability Computation Module

Figure 2 shows the block diagram of the HPCA Algorithm.

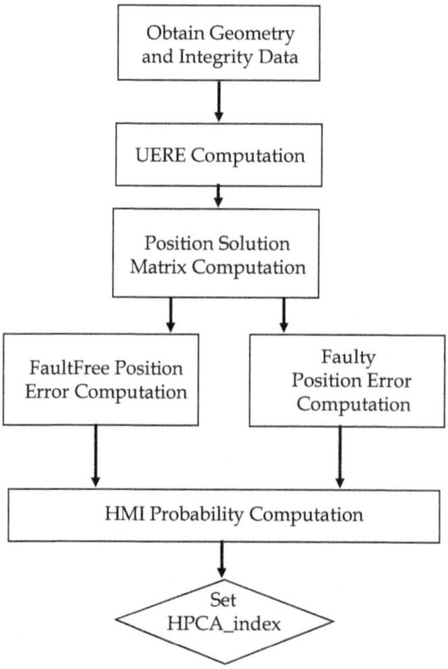

Fig. 2. HPCA algorithm

The "UERE Computation module" computes the predicted standard deviation of total pseudorange error on each signal from visible satellites. The principal formula is as follows:

$$\sigma_{u,RX}[i] = \sqrt{\sigma_{SISA}^2[i] + \sigma_{u,L}^2[i]} \tag{3}$$

It takes into account the signal in space as well as the local errors.

$\sigma_{SISA}[i]$ is the SISA value for the ith satellite used at user level; it is equal to the SISA value broadcasted in the navigation message increased by a factor of 1.1.

$$\sigma_{SISA}^2[i] = (1.1 \cdot SISA[i])^2 \tag{4}$$

$\sigma_{u,L}[i]$ is the predicted standard deviation of the local errors (troposphere, noise, multipath) on the ith signal. The values of the standard deviation of local errors can be read in the UERE table (Galileo satellites). Typical values are those reported in the following Table (F. Luongo et al., 2004). The computation is performed using the following interpolation function:

$$\sigma_{u,L}(i) = a + b \cdot e^{-10 \cdot EL_i} \tag{5}$$

Where A is a matrix that depends on the elevation angles, σ_i is the standard deviation value of the UERE for the ith satellite, while a and b can be computed by the following equation:

$$|ab|^T = (A^T \cdot A)^{-1} \cdot A^T \cdot \sigma_i \tag{6}$$

ID	01	02	03	04	05	06	07	08
Elevation [rad]	0.1745	0.2618	0.3491	0.5236	0.6981	0.8727	1.0472	1.5708
σ_i [m]	1.0300	0.7800	0.6700	0.6000	0.5800	0.5700	0.5600	0.5500

Table 2. UERE table

The predicted standard deviation of total pseudorange error ($\sigma_{u,RX}$) is considered an internal variable of the HMI Probability Computation Algorithm (HPCA).

The "Position Solution Matrix" module computes the position solution matrix. The main formula is reported:

$$K = \left(G^T W G\right)^{-1} G^T W \tag{7}$$

where K is the Position Solution Matrix, G is the Observation Matrix and W is the Weighting Matrix obtained by inverting the Covariance Matrix.

The "Fault Free Position Error" module computes the characteristic of the position error in fault free mode (fault free geometry: all the useable satellites). In particular, it evaluates the standard deviation of the distribution that overbounds the vertical position error and the variance of the distribution that overbounds the horizontal position error. The principal formulas are reported as follows:

$$\sigma_V = \sqrt{\sigma_{ZZ}^2} \tag{8}$$

$$\xi_H^2 = \frac{\sigma_{XX}^2 + \sigma_{YY}^2}{2} + \sqrt{\left(\frac{\sigma_{XX}^2 - \sigma_{YY}^2}{2}\right)^2 + \sigma_{XY}^4} \tag{9}$$

where σ_V is the standard deviation of the model (zero-mean normal CDF) used to overbound the vertical position error in fault free mode.

ξ_H^2 is the variance of the model used to overbound the horizontal position error (along the semi-mayor axis of the error ellipse) in fault free mode.

The $\sigma_{m,n}$ components are obtained using this general expression:

$$\sigma_{m,n} = \sum_{i=1}^{N} K[m,i] \cdot K[n,i] \cdot \sigma_{u,RX}^{2}[i] \tag{10}$$

where "i" indicates the ith satellite, "m" and "n" the reference axis: X, Y or Z.

The "Faulty Position Error" module computes the characteristic of the position error in faulty mode (faulty geometry: one single failure satellite). In particular, the standard deviation of the distribution that overbounds the vertical position error and the variance of the distribution that overbounds the horizontal position error are computed. The principal formulas are reported here:

$$\sigma_V = \sqrt{\sigma_{ZZ\_F}^{2}} \tag{11}$$

$$\xi_H^2 = \frac{\sigma_{XX\_F}^{2} + \sigma_{YY\_F}^{2}}{2} + \sqrt{\left(\frac{\sigma_{XX\_F}^{2} - \sigma_{YY\_F}^{2}}{2}\right)^2 + \sigma_{XY\_F}^{4}} \tag{12}$$

where σ_V is the standard deviation of the model (zero-mean normal CDF) used to overbound the vertical position error in faulty mode.

ξ_H^2 is the variance of the model used to overbound the horizontal position error (along the semi-mayor axis of the error ellipse) in faulty mode.

The σ_{m,n_F} components are obtained using the following general expression:

$$\sigma_{m,n\_F} = \sum_{i=1}^{N} K[m,i] \cdot K[n,i] \cdot \sigma_{u,RX}^{2}[i] + K[m,i_0] \cdot K[n,i_0] \cdot \left(\sigma_{SISMA}^{2}[i_0] - \sigma_{SISA}^{2}[i_0]\right) \tag{13}$$

where "i" indicates the ith satellite, "m" and "n" the reference axis: X, Y or Z.

The "HMI Probability" module computes the probability of HMI. The principal formulas are reported as follows:

$$P_{HMI} = P_{HMI,Fault-Free} + p_{fail} P_{HMI,Faulty} \tag{14}$$

$$P_{HMI,Fault-Free} = P_{HMI,Fault-Free,V} + P_{HMI,Fault-Free,H} \tag{15}$$

$$P_{HMI,Faulty} = P_{HMI,Faulty,V} + P_{HMI,Faulty,H} \tag{16}$$

3. EGNOS integrity

The GPS system is neither accurate nor reliable enough to be accepted as the only instrument of navigation for critical applications. One of the reasons is that no reliable and quick (within seconds) information can reach the user if any problems with the system occur. As a consequence, the GPS system cannot be used for landing approaches, for instance. Airplanes still have to use ILS-systems (Instrument Landing Systems) in case of poor visibility. But the installation and the maintenance of ILS-systems in every airport is expensive. With the SBAS systems, CAT I approaches (limited visibility) will be possible without additional ILS systems. For CAT III approaches (zero visibility), even the SBAS will not suffice, and ILS is still required.

EGNOS provides a European-wide, standardized and quality-assured augmentation service suitable for different fields of applications. Integrity is a key quality and safety parameter, and it alerts users when the system exceeds tolerance limits. EGNOS broadcasts wide-area differential corrections to improve accuracy, and alerts users within six seconds if something goes wrong (integrity).

The receiver combines satellite/user geometry information, with EGNOS-corrected pseudo-ranges, and internal estimates of the tropospheric delay to compute the user position. Ideally, the user would like to have the difference between the computed position and the true position - the true position error (PE) - to be less than the AL. However, since the true position is not known, the PE cannot be determined, and an alternative approach is required.

In fact, the receiver continuously estimates a predicted position error, known as the protection level (PL), for each position solution. The PL can be estimated using the UDRE and GIVE parameters and other local error-bound estimates. It is scaled for compatibility with the probability of non-integrity detection so that the PL should always be larger than PE.

Integrity assessments are based on PL and AL. A new PL is estimated for each computed position solution, then it is compared with the required AL, and an integrity alert is triggered if PL>AL. There is an underlying assumption, that PL>PE, when assessing integrity, and this corresponds to the "safe" zone to the left of the leading diagonal in Figure 3. In the nominal operation case, PL<AL and the system is available. If PL>AL for a particular operation, the EGNOS integrity cannot support the operation, and the system is unavailable.

There is also an "unsafe" zone to the right of the leading diagonal where PL<PE and the integrity assessment provide misleading information (Figure 3). The case at the bottom left corner of the diagram (PL<PE<AL) is also "safe," theoretically, because the AL has not been exceeded, but it should be noted that EGNOS also protects against these out-of-tolerance situations (ESA, 2005).

Different parameters, used in the XPL computation, must be elaborated by the ground segment (Roturier et al., 2001):

- the variance $\sigma^2_{UDRE,i}$ of a zero-mean normal distribution that describes the user differential range error (UDRE) for each ranging source after the application of fast and long-term corrections and excluding atmospheric effects and receiver errors;
- the variance $\sigma^2_{UIRE,i}$ of a zero-mean normal distribution that describes the L1 residual user ionospheric range error (UIRE) for each ranging source after ionospheric corrections have been applied. This variance is determined from the variance ($\sigma^2_{GIVE,i}$) of an ionospheric model based on the broadcast grid ionospheric vertical error (GIVE).
- the variance $\sigma^2_{local,i}$ of a zero-mean normal distribution that relates the pseudo range error due to local receiver noise and multipath;
- the variance $\sigma^2_{tropo,i}$ of a zero-mean normal distribution that defines the residual pseudo range error of a tropospheric correction model.

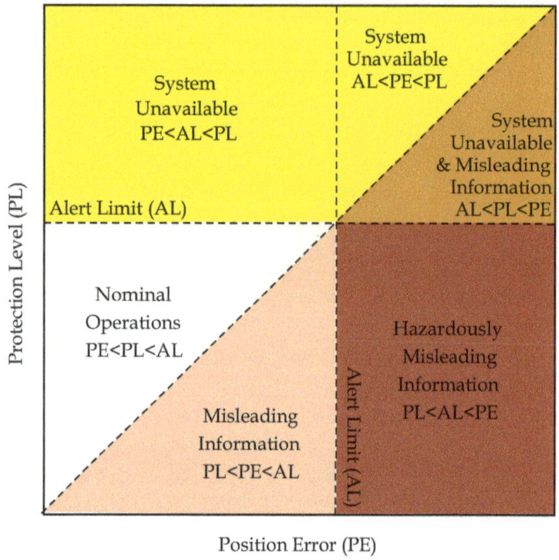

Fig. 3. EGNOS protection level

Assuming that the error in the range domain can be overbounded by a zero-mean Gaussian probability density function, the variance of that distribution is:

$$\sigma^2{}_i = \sigma^2{}_{i,flt} + \sigma^2{}_{UiRE,i} + \sigma^2{}_{air,i} + \sigma^2{}_{tropo,i} \tag{17}$$

where the variance $\sigma^2_{i,flt}$ may be easily derived from $\sigma^2_{UDRE,i}$. From Eq. (17) and for a given user/satellite geometry, it is quite simple to derive the vertical (horizontal) protection level VPL (HPL) equation by:

1. passing from the pseudo range variance domain to the position variance domain, noting that the integrity definitions are all in the position domain;
2. scaling the position domain variance to the integrity requirement. VPL (HPL) is scaled for compatibility with the probability of non integrity detection so that the VPL (HPL) should always be larger than PE.

Compared to the first step, the variance in the position domain residual error is a linear combination of $\sigma^2{}_i$ and is also representative of a zero-mean normal law.

$$\sigma^2{}_{Vposition} = \sum_{i=1}^{N} s^2{}_{V,i}\sigma^2{}_i \tag{18}$$

where $S_{V,i}$ are geometrical parameters.

The second step is achieved by multiplying the position domain variance by a factor k that propagates this variance to a level compatible with the integrity requirement.

$$VPL_{EGNOS} = K_V \sqrt{\sum_{i=0}^{N} s^2{}_{V,i}\sigma^2{}_i} \tag{19}$$

Integrity assessments are based on XPL and XAL: a new XPL is estimated for each computed position solution. It is compared with the required XAL, and an integrity alert is triggered if XPL > XAL.

4. RAIM integrity

The integrity of a navigation system can be checked by using external systems such as SBAS to monitor the correctness of the signals used to calculate position. One of the main drawbacks to this approach is the inherent delay that is introduced when an error is detected, due to the time taken to uplink the information on errors. This section will focus on internal monitoring, and in particular on RAIM. RAIM stands for Receiver Autonomous Integrity Monitoring and is used to denote a monitoring algorithm that uses nothing but the measurements of one particular navigation subsystem, usually a GPS receiver. Conventional RAIM algorithms are designed to protect users from a single satellite failure at a time. However, recent developments have shown that RAIM has the potential to provide integrity even in case of multiple failures for challenging flight categories such as LPV-200 and APV-II (Ciollaro, 2009).

Measurement information is used to compute a position. A test statistic is derived from this position computation. It gets passed to an error detector that will warn the user whenever something is wrong. The error detection procedure will have to obey navigation requirements, and it is important to determine the detection power (or 'error detectability'). It depends on the measurement quality and the configuration, and it is this detection power computation that monitors the system's integrity, determining whether the system has the ability to provide timely warnings when the system is in error. If this is not the case, it will inform the user that it might be unsafe to use the system. It should be noted that position computation algorithms always assume that noise on the measurements has a zero mean. An error or bias, as it is commonly called, is therefore defined as the non-zero mean of measurement noise.

4.1 Satellite slope

The slope, which relates the induced position error to the test statistic, can be calculated directly from geometry and is different for each satellite. The satellite with the largest slope is the most difficult to detect. It produces the largest position error for a given test statistic (Figure 4) (Ciollaro, 2009).

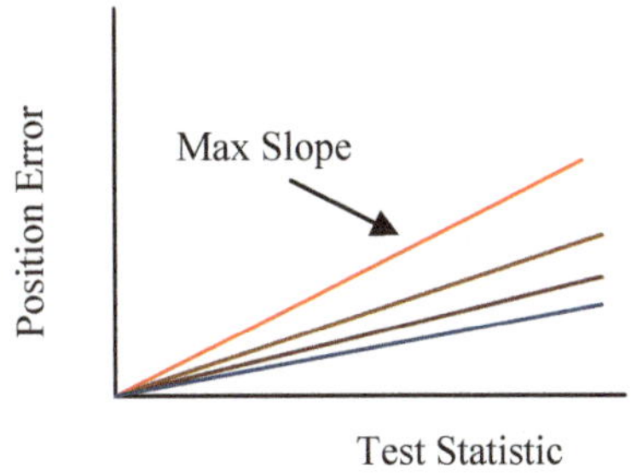

Fig. 4. Satellite slope

The slope is a geometric parameter that can be directly computed from the specific satellite-user geometry, based on the following equations, in the horizontal and vertical planes respectively:

$$Hslope_i = \frac{\sqrt{K_{1i}^2 + K_{2i}^2}\sigma_i}{\sqrt{1 - P_{ii}}} \tag{20}$$

$$Vslope_i = \frac{|K_{3i}|\sigma_i}{\sqrt{1 - P_{ii}}} \tag{21}$$

where $K = (G^T \cdot W \cdot G)^{-1} \cdot G^T \cdot W$ is the weighted pseudo-inverse of the design matrix, where W the inverse of the covariance matrix, while $P = G \cdot K$. The geometric contribution to the slope is given by the K and P matrices.

4.2 RAIM protection levels

The Protection Levels in the vertical and horizontal planes can be described by the following equations (Walter & Enge, 1995), for the vertical and horizontal cases, respectively:

$$VPL_{FD} = \max\{V_{slope}\}T(N, P_{fa}) + k(P_{md})\sigma_V \tag{22}$$

$$HPL_{FD} = \max\{H_{slope}\}T(N, P_{fa}) + k(P_{md})\sigma_H \tag{23}$$

where:

- V_{slope} and H_{slope} are the satellite error slope in the vertical and horizontal planes
- $T(N,P_{fa})$ is the test statistic threshold, and it is a function of the number of satellites (N) and the desired probability of false alarm (P_{fa}). Given the probability of false alarms, the threshold can be found by inverting the incomplete gamma function:

$$1 - P_{fa} = \frac{1}{\Gamma(a)} \int_0^{T^2} e^{-s} s^{a-1} ds \tag{24}$$

where a is the number of degrees of freedom divided by two, or, in terms of the number of measurements N and unknowns M:

$$a = \frac{N - M}{2} \tag{25}$$

- $k(P_{MD})$ is the number of standard deviations corresponding to the specified Probability of Missed Detection. The smaller the P_{MD} value, the higher the number of standard deviations should be considered, since longer tails for the Gaussian distribution should be taken into account.
- σ_V and σ_H are the standard deviations of the error in the position domain in the vertical and horizontal planes.

It should be noted that, when using RAIM, it is common to allocate the whole Integrity Risk, and so the whole Pmd is confined to only one plane (vertical or horizontal) according to the specific operation. For example, for LPV-200, the whole Integrity Risk is allocated to the vertical domain, since this is the most demanding requirement.

4.3 RAIM tst statistic

It is not possible to obtain a direct measurement of the position error. Therefore, the overall consistency of the solution has to be investigated (Walter & Enge, 1995). As long as there are more than four measurements, the system is overdetermined and cannot be solved accurately. This is why a least squares solution is performed in the first place. Since all of the conditions cannot be met realistically and exactly, there is always an error residual to the fit. Therefore, we need to be able to estimate the fit and assume that, if there is a good fit, the position error is most likely small.

An estimate of the ranging errors from the least squares fit and the basic measurement equation is given by:

$$\varepsilon_{wls} = y - G \cdot x_{wls} = (I - G \cdot K) \cdot y = (I - P) \cdot y \qquad (26)$$

where:

$$P = G \cdot K = G(G^T \cdot W \cdot G)^{-1} \cdot G^T \cdot W \qquad (27)$$

From these error estimates it is possible to define a scalar measure, defined as the Weighted Sum of the Squared Errors (WSSE):

$$WSSE = \varepsilon_{wls}^T \cdot W \cdot \varepsilon_{wls} = [(I - P) \cdot y]^T \cdot W \cdot [(I - P) \cdot y] \qquad (28)$$

which is equivalent to:

$$WSSE = y^T \cdot W \cdot (I - P) \cdot y \qquad (29)$$

The square root of WSSE plays the role of the basic observable, because it yields a linear relationship between a satellite bias error and the associated induced test statistic. The test statistic can be defined in both the horizontal and vertical planes.

Typically, a certain threshold, which depends on the required probability of false alarm, is selected. If the statistic exceeds that threshold, then the position fix is assumed to be unsafe. On the other hand, if the statistic is below the threshold, then the position fix is assumed to be valid.

The statistic-vertical error plane is thus broken up into four regions consisting of: normal operation points, missed detections, successful detections and false alarms. Ideally, there would never be any missed detections or false alarms. In reality, a certain number of missed detections and false alarms are allowed, based on the P_{md} and P_{fa} requirements, respectively.

5. Multisystem integrity

With the advent of Galileo, users will be provided with multiple signals coming from different satellite systems. This will improve position accuracy, because the number of satellites in view per user will be almost doubled. Moreover, the higher measurements redundancy will help guarantee a safer position and the detection of errors. This will also result in an improved availability as well as meet the requirements for more demanding flight categories. Therefore, it is necessary to introduce a base-line for a combined system, defining new parameters, a new integrity algorithm and possible ways to combine the two independent systems.

With the term "Multisystem", we intend the improvement of the accuracy and availability of the navigation solution using the combined Galileo and GPS signals. In this context, it is essential for the user to be able to take advantage of the integrity information coming from both Galileo and GPS satellite constellations, in order to prevent users from making errors that might represent an excessive risk. The multisystem integrity algorithm has to establish a link between the two generations of GNSS, defining the relation for integrating different integrity monitoring schemes (Pecchioni et al. 2007) (Ciollaro, 2009).

5.1 Definition of a new integrity algorithm (EGNOS + Galileo)

Two different approaches have been studied to define the new integrity algorithm. They represent two opposite ways of solving the problem of how to combine different integrity concepts: the first one has been called "One-System-Based Integrity," and the second one "Parallel Integrity" (Dore & Calamia, 2009).

The first approach is based on the use of only one algorithm for both systems with an a priori definition of integrity inputs. The integrity analysis can be made either by converting the EGNOS integrity message into an equivalent Galileo integrity message or vice versa, by using the inverse transformation from a Galileo to an EGNOS-like message. In the first case, known as Galileo-Based-Integrity-Algorithm (GBIA), the Galileo Integrity is used as a baseline; in the second case, called EGNOS-Based-Integrity-Algorithm (EBIA), the One-System Integrity is the EGNOS algorithm.

The second approach is based on the use of independent (parallel) algorithms, one for each System, and on an a posteriori integration of the integrity results. The integrity analysis can be performed by monitoring the values assumed by both the Integrity Risk and the Protection Level. If the IR is used as monitored variable, the scheme will be called IR-PIA; otherwise, if the monitored variable is the PL, the method will be called PL-PIA. It is worth noting that the computational load for the IR/PL conversion is expected to be higher than the PL/IR conversion, because an iterative method must be applied (Ciollaro, 2009).

5.2 Galileo based integrity algorithm

The approach chosen for this study is GBIA. The integrity data in fact arrives from the two systems, Galileo and EGNOS, and is implemented inside the Integrity Risk equation of Galileo, in order to estimate the HMI Probability.

Figure 5 shows the block diagram of a GBIA system. The fundamental block of this diagram is the EGNOS/Galileo converter, which has the aim of converting the EGNOS Integrity message into a message that can be used by the Galileo Integrity Algorithm.

The main functions implemented by the EGNOS-Galileo converter are the following (Ciollaro, 2009):

$$\sigma_{SISA,GPS} = f_{SISA,GPS}\sigma_{UDRE} \tag{30}$$

$$\sigma_{SISMA,GPS} = f_{SISMA,GPS}\sigma_{UDRE} \tag{31}$$

Taking into account the different integrity allocation between the Galileo concept, which implies the use of four failure mechanisms, and the EGNOS concept, based on a failure assumption, the contribution of the GPS satellite to IR computation is reduced only to the faulty free mode.

Then it is possible to assume $f_{SISMA,GPS} = 0$ and $f_{SISA,GPS} = 1$, that is:

$$\sigma_{SISA,GPS} = \sigma_{UDRE} \tag{32}$$

Moreover, in order to estimate the standard deviation of the error, the following equation can be used:

$$\sigma^2_{u,L,GPS} = \sigma^2_{UIRE} + (\sigma^2_{Air} + \sigma^2_{Tropo}) \tag{33}$$

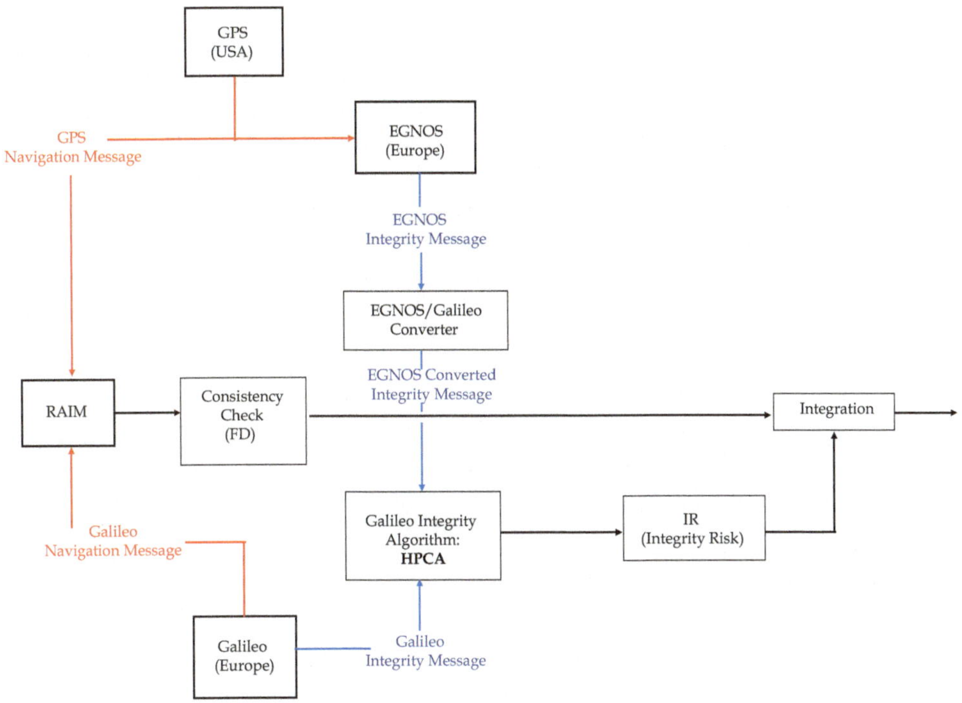

Fig. 5. Galileo-based integrity algorithm

5.3 Multisystem integrity (MSI) algorithm implementation

In this section, we describe a new proposed multisystem integrity algorithm. The algorithm merges integrity data originated by the Galileo and EGNOS systems and employs a Receiver

Autonomous Integrity (RAIM) technique (Weighted RAIM). One of the potential uses of this algorithm consists in the combination of the IR algorithm with the RAIM technique. RAIM is able to detect failures that have not been detected by the IR algorithm. In case of multiple failures, when the WRAIM technique fails, the IR algorithm triggers an alarm.

In this Section we describe the characteristics of this innovative algorithm, pointing out the reason for using the IR equation for the combined constellation Galileo/EGNOS and the reason for taking advantage of a RAIM technique. The EGNOS integrity equation provides a way to measure the integrity based on the incoming signal variances and the satellite geometry. The same is also true for the Galileo IR equation, obviously bearing in mind which data is the Galileo integrity data.

This algorithm is supposed to enable the user to take advantage of the data transmitted by the Galileo and EGNOS systems: the user receiver must consider a single and large constellation in order to strengthen the positioning algorithm and improve the accuracy. This idea is simply in need of the definition of a new integrity concept, which would be able to combine the techniques mentioned above.

5.3.1 IR equation

First of all, we have to explain why the protection level concept turns into the integrity risk concept in a Galileo environment. In an EGNOS domain, IR is the probability that the horizontal (vertical) PL exceeds the horizontal (vertical) AL without the user receiving any alarm whatsoever. This definition requires a clear distinction between the horizontal and vertical cases. Therefore, it is necessary to split IR into two a priori fixed quantities.

On the contrary, as far as a Galileo integrity equation is concerned, the users do not have to evaluate the horizontal and vertical protection levels, but the global IR directly, without making any strict allocations. In fact, different applications need distinct integrity requirements for the horizontal and vertical situations: for example, for a ship the vertical component of the error is not that important for a ship, but it is instead essential for a plane. This last observation leads us to choose the Galileo integrity equation to perform the multisystem integrity algorithm.

The first step in the definition of a new integrity algorithm concerning a combined constellation (Galileo+GPS), is the characterization of the equivalent elements belonging to the two navigation systems. In order to perform the test on the position solution, we opted for the relationship between σ_{SISA}^2 of Galileo and σ_{UDRE}^2 of EGNOS. First of all, these are quantities defined in the same domain SIS. Secondly, they are related to the same typology of error (clock and ephemeris).

The local contribution to the variance of the error in the SIS depends on the elevation angles of the satellite belonging to the two constellations considered. As mentioned before, in order to consider a single constellation composed by both GPS and Galileo SW, we have considered the variance of the error in the SIS as follows:

$$\sigma_i^2 = \sigma_{SISA/UDRE,i}^2 + \sigma_{u,L,i}^2 \tag{34}$$

where the first term, in the case of an EGNOS satellite, derives from Eq. 32; the second term instead represents the local error contribution and can be estimated via the following equation:

$$\sigma_{u,L,i} = a + b \cdot e^{-10El_i} \tag{35}$$

where a and b are parameters that depend directly on El_i (Luongo, 2004). This estimation was performed for both cases, the Galileo and GPS satellites.

The IR equation has been implemented by means of a numerical code in a computer. FF and FM, in Eq. 2, suggest the faulty and faulty free modes. In fact, the Galileo system assumes two separate scenarios: one in which the satellites are all set as use, and the other in which one of the satellites set as use is supposed not to be functioning. When you are in the faulty mode, in the case of Galileo satellites, the SISMA element comes out; in an EGNOS case, only the faulty free mode is instead expected and, because we could not find an equivalent for the Galileo SISMA in its navigation message, we are going to consider the following situation:

- Faulty free: for the Integrity Risk computation we consider all satellites in view, GPS and Galileo are set as OK.
- Faulty mode: the involved satellites are only those belonging to the Galileo constellation; hence the index of the sum concerns only those satellites.

5.3.2 Inputs of the implemented algorithm

The information available a priori for the new algorithm consists of two text files containing position (X,Y and Z components) and velocity (X,Y and Z components) of the SV belonging to the two constellations considered, and obtained through a constellation simulator.

Pseudoranges are obtained by the true satellite-user distance, adding a zero-mean Gaussian noise with variance depending on SISA and the elevation angles of the satellites.

Regarding the SISA and SISMA evaluation, we have considered actual values, adding a Gaussian noise:

$$\begin{aligned} SISA &= 0.87 + N(0, \sigma_{SISA}) \\ SISMA &= 0.7 + N(0, \sigma_{SISMA}) \end{aligned} \tag{36}$$

In this case, $\sigma_{SISA} = \sigma_{SISMA} = 0.01$, in order to simulate a sort of degradation on the signal received. We must also describe the behaviour of the positioning algorithm in the combined constellation case. Generally speaking, if we define X^k, Y^k and Z^k as the coordinates of the K-th satellite and X, Y and Z as the coordinates of the user position, we are able to compute the distance between the satellite and the user (d^k) and the pseudoranges (ρ^k) as follows (Misra & Enge, 2001):

$$d^k = \sqrt{(X^k - X)^2 + (Y^k - Y)^2 + (Z^k - Z)^2} \tag{37}$$

and

$$\rho_c^{(k)} = d^{(k)} + c\delta t_u + \tilde{\varepsilon}_\rho^{(k)} \tag{38}$$

where:

ε_ρ^k : residual error on k-th satellite. b : clock's offset.

Applying a linearization to the (38), we get the expression of the pseudo range model:

$$\Delta\underline{\rho} = G\Delta\underline{X} + \underline{\varepsilon}_{\rho} \tag{39}$$

The matrix G is named Design Matrix, and it consists of the linear coefficients obtained by the partial derivatives of the observation's equations with respect to the estimated coordinates. This matrix characterizes the user-satellite geometry. The number of the columns of G agree with the number of unknowns to be determined (X,Y,Z and b), while the rows equal the number of the available observations (number of satellites in view for both navigation systems). The union of the Galileo and the GPS constellations causes a change in the G matrix. The number of unknowns in fact become five, in order to compute the clock's offset for both systems. In order to estimate the user position's ($\Delta\tilde{X}$) we have to apply the weighted least mean square method to the pseudo range model, organizing the weight matrix (W) with the information contained in the navigation message sent by EGNOS or by the Galileo satellites (considering only SWs in view, or those with an elevation angle greater than 10°):

$$\Delta\tilde{X} = \left(G^T \cdot W \cdot G\right)^{-1} \cdot G^T \cdot W \cdot \Delta\rho \tag{40}$$

where G and W are two matrices of dimension N×5 and N×N respectively, with N representing the number of the satellites used in the positioning algorithm.

5.3.3 Outputs of the implemented algorithm

In this Section we will describe the characteristics of the implemented multisystem integrity algorithm. We will discuss the results of a few simulation tests organized by different typologies (with or without failure) and different durations, in order to test the validity of the proposed algorithm and confirm the expected results.

A peculiarity of this algorithm is the allocation of the Integrity Risk, valid for the computation of the P_{HMI}, and the P_{FA} (False Alarm Probability), required to estimate the RAIM statistics. The false alarm probability of RAIM and the Integrity Risk of Galileo are related to the time required for a specific flight operation. For example, in the case of safety of life applications, this time is equal to 150 seconds. Our study refers to these applications.

The proposed algorithm elaborates the position computation, the RAIM statistics and the IR equation in every second. It is therefore useful to refer to the probability mentioned above as to one second. In order to perform this conversion, we use the binomial distribution, obtaining the value of P_{HMI} and P_{FA}, both initially set[1] at 0.5×10^{-7}, referred to as one epoch (second).

The failures have been reproduced in two different ways:

- Introduction of a step function, at a given test epoch, on the pseudo range of a satellite in view.

[1] Equally split between the two integrity requirements from the initial value of $1 \times 10^{-7} / 150s$ as defined by the ICAO for the avionic integrity requirements.

- Bias on SISA and SISMA.

We chose the step function because it is able to characterize a lasting failure on a satellite. In fact, when we are looking at aeronautical applications, any failures lasting more than six seconds (TTA) are relevant. SISA results from the predictions on a satellite clock and ephemeris errors, and these error estimations are based on long term observations: SISA increases mark out long term failures. SISA derives from a large data batch, so the anomalous behaviour of just one sample is not relevant. On the other hand, pseudorange variations point out instantaneous failures. In case of failures, the new algorithm is able to protect users from:

- long term bias due to errors from the clock and ephemeris data (IR equation);
- long term bias and short term bias due to local errors (multipath, receiver noise) and errors caused by the SV, the SV payload and the navigation message (i.e. ephemeris data, clock) (RAIM algorithm).

5.3.3.1 No failure mode

In a "no failure" condition we are able to judge the behaviour of the new algorithm compared to the single constellation case, and we can also evaluate the performances offered by the code in term of probability of false alarm and missed detection.

Figure 6 illustrates the RAIM statistic in normal operations (Vertical case), without failure, and the correct functioning of this part of the algorithm. In this case the RAIM algorithm has been simulated independently from the IR algorithm, in order to estimate how it behaves with many samples in an epoch.

We tested the IR algorithm in the same way, for the two constellations and in absence of failures (Figure 7).

Figures 6 and 7 show that the RAIM statistic presents some samples that exceed the threshold. In particular, these samples do not exceed the VPL (Vertical Protection Level), therefore they are in the False Alarm zone. This tells us that the RAIM statistic presents a low probability of triggering an alarm, whenever it is not necessary (the main reason for this behaviour of the WRAIM could be seen in the largest sensibility to the outliers of this integrity algorithm). Instead, the IR algorithm has a lower false alarm probability than the previous case, consequently to the fact that the threshold is never exceeded, and the system does not trigger any alarms when the SIS is not affected by any bias.

5.3.3.2 Error on pseudoranges

We simulated the local error by adding a bias (fixed value) to the pseudoranges. Our intent was to emulate the contribution of some types of errors (i.e. multipath) that are not present in the SIS transmitted (local errors) and consequently are not detectable by the ground segment of EGNOS or Galileo, but only by a RAIM technique.

The pseudoranges are calculated by using the true distance between the satellites and the receiver, adding a Gaussian noise that depends on the variance σ_i^2 (Eq. 34). In addition to the noise, in order to simulate the malfunctioning in the biased case, by a certain epoch we added a fixed value to the range measurement. Since the IR algorithm is not able to detect these kinds of errors, we present the results of the WRAIM part of the proposed algorithm for this first model of failure.

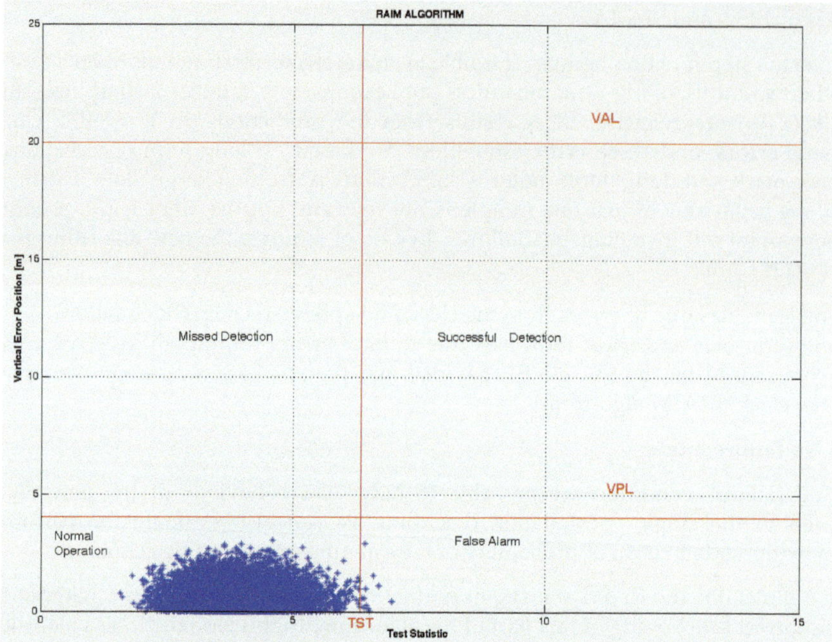

Fig. 6. WRAIM in faulty free condition

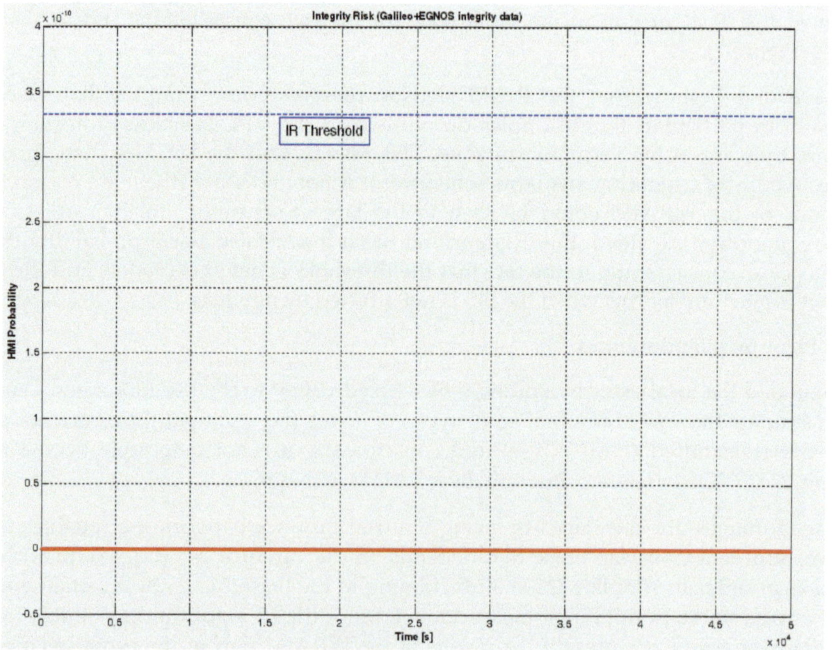

Fig. 7. IR algorithm in faulty free condition

Figures 9 and 10 show how the RAIM statistic behaves in the presence of a bias of 20 meters inserted from the 30-th epoch in a satellite belonging to the Galileo constellation, and a bias of 10 meters, from the 50-th epoch, in a GPS satellite. In both cases the RAIM statistic (green and blue curves) exceeds the Test Statistic Threshold by a probability of 100%. The Figures show the instantaneous behaviour of the RAIM. This way of representing the RAIM process

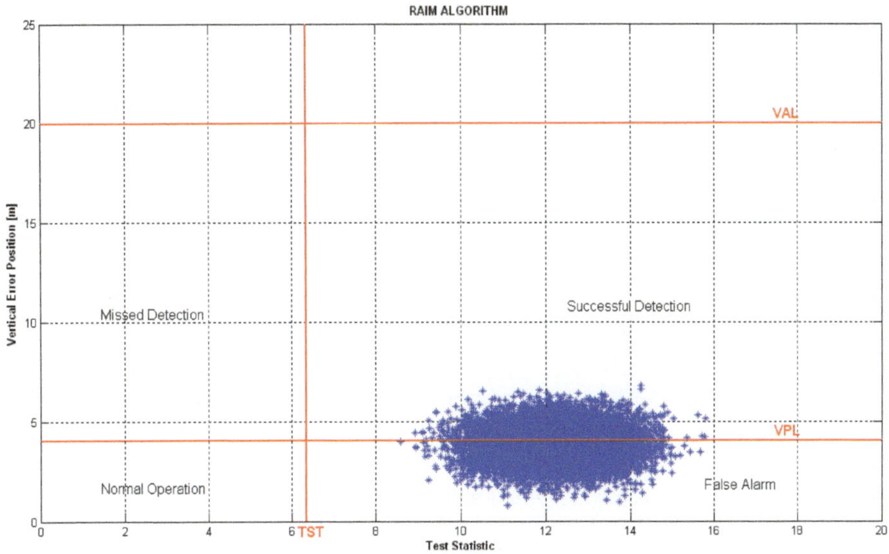

Fig. 8. RAIM and bias on pseudorange - satellite Galileo PRN 4 - RAIM algorithm

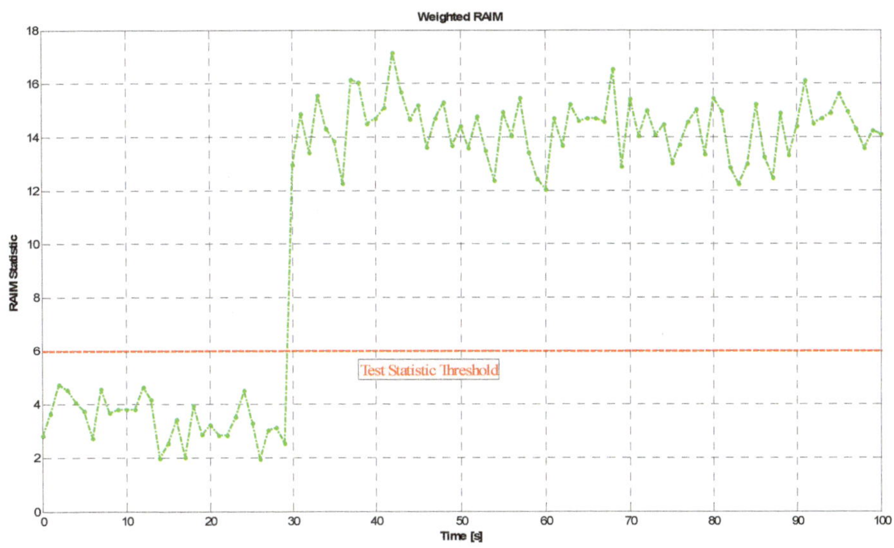

Fig. 9. RAIM and bias on pseudorange - satellite Galileo PRN 4 - Multisystem Integrity Algorithm

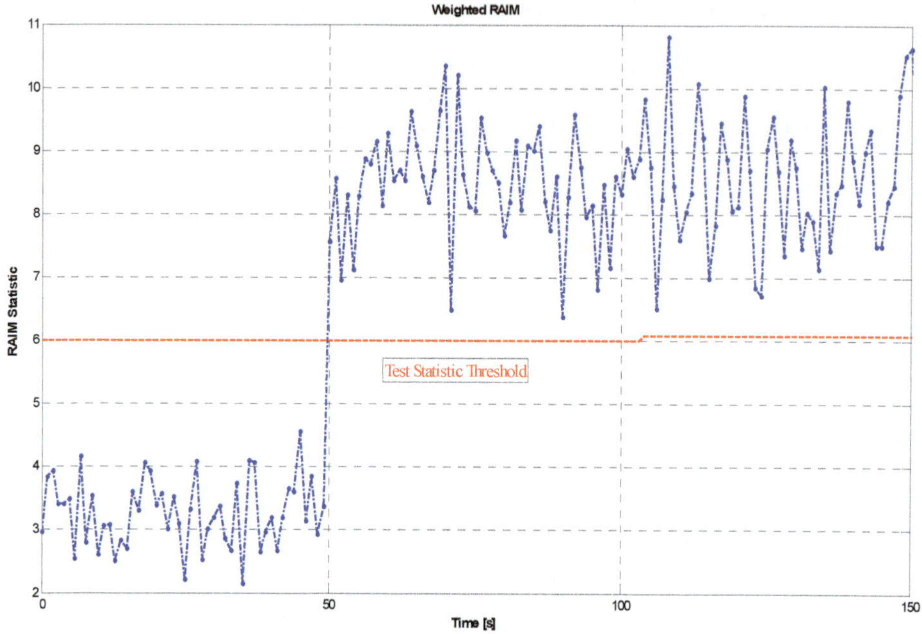

Fig. 10. RAIM and bias on pseudorange - satellite GPS PRN 8

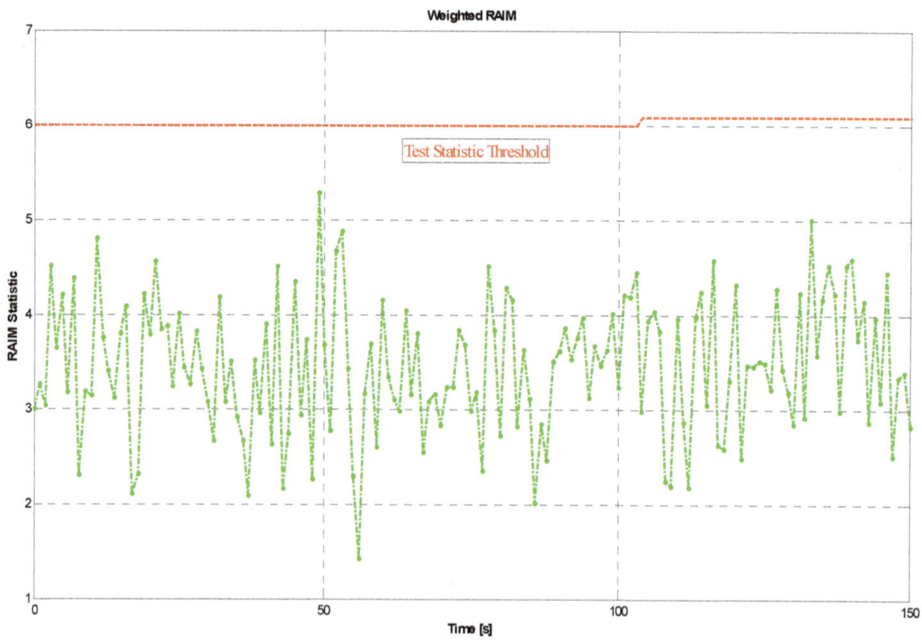

Fig. 11. RAIM and bias on all pseudoranges

is not typical; in fact, in the example depicted in Figure 6, different samples of one epoch[2] are considered in order to obtain an estimate of the proper functioning of the algorithm. Figure 8 shows this kind of analysis made for the biased case (Galileo satellite, Vertical case). The RAIM technique clearly detects the error in the pseudorange; in fact, the ellipse of point leaves the normal operation region exceeding the TST. The same result can be achieved also in the GPS biased case.

We can obtain different results by adding a bias on all pseudo ranges relative to all satellite in view. Through this kind of simulation we can reach the results shown in Figure 11.

In Figure 11, the RAIM statistic remains around zero value: the receiver assigns the 10 metres bias entirely to the two temporal unknowns, the GPS time clock's offset and the Galileo time clock's offset.

In conclusion, we figured out that the RAIM statistic is not able to detect the failures on more than one satellite at the same time. This is a limit for this algorithm, which leads us to conclude that the RAIM algorithm does not work properly when used as single integrity system.

5.3.3.3 Error on the SISA/SISMA value

We simulated the error on the signal in space by adding a bias on the standard deviation of the noise considered in the SISA and SISMA computation of two random satellites belonging to one of the two constellations considered. In this failure mode, SISA and SISMA values have been implemented as in Eq. 36, assuming the following value for the respective standard deviation:

$$\begin{aligned} \sigma_{SISA} &= 10 \\ \sigma_{SISMA} &= 7 \end{aligned} \tag{41}$$

Figure 12 shows the behaviour of the implemented algorithm, in particular the IR equation, when the bias on SISA and SISMA is considered in two Galileo satellites. In this case the algorithm triggers alarms with a probability of 2%.

Comparing this with a single constellation case (only a Galileo satellite), Figure 13 shows the behaviour of the Integrity Risk algorithm in the Galileo case, considering the same size of bias for the same satellite. What is clear from this comparison is the decrease of alarms (~10%, in the second case) triggered by the system achieved by using the combined constellations. This means that in a dual constellation the combined system provides a safe position for the user.

As described in the previous Section, since these disturbances are not related to a variation in the pseudoranges ($\Delta\rho$), we are not able to detect those errors through the RAIM statistic. The trend of the statistics is similar to that in Figure 11, in which the curve never exceeds the Threshold; for the sake of brevity we didn't report this picture.

In conclusion, Figure 14 shows the behaviour of the described algorithm when the bias is applied to the SISA value belonging to two GPS satellites; in this case the biased SISMA is

---

[2] the satellite configuration remains the same during the simulation; however, the noise added to the pseudoranges varies.

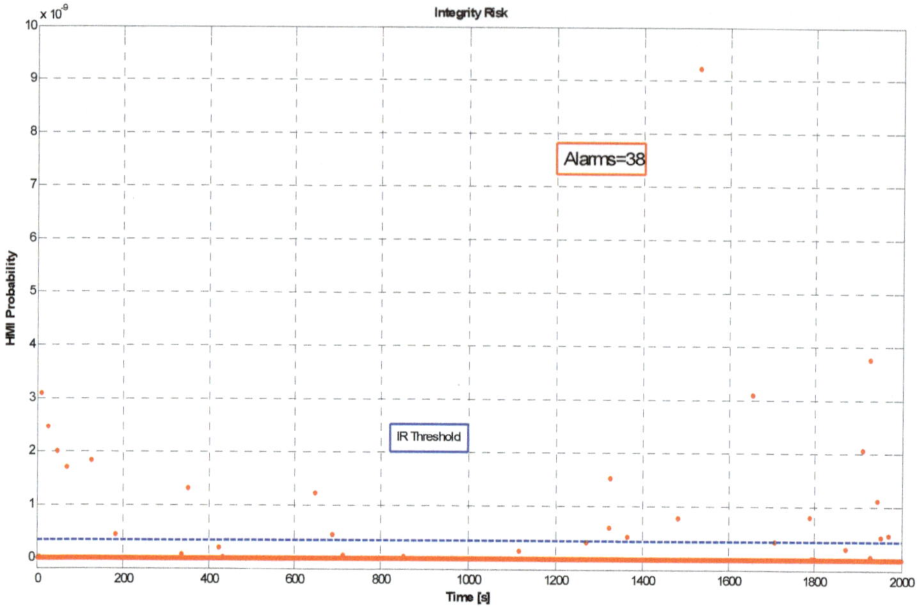

Fig. 12. IR algorithm combined constellation and bias on SISA and SISMA, satellites Galileo PRN15 and PRN22

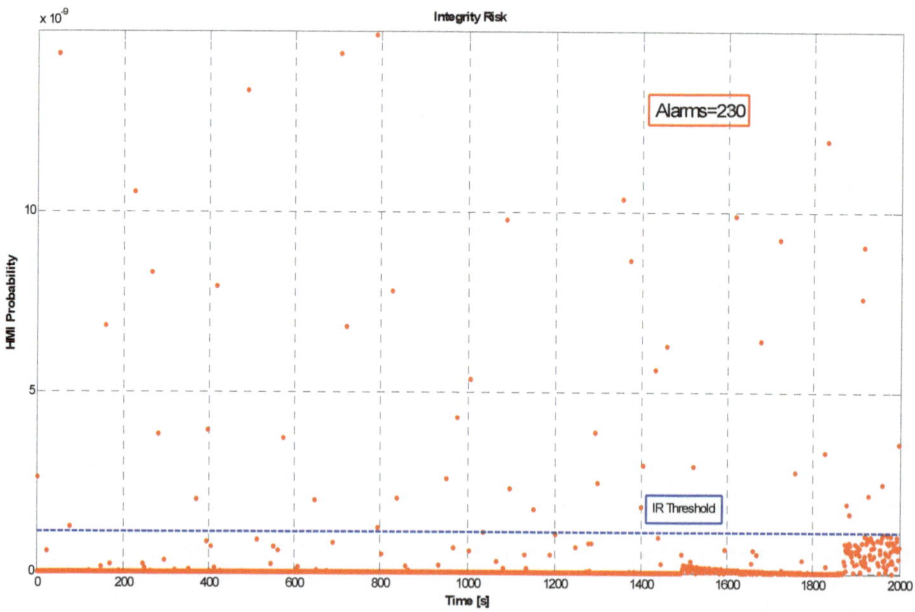

Fig. 13. Galileo IR algorithm and bias on SISA and SISMA, satellites Galileo PRN15 and PRN22

not present because this quantity is not broadcasted by the GPS satellite and, as we stated previously, the integrity data delivered by these satellites does not alter the Faulty section of the P_{HMI} equation. For this reason, in this context, the above-mentioned probability does not reach high values. Indeed, the statistics never exceed the Threshold.

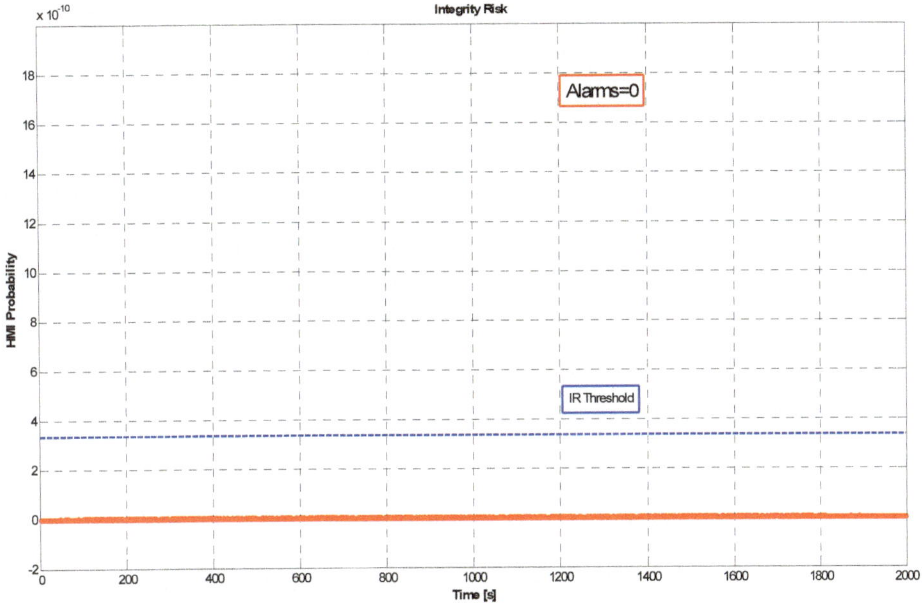

Fig. 14. IR algorithm combined constellation and bias on SISA, satellites GPS PRN3 and PRN10

6. RAIM evolution: ARAIM

An interesting development of the described study is the analysis of the new RAIM technique, the Advanced RAIM. ARAIM, proposed in 2010 by the GEAS (GNSS Evolutionary Architecture Study), could be considered as an evolution of the classical RAIM. This new solution takes advantage of the availability of different new navigation systems (i.e. Galileo) in order to improve receiver performances.

ARAIM is an extension of the single-frequency RAIM. Both are based on an airborne comparison of each satellite measurement to the consensus of the other available satellite measurements (GEAS, 2010). However, the differences between the two techniques are also important. ARAIM should be pursued for the worldwide vertical guidance of civil aircraft based on two or more GNSS constellations radiating at two ARNS/RNSS frequencies (L1 and L5). The main characteristic of the Advanced RAIM would support vertical guidance to decision heights of 200 feet (LPV-200), whereas single-frequency RAIM only supports LNAV guidance. As such, ARAIM must protect vertical errors at levels of 35 meters, while RAIM only needs to detect lateral errors of 200 meters or so. In addition, LPV-200 corresponds to a severe major hazard level (10-7), and LNAV is only major (10-5).

The proper functioning of this system does require some assistance from the ground; for example, the ISM (Integrity Support Message), which is a message developed using reference receivers on the ground, is communicated to the aircraft. The ISM message conveys the safety assertions associated with each of the underlying satellite systems to the sovereign responsible for a given airspace. These messages would contain performance estimates for each satellite to be used for navigation. ARAIM therefore uses a multiplicity of satellites in a dual-constellation environment to take responsibility for all faults that arise between dispatch and the completion of approach.

As described in the previous Section, one of the potential uses of the Multisystem Integrity algorithm is represented by the combination of the IR algorithm with the RAIM technique. ARAIM is still in a feasibility status, and a comparison—in order to test and verify the requirements and highlight the differences between the two approaches—between its results and Multi System Integrity cannot yet be performed. Indeed, the two systems have the same aim: to improve the reliability of the position solution provided by the system in particular conditions (LPV-200 for ARAIM), taking advantage of different navigation systems.

7. Conclusions

The totality of the tests made on the implemented code has been planned with the aim of characterizing the performances of the algorithm respectively in faulty free and in faulty mode. The use of the Galileo and EGNOS system as a single and augmented constellation allows us to develop the positioning algorithm and improve the position accuracy. Furthermore, the combination of the two SVs systems enables us to obtain some benefits from the RAIM point of view.

Our proposed solution starts from the integrity equation defined for the Galileo system and adapts it to the combined Galileo + EGNOS system, or rather, it combines the integrity data supplied separately by the two navigation systems, with the aim of computing the Hazardous Misleading Information Probability. We focused our attention on the IR equation: the implemented code reproduces the IR equation as it is presented in literature, that is, with the SISA values relative to Galileo and GPS satellites, and SISMA relative only to the Galileo ones, in faulty free and faulty mode, respectively. The results obtained testing the algorithm in the presence of failure have provided positive indications on the implemented IR equation: in these cases, the HMI probability increases with the value of the bias.

Alhough the IR protects the user against extended failure, whose effects revert on the SISE estimation, the RAIM technique could instead highlight instantaneous errors on the distances measured by a Galileo or a GPS satellite. RAIM and IR compensate each other, or rather, the RAIM indicates failure unperceived by the IR and vice versa; therefore the combination of the RAIM technique with the integrity equation has proved to be a good idea. This technique is based on a very different concept than protection levels and leads to different results. However, the Galileo integrity concept is more complete than the GPS/SBAS and RAIM integrity concepts and offers more protection from failures. However, this concept needs to be investigated further, in particular regarding the assumptions to be used for the error distributions and the parameters to be considered in the integrity equation. Indeed, although more complete, the new integrity concept introduced by Galileo is more complex and less intuitive than SBAS and RAIM protection level concept.

A possible development of the proposed algorithm could be the definition of SISMA analogous for the GPS satellites in order to contribute to the IR equation under faulty conditions.

The present study is only a preliminary analysis. In order to better evaluate the performances of the proposed algorithm, we need to use realistic data (i.e. pseudoranges measurements obtained through a real GNSS receiver) as inputs of the implemented code.

8. Acknowledgements

Some of the concepts illustrated in this Chapter have been developed by the authors and their collaborators at the University of Florence, also within the following projects:

- GIRASOLE (Galileo Safety of Life Receivers Development), March 2005 - September 2006, financed by the GSA (European GNSS Agency) under the contract GJU/05/2415/CTR/GALILEOSOL;
- SWAN (Sistemi software per Applicazioni di Navigazione), December 2007 - June 2009, financed by the Italian Space Agency (ASI) under the contract DC-IPC-2006-160;
- PEGASUS (Platform of Enhanced Gnss receiver for Application in Sol User Segment), December 2010 - December 2011, financed by the Italian Space Agency (ASI) under the contract. ASI I/024/10/0.

9. References

G. Dore, M. Calamia, "Evolution of Integrity Concept : from Galileo to Multisystem", ENC-GNSS 2009, Naples (Italy), May 2009.

V. Oehler, F. Luongo, J. P. Boyero, R. Stalford, H. L. Trautenberg, J. Hahn, F. Amarillo, M. Crisci, B. Schalarmann, J. F. Flamand, "The Galileo Integrity Concept", ION GNSS 17th International Technical Meeting of the Satellite Division, 21-24 Sept. 2004.

C. Pecchioni, M. Ciollaro, M. Calamia, "Combined Galileo and EGNOS Integrity Signal: a multisystem integrity algorithm", 2nd Workshop on GNSS Signals & Signal Processing, Apr. 2007.

T. Walter and P. Enge, "Weighted RAIM for Precision Approach", Stanford University, 1995.

C. Pecchioni, "L'integrity nei sistemi combinati di navigazione satellitare: confronti, algoritmi e verifiche", Università degli Studi di Firenze, Sept. 2006.

P. Misra and P. Enge, "Global Positioning System, Signal, Measurements and Performance", Ganga-Jamuna Press, 2001.

M. Ciollaro, "GNSS Multisystem Integrity for Precision Approaches in Civil Aviation", Università degli Studi di Napoli "Federico II", Feb. 2009.

"Galileo Integrity User Equations - Working Paper", GAL-TNO-GLI-SYST-I/0630.

F. Luongo, V. Oehler and R. Stalford, "HPCA Input/Output Test Data", Galileo Industries, Apr. 2004.

B. Roturier, E. Chatre and J. Ventura-Traveset, "The SBAS Integrity Concept Standardised by ICAO. Application to EGNOS", GNSS 2001, May 2001.
I. Martini, "Analisi delle prestazioni degli algoritmi di integrità del sistema Europeo di navigazione satellitare Galileo", Università degli studi di Firenze, 2006.
J. Rife, S. Pullen, B. Pervant and P. Enge, "Paired Overbounding and Application to GPS Augmentation", Stanford University & Illinois Institute of Technology, 2004.
ESA-DEUI-NG-TN/01331, "Galileo Integrity Concept", issue 1, 5 July 2005.
ESA, "EGNOS Fact Sheet - 3. Integrity Explained", May 2005.
GEAS Panel, "Phase II of the GNSS Evolutionary Architecture Study", Feb. 2010.

4

Unambiguous Processing Techniques of Binary Offset Carrier Modulated Signals

Zheng Yao
Tsinghua University
China

1. Introduction

In recent years, applications of global navigation satellite systems (GNSS) are developing rapidly. The growing public demand for positioning and location services has generated higher requirements for system performance. However, the performance of the traditional GPS is constrained by its inherent capability. In order to cope with both the civil and military expectations in terms of performance, several projects are launched to promote the next generation of GNSS (Hegarty & Chatre, 2008). GPS is undergoing an extensive modernization process (Enge, 2003), while the European satellite system, Galileo, is also under construction. In addition, Russia is restoring their GLONASS (Slater et al., 2004), and China is in the midst of launching Compass (Gao et al., 2007).

Based on the experience gained during the traditional GPS design and operation, signal structures of these new navigation systems have been well designed (ARINC, 2005, 2006). A large number of modifications have been made intended to address the main weakness of traditional GPS, and to enhance its inherent performance. The accuracy and reliability of those modernized signals and the compatibility between new signals and already-existing signals have been simultaneously taken into account in the design.

Binary offset carrier (BOC) (Betz, 2001) and multiplexed binary offset carrier (MBOC) modulations (Hein et al., 2006) have been chosen as the chief candidate for several future navigation signals, for example, GPS L1C, GPS M-code, and Galileo open service (OS) signals. BOC modulation is a square-wave modulation scheme. It moves signal energy away from the band center and thus achieves a higher degree of spectral separation between BOC modulated signals and other signals which use traditional binary phase shift keying (BPSK) modulation, such as the GPS C/A code, in order to get a more efficient sharing of the L-band spectrum. Besides, many studies (Avila-Rodriguez, et al., 2007; Betz, 2001; Hein, et al., 2006) show that BOC modulation also provides better inherent resistance to multipath and narrowband interference.

However, despite these advantages, some problems remain with the use of BOC modulation. According to the theory of matched filtering (Proakis, 2001), when the waveform of the local signal is as same as the received one, the output of the correlator has the highest signal-noise-ratio (SNR). For this reason, in traditional GPS receivers, both of the acquisition and tracking are based upon the auto-correlation function (ACF) of the received signals. Nevertheless, because of the square-wave modulated symbol, a BOC modulated signal has a sawtooth-like, piecewise linear ACF which has multiple non-negligible side peaks along with the main peak. Since there are significant amount of signal energy located at side peaks of BOC ACF, in acquisition stage, under the influence of noise it is quite likely that one of side peak magnitudes exceeds the main peak, and false acquisition will happen. If false acquisition occurs, the code tracking loop will initially lock on the side peak. Similarly, due to the side peaks of ACF, in code tracking loop, the discriminator characteristic curve of a BOC modulated signal has multiple stable false lock points. Once the loop locks on one of the side peaks, it would result in intolerable bias in pseudorange measurements, which is unacceptable for GNSS aiming to provide accurate navigation solution. This problem is reputed as the ambiguity problem for BOC modulated signal acquisition and tracking. And in order to employ BOC modulated signals in the next generation GNSS, solutions have to be found to minimize this bias threat.

In this Chapter, the ambiguity problem of BOC modulated signals as well as its typical solutions is systematically described. An innovative design methodology for future unambiguous processing techniques is also proposed. Some practical design examples on this methodology are also given to show the practicality and to provide reference to further algorithm development.

The rest of the Chapter is organized as follows. In Section 2, the concept and some main characteristics of BOC modulated signals are given, and the ambiguity problem is also described. In Section 3, some existing representative solutions to ambiguity problem are reviewed. Then in Section 4, we present a parameterized chip waveform pattern, and on this basis, give the analytic design framework for side-peak cancellation (SC) based unambiguous BOC signal processing algorithm development. As two application examples of the proposed design framework, the design process of an SC unambiguous acquisition algorithm as well as an SC unambiguous tracking loop is described in Section 5 and Section 6, respectively. Finally, some conclusions are drawn in Section 7.

2. BOC modulated signals

2.1 Definitions and main characteristics

In order to take advantage of the frequent phase inversions in the spreading waveform to realize the precise ranging, and to obtain excellent multiple access capability, the majority of GNSS employ direct sequence spread spectrum (DSSS) technique. DSSS can be regarded as an extension of binary phase shift keying (BPSK). The transmitting signal yielded by this technique can be expressed as the product of the un-modulated carrier, data $d(t)$, as well as the baseband spreading signal $g(t)$, that is

$$s(t) = A_s d(t) g(t) \cos(2\pi f_0 t + \theta) \tag{1}$$

where A_s is the amplitude of signal, f_0 is the carrier frequency in Hz, and θ is the carrier phase in radians. The baseband spreading signal $g(t)$ can be further represented as

$$g(t) = \sum_{i=-\infty}^{\infty} (-1)^{c_i} p(t - iT_c) \tag{2}$$

where c_i is the spreading sequence of binary digits {0,1}, $p(t)$ is spreading symbol, and T_c is the period of the modulated symbol. In conventional GPS, both of C/A code signal and P(Y) code signal use BPSK-R(n) modulation whose spreading symbol is the energy normalized rectangular pulse with the lasting time $T_c = 1/(n \times 1.023\text{MHz})$:

$$p_{\text{BPSK-R}}(t) = \begin{cases} \frac{1}{\sqrt{T_c}}, & 0 \le t < T_c \\ 0, & \text{others} \end{cases} \tag{3}$$

In principle, the spreading symbol of DSSS signals can be any shape. BOC modulated signal is a variant of basic DSSS signal. The baseband BOC modulated signal can be regarded as the result of multiplying the BPSK-R signal with a sub-carrier which is equal to the sign of a sine or a cosine waveform:

$$s_{\text{BOC}}(t) = s_{\text{BPSK-R}}(t)\text{sgn}\left[\sin(2\pi f_s t + \phi)\right] \tag{4}$$

where sgn($\cdot$) is sign function, f_s is the sub-carrier frequency, and ϕ is the phase of sub-carrier. Two common values of ϕ are 0 or $\pi/2$, for which the resultant BOC signals are referred to as sine-phased BOC or cosine-phased BOC, respectively. In this Chapter, we focus on sine-phased case. For information on cosine-phased BOC signal unambiguous processing, see (Lohan et al., 2008). Using the terminology from (Betz, 2001), a sine phased BOC modulated signal is denoted as BOC$_s$(m,n), where m means the ratio of the square wave frequency f_s to 1.023 MHz, and n represents the ratio of the spreading code rate f_c to 1.023 MHz. m and n are constrained to positive integer $m \ge n$, and the ratio $M = 2m/n$ is referred to as BOC-modulation order, which is constrained to positive integer.

Under the assumption that the spreading sequence has an ideal correlation characteristic, the power spectrum density (PSD) of BOC$_s(f_s, f_c)$ can be expressed as (Betz, 2001)

$$S_{\text{BOC}_s}(f) = \begin{cases} T_c \frac{\sin^2(\pi f T_c)}{(\pi f T_c)^2} \tan^2\left(\frac{\pi f}{2 f_s}\right), & M \text{ even} \\ T_c \frac{\cos^2(\pi f T_c)}{(\pi f T_c)^2} \tan^2\left(\frac{\pi f}{2 f_s}\right), & M \text{ odd} \end{cases} \tag{5}$$

It can be seen that due to the effect of subcarrier, BOC modulated signals symmetrically split the main energy component of the signal spectrum and move them away from the band center, so that they have a higher degree of spectral separation with other BPSK-R modulated signals on the same carrier frequency. Moreover, as noted in (Betz, 2001), BOC modulated signals have greater root-mean-square (RMS) bandwidth compared with traditional BPSK signals with the same spreading code frequency. The greater the RMS bandwidth is, the better the inherent ability to mitigate white Gaussian noise and narrowband interference during tracking will be. Consequently, with same f_c, BOC modulation provides better resistance to thermal noise and narrowband interference than BPSK-R modulation theoretically. However, the ambiguity of the autocorrelation function of sine-BOC modulated signal induces a risk of biased measures in code synchronization.

2.2 Ambiguous problem

The difference of the spreading chip waveforms between BPSK-R signals and BOC signals leads their difference in ACF shapes, and thus makes the distinction of acquisition and tracking performance. The ACF of BPSK-R modulated signals is a triangle, but BOC signals have sawtooth-like, piecewise linear ACF. The normalized BOC(m,n) ACF without filter can be expressed as (Yao, 2009)

$$R_{BOC}(\tau)=\begin{cases}(-1)^{k+1}\left[\left(-\frac{2k^2-2k}{M}+2k-1\right)-(2M-2k+1)|\tau|\right], & |\tau|\le T_c \\ 0, & \text{others}\end{cases} \quad (6)$$

where $k=\lceil M\cdot|\tau|\rceil$, and $\lceil x\rceil$ means the smallest integer not less than x. In Figure 1, the normalized ACF envelopes of BPSK-R(1) signal and BOC(2,1) signal are drawn.

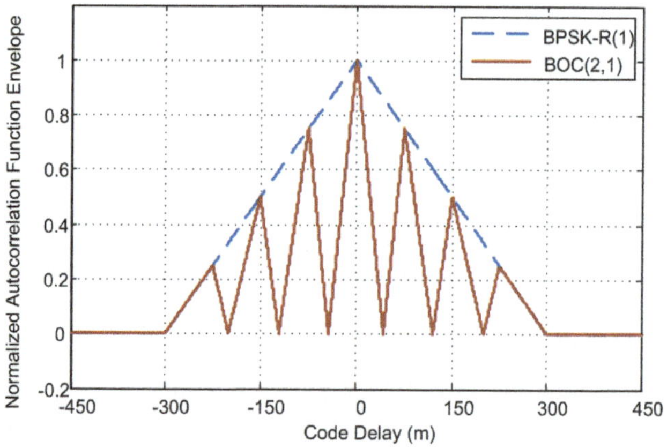

Fig. 1. BPSK-R(1) and BOC(2,1) Normalized ACF Envelopes

From Figure 1 we can see that these two signals have the same spreading chip rate 1.023 MHz, but their ACFs have entirely different shapes. Compared with the triangular ACF of BPSK-R(1) signal, that of BOC(2,1) signal has a sharper main peak, which means better tracking accuracy in thermal noise. However, the ACF of BOC signal has multiple side peaks within $\tau=\pm1$ chips. At the acquisition and tracking stages, these side peaks could be mistaken for the main peak.

When the traditional acquisition and tracking algorithms are employed to process BOC($2n$,n) signal, the shapes of statistic and the discriminator curve are shown in Figure 2(a) and Figure 2(b), respectively.

The traditional acquisition and tracking of DSSS signal have been very well discussed (Ziemer & Peterson, 1985). From Figure 2(a) we can see that since there are significant amount of signal energy located at side peaks of BOC ACF, under the influence of noise it is quite likely that one of side peak magnitudes exceeds the main peak, and false acquisition will happen. For M-order BOC signal, the energy ratio between the i-th largest side peak and the main peak is

$$\xi_i = \left(\frac{M-i}{M}\right)^2 \tag{7}$$

For $M = 2$, side peaks are 6 dB weaker than the main peak. But for $M = 6$, the gap between the largest side peak and the main peak is only 2.5 dB. With increase of M, the difference between the maximum side peak and the main peak decreases, while the false acquisition probability increase.

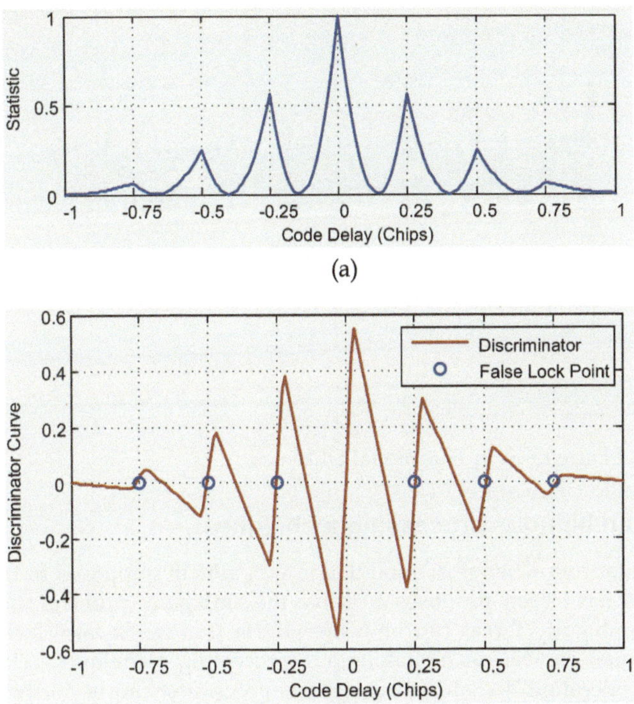

(a)

(b)

Fig. 2. (a) The Statistic in Acquisition Stage and (b) The Discriminator Curve in Tracking Stage of BOC(2*n*,*n*) Signal

It can be seen from Figure 2(b) that when using a traditional narrow early-minus-late (NEML) tracking loop (Van Dierendonck et al., 1992) with the early-late separation Δ, the discriminator characteristic curve of BOC(*m*,*n*) signal has a smaller linear domain than the one of the BPSK-R(*n*) signal. Besides, the discriminator characteristic curve of a BOC(*m*,*n*) signal has $2M-2$ stable false lock points which are due to the side peaks of the autocorrelation function. If a false acquisition occurs, in tracking stage, the code tracking loop will initially lock on a false lock point. Even if there is no false acquisition, the false lock can result from high noise, jitter, or short loss of lock.

Figure 3 shows an example of false lock caused by the excessive initial code delay bias in BOC(2*n*,*n*) signal tracking. The C/N_0 in this example is 45 dB-Hz, and the predetection integration time is 1 ms, with the early-late separation 0.1 chips. It can be seen that with the initial code delay bias of 0.14 chips, the loop locks on a false lock point. Although the output of discriminator hovers around zero and DLL maintains lock, the true code phase measurement bias is 0.25 chips, which corresponds to 75 m for BOC(2,1) signal. Such a considerable bias is unacceptable in most of the positioning applications.

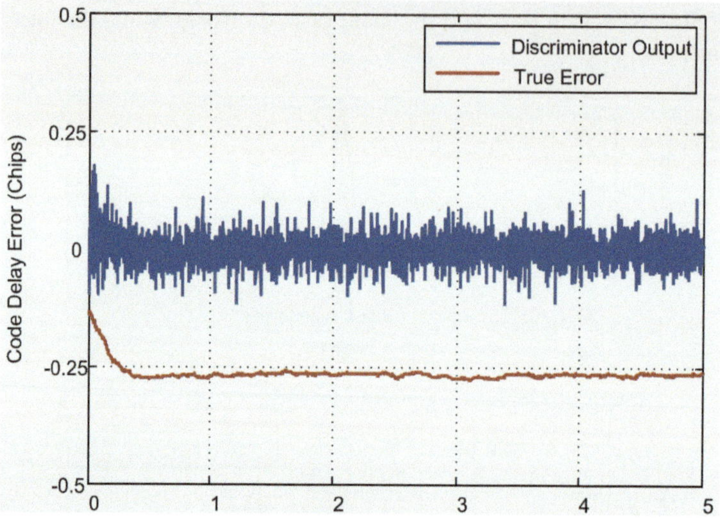

Fig. 3. Example of False Lock in Traditional DLL

3. Existing unambiguous processing techniques

During the decade from when BOC modulation was initially proposed to the present time, several solutions have been proposed to solve the ambiguity problem. In summary, the elimination of ambiguity threat can be achieved via two ways: false lock detection and recovery technique, as well as unambiguous processing techniques. More specifically, considering the operation domain, unambiguous processing can be further classified into frequency-domain processing and time-domain processing.

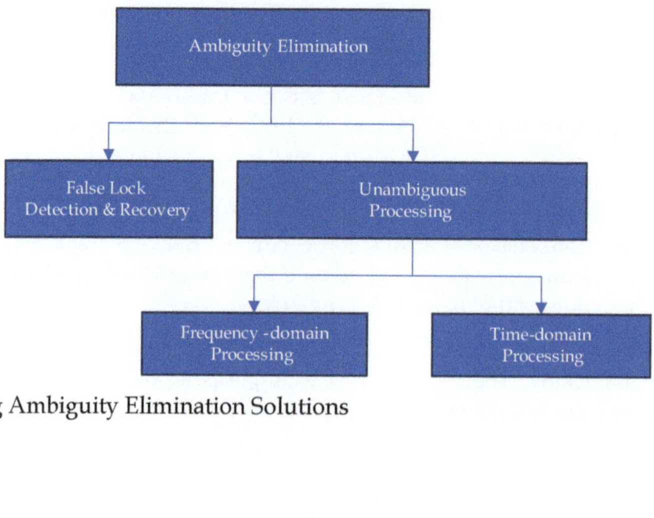

Fig. 4. Existing Ambiguity Elimination Solutions

3.1 False lock detection & recovery

False lock detection and recovery technique does not remove ambiguity but rather checks false lock. The most representative detection and recovery technique is referred to as bump-jumping technique (Fine & Wilson, 1999). This technique employs the traditional ambiguous code tracking loop and constantly check whether this loop is locked on the main peak of BOC ACF. To do so, bump-jumping technique uses two additional correlators located at the theoretical location of the two highest side peaks, as shown in Figure 5.

These two correlators are referred to as very early (VE) and very late (VL) correlators. By measuring and comparing the magnitude of the outputs of these two correlators and the prompt one, bump-jumping technique determines whether the false lock happens. It can be seen from Figure 5 that ignoring the effect of noise, when the code loop locks on the main peak, the magnitude of prompt correlator output is the greatest. And if either VE or VL correlator output is the largest, it means that tracking might be biased, and the loop will "jump" in the appropriate direction.

When locked on the main peak, this technique has high tracking accuracy. However, since it is based on the comparing of the main and side peaks magnitudes, the detection may have a high probability of false alarm when the signal-to-noise ratio (SNR) is low. In (Fine & Wilson, 1999), two up/down counter mechanisms are employed to reduce this false alarm probability. After each comparison, if one of the magnitudes of VE and VL correlator outputs exceeds that of the prompt one, the corresponding counter is incremented by one, otherwise the corresponding counter is decremented by one. The counter is not decremented below 0 or incremented above the preset threshold *N*. When the counter reaches the threshold, the loop jumps to the highest peak. By using this counter mechanism, the false alarm probability can be reduced effectively. However, the response time is also increased. Once the false lock happen, this technique needs time to detect and recover from false lock, so it might be intolerable for some critical applications such as aircraft landing.

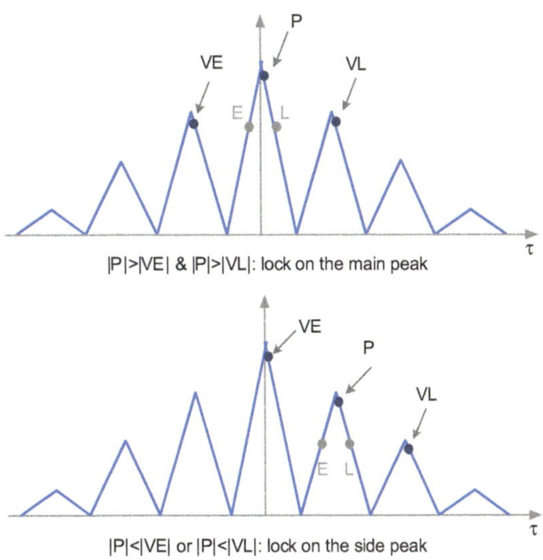

Fig. 5. Bump-jumping technique

3.2 Frequency-domain unambiguous processing

Frequency-domain processing techniques are represented by sideband techniques. Sideband technique considers the received BOC signal as the sum of two BPSK signals with carrier frequency symmetrically positioned on each side of the BOC carrier frequency. Thus each side lobe is treated independently as a BPSK signal, which provides an unambiguous correlation function and a wider S-curve steady domain.

The earliest sideband technique was described in (Fishman & Betz, 2000). As shown in Figure 6, the single sideband technique uses only one of the sidebands (either upper or lower) of BOC modulated signal. Both the received signal and the local BOC modulated baseband signal are filtered. Only the upper or lower sidebands of the received and local signals are retained. The shape of the correlation function of these two filtered signals is close to that of two BPSK-R signals. Therefore this correlation function can be used instead of BOC ACF in acquisition and tracking. The double sideband technique uses both the upper and lower sideband of BOC modulated signal. These two sidebands are processed separately before the output of correlators, and then the correlation values are added non-coherently. Compared with single sideband technique, double sideband technique suffers lower non-coherent correlation losses. However, it requires twice the sideband-selection filter number of single sideband technique.

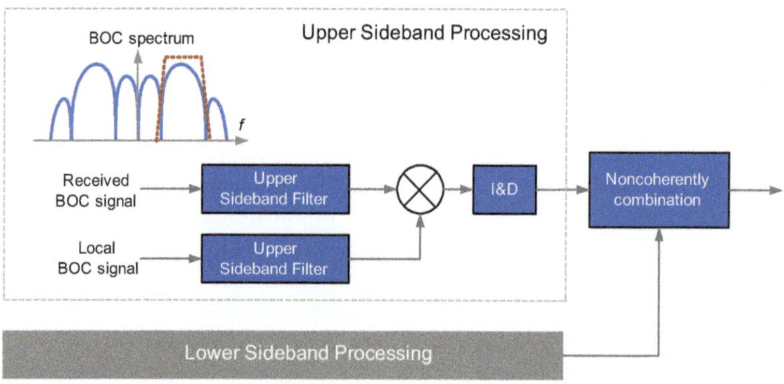

Fig. 6. Block diagram of sideband technique

BPSK-like method (Martin et al., 2003) is another frequency-domain unambiguous processing technique. This method is also based on the consideration of the BOC spectrum as the sum of two BPSK spectrum shifted by $\pm f_s$. The main difference compared with the method described above is the fact that only one low-pass filter is employed for the received signal. As shown in Figure 7, the filter bandwidth includes the two principal lobes of the spectrum. Another difference is that, the local signal is not the filtered BOC-modulated baseband signal but the BPSK-R signal, shifted by the sub-carrier frequency f_s. The BPSK-like technique can also be either single or double sideband, according to whether both the sidebands are used and combined non-coherently or only one sideband is used.

The original BPSK-like method can only be used for sine-phased BOC modulations with even BOC order. In (Burian et al., 2006), a modified version of BPSK-like method is proposed to extend BPSK-like method to BOC signals with odd order.

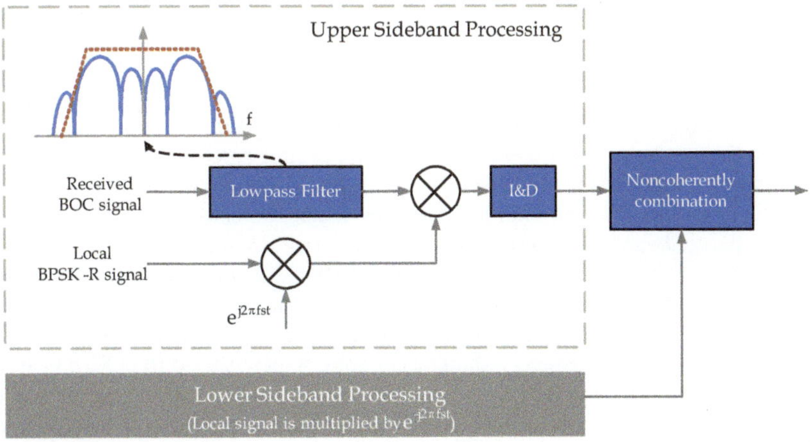

Fig. 7. Block diagram of BPSK-like method

Although the correlation functions in sideband techniques do not present any side peak, meaning that they are fully unambiguous, this kind of methods suffers from some drawbacks. The principle defect is that this kind of methods completely removes all of the advantages of BOC signal tracking in terms of Gaussian noise and multipath mitigation, since it causes the Gabor bandwidth of the received signal to approach that of the BPSK-R signal. Moreover, two side lobes are combined in non-coherent mode, which introduces correlation losses into the process. It seems that sideband techniques are not appropriate in terms of tracking. However, the correlation functions in this kind of methods have wide main correlation peak, which allows using longer code delay step in acquisition to reduce mean acquisition time. Therefore, sideband techniques can be attractive options in BOC modulated signal acquisition.

3.3 Time-domain unambiguous processing

Time-domain processing techniques are also referred to as side-peaks cancellation (SC) techniques which solve the ambiguity problem by taking advantage of the geometrical property of correlation functions (CF). The basic idea of SC techniques is using synthesized correlation function (SCF) instead of BOC ACF in acquisition and tracking. CF between the received BOC signal and some local auxiliary signals whose chip waveforms may be different from the received one are combined linearly or non-linearly to form the SCF with no side-peak. SC methods are flexible. Due to different auxiliary signal chip waveforms and combination modes, SC methods differentiate from each other greatly.

The first side-peaks cancellation technique is proposed in (Ward, 2003). This approach removes the ambiguities of the correlation function, but one drawback is that this method destroys the sharp peak of the correlation function. For accurate tracking, preserving a sharp peak of the correlation function is a prerequisite. An innovative unambiguous tracking technique, which is referred to as autocorrelation side-peak cancellation technique (ASPeCT), is described in (Julien et al., 2007). This technique uses ten correlation channels, completely removing the side peaks from the correlation function and keeping the sharp main peak. However, this technique has some limitations, for it is only applicable to sine-BOC(n,n) signals. Some other side-peaks cancellation methods have been proposed recently (Dovis, et al., 2005; Fante, 2003; Musso et al., 2006; Nunes et al., 2007).

However, the design of SC algorithms is still scarce of uniform theoretical frame and analytical method. There is no easy handling design method for SC algorithms development. The key of SC methods is the selection of local auxiliary signal chip waveforms. Due to lack of mathematical analysis tools, the selection of local signal chip waveforms is mainly based on intuition and trail-and-error. In new SC algorithm design, the shape of auxiliary signal waveforms is limited by the imagination of the designer, thus concentrating on some common shapes such as rectangular pulse, square wave with sine phase or cosine phase, and return to zero (RZ) code wave. When the chip waveform of the received signal is simple, for instance, Manchester code which is used in BOC(n,n) signals, it is easy to find a corresponding auxiliary chip waveform by using trail-and-error method. However, when the chip waveform of target signal gets more complicated, the design process becomes tough, and a mathematical analysis method is needed.

In the next section, a SC analytic design framework is presented. In this framework, the local auxiliary signal chip waveform can be designed under this framework by means of mathematic analysis so that the waveform shape selection can be more flexible.

4. SC analytic design framework

4.1 SCS waveform

The main difficulty of SC method design is how to select the spreading code chip waveform of local signals. It is desired to define a parameterized local signal model the chip waveform of which has a high degrees of freedom and is easy to generate in receivers to provide more opportunities for waveform optimization. Although there are few investigations about general local signal model for receiver designers since most of the signal receiving techniques are based on matched correlator in GNSS, it is interesting to note that some generalized waveform models are proposed for satellite signal design in order to offer degrees of freedom for shaping the signal spectrum, such as the binary coded symbols (BCS) (Hegarty et al., 2004). The advanced idea can be instructive for SC algorithm design.

For BCS signals, in order to ensure constant modulus, the envelope of $p(t)$ is restricted to 1. However, when considering auxiliary chip waveform in SC techniques, since local signals do not relate to amplifying and transmitting, they do not need to satisfy the request of constant modulus but their chip waveform should be easy to generate. Therefore, we expand the definition of BCS signal, restricting the chip waveform to being real-valued and having normalized energy. The chip waveform is divided into M segments, each with equal length $T_s = T_c / M$, and in each segment the level remains constant.

Since such waveform looks like steps, for expressional simplicity, we call this kind of chip waveform the step-shape code symbol (SCS) waveform, and call the signal which uses this waveform the SCS signal hereafter. Sticking with the terms used for BOC signals, M is referred to as the order of SCS signal. Some examples of SCS waveforms are shown in Figure 8.

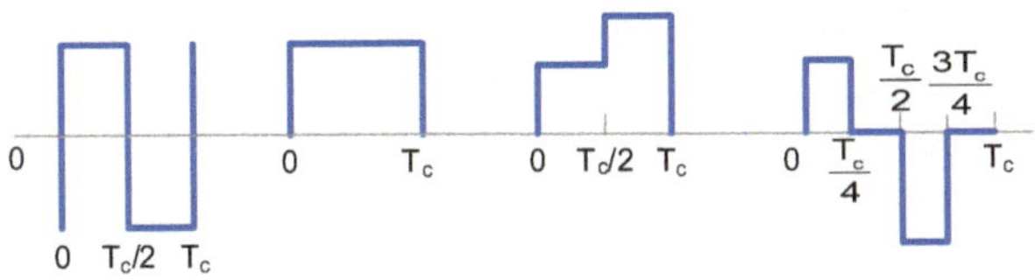

Fig. 8. Examples of SCS Waveforms

4.2 Vector representation

Any SCS waveform can be represented by a coordinate vector. Given a pair of T_c and T_s, we can construct a set of function $\{\psi_k(t) : k = 0,1,\cdots,M-1\}$, where

$$\psi_k(t) = \begin{cases} \frac{1}{\sqrt{T_c}}, & kT_s \le t < (k+1)T_s \\ 0, & \text{others} \end{cases} \tag{8}$$

As one can easily confirm, these M functions are orthogonal to each other and any SCS chip waveform $p_{SCS}(t)$ with the same T_c and M can be written as a linear combination of $\{\psi_k(t)\}$, that is

$$p_{SCS}(t) = \sum_{k=0}^{M-1} \psi_k(t) \cdot d_k \tag{9}$$

where d_k is the projection of $p_{SCS}(t)$ onto $\psi_k(t)$, i.e.

$$d_k = M \int_0^{T_c} p_{SCS}(t) \psi_k(t) dt \tag{10}$$

Therefore, each chip symbol $p_{SCS}(t)$ corresponds to a coordinate vector

$$\boldsymbol{d} = [d_0, d_1, \cdots, d_{M-1}]^{\mathrm{T}} \tag{11}$$

in the space spanned by $\psi_k(t)$ for $k = 0,1,\cdots,M-1$. To meet the energy normalization condition of the SCS chip waveform, each vector $\boldsymbol{d}$ must satisfy

$$\frac{1}{M}\|\boldsymbol{d}\|^2 = \frac{1}{M}\boldsymbol{d}^{\mathrm{T}}\boldsymbol{d} = 1 \tag{12}$$

Given the spreading chip rate f_c and the vector $\boldsymbol{d}$, the chip waveform $p(t)$ is determined. Borrowing from the notation of BCS signals (Hegarty et al., 2004), we call the vector $\boldsymbol{d}$ shape vector, and use the notation $p(t; \boldsymbol{d}, f_c)$ to denote a SCS signal whose shape vector is $\boldsymbol{d}$ and the chip rate is f_c. If it is understood from context, we will omit f_c from the notation.

It can be seen that the chip waveforms employed in most of the modulations in satellite navigation such as BPSK-R, BOC with even order, and BCS are special cases of SCS waveforms. Besides, almost all the auxiliary signal chip waveforms used in SC algorithms also belong to this family. When d=1, $p(t;\boldsymbol{d})$ degenerates to the rectangular pulse, and when $\boldsymbol{d} = [1,-1,\cdots,1,-1]^{\mathrm{T}}_{2\ell\times 1}$, $p(t;\boldsymbol{d})$ is the chip waveform of a sine phased BOC signal with the order $M = 2\ell$. Note that in an odd order SCS signal, the chip waveform is time-varying. For example, for sin-BOC signal with $M = 3$, the shape vector of the spreading chip is $(1,-1,1)^T$ in the time interval $t \in [2nT_c, (2n+1)T_c)$, while it is $(-1,1,-1)^T$ in $t \in [(2n-1)T_c, 2nT_c)$. We assume that M is even hereafter.

4.3 CCF of SCS signals

All the SC techniques are based on the shapes of CCF between the received signal and the local signal. Here we consider the CCF of two SCS signals which have the same chip rate f_c, spreading sequence $\{c_i\}$, and the order M, while the chip waveform are difference.

By using (2) and (9), a SCS baseband signal can be expressed as

$$g(t;\boldsymbol{d}) = \sum_{n=-\infty}^{+\infty} \sum_{k=0}^{M-1} (-1)^{c_n} d_k \psi_k (t - nMT_s) \tag{13}$$

The CCF of two SCS signals is

$$\begin{aligned} R_{gg'}(\tau;\boldsymbol{d},\boldsymbol{d}') &= \frac{1}{T}\int_0^T g(t) g'(t+\tau) dt \\ &= \frac{1}{T} \sum_n \sum_m \sum_{k=0}^{M-1} \sum_{q=0}^{M-1} (-1)^{c_n + c_m} d_k d'_q \int_0^T \psi_k (t - nMT_s) \psi_q (t - mMT_s + \tau) dt \end{aligned} \tag{14}$$

where $T = NT_c$ is the period of the spreading sequence. The integral in (14) is nonzero only when $\psi_k(t - nMT_s)$ and $\psi_q(t - mMT_s + \tau)$ have the overlapping parts. The delay τ can be expressed as the summation of three parts $\tau = aT_c + bT_s + \varepsilon$, where a is an integer, $b = 0, 1, \cdots, M-1$, and $\varepsilon \in [0, T_s)$. And after some algebraic simplifications, (14) can be rewritten as (Yao & Lu, 2011)

$$\begin{aligned} R_{gg'}(\tau) &= R_{gg'}(aT_c + bT_s + \varepsilon) \\ &= R_c(a)\left[r_b\left(1 - \frac{\varepsilon}{T_s}\right) + r_{b+1}\left(\frac{\varepsilon}{T_s}\right)\right] + R_c(a+1)\left[r_{b-M}\left(1 - \frac{\varepsilon}{T_s}\right) + r_{b-M+1}\left(\frac{\varepsilon}{T_s}\right)\right] \end{aligned} \tag{15}$$

where

$$R_c(a) \triangleq \frac{1}{N} \sum_{n=0}^{N-1} (-1)^{c_n + c_{n+a}} \tag{16}$$

and the aperiodic cross-correlation function (ACCF) of $\boldsymbol{d}$ and $\boldsymbol{d}'$ is

$$r_b(\boldsymbol{d},\boldsymbol{d}') = \begin{cases} \dfrac{1}{M} \sum\limits_{i=0}^{M-1-b} d_i d'_{b+i}, & 0 \le b \le M-1 \\ \dfrac{1}{M} \sum\limits_{i=0}^{M-1-b} d_{i-b} d'_i, & 1-M \le b < 0 \\ 0, & |b| \ge M \end{cases} \tag{17}$$

By using (15) one can derive other expression form of an M-order BOC signal ACF, which is

$$R_{\mathrm{BOC}}(\tau)=\begin{cases}(-1)^{k+1}\left[\frac{\tau(2M-2k-1)}{T_c}-\frac{2(M-1)k-2k^2+M}{M}\right], & \frac{kT_c}{M}\le\tau<\frac{(k+1)T_c}{M}\\ (-1)^{k+1}\left[\frac{\tau(2k-1)}{T_c}+\frac{(M-k)(2k-1)-k}{M}\right], & \frac{(k-M)T_c}{M}\le\tau<\frac{(k-M+1)T_c}{M}\\ 0, & \text{others}\end{cases} \quad (18)$$

where $k=0,1,\cdots,M-1$. And the CCF between an M-order BOC signal and a SCS signal is

$$R_{\mathrm{B/L}}(\tau;\boldsymbol{d}_L)=\begin{cases}\left(\frac{\tau M-kT_c}{T_c}\right)\left(r_{k+1}-r_k\right)+r_k, & \frac{kT_c}{M}\le\tau<\frac{(k+1)T_c}{M}\\ \left(\frac{\tau M-kT_c+MT_c}{T_c}\right)\left(r_{k-M+1}-r_{k-M}\right)+r_{k-M}, & \frac{(k-M)T_c}{M}\le\tau<\frac{(k-M+1)T_c}{M}\\ 0, & \text{others}\end{cases} \quad (19)$$

where $\boldsymbol{d}_L$ is the shape vector of SCS signal, and

$$r_k=\begin{cases}\frac{1}{M}\sum_{i=0}^{M-1-k}(-1)^i d_{k+i}, & 0\le k\le M-1\\ \frac{1}{M}\sum_{i=0}^{M-1-k}(-1)^{i-k} d_i, & 1-M\le k<0\\ 0, & |k|\ge M\end{cases} \quad (20)$$

Fig. 9 shows a schematic diagram of $R_{\mathrm{B/L}}(\tau;\boldsymbol{d}_L)$. Note that within ($-T_c,T_c$) the correlation function is piecewise linear between kT_s and $(k+1)T_s$, and $R_{\mathrm{B/L}}(kT_s)=r_k$, for $k\in[-M+1,M-1]$ and $k\in\mathbb{Z}$.

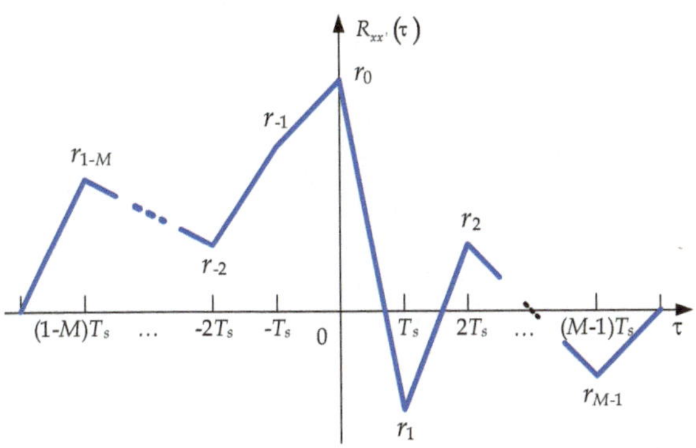

Fig. 9. Schematic diagram of correlation function $R_{\mathrm{B/L}}(\tau;\boldsymbol{d}_L)$.

4.4 SC algorithm design process under SC framework

From (15) and (17), it can be seen that the CCF of two SCS signals mostly depends on the ACCF of their chip waveforms when these two signals have the same spreading sequences. In SC algorithm design, since the chip waveform shape of the received signal is known, CCF entirely depends on the shape of local signal spreading chip waveform. Note that each SCS chip waveform corresponds to an unique point in M-dimensional space whose coordinate is $(d_0, d_1, \cdots, d_{M-1})$, so by changing the value of d_k, one can adjust the shape of CCF. That is, CCF is a function of $\boldsymbol{d}$.

After building the relation between the shape of CCF and the value of a vector, the search for good chip waveform can be equivalent to an optimization problem which can be formulated as

$$\begin{cases} \min f(\boldsymbol{d}_1, \boldsymbol{d}_2, \cdots, \boldsymbol{d}_n); \\ s.t. \quad g_i(\boldsymbol{d}_1, \boldsymbol{d}_2, \cdots, \boldsymbol{d}_n) \geq 0, i = 1, \cdots, m \end{cases} \tag{21}$$

where f is the objective function, g_i is the constraint function, m is the number of constraints, and $\boldsymbol{d}_1, \cdots, \boldsymbol{d}_n$ are shape vectors of local signals. Then the development of SC algorithm becomes the solving of an optimization problem with a set of inequalities constraints and can be achieved through four steps (Yao & Lu, 2011), as shown in Figure 10.

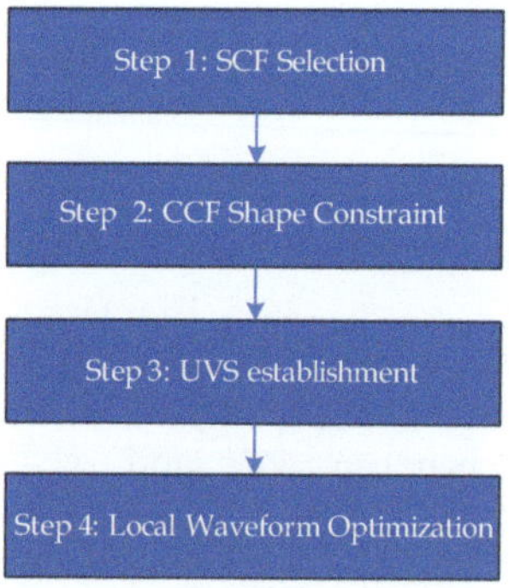

Fig. 10. SC Algorithm Design Process under SC Framework

The first step is SCF selection. In SC algorithms, the SCF $\tilde{R}$ is used instead of ACF in acquisition and tracking. The choice of SCF determines the combination mode of CCFs, and provides for the number of local auxiliary signals. The more local signals are used to form SCF, the more flexible the shape control is, but the more correlators are needed as well.

The second step is CCF shape constraint. Once the SCF is determined, the shape of each CCF employed in SCF can be restricted. Since CCF is piecewise linear between its values at integer multiples of T_c / M, and $R_{gg'}(kT_c / M) = r_k$, the constraint of the i'th CCF shape can be translated into the restriction on values of those $r_k^{(i)}$.

The third step is referred to as unambiguous vector subset (UVS) establishment. Using the corresponding relationship between $r_k^{(i)}$ and the shape vector $\boldsymbol{d}_i$ of local signal, one can further translate the restriction on the value of $r_k^{(i)}$ into the limitation of $\boldsymbol{d}_i$. In most cases, the values of each $r_k^{(i)}$ which makes the SCF unambiguous are not unique, thus the feasible solution of $\boldsymbol{d}_i$ is not a fixed point. The feasible region of $\boldsymbol{d}_i$ is denoted as $\mathcal{S}_i \subset \mathbb{R}^M$ and is called UVS of $\boldsymbol{d}_i$.

The final step is local waveform optimization. As the UVS has been established, with an optimization object, (21) can be rewritten as

$$\begin{cases} \min f(\boldsymbol{d}_1, \boldsymbol{d}_2, \cdots, \boldsymbol{d}_n); \\ s.t. \quad \boldsymbol{d}_i \in \mathcal{S}_i, i = 1, \cdots, n \end{cases} \tag{22}$$

and the final step is to find n optimal shape vectors $\boldsymbol{d}_1^{(\mathrm{opt})}$, $\boldsymbol{d}_2^{(\mathrm{opt})}$,..., $\boldsymbol{d}_n^{(\mathrm{opt})}$ from UVS, which correspond to the optimal local chip waveforms.

Note that usually at different processing stages, the optimization objects are difference. For example, in acquisition, the optimization objects may be the maximum SNR or the widest SCF main peak, while at tracking stage it may be the ability of multipath rejection, the greatest slope or the widest linear range of the discriminator curve, or even some compromises between them. In next two sections, we will give two examples of SC algorithm design under the steps described above. The design process of an SC unambiguous acquisition algorithm as well as an SC unambiguous tracking loop is described respectively.

5. GRASS technique

Under the analytic design framework described above, an unambiguous acquisition technique named General Removing Ambiguity via Side-peak Suppression (GRASS) technique is developed. This technique is suitable for generic sin-BOC(kn,n) signals and it is convenient to implement. The detailed performance analysis of this technique can be found in (Yao et al., 2010a). This section puts its emphasis on the design process of this technique.

5.1 Step 1 - SCF selection

Theoretically, when the number of local auxiliary signals is unlimited, SCF can be shaped into any desired forms. However, from the view of engineering, the more local signals are used, the more correlators are needed in a receiver which is directly related to the complexity and power consumption. Moreover, the noncoherent combination of too much correlator results may aggravate SNR deterioration. Therefore, in our design, the number of local auxiliary signals should be as few as possible.

Since the signal acquisition is a process of searching pronounced energy peak in a 2-dimentional space, the requirement to the shape of SCF in acquisition is relatively generous compared to code tracking. A SCF having main peak without positive side peak is enough. Therefore in GRASS technique only one local auxiliary SCS signal with a matched BOC signal is employed to suppress the side peaks of BOC ACF in noncoherent mode. The SCF used is as follow:

$$\tilde{R}(\Delta\tau) = R_{\mathrm{B}}^2(\Delta\tau) - \alpha R_{\mathrm{B/L}}^2(\Delta\tau) \tag{23}$$

where R_{B} is the ACF of BOC signal, $R_{\mathrm{B/L}}$ is the CCF between the received BOC signal and the local SCS signal, and α is the weight coefficient. It can be seen that (23) is similar with the SCF used in (Julien et al., 2007) in form. However, as shown later, GRASS technique is not only suitable for BOC(n,n) signals but also for other BOC(kn,n) signals.

5.2 Step 2 - CCF shape constraint

The objective is to keep the main peak of BOC ACF envelop while remove all the positive side peaks (the negative side peaks do not interfere with the statistical test since only positive values could pass the threshold). In consideration of the shape of BOC ACF, it is desirable that the envelop of $R_{B/L}$ be zigzag and symmetric with respect to $\tau = 0$. Moreover, $R^2_{B/L}(0)$ should be zero in order to ensure that the magnitude of main peak is unaffected after the subtracting.

As explained in the previous section, the above constraints of the CCF shape can be translated into the restriction on r_k via (19). The constraint $R^2_{B/L}(0) = 0$ is equivalent to

$$r_0 = 0 \tag{24}$$

and the axial symmetry of $|R_{B/L}|$ means

$$|r_i| = |r_{-i}| \tag{25}$$

The requirement of zigzag shape can be realized through making adjacent r_k have opposite sign, that is

$$\begin{cases} r_i r_{i+1} < 0, & i > 0 \\ r_i r_{i-1} < 0, & i < 0 \end{cases} \tag{26}$$

Actually, under the restrictions of (24) and (26), (25) can be simplified to

$$r_i = -r_{-i} \tag{27}$$

because by (20) and (24) we have

$$\begin{aligned} |r_{M/2} + r_{-M/2}| &= \frac{1}{M}\left|\sum_{i=0}^{M/2-1} (-1)^i d_{i+M/2} + \sum_{i=0}^{M/2-1} (-1)^{i-M/2} d_i\right| \\ &= \frac{1}{M}\left|\left(\sum_{i=0}^{M/2-1} + \sum_{i=M/2}^{M-1}\right)(-1)^{i-M/2} d_i\right| \\ &= \frac{1}{M}\left|\sum_{i=0}^{M-1} (-1)^i d_i\right| = |r_0| = 0 \end{aligned} \tag{28}$$

so that $r_{M/2} = -r_{-M/2}$. Then using (26), we obtain (27).

5.3 Step 3 - UVS establishment

Substituting (20) into (24), (26) and (27), after some straightforward algebraic simplification, we have the set of inequalities constraints on the elements of $\boldsymbol{d}_L$:

$$\begin{cases} \sum_{i=0}^{k} (-1)^i d_i > 0, & k = 0,1,\cdots,\frac{M}{2}-1 \\ d_k = d_{M-1-k}, & k = 0,1,\cdots,\frac{M}{2}-1 \\ \sum_{i=0}^{M-1} d_i^2 = M \end{cases} \tag{29}$$

Note that the last term in (29) is the energy normalization constraint of SCS waveform. So that the UVS can be represented as

$$\mathcal{S}=\left\{ \begin{array}{c} \sum_{i=0}^{M-1} d_i^2 = M \\ \boldsymbol{d}_L \in \mathbb{R}^M : \sum_{i=0}^{k} (-1)^i d_i > 0 \\ d_k = d_{M-1-k}, k=0,1,\cdots,\frac{M}{2}-1 \end{array} \right\} \tag{30}$$

With an appropriate weight coefficient α, all the undesired positive side peaks of R_B^2 can be canceled by subtraction. From (18), the coefficient α must satisfies

$$\alpha|r_k| \geq |R_B(kT_c / M)| = \frac{M-k}{M} \tag{31}$$

for $k=1,2,\cdots,M-1$, or

$$\alpha \geq \max_{k\neq 0} \frac{M-k}{M|r_k|}. \tag{32}$$

5.4 Step 4 - Local waveform optimization

So far the effect of thermal noise has not been considered. In fact, the coefficient α amplifies noise components in $R_{B/L}$. Under a given pre-correlation SNR, the larger α is, the lower SNR in the SCF is. Therefore, from the viewpoint of sensitivity, it is desired that α be as small as possible. So that with a given $\boldsymbol{d}_L$ the optimum α is

$$\alpha = \max_{k\neq 0} \frac{M-k}{M|r_k(\boldsymbol{d}_L)|} \tag{33}$$

and in UVS, the optimum $\boldsymbol{d}_L$ is the one minimizing α, that is

$$\boldsymbol{d}_{\text{opt}} = \arg\min_{\boldsymbol{d}_L \in \mathcal{S}} \alpha = \arg\min_{\boldsymbol{d}_L \in \mathcal{S}} \max_{k\neq 0} \frac{M-k}{M|r_k|} \tag{34}$$

It can be proved that the explicit expression of (34) is

$$\boldsymbol{d}_{\text{opt}} = [d_0, d_1, \cdots d_{M-1}]^{\text{T}} \tag{35}$$

where

$$\begin{cases} d_0 = d_{M-1} = \frac{M-1}{\sqrt{2M-3}} \\ d_i = d_{M-i+1} = \frac{(-1)^{i-1}}{\sqrt{2M-3}}, \left(i=1,2,\cdots,\frac{M}{2}-1\right) \end{cases} \tag{36}$$

and $\alpha_{\min} = 2M-3$.

Figure 11 (a)-(c) depict the optimum local SCS waveforms for BOC(n,n), BOC($2n$,n), and BOC($3n$,n) signals respectively. For BOC(n,n) signals, $M=2$. So $\boldsymbol{d}_{\text{opt}} = (1,1)^{\text{T}}$ and $\alpha_{\min}=1$. It can be found that the SC method proposed in (Julien, et al., 2007) is equivalent to this case. The local symbol is simply a rectangular pulse. For BOC($2n$,n), $\boldsymbol{d}_{\text{opt}} = \sqrt{\frac{9}{5}}\left(1,\frac{1}{3},\frac{1}{3},1\right)^{\text{T}}$ and $\alpha_{\min}=5$. For BOC($3n$,n), $\boldsymbol{d}_{\text{opt}} = \frac{1}{3}(5,1,-1,-1,1,5)^{\text{T}}$ and $\alpha_{\min}=9$. It can be seen that the complexity of the chip shape increases as the BOC-modulation order is raised. For $M=4$ or higher, the optimum local SCS waveform is hard to obtain by geometrical intuition. Figure 12, 13, and 14 show the envelop of R_B, $R_{B/L}$ and the SCF for $M=2$, 4, 6, respectively.

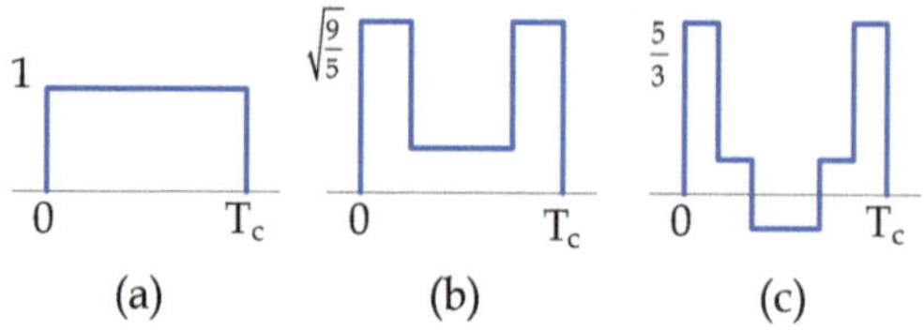

Fig. 11. Optimum Local SCS Waveforms for (a) M=2, (b) M=4, (c) M=6

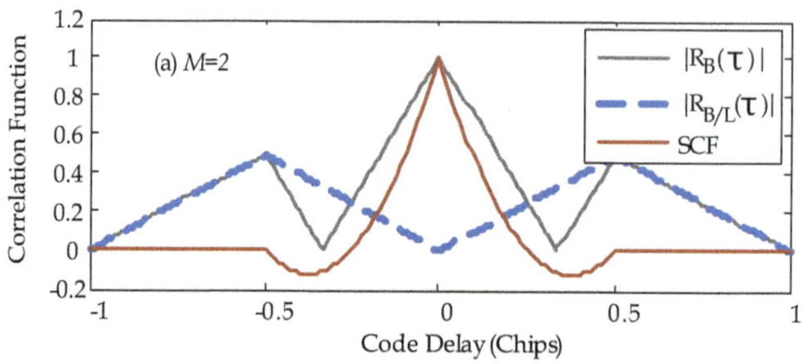

Fig. 12. The Envelops of R_B, $R_{B/L}$ and the SCF for M=2

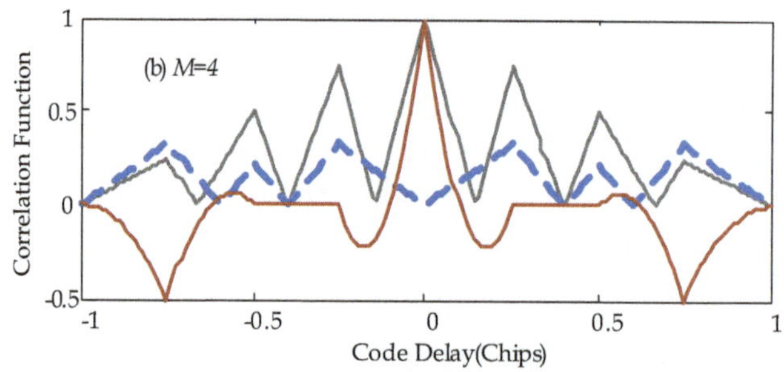

Fig. 13. The Envelops of R_B, $R_{B/L}$ and the SCF for M=4

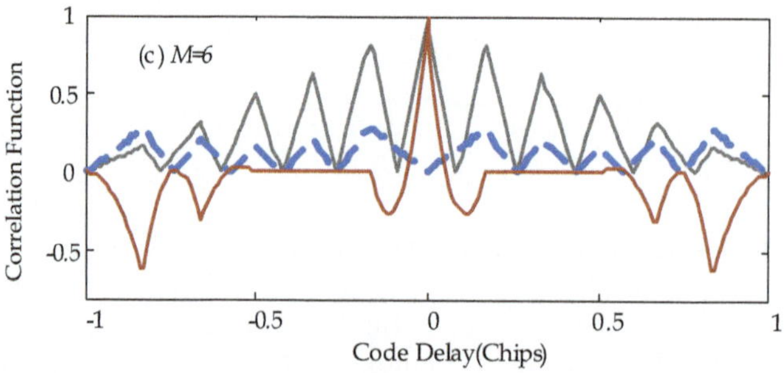

Fig. 14. The Envelops of R_B, $R_{B/L}$ and the SCF for M=6

From the figures it can be seen that with any M, no major positive side peak exists in SCF. Although there are some pits on the SCF, their magnitudes are all below zero, so they bring no threat to the acquisition.

Although all examples shown above are for sin-BOC signals, it is easy to demonstrate that GRASS technique can also be applied to cos-BOC signals by simply replacing basis function (8) with

$$\psi'_k(t)=\begin{cases}\frac{1}{\sqrt{T_c}}\operatorname{sgn}\left\{\sin\left[2\pi f_s\left(t-kT_s\right)\right]\right\}, & kT_s\leq t<(k+1)T_s\\ 0, & \text{others}\end{cases} \tag{37}$$

6. PUDLL

As another example of the application of SC design framework, in this section we show the design process of an unambiguous code tracking technique named pseudo-correlation-function-based unambiguous delay lock loop (PUDLL) (Yao et al., 2010b) which is applicable to any BOC(*kn*,*n*) signals. At tracking stage, because the discriminator characteristic curve is based on the first derivative of SCF, the requirement to the shape of SCF is more stringent than in acquisition.

6.1 Step 1 – SCF selection

It is desired that the SCF used in code tracking has an ideal triangular main peak with no side peak. Since the CCF between the received BOC signal and the local SCS signal is piecewise linear, utilizing this characteristic, and using the absolute-magnitude operation to change the direction of lines on one side of the zero crossing point, following by the linear combination, it is possible to obtain the SCF without any side peak. Therefore, in this example, the SCF is chosen as

$$\tilde{R}(\tau)=\left|R_1(\tau)\right|+\left|R_2(\tau)\right|-\left|R_1(\tau)+R_2(\tau)\right| \tag{38}$$

where R_1 and R_2 are CCFs between BOC signal and two local SCS signals $g_1(t;\boldsymbol{d}_1)$ and $g_2(t;\boldsymbol{d}_2)$, respectively, in which $\boldsymbol{d}_1$ and $\boldsymbol{d}_2$ are the shape vectors of g_1 and g_2 respectively.

6.2 Step 2 – CCF shape constraint

A sufficient condition for (38) being symmetric with respect to $\tau=0$ is

$$\left|R_1(\tau)\right|=\left|R_2(-\tau)\right|. \tag{39}$$

In fact, the only difference between $R_1(\tau)=R_2(-\tau)$ and $R_1(\tau)=-R_2(-\tau)$ lies in the polarity of the local SCS chip waveform. Without loss of generality, assume $R_1(\tau)=-R_2(-\tau)$, so that at each endpoint of CCF segment

$$r'_i=-r_i \tag{40}$$

where $r_i=R_1(iT_c/M)$ and $r'_i=R_2(iT_c/M)$.

From (38), it can be proved that $\tilde{R}$ is also piecewise linear. In order to shape the SCF into an ideal triangle, one have to make all of the ending points of lines in $\tilde{R}$ be zero except the central one.

To ensure the triangular shape, SCF must satisfy the following request:

$$\begin{cases} \tilde{R}(0) \neq 0 \\ \tilde{R}(kT_c / M) = 0, (k \neq 0) \end{cases} \tag{41}$$

where the first term is equivalent to

$$r_0 = 0 \tag{42}$$

and using (38) and (40), the second term can be simplified as

$$r_k r_{-k} \leq 0 \tag{43}$$

for $k \neq 0$.

The constraints (42) and (43) are necessary but not sufficient, since the absolute-magnitude operation introduces additional endpoints at the zero crossing points in R_1 and R_2. If R_1 has a zero crossing point τ_0 within the interval $\left[kT_c / M, (k+1)T_c / M\right]$, $(k > 0)$, easily proved, it must be

$$\tau_0 = \frac{kT_c}{M} + \frac{|r_k|}{|r_k| + |r_{k+1}|}\frac{T_c}{M} \tag{44}$$

From (43) we know that R_2 must have a zero crossing point within the same interval, which is

$$\tau_0' = \frac{kT_c}{M} + \frac{|r_{-k}|}{|r_{-k}| + |r_{-k-1}|}\frac{T_c}{M} \tag{45}$$

In order to eliminate the inclined lines on both sides of the zero crossing point, we must have $\tau_0 = \tau_0'$, which can be simplified as

$$r_k r_{-k-1} = r_{-k} r_{k+1} \tag{46}$$

for $k > 0$.

6.3 Step 3 – UVS establishment

The necessary and sufficient conditions for SCF being triangular are (40), (42), (43), and (46). From (40), we obtain that $\boldsymbol{d}_1$ and $\boldsymbol{d}_2$ are mirror images of each other, i.e.

$$d_k' = d_{M-k-1} \tag{47}$$

where d_k and d_k' are the entries of $\boldsymbol{d}_1$ and $\boldsymbol{d}_2$, respectively. When $M = 2$, by using the relationship (20), the UVS can be represented as

$$\mathcal{S} = \left\{ \boldsymbol{d}_1 \in \mathbb{R}^2 : d_1^2 + d_2^2 = 2, d_0 > d_1 \geq 0 \right\} \tag{48}$$

In the case $M = 4$ (Yao, 2008), UVS can be expressed as

$$\mathcal{S} = \left\{ \boldsymbol{d}_1 \in \mathbb{R}^4 \left| \left\{ \begin{array}{c} d_0 > d_3 \geq 0 \\ d_1 = d_2 = 0 \end{array} \right\} \cup \left\{ \begin{array}{c} 0 \leq d_3 \leq d_0 \leq d_1 \\ d_3 \leq d_2 \\ -d_3 \leq d_1 - d_2 \leq d_0 \\ d_0 - d_1 + d_2 - d_3 \neq 0 \end{array} \right\} \right. \right\} \tag{49}$$

For larger M, as the degrees of freedom of $\boldsymbol{d}_1$ increase, the explicit expressions for the constraints on $\boldsymbol{d}_1$ become complex and hard to derive. The full UVS can be obtained through the use of numerical method. However, it is easy to verify that any element in one of the subset of UVS

$$\mathcal{S}' = \left\{ \boldsymbol{d}_1 \in \mathbb{R}^M \left| \left\{ \begin{array}{l} d_0 > d_{M-1} \geq 0 \\ d_i = 0, i = 1, 2, \cdots, M-2 \end{array} \right\} \right. \right\} \tag{50}$$

can make a triangular SCF for all M even.

6.4 Step 4 – Local waveform optimization

Under the energy normalization restriction (12), $\boldsymbol{d}_1$ in (50) has one degree of freedom. So by defining $\kappa = d_{M-1} / d_0$, the shape vector in $\mathcal{S}'$ can be expressed as

$$\boldsymbol{d}_1 = \left(\sqrt{\tfrac{M}{1+\kappa^2}}, 0, \cdots, 0, \kappa \sqrt{\tfrac{M}{1+\kappa^2}} \right)^{\mathrm{T}} \tag{51}$$

Utilizing (51), (19), and (38), without considering front-end filtering, we can obtain the expression of $\tilde{R}$

$$\tilde{R}(\tau;\kappa) = \begin{cases} \dfrac{M(2\kappa-4)|\tau| + 2(1-\kappa)T_c}{\sqrt{M(1+\kappa^2)}T_c}, & |\tau| < \dfrac{(1-\kappa)T_c}{M(2-\kappa)} \\ 0, & \text{others} \end{cases} \tag{52}$$

the base line half width of which is

$$w(\kappa) = \frac{(1-\kappa)T_c}{M(2-\kappa)} \tag{53}$$

and the height of the peak is

$$h(\kappa) = \frac{2(1-\kappa)}{\sqrt{M(1+\kappa^2)}} \tag{54}$$

Figure 15 (a) and (b) show some SCFs with different κ for BOC(n,n) and BOC($2n$,n) signals, respectively. Figure 16 depicts the discriminator characteristic curve of the early-minus-late power (EMLP) loop which uses SCF instead of BOC(n,n) ACF.

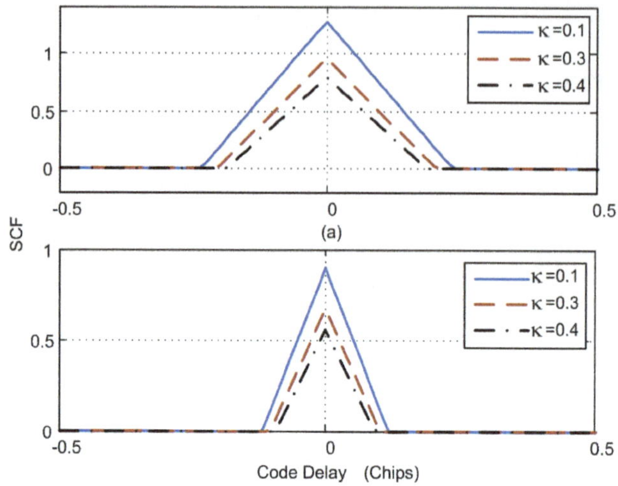

Fig. 15. SCF for (a) BOC(n,n) and (b) BOC($2n$,n) Signals, with Different κ

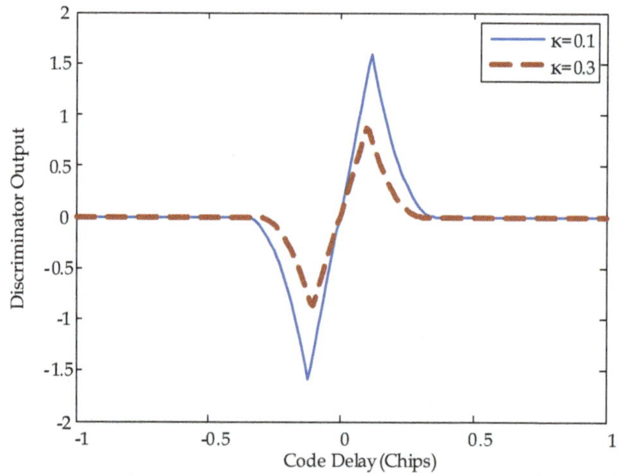

Fig. 16. The Discriminator Characteristic Curve for BOC(n,n) Signals, with Different κ

It can be noted that by using SCF, this technique completely removes the false lock points. And it can also be seen that both shapes of SCF and discriminator characteristic curve are functions of κ. Consequently, changing the value of κ can adjust the linear range of the discriminator and the slope of the curve, thus changes the multipath and thermal noise mitigation performances of the tracking loop (Yao, Cui, et al., 2010). With different optimization objective, the optimum κ and the optimum chip waveform are not the same.

7. Conclusion

In this Chapter, the ambiguity problem of BOC modulated signals which have been chosen as the chief candidate for several navigation signals in the next generation GNSS as well as its solutions is systematically described. An innovative design methodology for future unambiguous processing techniques is also proposed. Under the proposed design framework, the development of SC algorithm becomes the solving of an optimization problem with a set of inequalities constraints and can be achieved through four steps.

As two practical examples, the design process of an SC unambiguous acquisition algorithm as well as an SC unambiguous tracking loop is described respectively to demonstrate the practicality of the proposed framework and to provide reference to further SC algorithm development. Although the optimization objects are difference in these two algorithms, the methods of analysis and the design steps under the analytic design framework are unified. It is proved that both of these two algorithms can completely eliminate the ambiguity problem in acquisition and tracking. Moreover, these two algorithms have outstanding compatibility. Both of them are suitable for generic even-order sine-BOC signal.

Future works will focus on the development of new algorithms under the proposed framework. Considering the complexities, both of the example algorithms in this Charpter employ only two local signals. In fact, by using more local signals, degrees of freedom in design can be further increased and the shape control will be more flexible. Besides, this analytic design framework can be used not only on unambiguous algorithm development but also on finding good local waveform to resist the effect of multipath.

8. References

ARINC. (2005). NAVSTAR GPS space segment/navigation L5 User interfaces. In *IS-GPS-705*.

ARINC. (2006). Navstar GPS space segment/user L1C interfaces. In *IS-GPS-800*. El Segundo, CA, US.

Avila-Rodriguez, J.-A., Hein, G. W., Wallner, S., Issler, J.-L., Ries, L., Lestarquit, L., Latour, A. d., Godet, J., Bastide, F., Pratt, T., & Owen, J. (2007). The MBOC modulation: the final touch to the Galileo frequency and signal plan, *Proceedings of ION GNSS 20th International Technical Meeting of the Satellite Division*, pp. 1515-1529, Fort Worth, TX, US, 2007.

Betz, J. W. (2001). Binary offset carrier modulations for radionavigation, *Navigation: J. Inst. Navig.*, 48(4), 227-246.

Burian, A., Lohan, E. S., & Renfors, M. (2006). BPSK-like methods for hybrid-search acquisition of Galileo signals, *Proceedings of IEEE ICC 2006*, pp. 5211-5216, Istanbul, Turkey, 2006.

Dovis, F., Mulassano, P., & Presti, L. L. (2005). A novel algorithm for the code tracking of BOC(*n*,*n*) modulated signals, *Proceedings of ION GNSS 2005*, pp. 152-155, Long Beach, CA, 2005.

Enge, P. (2003). GPS modernization: capabilities of the new civil signals. *Proceedings of Australian International Aerospace Congress*, pp. 1-22. Brisbane, 2003.

Fante, R. L. (2003). Unambiguous tracker for GPS binary-offset-carrier signals, *Proceedings of the 59th Annual Meeting of The Institute of Navigation and CIGTF 22nd Guidance Test Symposium*, pp. 141-145, Albuquerque, NM, US, 2003.

Fine, P., & Wilson, W. (1999). Tracking algorithm for GPS offset carrier signals. *Proceedings of ION NTM 1999*, pp. 671-676, San Diego, CA, US, 1999.

Fishman, P., & Betz, J. W. (2000). Predicting performance of direct acquisition for the M-code signal. *Proceedings of ION NTM 2000*, pp. 574-582, Anaheim, CA, US, 2000.

Gao, G. X., Chen, A., Lo, S., Lorenzo, D. d., & Enge, P. (2007). GNSS over China - the Compass MEO satellite codes. *Inside GNSS*, 2007, 2(5), pp. 36-43.

Hegarty, C. J., Betz, J. W., & Saidi, A. (2004). Binary coded symbol modulations for GNSS. *Proceedings of ION 60th Annual Meeting*, pp. 56-64, Dayton, OH, US, 2004.

Hegarty, C. J., & Chatre, E. (2008). Evolution of the global navigation satellite system (GNSS). *Proceedings of IEEE*, 96(12), pp. 1902-1917.

Hein, G. W., Avila-Rodriguez, J.-A., Wallner, S., Pratt, A. R., Owen, J., Issler, J.-L., Betz, J. W., Hegarty, C. J., Lenahan, L. S., Rushanan, J. J., Kraay, A. L., & Stansell, T. A. (2006). MBOC: the new optimized spreading modulation recommended for GALILEO L1 OS and GPS L1C. *Proceedings of IEEE/ION PLANS 2006*, pp. 883-892, San Diego, CA, US, 2006.

Julien, O., Macabiau, C., Cannon, M. E., & Lachapelle, G. (2007). ASPeCT: unambiguous sine-BOC(n,n) acquisition/tracking technique for navigation applications, *IEEE Transactions on Aerospace and Electronic Systems*,Vol. 43, No.1, pp. 150-162.

Lohan, E. S., Burian, A., & Renfors, M. (2008). Low-complexity unambiguous acquisition methods for BOC-modulated CDMA signals, *Int. J. Commun. Syst. Network*, 26 (2008), pp. 503–522.

Martin, N., Leblond, V., Guillotel, G., & Heiries, V. (2003). BOC(x,y) signal acquisition techniques and performances. *Proceedings of ION GPS 2003*, pp. 188-198, Portland, OR, US, 2003.

Musso, M., Cattoni, A. F., & Regazzoni, C. S. (2006). A new fine tracking algorithm for binary offset carrier modulated signals. *Proceedings of ION GNSS 2006*, pp. 834-840, Fort Worth, TX, US, 2006.

Nunes, F., Sousa, F., & Leitao, J. (2007). Gating functions for multipath mitigation in GNSS BOC signals, *IEEE Transactions on Aerospace and Electronic Systems*, Vol. 43, No. 3 (2007), pp. 941-964.

Proakis, J. G. (2001). *Digital Communications*, Boston: McGraw-Hill Companies, Inc.

Slater, J. A., Weber, R., & Fragner, E. (2004). The IGS GLONASS pilot project - transitioning an experiment into an operational GNSS service. *Proceedings of ION GNSS 2004*, pp. 1749 - 1757, Long Beach, CA, US, 2004.

VanDierendonck, A. J., Fenton, P., & Ford, T. (1992). Theory and performance of narrow correlator spacing in GPS receiver, *Navigation: J. Inst. Navig.*, Vol. 39, pp. 115 - 124.

Ward, P. W. (2003). A design technique to remove the correlation ambiguity in binary offset carrier (BOC) spread spectrum signals. *Proceedings of ION AM 2003*, pp. 146-155, Albuquerque, NM, US, 2003.

Yao, Z. (2008). A new unambiguous tracking technique for sine-BOC(2*n*,*n*) signals. *Proceedings of ION GNSS 2008*, pp. 1490-1496, Savannah, GA, US, 2008.

Yao, Z. (2009). *Code Synchronization and Carrier Tracking Algorithms for New Generation of GNSS*. PhD Thesis, Tsinghua University, Beijing.

Yao, Z., Cui, X., Lu, M., & Feng, Z. (2010b). Pseudo-correlation-function-based unambiguous tracking technique for sine-BOC signals. *IEEE Transactions on Aerospace and Electronic Systems*, Vol. 46, No. 4, pp. 1782-1796.

Yao, Z., & Lu, M. (2011). Side-peaks cancellation analytic design framework with applications in BOC signals unambiguous processing, *Proceedings of ION ITM 2011*, pp. 775-785, San Diego, CA, US, 2011.

Yao, Z., Lu, M., & Feng, Z. (2010a). Unambiguous sine-phased binary offset carrier modulated signal acquisition technique, *IEEE Transactions on Wireless Communications*, Vol. 9, No. 2, pp. 577-580.

Ziemer, R. E., & Peterson, R. L. (1985). *Digitial Communications and Spread Spectrum Systems*. New York: Macmillan Publishing Company.

Part 2

GNSS Navigation and Applications

5

GNSS in Practical Determination of Regional Heights

Bihter Erol and Serdar Erol
Istanbul Technical University, Civil Engineering Faculty
Department of Geomatics Engineering
Turkey

1. Introduction

Describing the position of a point in space, basically relies on determining three coordinate components: the Cartesian coordinates (X, Y, Z) in rectangular coordinate system or latitude, longitude and ellipsoidal height (φ, λ and h) in ellipsoidal coordinate system, referred to any given reference ellipsoid. Today, of course, global navigation satellite systems (GNSS) is the best and most popular method for determining φ, λ and h, directly. The instantaneous determination of position and velocity on a continuous base, and the precise coordination of time are included in the objectives of GNSS, and positioning with GNSS base on ranging from known positions of satellites in space to the unknown positions on the earth or in space. Besides the geometrically described coordinates however, the natural coordinates, the astrogeodetic latitude, longitude and orthometric height (Φ, Λ, H), which directly refer to the gravity field of the earth, are preferable to take for many special purposes. In particular the orthometric heights above the geoid are required in many applications, not only in all earth sciences, but also in other disciplines such as; cartography, oceanography, civil engineering, hydraulics, high-precision surveys, and last but not least geographical information systems. Traditionally, these heights are determined by combining geometric levelling and gravity observations with millimetre precision in smaller regions. This technique, however, is very time consuming, expensive and makes providing vertical control difficult, especially in mountainous areas which are hard to access. Another disadvantage is the loss of precision over longer distances since each height system (regional vertical datum) usually refers to a benchmark point close to the sea level, which is connected to a tide gauge station representing the mean sea level (Hofmann-Wellenhof & Moritz, 2006).

In order to counteract these drawbacks of levelling, GNSS introduces a revolution also in the practical determination of the heights in regional vertical datum depending on the basic relation $H = h - N$ among the heights. This equation relates the orthometric height H (above the geoid), the ellipsoidal height h (above the ellipsoid), and the geoidal undulation N, as such, when the h is provided by GNSS and N exists from a reliable and precise digital geoid map, the orthometric height H can then be obtained immediately. This alternative technique for the practical determination of H is called GNSS levelling. In the recent decades the wide and increasing use of GNSS in all kinds of geodetic and surveying applications demands modernization of vertical control systems of countries. The current position is that, the most developed countries are concentrating efforts on establishing a dynamic geoid based vertical datum accessible via GNSS positioning (see e.g. Rangelova et al., 2010). Besides enabling the accurate determination of most up-to-date geoidal heights under the effects of secular dynamic changes of the earth for GNSS levelling purposes, it is envisioned that this new datum concept, will also provide a compatible vertical datum with global height system, which is crucial for studies related to large scale geodynamics and geo-hazards processes.

The accurate determination of orthometric heights via GNSS levelling requires a centimetre(s) accuracy of the geoid model. The level of achievable accuracy of the models varies depending on the computational methodology (assumptions used) and available data within the region of interest (Featherstone et al., 1998; Fotopoulos, 2003; Fotopoulos et al., 2001; Erol, 2007; Erol et al., 2008). The regional models provide better accuracies in comparison to global models. However, for many parts of the globe a high precision regional geoid model is not accessible usually due to lack of data. In these cases, depending on the required accuracy level of the derived heights, one may resort to applying global geopotential model values. An alternative way to determining discrete geoid undulation values is the geometric approach. The approach, which works well in relatively small areas, utilises the relationship between the GNSS ellipsoidal and regional orthometric heights at the known points to interpolate new values. In determining orthometric height with GNSS levelling, apart from consideration for the error budgets of each height data (h, H, N), it will also be necessary to take into account the systematic shifts and datum differences among these data sets, which also restrict the precision of determination. Since the regional vertical datum is not necessarily coincident with the geoid surface, the discrepancies between the regional vertical datum and the geoid surface are preferably accounted for using a special technique allowing for an improved computation of the regional heights with GNSS coordinates (Fotopoulos, 2003; Erol, 2007).

This chapter aims to review the geoid models for GNSS levelling purposes in Turkey and mapping the progress of the global and regional geoid models in Turkish territory. In this respect the study consists of two parts; the first part provides validation results of the recently released eight global geopotential models from satellite gravity missions namely; EGM96, EGM08 (of full expansion and up to 360 degree), EIGEN-51C, EIGEN-6C, EIGEN-6S, GGM03C, GGM03S and GOCO02S, as well as two Turkish regional geoid models TG03 and TG09, based at 28 homogeneously distributed reference benchmarks with known ITRF96 coordinates and regional orthometric heights. The validations consist of comparison of the geoid undulations between the used models and the observed height data (h, H). It should be noted that the results from the validations were evaluated against the reported precisions of the models by the responsible associations.

The second part of this chapter focuses on the determination and testing of the geoid models using geometrical approach in small areas and their assessments in GNSS levelling. In the numerical evaluation, two geodetic networks were used. Each network had 1205 and 109 reference benchmarks with known ITRF96 positions and regional vertical heights, established in these neighbour local areas. Since the topographical character, distribution and density of the reference benchmarks at each area were totally different, these networks provided an appropriate test bed for the local geoid evaluations. In the coverage of the second part, each network was evaluated independently. A group of modelling algorithms were run using the reference benchmarks and tested at the independent test benchmarks of each network. The applied modelling techniques for local geoids (ranging from simple to more comlex methods) including, the multivariable polynomial regression and artificial neuro-fuzzy inference systems (ANFIS). In the light of these conclusions, the roles of topography of the area of interest, the distribution and density of the reference benchmarks, and computation algorithm used in the precise determination of the geoid model and therefore the accuracy of regional heights from the GNSS levelling, were investigated. In addition to the investigation and review on local geoids, local improvement of the recent Turkish regional geoid using 31 reference benchmarks of Çankırı GNSS/levelling networks has also been included. The next section has been structured accordingly to report on these areas.

The outline for this chapter is as follows; the first section provides background information regarding the geoid models, the height data used to conduct the research are also presented and explained. As special emphasis has been given to the error sources affecting the used heights (h, H) and thus the accuracy of the geoid model (N), information relating to the global geoid models (EGM96, EGM08, EIGEN-51C, EIGEN-6C, EIGEN-6S, GGM03C, GGM03S and GOCO02S) and Turkish regional geoid models (TG03, TG09) has also been included in this section. In addition to an overview of the aforementioned models, a list of related references where they have been used in previous studies is provided for further reading purposes. Furthermore the numerical validations of the explained models are included within a sub-heading and validation results appropriately presented as graphics and tables.

The second section is focused on the methodology, and the theoretical background of the applied methods used in the calculation of the local geoid models. The local improvements of regional models are also summarized, and corresponding literature provided. The merits and limitations of each method are also referred to in this section. The outlined methodology was implemented using the three test network's data, in order to test the computation algorithms and demonstrate the role of the data and topographical patterns in geometrical modelling of the local geoids and local improvements of regional geoids. The findings are presented as graphics and tables.

The last section summarises the main conclusions of this research and some practical considerations for modernizing vertical control, as parallel to GNSS development are presented. This section therefore essentially focuses on how to evaluate the achievable accuracy of GNSS levelling. A brief discussion outlines some of the key concepts for providing users of GNSS with the proper information to transform ellipsoidal heights to heights associated with a regional vertical datum. To conclude the chapter, recommendations for future work in this area are also provided.

2. Global and regional geoid models: Methodology and data

GNSS ellipsoidal heights are purely geometric definitions and do not refer to an equipotential surface of the earth's gravity field, as such they cannot be used in the same way as conventional heights derived from levelling in many applications. In order for GNSS derived ellipsoidal heights to have any physical meaning in application, they must be transformed to orthometric heights referring to mean sea level (geoid). This transformation is applied using the geoidal heights (N) from a geoid model that must be known with sufficient accuracy (Fotopoulos et al., 2001; Fotopoulos, 2005). The computation methods of geoid models are many (Schwarz et al., 1987; Featherstone, 1998; Featherstone, 2001; Hirt and Seeber, 2007; Erol et al., 2008; Erol et al., 2009). The most commonly used methods for geoid surface construction are described in textbooks like Heiskanen & Moritz (1967), Vaniček & Krakiwsky (1986), Torge (2001). The so-called remove-restore (R-R) procedure is one of these methods; where a global geopotential model and residual topographic effects are subtracted (and later added back) (see Equations 1 and 2). The smooth resulting data set is then suitable for interpolation or extrapolation using for example least squares collocation with parameters (Sideris, 1994). According to R-R method, the reduced gravity anomaly is:

$$\Delta g = \Delta g_{FA} - \Delta g_{GM} - \Delta g_{H} \tag{1}$$

and the computed geoid height is:

$$N = N_{GM} - N_{\Delta g} - N_{ind} \tag{2}$$

where Δg_{GM} is the effect of the global geopotential model on gravity anomalies, Δg_H is the terrain effect on gravity, $N_{\Delta g}$ is the residual geoid height, which is calculated using Stokes integral (see Equation 3), N_{ind} is the indirect effect of the terrain on the geoid heights and N_{GM} is the contribution of the global geopotential model (expressed by the Equation 4), (Heiskanen & Moritz, 1967; Sideris, 1994). The residual geoid height, computed from Stokes's equation is;

$$N_{\Delta g} = \frac{R}{4\pi} \iint_{\sigma} \Delta g S(\Psi) \, d\sigma \tag{3}$$

where σ denotes the Earth's surface, Δg is the reduced gravity anomaly (Equation 1) and $S(\Psi)$ is the Stokes kernel function where Ψ is the spherical distance between the computation and running points (Haagmans et al., 1993; Sideris; 1994).

The global geopotential model derived geoid height using spherical harmonic coefficients, $\bar{C}_{\ell m}$ and $\bar{S}_{\ell m}$, is;

$$N_{GM} \approx R \sum_{\ell=2}^{\ell \max} \sum_{m=0}^{\ell} \bar{P}_{\ell m}(\sin\theta)\left[\bar{C}_{\ell m} \cos m\lambda + \bar{S}_{\ell m} \sin m\lambda\right] \tag{4}$$

where R is the mean radius of the Earth, (θ, λ) are co-latitude and longitude of the computation point, $\bar{P}_{\ell m}$ are fully normalized Legendre functions for degree ℓ and order m, and ℓ_{max} is the maximum degree of the global geopotential model (Heiskanen & Moritz, 1967).

Following the Equation 2, it is obvious that the accuracy of the computed geoid heights depends on the accuracy of the three height components, namely N_{GM}, $N_{\Delta g}$ and N_H (Fotopoulos, 2003). The global geopotential model not only contributes to the long wavelength geoid information but also introduces long-wavelength errors that originate from insufficient satellite tracking data, lack of terrestrial gravity data and systematic errors in satellite altimetry. The two main types of errors can be categorized as either omission or commission errors. Omission errors occur from the truncation of the spherical harmonic series expansion (Equation 4), which is available in practice ($\ell_{max} < \infty$). The other major contributing error type is due to the noise in the coefficients themselves and termed as commission errors. As the maximum degree ℓ_{max}, of the spherical harmonic expansion increases, so does the commission error, while the omission error decreases. Therefore, it is important to strike a balance between the various errors. In general, formal error models should include both omission and commission error types in order to provide a realistic measure of the accuracy of the geoid heights computed from the global geopotential model. In the following section, recently released global geopotential models using the data from low earth orbiting missions such as CHAMP, GRACE and GOCE are exemplified and their performances in Turkish territory investigated. Parallel to the improvements in techniques, the new global geopotential models derived by incorporating the satellite data from these missions are quite promising (Tscherning et al., 2000; Fotopoulos, 2003).

The other errors in the budget contributing to the $N_{\Delta g}$ component stem from the insufficient data coverage, density and accuracy of the local gravity data. Obviously, higher accuracy is implied by accurate Δg values distributed evenly over the entire area with sufficient spacing, however there are some systematic errors such as datum inconsistencies, which influence the quality of the gravity anomalies too. The shorter wavelength errors in the geoid heights are introduced through the spacing and quality of the digital elevation model used in the computation of N_H. Improper modelling of the terrain is especially significant in mountainous regions, where terrain effects contribute significantly to the final geoid model. This is in addition to errors relating to the approximate values of the vertical gravity gradient (Forsberg, 1994). Improvements in geoid models according to the computation of N_H, will be seen through the use of higher resolution (and accuracy) digital elevation data, especially in mountainous regions.

2.1 Testing global geoid models

The global geopotential model used as a reference in the R-R technique has the most significant error contribution in the total error budget of the computed regional geoid models. Therefore employing an appropriate global model in R-R computations is of primary importance. Likewise, in areas where regional models exist, they should be used as they are more accurate compared to global models. However, many parts of the globe do not have access to a regional geoid model, usually due to lack of data. In these cases, one may resort to applying global geopotential model values (Equation 4) that best fit the gravity field of the region. Determining the optimal global model for either, using the base model in R-R construction of the regional geoid or estimating the geoid undulations in the region with a relatively low accuracy, it will be necessary to undertake a comparison and validation of the models with independent geoid and gravity information, such as GNSS/levelling heights and gravity anomalies (Gruber, 2004; Kiamehr & Sjöberg, 2005; Merry, 2007).

The global geopotential models are mainly divided into three groups based on the data used in their computation, namely satellite-only (derived from the tracking of artificial satellites), combined (derived from the combination of a satellite-only model with terrestrial and/or airborne gravimetry, satellite altimetry, topography/bathymetry) and tailored (derived by refining existing satellite-only or combined global geopotential models using regional gravity and topography data) models. Satellite-only models are typically weak at coefficients of degrees higher than 60 or 70 due to several factors, such as the power-decay of the gravitational field with altitude, modelling of atmospheric drag, incomplete tracking of satellite orbits from the ground stations etc. (Rummel et al., 2002). Although the effects of some of these limitations on the models decreased after the dedicated satellite gravity missions CHAMP, GRACE and GOCE (GGM02, 2004; GFZ, 2006; GOCE, 2009), the new satellite-only models still have full power until a certain degree, however rapidly increasing errors make their coefficients unreliable at high degrees (see e.g. Tapley et al., 2005; ICGEM, 2005). Whilst, the application of combined models reduce some of the aforementioned limitations, the errors in the terrestrial data effectively remain the same.

Theoretically, the observations, used in computation of the global models, should be scattered to the entire earth homogenously, but it is almost impossible to realise this exactly. As such, accuracy of quantities computed via global geopotential models, such as geoid undulation (Equation 4), is directly connected to the quality and global distribution of gravity data as well as to the signal power of satellite mission. The distribution and the availability of quality gravity data therefore plays a major role in the global model-derived values in different parts of the Earth. It may however be argued that, the various models may not be as good as they are reported to be, otherwise the differences between them should not be so great as they are (Lambeck & Coleman, 1983). As such, validating the models in local scale with in situ data before using them with geodetic and geophysical purposes is highly important (Gruber 2004). In this manner, Roland & Denker (2003) evaluated the fit of some of the global models to the gravity field in Europe using external data such as GPS/levelling and gravity anomalies. Furthermore, Amos & Featherstone (2003) included astrogeodetic vertical deflections at the Earth surface in the external data for validating the EGMs at that date in Australia. Similar evaluations were also undertaken by Kiamehr & Sjöberg (2005), Abd-Elmotaal (2006), Rodriguez-Caderot et al (2006), Merry (2007) and Sadiq & Ahmad (2009) in Iran, Egypt, Southern Spain, Southern Africa, Pakistan, respectively. Satellite altimeter data and orbit parameters were also used by Klokočník et al (2002) and Förste et al (2009) in comparative assessments of the EGMs. Erol et al (2009), Ustun & Abbak (2010) and Yılmaz & Karaali (2010) provided some specific results on spectral evaluation of global models and on their local validations using terrestrial data in territory of Turkey. Motivated research conducted by Lambeck&Coleman (1983) and Gruber (2004), we tested some of the recent global geopotential models having various orders of spherical harmonic expansion for Turkish territory, the results of which have been recorded later in this chapter. The listed global geopotential models in Table 1 were validated at 28 GNSS/levelling benchmarks, homogenously distributed over the country. The table provides the maximum degrees of the harmonic expansions, the data contributed for developing the models and also the principle references for further reading on these models. The reference data for validations are included by Yılmaz & Karaali (2010), hence the results from the models evaluated in both studies are comparable (see Figure 1 for the distribution of the benchmarks).

In evaluations, the geoid heights derived from the models (Equation 4) were compared with observations at the benchmarks, and the statistics of comparisons (see Table 2) were investigated. In the validation results, superiority of ultra-high resolution models EIGEN-6C (ℓ_{max} = 1420) and EGM08 (ℓ_{max} = 2190) in representing the gravity field in the region is naturally obvious given that these models comprise information relating to full content of gravity field spectrum. Considering the ±16.3 cm and ±17.9 cm accuracies of EIGEN-6C and

EGM08 in terms of root mean square errors of geoid heights, these models can be employed to obtain regional orthometric heights from GNSS heights for the applications that require a decimetre level accuracy in heights.

*Model	Degree	Data	Citation
EGM96	360	Satellite, gravity, altimetry	Lemoine et al., 1998
EGM08[a]	360	GRACE, gravity, altimetry	Pavlis et al., 2008
EGM08[b]	2190	GRACE, gravity, altimetry	Pavlis et al., 2008
EIGEN-6C	1420	GOCE, GRACE, LAGEOS, gravity, altimetry	Förste et al., 2011
EIGEN-6S	240	GOCE, GRACE, LAGEOS	Förste et al., 2011
EIGEN-51C	359	GRACE, CHAMP, gravity, altimetry	Bruinsma et al., 2010
GGM03C	360	GRACE, gravity, altimetry	Tapley et al., 2007
GGM03S	180	GRACE	Tapley et al., 2007
GOCO02S	250	GOCE, GRACE	Goiginger et al., 2011

* Related to the global geopotential models that were used in the study: *i*-) The adopted reference system is GRS80, *ii*-) The applied models are in tide free system, *iii*-) Zero degree terms were included in computations, *iv*-) The model coefficients are available from ICGEM (2011).

Table 1. Validated global geopotential models in the study

Some other conclusions drawn from the statistical inspection of the validation results that EGM08 provided improved results compared to its previous version EGM96 in the study region (compare the statistics of EGM96 and EGM08[a] in Table 2). Among the satellite only models EIGEN-6S fits best, and as such, can be recommended as a reference model for a future regional geoid of Turkey with R-R technique.

Model	ℓ_{max}	Type	min.	max.	mean	std. dev.
EGM96	360	Combined	-183.1	336.5	38.2	156.3
EGM08[a]	360	Combined	-105.0	47.6	-18.1	36.4
EGM08[b]	2190	Combined	-58.6	27.0	-4.5	17.3
EIGEN-6C	1420	Combined	-41.9	28.3	-4.1	15.8
EIGEN-6S	240	Satellite only	-77.5	85.0	-9.7	43.2
EIGEN-51C	359	Combined	-126.2	50.5	-21.8	38.9
GGM03C	360	Combined	-151.2	213.0	-2.4	76.3
GGM03S	180	Satellite only	-394.3	331.4	-18.5	198.1
GOCO02S	250	Satellite only	-87.2	90.9	-8.6	43.5

Table 2. Statistics of the geoid height differences between global models and observations (in centimetre)

The geoid height differences of EIGEN-6C and EIGEN-6S global models from the observed geoid heights at the reference benchmarks are illustrated in Figures 2 and 3, respectively. These differences can be compared and interpreted considering the topographic map of Turkey in Figure 1.

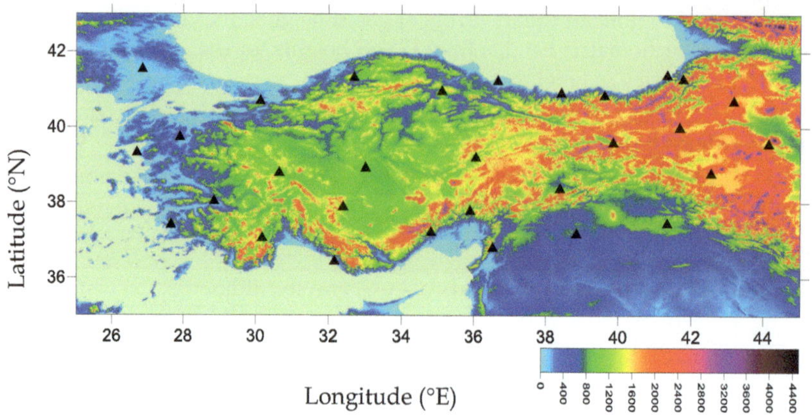

Fig. 1. Topographic map of Turkey and validation benchmarks (units metre) (using GTOPO30 data (USGS, 1997))

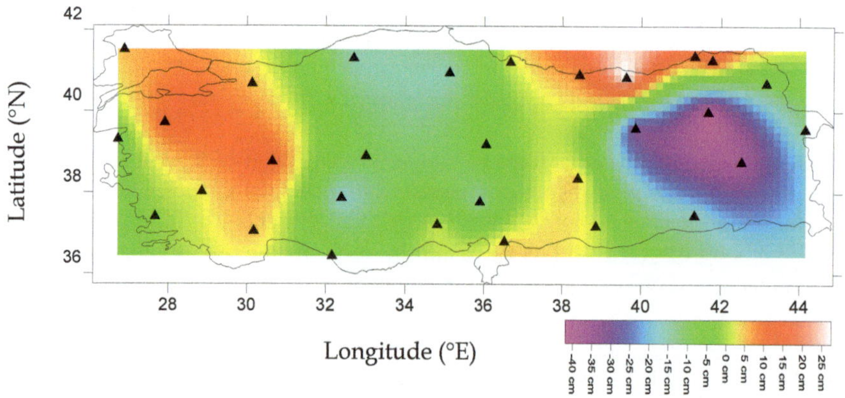

Fig. 2. Geoid height differences between EIGEN-6C model and GNSS/levelling observations

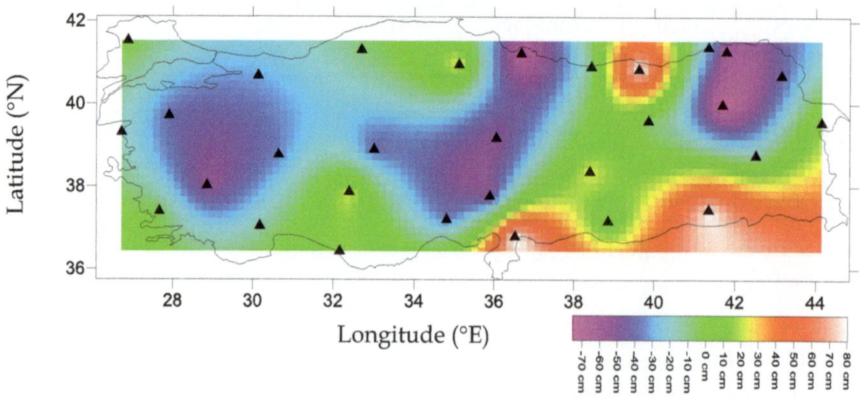

Fig. 3. Geoid height differences between EIGEN-6S model and GNSS/levelling observations

2.2 Regional geoid models in Turkey

In Turkey, various regional geoid models have been computed with different methods, since the 1970's (see e.g. Ayan, 1976; Ayhan, 1993; Ayhan et al., 2002; TNUGG, 2003; TNUGG, 2011), along with the technologic advances and increasing use of GNSS techniques in 1990's, modernization of national geodetic infrastructure, including the vertical datum definition, was required. As a consequence of these developments the geodetic control network was re-established in ITRF96 datum by Turkish Ministry of National Defence-General Command of Mapping between 1997 and 2001, a geoid model (TG99A) as a height transformation surface from GNSS to the regional vertical datum was released in 2000 (Ayhan et al., 2002). Turkey regional geoid model TG99A was gravimetrically determined and fitted to the regional vertical datum at homogeneously distributed GNSS/levelling benchmarks throughout the country. The absolute accuracy of TG99A model is reported between ±12 cm and ±25 cm, however the performance of the model decreases from the central territories through the coastline and boundaries of the country (Ayhan et al., 2002). An updated version of TG99A was released by General Command of Mapping in 2003 (TG03) (TNUGG, 2003). TG03 was computed with R-R method and Least Squares Collocation using terrestrial gravity data in 3-5 km density over the country (at Potsdam gravity datum), marine gravity data (acquired with shipborne and satellite altimetry), terrain based elevation model in 450 m x 450 m resolution and reference global model EGM96, and fitted to the regional vertical datum at 197 high order GNSS/levelling benchmarks (TNUGG, 2003). The accuracy of TG03 is reported as ±8.8 cm by TNUGG (2003) this revealed good improvement when compared the previous TG model.

Release of the Earth Gravitational Model 2008 (EGM08), the collection of new surface gravity observations (~266000), the advanced satellite altimetry-derived gravity over the sea (DNSC08), the availability of the high resolution digital terrain model (90 m resolution) and increased number of GNSS/levelling benchmarks (approximately 2700 benchmarks cover the entire country) enabled the computation of a new regional geoid model for Turkey in 2009, hence TG09 was released by General Command of Mapping as successor of TG03 (TNUGG, 2011). In computations, the quasi geoid model was constructed first using R-R procedure based on EGM08 and RTM reduction of surface gravity data and since the Helmert orthometric heights are used for vertical control in Turkey, the quasi geoid model was then converted to the geoid model. Ultimately, the hybrid geoid model TG09 was derived with combining the gravimetric geoid model and GNSS/levelling heights to be used in GNSS positioning applications. In the test results of TG09 with GNSS/levelling data, the accuracy of the model is reported as ±8.3 cm by TNUGG (2011). This result does not signify much improvement when comparing the TG03.

This section examines the published accuracies of TG03 and TG09 models at 28 GNSS/levelling benchmarks used in the validation of global geopotential models in the previous section. With this purpose in mind, the derived geoid heights at the benchmarks were compared with observations and in the results: TG03 model revealed ±10.5 cm standard deviation with minimum -10.1 cm, maximum 28.9 cm and mean 7.3 cm geoid height differences, whereas the TG09 model has ±9.2 cm standard deviation with minimum -11.3 cm, maximum 36.7 cm and mean of 10.5 cm in geoid height differences. The distribution of geoid height residuals versus the numbers of point are given in histograms in Figure 4. The geoidal height differences for TG03 and TG09 models are illustrated in Figures 5 and 6, respectively.

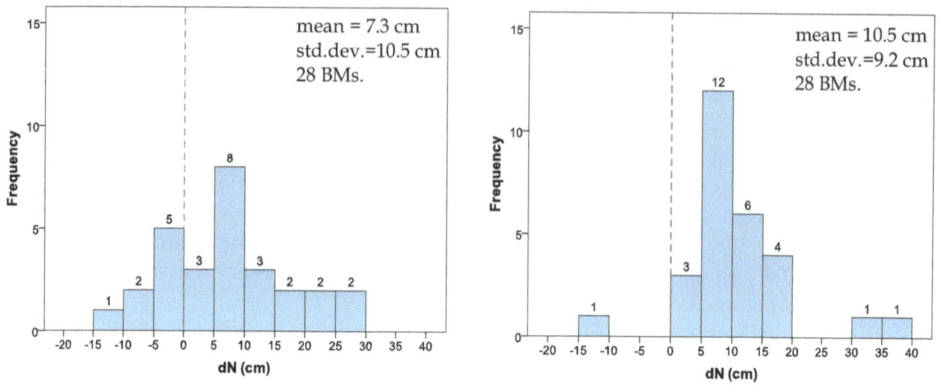

(a) (b)

Fig. 4. Validation results of (a) TG03 and (b) TG09 models: geoid height differences (in cm) versus reference benchmark numbers

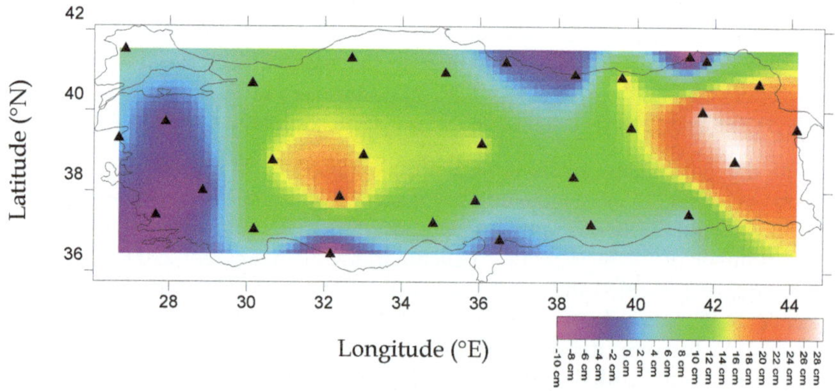

Fig. 5. Geoid height differences between TG03 and GNSS/levelling observations

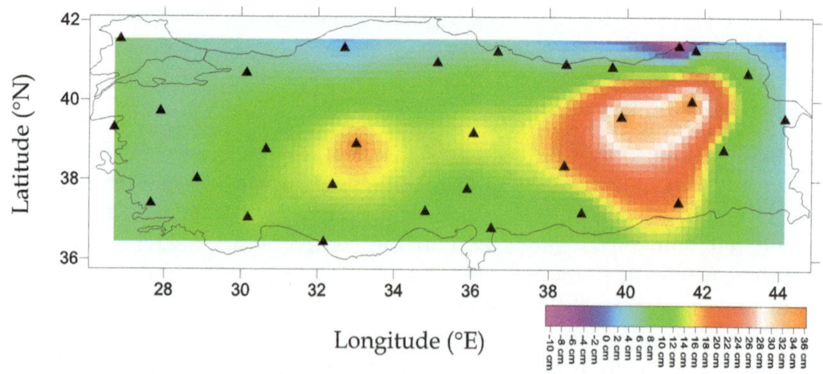

Fig. 6. Geoid height differences between TG09 and GNSS/levelling observations (TG09 data were used from Yılmaz&Karaali (2010))

3. Local GNSS/levelling geoids

Among the computation methods of geoid models (see e.g. Schwarz et al., 1987; Featherstone, 1998; Featherstone, 2001; Hirt and Seeber, 2007; Erol et al., 2008; Erol et al., 2009), geometric approach that GNSS and orthometric heights (h and H, respectively) can be used to estimate the position of the geoid at discrete points (so called geoid reference benchmarks) through a simple relation between the heights ($N \approx h\text{-}H$) provides a practical solution to the geoid problem in relatively small areas (typically a few kilometers) (Featherstone et al., 1998; Ayan et al., 2001, 2005; Erol and Çelik, 2006). This method addresses the geoid determination problem as "describing an interpolation surface depending on the reference benchmarks" (Featherstone et al., 1998; Erol et al., 2005; Erol & Çelik, 2006; Erol et al., 2008). The approximate equality in the equation arises due to the disregard for the deflection of the vertical that means the departure of the plumbline from the ellipsoidal normal (Heiskanen and Moritz, 1967). However the magnitude of error steming from this oversight is fairly minimal and therefore acceptable for the height transformation purposes (Featherstone, 1998).

The data quality, density and distribution of the reference benchmarks have important role on the accuracy of local GNSS/levelling geoid model (Fotopoulos et al., 2001; Fotopoulos, 2005; Erol & Çelik, 2006; Erol, 2008, 2011). There are certain criteria on the geoid reference benchmark qualities and locations, as described in the regulations and reference books (LSMSDPR, 2005; Deniz&Çelik, 2007) that will be mentioned in the text that follows. On the other hand using an appropriate surface approximation method in geoid modelling with geometrical approach is also critical for the accuracy of the model. The modelling methods are various but those most commonly employed among are; polynomial equations (of various orders) (Ayan et al, 2001; Erol, 2008; Erol, 2011), least squares collocation (Erol and Çelik, 2004), geostatistical kriging (Erol and Çelik, 2006), finite elements (Çepni and Deniz, 2005), multiquadric or weighted linear interpolation (Yanalak and Baykal, 2001). In addition to these classical methods, soft computing algorithms such as artificial neural networks (either by itself, see e.g. Kavzaoğlu and Saka (2005) or as part of these classical statistical techniques, e.g. Stopar et al. (2006)), adaptive network-based fuzzy inference systems (ANFIS) (Yılmaz and Arslan, 2008) and wavelet neural networks (Erol, 2007) were also evaluated by researchers in the most recent investigations on local geoid modelling.

3.1 Case studies: Istanbul and Sakarya local geoids

In this section, we discuss and explain the handicaps and advantages of geometric approach and local geoid models from the view point of transformation of GNSS ellipsoidal heights. This includes two case studies: Istanbul and Sakarya local geoids, using polynomial equations and ANFIS methods.

3.1.1 Data

One of the case study areas, Istanbul, is located in the North West of Turkey (between 40°30' N - 41°30' N latitudes, 27°30' E - 30°00' E longitudes, see Figure 7). The region has a relatively plain topography and elevations vary between 0 and 600 m. The GNSS/levelling network (Istanbul GPS Triangulation Network 2005, IGNA2005) was established between 2005 and 2006 as a part of IGNA2005 project (Ayan et al., 2006), and the measurement campaigns and data processing strategies adopted to compute benchmark coordinates satisfy the criteria of LSMSDPR (2005), on determination and use of local GNSS/levelling geoids. Accordingly the geoid reference benchmarks must be the common points of C1, C2 and C3 order GNSS benchmarks and high order levelling network points. Thus the GNSS observations of IGNA2005 project were carried out using dual frequency GNSS receivers, with observation durations of at least 2 hours for C1 type network points (for the baselines 20 km in length), and between 45 and 60 minutes for the C2 type network points (for the baselines 5 km in length). The recording interval was set 15 seconds or less during the campaigns. The GNSS coordinates of network benchmarks were determined in ITRF96 datum 2005.000 epoch with ±1.5 cm and ±2.3 cm of root mean square errors in the two dimensional coordinates and heights, respectively (Ayan et al., 2006). The levelling measurements were done simultaneously during the GNSS campaigns and Helmert orthometric heights of geoid reference benchmarks in Turkey National Vertical Control Network 1999 (TUDKA99) datum (Ayhan and Demir, 1993) were derived. Total number of the homogenously distributed reference benchmarks is 1205 with the density of 1 benchmark per 20 km^2 in the network (see Figure 7).

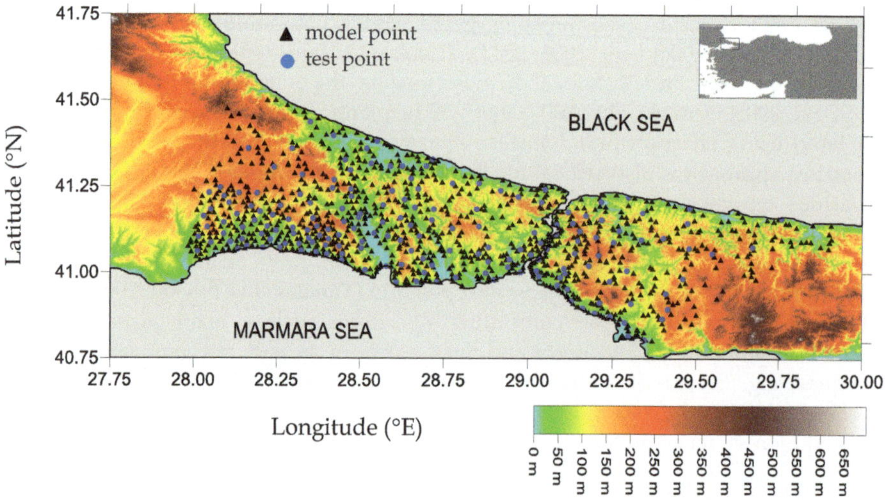

Fig. 7. Geoid reference benchmarks in Istanbul (topographic data from SRTM3 (USGS, 2010)

The second case study on determining local GNSS/levelling geoids was carried out in the Sakarya region situated in the East of Marmara sea and Izmit Gulf (between 40°30' N - 41°30' N latitudes, 28°30' E - 31°00' E longitudes). The GNSS/levelling network was established during the Geodetic Infrastructure Project of the Marmara Earthquake Region Land Information System (MERLIS) in 2002 (Çelik et al., 2002), and overlap with IGNA2005 network. Compared to the Istanbul area, the topography in Sakarya is quite rough and the elevations are between 0 m and 2458 m. The GNSS and levelling observations, and data processes were executed according to the regulation of the project. After the adjustment of GNSS network, the accuracies of ±1.5 cm and ±3.0 cm for the horizontal coordinates and ellipsoidal heights were derived. During the GNSS campaign of the MERLIS project, precise levelling measurements were undertaken, simultaneously, and in the adjustment results of levelling observations the relative accuracy of Helmert orthometric heights is reported as 0.2 ppm by Çelik et al. (2002). The GNSS coordinates are in ITRF96 datum, while the orthometric heights are in TUDKA99 datum.

The distribution of the 109 GNSS/levelling benchmarks is homogenous but rather sparse, and given the rough topography of the region, the coverage of the benchmarks cannot characterize the topographic changes well. The reference point density is 1 benchmark per 165 km^2. Figure 8 shows the reference network benchmarks on the topographic map of the area.

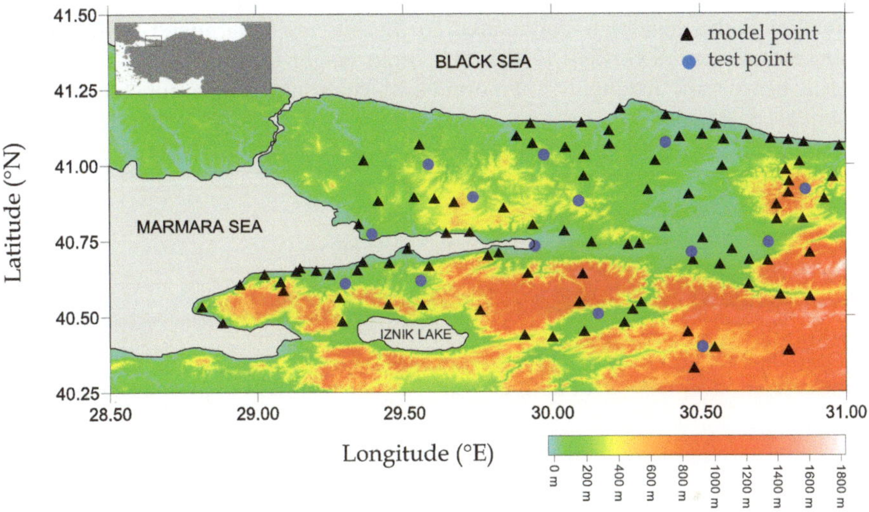

Fig. 8. Geoid reference benchmarks in Sakarya (topographic data from SRTM3 (USGS, 2010)

3.1.2 Methods

Since the computation algorithms, applied for local GNSS/levelling geoid determination in the study, are not able to detect potential blunders in data sets, the geoid heights derived from the observations at the benchmarks were statistically tested and the outliers were cleaned before modelling the data (see Erol (2011) for a case study in screening the reference data before geoid modelling). After removing the outliers from data sets, in Istanbul data, uniformly distributed 200 points of 1205 reference benchmarks (approximately 16% of the entire data) were selected to form the test data, and the remaining 1005 benchmarks were used in computation of the geoid. Similarly, in Sakarya, 14 of the 109 data points (nearly 13% of all data) having homogenous distribution were selected and used for external tests of the geoid model. The model and test points are distinguished with different marks on Figures 7 and 8. The theoretical review of the applied surface interpolation methods and comparisons of their performances by means of the test results are provided in the next section.

3.1.2.1 Polynomials

The polynomial equation for representing a local geoid surface based on the discrete reference benchmarks with known geoid heights in the closed form is:

$$N(u,v) = \sum_{m=0}^{l} \sum_{n=0}^{l-m} a_{mn} u^m v^n \tag{5}$$

where a_{mn} are the polynomial coefficients for $m,n = 0$ to l, which is the order of polynomial. u and v represent the normalized coordinates, which are obtained by centring and scaling the geodetic coordinates φ and λ. In the numerical tests of this study, the normalized coordinates were obtained by $u = k\,(\varphi - \varphi_o)$ and $v = k\,(\lambda - \lambda_o)$ where φ_o and λ_o are the mean latitude and longitude of the local area, and the scaling factor is $k = 100/\rho°$.

In Equation 5, the unknown polynomial coefficients are determined with least squares adjustment solution. According to this, the geoid height (Ni) and its correction (Vi) at a reference benchmark having (u, v) normalized coordinates as a function of unknown polynomial coefficients is:

$$\begin{aligned} N_i + V_i = {} & a_{00} + a_{10}u + a_{11}v \\ & + a_{20}u^2 + a_{21}uv + a_{22}v^2 \\ & + a_{30}u^3 + a_{31}u^2v + a_{32}uv^2 + a_{33}v^3 \\ & + a_{40}u^4 + a_{41}u^3v + a_{42}u^2v^2 + a_{43}uv^3 + a_{44}v^4 \\ & \ldots \end{aligned} \tag{6}$$

and the correction equations for all reference geoid benchmarks in matrix form is:

$$\begin{bmatrix} N_1 \\ N_2 \\ . \\ . \\ N_i \end{bmatrix} + \begin{bmatrix} V_1 \\ V_2 \\ . \\ . \\ V_i \end{bmatrix} = \begin{bmatrix} 1 & u_1 & v_1 & \ldots \\ 1 & u_2 & v_2 & \ldots \\ . & . & . & \ldots \\ . & . & . & \ldots \\ 1 & u_i & v_i & \ldots \end{bmatrix} \begin{bmatrix} a_{00} \\ a_{10} \\ . \\ . \\ a_{mn} \end{bmatrix} \tag{7a}$$

$$N + V = AX \tag{7b}$$

and the unknown polynomial coefficients (a_{mn} elements of the X vector, see Equations 7a and 7b):

$$X = \left(A^T A\right)^{-1} A^T \ell \tag{8}$$

and the cofactor matrix of X

$$Q_{XX} = \left(A^T A\right)^{-1} \tag{9}$$

are calculated. In the equations A is coefficients matrix and ℓ is the vector of observations that the elements of the vector are the geoid heights ($N_{GNSS/levelling}$).

One of the main issues of modelling with polynomials is deciding the optimum degree of the expansion, which is critical for accuracy of the approximation as well and its decision mostly bases on trial and error (Erol, 2009). Whilst the use of a low-degree polynomial usually results in an insufficient or rough approximation of the surface, unnecessarily use of a higher degree function may produce an over fitted surface that may reveal unrealistic and optimistic values at the test points. Another critical phase of determining polynomial surface is selecting the significant parameters and hence ignoring the insignificant ones in the model that this decision also bases on statistical criteria. After calculating the polynomials with least squares adjustment, the statistical significance of the model parameters can be analyzed using F-test with the null hypothesis $H_o : X_i = 0$ and the alternative hypothesis $H_1 : X_i \neq 0$ (Draper and Smith, 1998). The F-statistic is used to verify the null hypothesis and computed as a function of observations (Dermanis and Rossikopoulos, 1991):

$$F = \frac{X_i^T Q_{X_i X_i}^{-1} X_i}{t\hat{\sigma}^2} \tag{10}$$

where $\hat{\sigma}^2$ is a-posteriori variance, t is the number of tested parameters. The null hypothesis is accepted if $F \leq F_{t,r}^{\alpha}$, where $F_{t,r}^{\alpha}$ is obtained from the standard statistical tables for a confidence level α and degrees of freedom r that means the tested parameters are insignificant and deleted from the model. If the contrary is true and $F > F_{t,r}^{\alpha}$ is fulfilled, then the parameters remain in the model. After clarifying the optimal form of a polynomial model with significance tests of parameters, the performance of the calculated model is tested empirically, considering the geoid residuals at the benchmarks of the network. The tests are repeated with the polynomials in varying orders and hence an appropriate order of polynomial is determined for the data depending on the comparisons of test results.

3.1.2.2 Adaptive network based fuzzy inference system

ANFIS is an artificial intelligence inspired soft computing method that is first purposed in the late 1960's depending on fuzzy logic and fuzzy set theory introduced by Zadeh (1965). After that this method was used in various disciplines for controlling the systems and modelling non-stationary phenomena, and recently applied in geoid determination, as well (see e.g. Ayan et al, 2005; Yılmaz and Arslan, 2008). The computation algorithm of the method mainly bases on feed-forward adaptive networks and fuzzy inference systems. A fuzzy inference system is typically designed by defining linguistic input and output variables and an inference rule base. Initially, the resulting system is just an approximation for an adequate model. Hence, its premise and consequent parameters are tuned based on the given data in order to optimize the system performance and this process bases on a supervised learning algorithm (Jang, 1993).

In computations with ANFIS, depending on the fuzzy rule structures, there are different neural-fuzzy systems such as Mamdani, Tsukamoto and Takagi-Sugeno (Jang, 1993). Tung and Quek (2009) can be referred for a review on implementation of different neural-fuzzy systems. In Figure 9 a two input, two-fuzzy ruled, one output type 3 fuzzy model is illustrated. In this example Takagi-Sugeno's fuzzy if-then rules are used and the output of each rule is a linear combination of input variables plus a constant term, and the final output is a weighted average of each rule's output.

In the associate fuzzy reasoning in the figure and corresponding equivalent ANFIS structure:

Rule 1: if x is A_1 and y is B_1; then $f_1 = p_1 x + q_1 y + r_1$

Rule 2: if x is A_2 and y is B_2; then $f_2 = p_2x + q_2y + r_2$

where the symbols A and B denote the fuzzy sets defined for membership functions of x and y in the premise parts. The symbols p, q and r denote the consequent parameters of the output functions f (Takagi and Sugeno, 1985; Jang, 1993; Yılmaz, 2010). The Gaussian function is usually used as input membership function $\mu_i(x)$ (see Equation 11) with the maximum value equal to 1 and the minimum value equal to 0:

$$\mu_i(x) = \exp\left[-\left(\frac{x-b_i}{a_i}\right)^2\right] \tag{11}$$

where a_i, b_i are the premise parameters that define the gaussian-shape according to their changing values. Yılmaz and Arslan (2008) apply various membership functions and investigate the effect of the each function on the approximation accuracy of the data set.

In the associated ANFIS architecture of Figure 9, the functions of the layers can be explained as such that in *Layer 1*, inputs are divided subspaces using selected membership function, in *Layer 2*, firing strength of a rule is calculated by multiplying incoming signals, in *Layer 3*, the firing strengths are normalised and in *Layer 4*, the consequent parameters (p_i, q_i, r_i) are determined and finally in *Layer 5*, the final output is obtained by summing of all incoming signals.

Using the designed architecture, in the running steps of the ANFIS, basically, it takes the initial fuzzy system and tunes it by means of a hybrid technique combining gradient descent back-propagation and mean least-squares optimization algorithms (see Yılmaz and Arslan, 2008). At each epoch, an error measure, usually defined as the sum of the squared difference between actual and desired output, is reduced. Training stops when either the predefined epoch number or error rate is obtained. The gradient descent algorithm is mainly implemented to tune the non-linear premise parameters while the basic function of the mean least-squares is to optimize or adjust the linear consequent parameters (Jang, 1993; Takagi and Sugeno, 1985).

After determination of the local geoid model using either of the methods, the success of the method can be assessed using various statistical measures such as the coefficient of determination, R^2, and the root mean square error, RMSE, of geoidal heights at the reference benchmarks:

$$R^2 = 1 - \frac{\sum_{i=1}^{j}\left(\ell - \hat{\ell}_i\right)^2}{\sum_{i=1}^{j}\left(\ell_i - \overline{\ell}\right)^2} \tag{12}$$

$$RMSE = \sqrt{\frac{\sum_{i=1}^{j}\left(\ell_i - \hat{\ell}_i\right)^2}{j}} \tag{13}$$

where $\hat{\ell}_i$ is the geoid height computed with the polynomial or ANFIS N_{model}, and $\overline{\ell}$ is the mean value of observations, and j is the number of observations (Sen and Srivastava, 1990). Coefficient of determination indicates how closely the estimated values ($\hat{\ell}$) from an approximation model corresponds to the actual data (ℓ), and takes values between 0 and 1 (or represented as percentage, and the closer the R^2 is to 1, the smaller the residuals and hence the better the model fit).

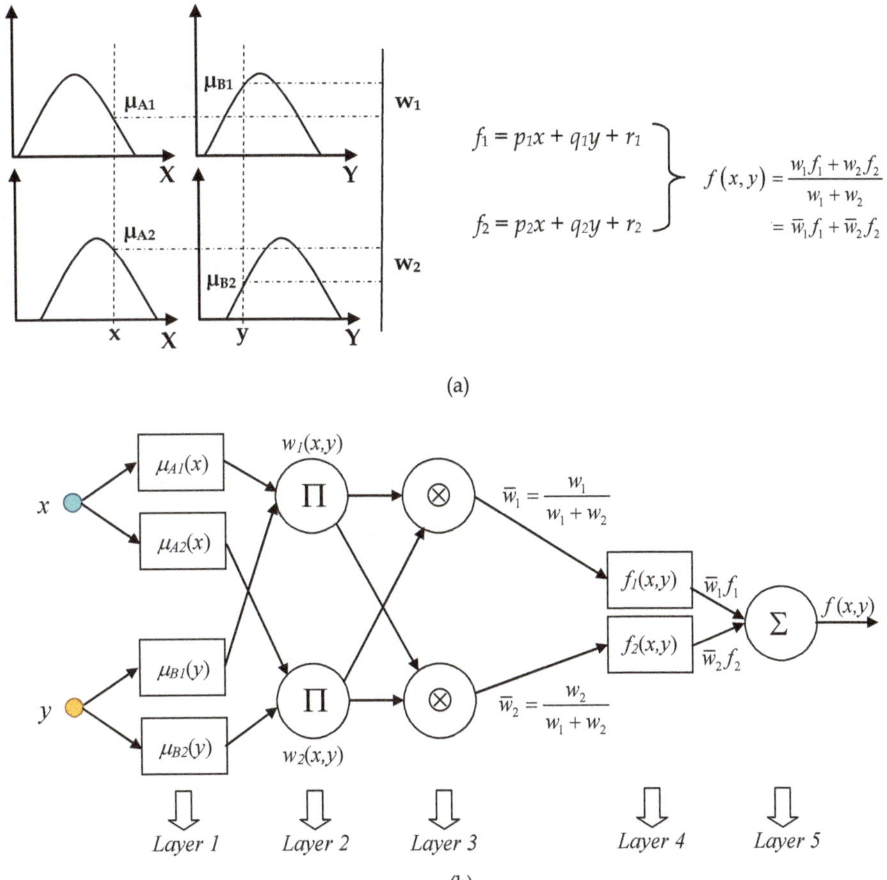

Fig. 9. (a) type 3 fuzzy reasoning, (b) a simple two-input, two-rule and single-output ANFIS structure (Jang, 1993)

3.1.3 Test results

In the results of the tests, repeated with the varying polynomial orders from first to sixth order, a 5th and 4th order polynomial models (having 21 and 15 coefficients) were determined as optimal for the Istanbul and Sakarya data, respectively. The significance tests of the polynomial parameters revealed the final forms of the models. Evaluation of these polynomials at the reference and test benchmarks, separately, in Istanbul and Sakarya regions, revealed the statistics in Tables 3 and 4. As is seen from the Table 3 for Istanbul area, the accuracy of the fifth order polynomial in terms of RMSE of geoid heights at the test points is ±4.4 cm with a coefficient of determination of 0.992. The geoid height differences of the polynomial model and observations at the benchmarks are mapped in Figure 10a. The test statistics of the polynomial model for Sakarya local geoid are summarized in Table 4 that the evaluation of the model at the independent test points revealed an absolute accuracy of ±20.4 cm in terms of RMSE of the geoid heights. Although the qualities of employed reference data in computations of both local geoid models are comparable (see section 3.1.1), the polynomial surface model revealed much improved results in Istanbul territory than Sakarya. The reasons of low accuracy in local geoid model of Sakarya territory can be told as sparse and non-homogeneous distribution of geoid reference benchmarks and rough topographic character of the territory that makes difficult to access for height measuring. Hence the GNSS/levelling benchmarks whose density and distribution are very critical indeed for precise modelling of the local geoid, are not characterize sufficiently the topographic changes and mass distribution in Sakarya (compare point distribution versus topography in Figure 8). Figure 10b shows the geoid height differences of the polynomial model and observations at the benchmarks for Sakarya.

	5th order polynomial		ANFIS		TG03
	Reference BMs	Test BMs	Reference BMs	Test BMs	
Minimum	-11.2	-11.5	-10.5	-9.7	-32.5
Maximum	11.4	11.5	12.4	9.5	30.0
Mean	0.0	0.0	0.0	0.0	-0.3
RMSE	4.2	4.4	3.6	3.5	10.8
R^2	0.993	0.992	0.996	0.995	0.960

Table 3. Statistical comparison of applied approximation techniques in Istanbul local geoid (units in centimetre, R^2 unitless)

	4th order polynomial		ANFIS		TG03
	Reference BMs	Test BMs	Reference BMs	Test BMs	
Minimum	-52.0	-36.3	-39.7	-35.4	-53.8
Maximum	82.7	24.1	42.1	19.0	64.3
Mean	-0.3	-7.5	0.0	-11.0	-4.4
RMSE	22.7	20.4	12.0	18.9	18.6
R^2	0.923	0.905	0.978	0.913	0.945

Table 4. Statistical comparison of applied approximation techniques in Sakarya local geoid (units in centimetre, R^2 unitless)

Nonlinear regression structure of ANFIS and its resulting system, based on tuning the model parameters according to local properties of the data may reveal improved results of surface fitting. However one must be careful whilst working with soft computing approaches and pay attention for choosing appropriate design of architecture with optimal parameters such as: (e.g. in ANFIS) the input and rule numbers, type and number of membership functions, efficient training algorithm. Since the prediction capabilities of these algorithms vary depending on adopted architecture, use of unrealistic parameters may reveal optimistic results but, at the same time, produce an over fitted surface model that should be avoided in geoid modelling. While modelling with ANFIS, deciding an optimal architecture for the system is based on trial and error procedure.

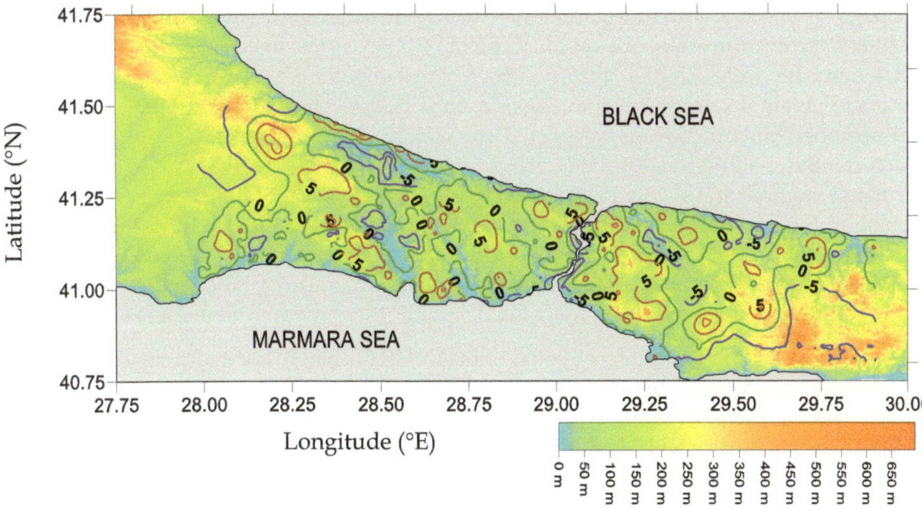

(a)

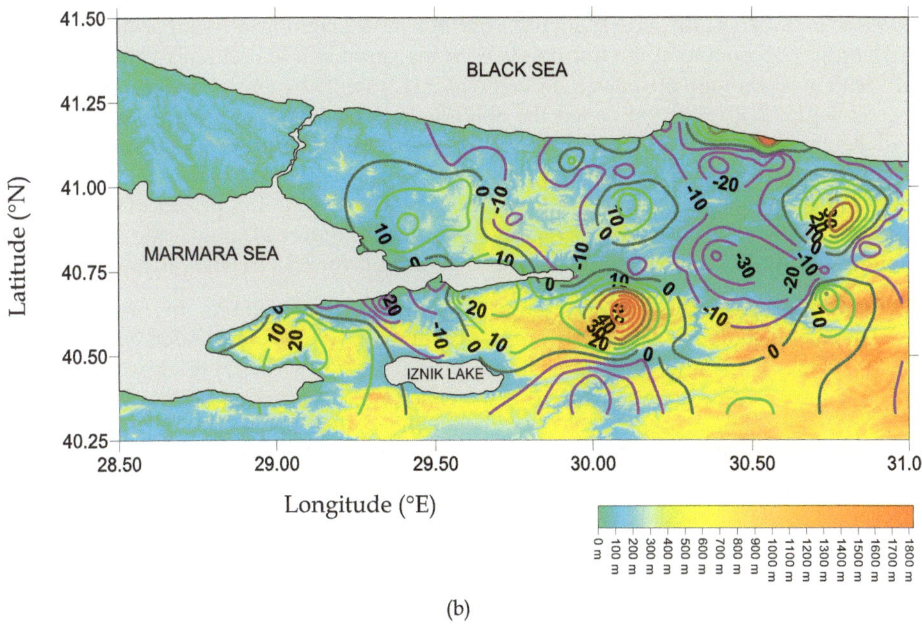

(b)

Fig. 10. Geoid height differences of polynomial models and observations in centimetre ($\Delta N = N_{GNSS/lev.} - N_{poly.}$): (a) Istanbul, (b) Sakarya

In modelling Istanbul and Sakarya local geoids using the ANFIS approach, training data (the geoid reference benchmarks) were used to estimate the ANFIS model parameters, whereas test data were employed to validate the estimated model. The input parameters are the geographic coordinates of the reference benchmarks, and the output membership functions are the first order polynomials of the input variables. As the number of the output membership functions depends on the number of fuzzy rules, in computations, the latitudes and longitudes were divided into 5 subsets to obtain 5 x 5 = 25 rules in Istanbul, and 4 subsets to obtain 4 x 4 = 16 rules in Sakarya. In both case studies, we adopted the Gaussian type membership function as suggested by Yılmaz (2010). After determining the ANFIS structure, the parameters of both the input and output membership functions were calculated according to a hybrid learning algorithm as a combination of least-squares estimation and gradient descent method (Takagi & Sugeno, 1985). Using the determined ANFIS model parameters for Istanbul and Sakarya data, separately, the geoid heights both at the reference and test benchmarks were calculated. In addition, the statistics of the geoid height differences between the model and observations were investigated in each local area.

In the test results for Istanbul local geoid with ANFIS (Table 3), the geoid height residuals at the test benchmarks vary between -9.7 cm and 9.5 cm with a standard deviation of ±3.5 cm. As the basic statistics in Table 3 provides a comparison between the performances of two methods in Istanbul, ANFIS has a 20% improvement in terms of RMSE of geoid heights comparing the 5th order polynomial model. As the RMSE of the computed geoid heights for the reference benchmarks and the test benchmarks are close values, we can say that the composed ANFIS structure is appropriate for modelling the Istanbul data. The coefficient of determination (R^2), as the performance measure of ANFIS model is 0.996.

However, in Sakarya, the ANFIS method did not reveal significantly superior results from the 4th order polynomial at the test points with the geoid height residuals between -35.4 cm and 19.0 cm with root mean square error of ±18.9 cm. The improvement of the model accuracy with ANFIS method versus the polynomial is around 7%, considering the RMSE of geoid heights. On the other hand ANFIS revealed much improved test statistics at the reference benchmarks than the polynomial. The inconsistency, observed between the evaluation results at the reference and test benchmarks for ANFIS model may indicate an inappropriateness of this model for Sakarya data. Figure 11 maps the geoid height differences of ANFIS model and observations at the benchmarks in Istanbul and Sakarya.

In addition to the evaluation of surface approximation methods in modelling local GNSS/levelling geoids in case study areas, TG03 model was also evaluated at the reference geoid benchmarks. The statistics of geoid height differences with 0.3 cm mean and ±10.8 cm standard deviation for Istanbul, confirms the reported accuracies of the model by TNUGG (2003) and Kılıçoğlu et al. (2005). Conversely, the validation results of TG03 model in Sakarya GNSS/levelling benchmarks revealed the differences of geoid heights with -4.4 cm mean and ±18.6 cm standard deviations. Considering these validation results, although the performance of TG03 model seems low by means of RMSE of geoid heights, they revealed approximately 44% of improvement when comparing to the performance of previous Turkish regional geoid TG99A in the same region (see the results of TG99A validations in Sakarya region by Kılıçoğlu&Fırat (2003)).

In the conclusion of this section, the Istanbul and Sakarya local GNSS/levelling geoid models by ANFIS approach can be observed in the maps depicted in Figures 12 and 13.

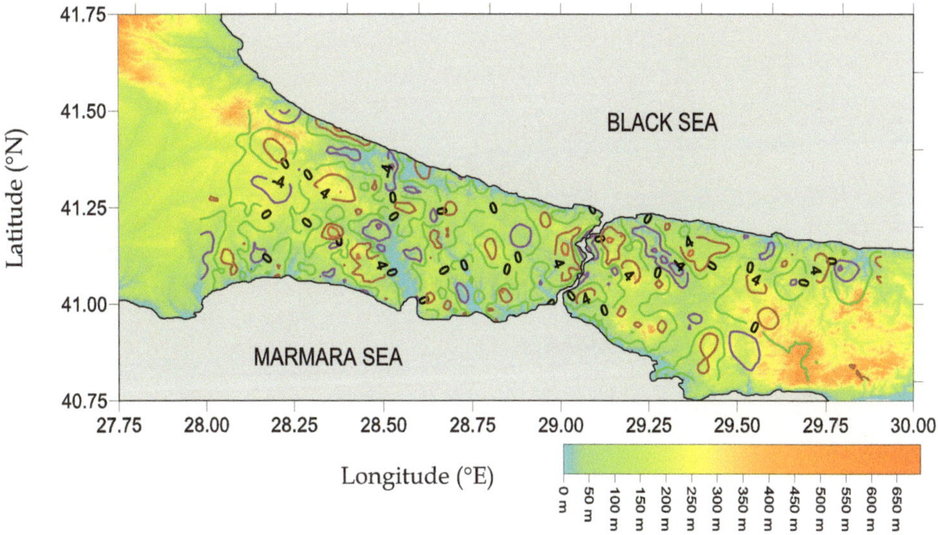

(a)

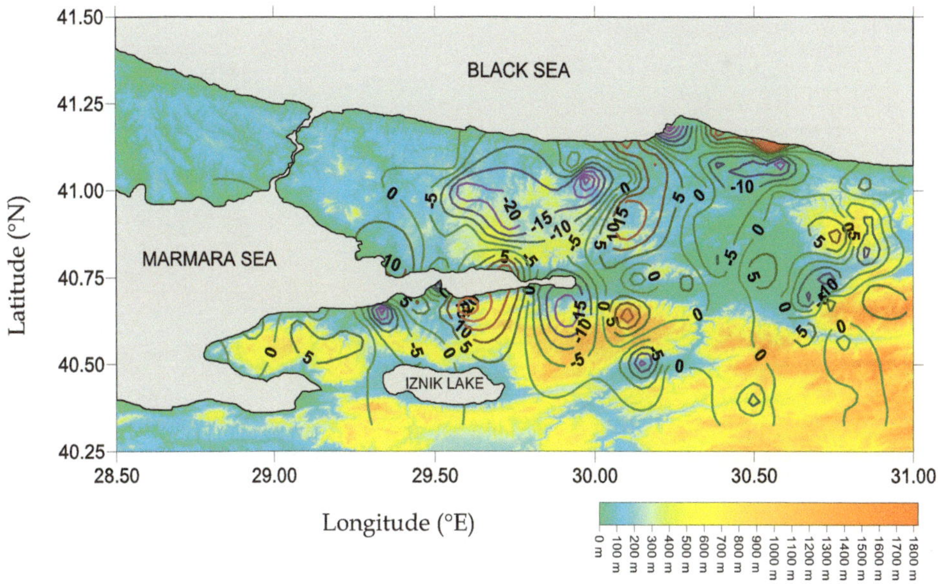

(b)

Fig. 11. Geoid height differences of ANFIS models and observations in centimetre ($\Delta N = N_{GNSS/lev.} - N_{ANFIS}$): (a) Istanbul, (b) Sakarya

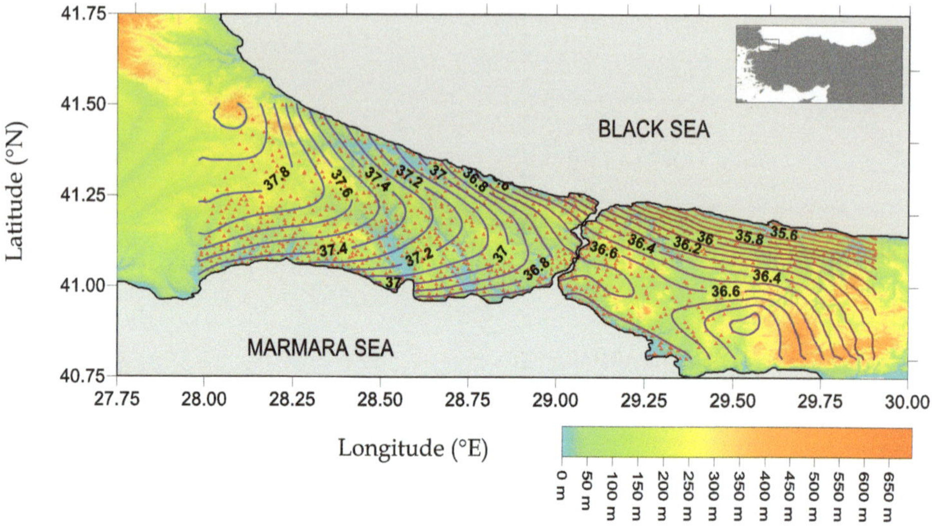

Fig. 12. Istanbul local GNSS/levelling geoid with ANFIS model

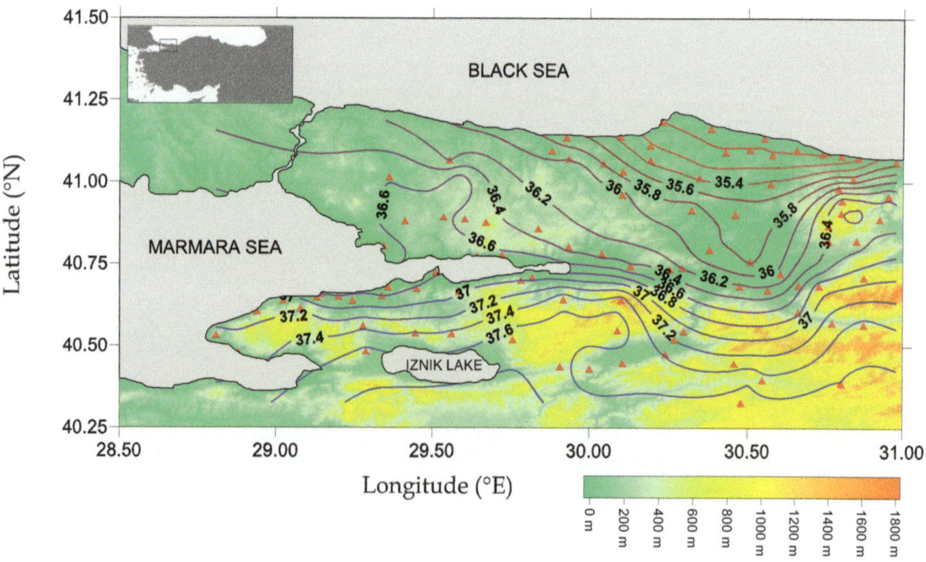

Fig. 13. Sakarya local GNSS/levelling geoid with ANFIS model

3.2 Local improvement of regional geoids

Besides the local GNSS/levelling geoid models, using locally improved regional geoid model with local GNSS/levelling data also provides an applicable solution for transformation of GNSS heights into regional vertical datum. Theoretically, the fundamental relationship between heterogeneous heights: $h_{GNSS} - H_{levelling} - N_{model} = 0$ should have been satisfied. However, because of physical realities and computational factors that cause discrepancies among the heights, this equation cannot be realised at all in real world. As such, this naturally affects the precision of transformation among the heights in practice. Dealing with these disturbing factors, especially the element caused by the systematic errors and datum inconsistencies as a part of geoid modelling, will reduce the discrepancies among the three heights and hence improve the transformation precision of GNSS ellipsoidal heights. As part of this chapter we therefore explain two methods, which are aimed at minimizing the systematic differences of three heights in terms of optimal combination of the heights, for the improvement of regional geoid models with limited reference data in local areas. In the first approach, the height discrepancies are modelled with a parametric equation, so called corrector surface model, which absorbs inconsistencies of the height sets and allow a direct transformation of GNSS heights to the regional vertical datum. The second method consists of the least squares adjustment of the orthometric height differences, which are derived from ellipsoidal heights and regional geoid model, on the base vectors. Hence the orthometric heights of the new points are derived using the adjusted orthometric height differences. Brief descriptions of these height combination approaches with formulations can be found in the sections below.

3.2.1 Corrector surface model

The corrector surfaces, determined according to combination of GNSS derived heights, orthometric heights from the vertical datum and a gravimetric based geoid model, provides an efficient and practical option to precise GNSS levelling in a local area (see e.g., Featherstone, 1998; Kotsakis & Sideris, 1999; Fotopoulos, 2003). The main idea of modelling the corrector surface is to make the regional model estimate of the geoid coincident with the valid vertical datum at GNSS/levelling benchmarks hence minimising the errors in the regional geoid model and the observed heights at the benchmarks. This provides a practical solution for GNSS users in order to accomplish a direct transformation from GNSS derived ellipsoidal heights to orthometric heights, based on local vertical datum.

Determining an optimal parametric model for discrepancies of three heights follows the similar steps as explained in section 3.1.2 for local GNSS/levelling geoid modelling. These steps basically include: determining an appropriate type for model, selecting the optimum extent (form) of the model, and finally assessing the performance of determined model. Accordingly, although one can find numerous models suggested in the literature for realizing corrector surfaces, selecting procedures of the parametric model is mostly arbitrary and based on comparison of statistical test results that measure the accuracy and numerical stabilities of the various models.

General expression of the discrepancies between GNSS/levelling derived geoid heights and geoid heights from the regional geoid model as a function of geodetic position:

$$h_{GNSS} - H_{lev.} - N_{model} - F(\varphi,\lambda) = 0 \qquad (14)$$

that $F(\varphi, \lambda)$ function can be presented in various forms in different levels of complexity (e.g. having elements as only a bias, a bias and a tilt, or higher order polynomials), and multiple regression equations generally as low-order polynomials (similar with Equation 5,

$F(\varphi,\lambda)=\sum_{m=0}^{l}\sum_{n=0}^{l-m}a_{mn}u^{m}v^{n}$), and four or five parameter similarity transformation equations (see Equations 15 and 16, respectively) are generally used.

$$F(\varphi,\lambda)=a_0+a_1\cos\varphi\cos\lambda+a_2\cos\varphi\sin\lambda+a_3\sin\varphi \tag{15}$$

and five parameter similarity transformation as an extended version of Equation 15:

$$F(\varphi,\lambda)=a_0+a_1\cos\varphi\cos\lambda+a_2\cos\varphi\sin\lambda+a_3\sin\varphi+a_4\sin^2\varphi \tag{16}$$

The coefficients of the parametric models are calculated using least squares adjustment method as described in section 3.1.2.1 with Equations 6-9. The appropriateness of the models are comparable according to the results of empirical tests, and RMSE of the height differences and coefficient of determinations (see Equations 12 and 13), are two of these statistics which provides useful hints on the compatibility of the parametric models as corrector surfaces. Hence the geoidal height at a new point can be determined with better precision as the summation of geoid height derived from the regional model and residual δN_{CS} from the corrector surface model as $N=N_{TG03}+\delta N_{CS}$.

3.2.2 Adjustment of the derived orthometric height differences on the baselines

The second method combines the height differences, which are derived from GNSS ellipsoidal heights (Δh) and regional geoid model (ΔN_{model}), in the least squares adjustment algorithm (Mikhail & Ackermann, 1976) and derives the adjusted orthometric height differences for the baselines between the reference GNSS/levelling benchmarks and new points according to following formulation:

$$\Delta H=\Delta h-\Delta N \tag{17}$$

where ΔH is the orthometric height difference for the baseline between the reference GNSS/levelling benchmark and new computation point, Δh is the ellipsoidal height difference derived from GNSS heights for the same baseline and finally ΔN is the geoid height difference of the baseline derived from the regional geoid model. In the adjustment computations that the orthometric heights of the reference benchmarks are set as 'known' to constrain the system, ΔH values are the observations. According to functional model of adjustment:

$$\Delta H+v=H-H^{*} \tag{18}$$

where H and H^{*} are approximate and precise orthometric heights of new and reference benchmarks, respectively. And the residual for the orthometric height difference of the baseline:

$$v=-H^{*}+H-\Delta H \tag{19}$$

and the residuals for all reference benchmarks set the matrix system:

$$v=AX-\ell \tag{20}$$

where the observations matrix is $\ell = \Delta h - \Delta N$, A is the coefficients matrix, and X consists the unknown parameters. The a-priori root mean square error of ΔH of a baseline of S km is $m = m_0\sqrt{S_{(km)}}$ that m_0 is the a-priori RMSE of unit observation. The unknown parameters from the solution of matrix system in Equation 20 is calculated as

$$X = \left(A^T PA\right) A^T P\ell \tag{21}$$

where P includes the weights of ΔH observations. Hence the adjusted orthometric height differences are:

$$\Delta H^* = \Delta H + v \tag{22}$$

The success of the method can be assessed at the test points where GNSS and levelling observations exist, and in the evaluations the orthometric heights of the test points are compared with their observed orthometric heights.

Furthermore, combining the height sets using the method of least squares, weights of each set are essential to correctly estimate the unknown parameters. Improper stochastic modelling can lead to systematic deviations in the results. Therefore, for the purpose of estimating realistic and reliable variances of the data sets, and therefore constructing the appropriate a−priori covariance matrix of the observations, variance component estimation techniques can be included in combining algorithms of the heights. Numerous solution algorithms suggested for variance component estimation problems can be found in various literature published on the subject however, Rao's Minimum Norm Quadratic Unbiased Estimation is commonly used one of these methods (Rao, 1971.). Sjöberg (1984), Fotopoulos (2003) and Erol et al. (2008) can be referred to for further readings and practicing variance component estimation techniques in the adjustment.

3.2.3 Case study: Local Çankırı geoid

Suggested data combination methods related to local improvement of regional geoids are exemplified and tested in a numerical case study in this title. These results are also included by Erol et al. (2008) to provide a detailed investigation on local performances of the various regional models and their improvement capabilities. The local area covers 154 km x 198 km, and the number of reference benchmarks used in the tests is 31. The GNSS positions of the benchmarks were determined with static measurements using dual frequency GNSS receivers. The accuracies of the latitudes and longitudes in ITRF96 datum is ±1.5 cm, and for the accuracy of ellipsoidal heights is reported as ±3.0 cm (Erol et al., 2008). The adjustment of levelling observations revealed the orthometric heights of the benchmarks with ±2.5 cm in TUDKA99 datum. As can be seen in Figure 14, the benchmarks have quite poor density and non-homogeneous distribution over the area. The approximate density of the benchmarks is 1 point per 900 km^2. When the poor density of the benchmarks and rough topographic pattern of the area (the heights of the region change between 41 m and 2496 m) are considered, alongside the levelling technique the regional geoid model or its locally improved version can be applied to obtain regional orthometric heights from GNSS. As a result of this the density and distribution of the reference benchmarks do not allow determination of local GNSS/levelling geoid. According to Large Scale Map and Spatial Data Production Regulation of Turkey, legalized by July 2005, the density of the geoid reference benchmarks must be at least 1 benchmark per 15 km^2 for determination of precise local geoid with geometric approach (LSMSDPR, 2005; Deniz&Çelik, 2008), however with the purpose of testing and local improvement of the regional geoid, the density of the reference GNSS/levelling benchmarks are foresighted to be at least 1 benchmark per 200 km^2 by the regulation.

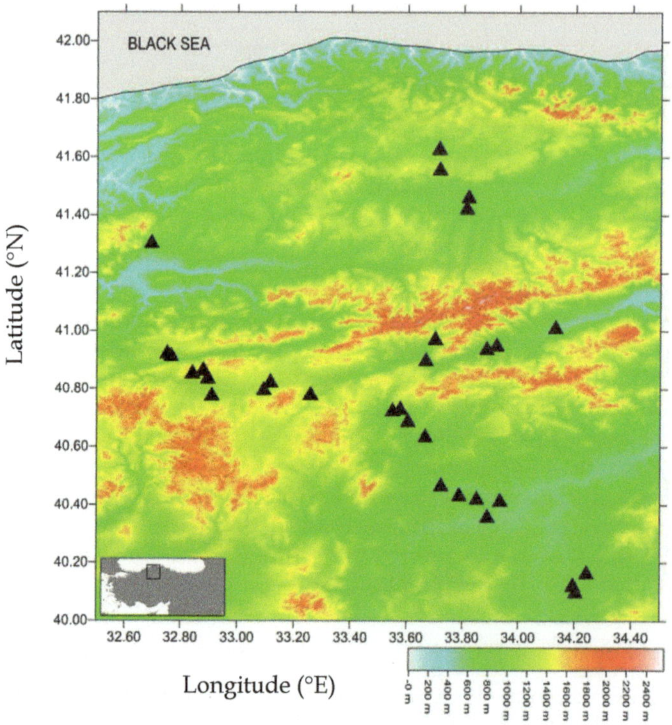

Fig. 14. Çankırı geoid reference benchmarks on topography (Erol et al., 2008)

In respect of the case study carried out with Çankırı local GNSS/levelling network by Erol et al. (2008), Turkish regional geoid TG03 (TNUGG, 2003; Kılıçoğlu et al., 2005) was tested at 31 GNSS/levelling benchmarks, and refined by combining the GNSS/levelling heights using least squares adjustment (LSA) of height differences derived from GNSS ellipsoidal heights and TG03 geoid undulations on the baselines, and simple corrector surface model (CS) with only a bias and a tilt. The performances of the refinement methods were also compared in terms of geoid height residuals at the 9 test points of 31 benchmarks. In addition, LSA of the geoid height differences on the baselines approach was applied with estimated variance components of each height sets, using iterative Minimum Quadratic Unbiased Estimation. The performances of TG03 and its refined versions were also compared with local GNSS/levelling geoid model which were determined with GNSS/levelling heights at 22 reference benchmarks using a 2nd order polynomial equation in the study. Considering the reported results, the accuracy of TG03 model is ±26.2 cm in terms of RMSE of geoid heights and the mean of geoid height differences at the benchmarks is 19.3 cm. When the TG03 is refined with LSA of orthometric height differences on the 58 baselines among the 22 reference and 9 test benchmarks, the accuracy of the refined TG03 model (version 1) is ±15 cm in terms of the RMSE of geoid heights at 9 test points. Hence the improvement of the model is approximately 42%. The refined version of TG03 (version 2) using CS fitting revealed ±19.2 of RMSE of geoid heights at the test points. The third version of refined TG03 was computed using LSA of geoid height differences on the baselines with estimated variance information from iterative MINQUE algorithm, and the internal accuracy of the computed geoid height values having ±4.9 cm RMSE at 31 points were obtained. As expected the 2nd order polynomial type local GNSS/levelling geoid model revealed the worst results with ±46.6 cm RMSE of geoid heights at the test benchmarks. All the results can be compared using the summary statistics at Table 5. For further reading on the applied methods for TG03 local refinement in Çankırı area and the associated case study, Erol et al. (2008) can be referred to.

	refining method		min.	max.	mean	RMSE
TG03	-		-10.8	60.4	19.3	26.2
refined TG03_ver.1	LSA of ΔH on Baseline		-28.7	23.1	0.0	15.0
refined TG03_ver.2	CS fit., 1st order polynomial	model	-25.4	43.3	0.0	17.6
		test	-45.4	28.5	0.0	19.2
local geoid model	2nd order polynomial		-120.1	85.7	0.0	46.6

Table 5. Statistical comparison of TG03 and its refined versions in Çankırı (units in centimetre) (Erol et al., 2008)

4. Summary of results and remarks

This chapter compares geoid models from various scales in Turkish territory and aims to provide a road map to GNSS users in practice, with regards to how to choose, compute and use of the geoid model as a tool for transformation of GNSS ellipsoidal heights to the regional vertical datum. As the traditional levelling techniques for obtaining precise height information are left aside, the improved accuracy of the geoid, as a modern technique for vertical control called, known as GNSS(-geoid) levelling, can be contemplated as an alternative for practical height applications. In the numerical evaluations, presented as part of this chapter, the recently released global geoid models, which include the data by the latest gravity field satellite missions, CHAMP, GRACE and GOCE, were tested against the terrestrial data. The results from this indicate the absolute accuracies of the two ultra-high resolution combined global geopotential models, EGM08 ($\ell_{max} = 2190$) and EIGEN-6C ($\ell_{max} = 1420$) in Turkey were calculated around ±17 cm, which means that these global models can directly be used for GNSS levelling in small scale map production and applications that requires regional orthometric heights with decimetre accuracy. A comparison on validation results of satellite only global models put EIGEN-6S and GOCO02S forward that these models were calculated using GOCE and GRACE missions' data until 240, 250 maximum degrees of expansion, with ±44.0 cm absolute accuracy at the test points. Comparing these models, performance of the GGM03S ($\ell_{max} = 180$), the GRACE only model, stayed rough in representation of the local gravity field in the region. Therefore in modelling the regional hybrid geoid, EIGEN-6S and GOCO02S may provide better performances.

In the content of numerical tests, beside the global models, the most recent regional geoids TG03 and TG09 were also validated against the GNSS/levelling heights at the test benchmarks. The validation results showed that although TG09 model provided approximately 12% improvement comparing to TG03 in terms of accuracy of geoid heights, the absolute accuracy of regional geoid models is not yet below 10 cm. This indicates that the regional geoid models remain insufficient to be applied for GNSS levelling purposes in large scale map production and applications that require centimetre level accuracy in heights. Since the lack of a 5-centimetre or higher precision regional geoid in the country; local geoid models, as an alternative solution for height transformation problems, are determined and used. This chapter presents examples of local geoid modelling using geometric approach, at the two case study areas, Istanbul and Sakarya situated in the north east of Turkey, which have precise GNSS/levelling data. Also from the test results of the computed local geoids, it is obvious that the topographic character of the local area, the quality of GNSS and levelling data, the density and distribution of the geoid reference benchmarks are very critical for the accuracy and reliability of the local geoid model. As such the design of the geoid reference network and data acquisition needs to be planned in a specific manner. Applied methodology for modelling the local geoid is another critical parameter that affects the final accuracy. In the numerical tests, the Istanbul and Sakarya local geoids were computed using classical polynomial type multi regression equations and ANFIS method. In Istanbul a fifth order polynomial equation fitted the best the reference geoid data, where as in Sakarya a fourth order polynomial was decided as an optimal model. Evaluation of the polynomial models at the test benchmarks revealed ±4.4 cm and ±20.4 cm absolute accuracies in Istanbul, and Sakarya, respectively. When the topographies and densities of the benchmarks in both local areas are compared, the difference between the accuracies of the polynomial representations of two local geoids can be understood (Figure 7 vs. Figure 8). On the other hand the ANFIS approach provided marked improvements in results with ±3.5 cm and ±18.9 cm accuracies, in Istanbul and Sakarya. TG03 regional geoid model has ±10.8 cm and ±18.6 cm accuracies in Istanbul and Sakarya. When comparing the regional model, in Istanbul, a local geoid model provides much better accuracy but in Sakarya many of the local geoid model solutions did not provide a better alternative to regional geoid for GNSS levelling purposes. The numerical tests on the local geoid modelling also provided an opportunity to compare the two surface approximation techniques. Hence it is concluded that although, ANFIS has a developed computation algorithm and potential to provide more improved results, it has handicaps from a practical point of view: the prediction capability of this method varies depending on the adopted architecture and it is too sensitive to the selection of the reference/test points. Therefore, while geoid modelling with ANFIS, one must be very carefully to employ the appropriate architecture and to decide reference and test data. Otherwise too optimistic and unrealistic statistics can appear with an over fitted surface model.

In the final part of this chapter, local improvement of geoid models is provided as another alternative solution to GNSS levelling. In the case study, local improvement of the TG03 in Çankırı using precise GNSS/levelling data, by corrector surface fitting and adjustment of derived orthometric height differences on the baselines, is presented. The accuracy of TG03 model in the region is ±26.2 cm. Applying least squares adjustment of height differences derived from GNSS and TG03 on the baselines approach provided 42% improvement in the model and the RMSE of the orthometric heights derived from the improved version of TG03 is reported as ±15.0 cm.

5. Conclusion

Numerous advantages of GNSS techniques from a practical perspective and its high precision in geodetic positioning make this satellite based positioning systems on service in a very large spectrum of applications, ranging from routine engineering surveys to scientific researches. On the other hand, the reference system definition of GNSS coordinates separates the geometry from the Earth gravity field, and therefore developing a solution for transition between the ellipsoidal and natural coordinates, especially in heights, constitutes a challenge for geodesists to be solved by a combination of terrestrial and GNSS data in the recent years. As a reflection of advances in computation techniques and improved data resolutions and accuracies, the precisions of geoid models increase and hence GNSS levelling, as a new concept in vertical control, become a consideration as a viable alternative for practical height determination. All these developments lead modernization of geodetic infrastructures in the national and consequently global scale, and cause leaving the traditional onerous surveying techniques aside as a means for obtaining heights. Today, in many countries, the new vertical datum definition is based solely on the geoid and vertical control is provided via GNSS levelling with a precise geoid model (see e.g. Rangelova et al., 2010).

In the light of recent developments on GNSS techniques and their tremendous impacts on definitions of the reference systems and hence geodetic infrastructures, this chapter reviewed the principle geoid models and widely used methodologies for practical determination of regional heights using GNSS. With this purpose, the evaluations on global models validated the improvement of the long and medium wavelength information of the gravity field, as a result of the current state of technologies with modernized GNSS, as well as new LEO missions for dedicated gravity field research (i.e., CHAMP, GRACE, GOCE). The improvements on global models as well as the terrestrial data qualities contribute also to the regional geoid models by reducing their errors in the total budget of hybrid geoid representation. However, according to results drawn from this study, the accuracy of regional geoid model of Turkey is insufficient yet for deriving regional orthometric heights with centimetre precision from GNSS levelling, and therefore local solutions such as modelling local geoid with geometric approach or improving the regional geoid model with local terrestrial data are still required for providing heights with an accuracy under 5 centimetres. Although the local geoids provide high accuracies, there are handicaps related to their determination and use. The determination of local geoid models requires specifically acquired reference data, having good quality and adequate distribution representing the topography well, and an appropriate modelling algorithm, fitting the data. One of the disadvantages related with the use of local geoid models is that they can be applied only in the limited area with high precision and so are not suitable for extrapolation. These local solutions do not contribute to a unified vertical datum definition in the country. In this manner the importance of a precise and reliable regional geoid model in the concept of GNSS levelling for practical determination of precise regional heights is obvious. In Turkey, geoid modelling efforts as a part of modernization of geodetic infrastructure continue, and with the enhanced data qualities, a precise regional geoid model with its time dependent variations for GNSS levelling purposes will be possible in the near future.

6. Acknowledgment

GNSS/levelling data, used in local geoid modelling, were provided by Istanbul GPS Levelling Network-2005 and Geodetic Infrastructure of Marmara Earthquake Region Land

Information System (MERLIS) projects by Istanbul Technical University, Geodesy Division. Validation of global geopotential models were done using the reference data which were published and used in the same purpose by Yılmaz & Karaali, Sci. Res. Essays 5(2010). Global geopotential models were used from International Centre for Global Earth Models of German Research Centre for Geosciences (GFZ). The validations and modelling computations were carried out using MATLAB ver.7.11. Special thanks go to Dr. R.N. Çelik, for his contributions on local geoid modelling in this study.

7. References

Abd-Elmotaal, H.A. (2006). High-Degree Geopotential Model Tailored to Egypt, In: *Gravity Field of the Earth,* A. Kılıçoğlu & R Forsberg (Eds.), 187-192, Map Journal, International Gravity Field Service, Turkey

Amos, M.J. & Fearhetstone, W.E. (2003). Comparisons of Global Geopotential Models with Terrestrial Gravity field over New Zealand and Australia. *Geomatics Research Australasia,* Vol.78 pp. 67-84

Ayan T. (1976). *Astrogeodätische Geoidberechnung für das Gebiet der Türkei,* PhD Thesis, Karlsruhe University, Karlsruhe, Germany [in German]

Ayan, T., Deniz, R., Çelik, R. N., Denli, H., Özlüdemir, M.T., Erol, S., Özöner, B., Akyılmaz, O. & Güney, C. (2001). *Izmir Geodetic Reference System–2001 (IzJRS 2001)* (Report ID-ITU 2000/2294), Istanbul Technical University, Turkey, 152 pp. [in Turkish]

Ayan, T., Deniz, R., Arslan, E., Çelik, R.N., Denli, H.H., Akyılmaz, O., Özşamlı, C., Özlüdemir, M.T., Erol, S., Erol, B., Acar, M., Mercan, H. & Tekdal, E. (2006). *Istanbul GPS Triangulation Network (IGNA) 2005-2006 Re-measurements and Data Processing* (Report ID-2005/3123), Volume 1, Istanbul Technical University, Turkey, 186 pp. [in Turkish]

Ayhan, M.E. (1993). Geoid determination in Turkey. *Bulletin Geodesique,* Vol.67, pp. 10–22

Ayhan, M.E. & Demir C., 1992. Turkish National Vertical Control Network-1992 (TNVCN-92). *Map Journal,* Vol.109, pp. 22–42.

Ayhan, M. E., Demir, C., Lenk, O., Kılıçoğlu, A., Aktuğ, B., Açıkgöz, M., Fırat, O., Şengün, Y. S., Cingöz, A., Gürdal, M. A., Kurt, A. I., Ocak, M., Türkezer, A., Yıldız, H., Bayazıt, N., Ata, M., Çağlar, Y. & Özerkan, A. (2002). Turkish National Fundamental GPS Network 1999 (TFGN-99). *Map Journal Special Issue,* Vol.16, pp. 47-50

Bruinsma S.L., Marty, J.C., Balmino, G., Biancale, R., Foerste, C., Abrikosov, O. & Neumayer, H. (2010). GOCE Gravity Field Recovery by Means of the Direct Numerical Method, *Proceedings of ESA Living Planet Symposium 2010,* Bergen, Norway, June - July 2010

Çelik, R.N., Ayan, T. & Erol, B. (2002). *Geodetic Infrastructure Project of Marmara Earthquake Region Land Information System (MERLIS)* (Report ID- ITU 2002/06/20), Istanbul Technical University, Istanbul

Çepni, M.S. & Deniz, R. (2005). Examination of Continuity on Geodetic Transformations. *ITU Journal/d,* Vol.4, No.5, pp. 43-54 [in Turkish]

Deniz, R. & Çelik, R.N. (Eds.). (2008). *Explanations and Examples Book of Large Scale Map and Spatial Data Production Regulation (legalized in 15 June 2005),* Chamber of Surveying Engineers, Ankara, Turkey, 86pp [in Turkish] 30.07.2011, Available from http://www.hkmo.org.tr/resimler/ekler/2CO1_db11d259a9db7fb_ek.pdf

Dermanis, A. & Rossikopoulos, D. (1991). Statistical Inference in Integrated Geodesy, *Proceedings of IUGG XXth General Assembly, International Association of Geodesy,* Vienna, August 1991

Draper, N.R. & Smith, H. (1966). *Applied Regression Analysis*, John Wiley &Sons, Inc., USA

Erol, B. (2007). *Investigations on Local Geoids for Geodetic Applications,* PhD Thesis, Institute of Science and Technology, Istanbul Technical University, Turkey

Erol, B. (2011). An automated height transformation using precise geoid models. *Scientific Research and Essays,* Vol.6, No.6, pp. 1351-1363

Erol, B. & Çelik, R.N. (2004). Precise Local Geoid Determination to Make GPS Technique More Effective in Practical Applications of Geodesy, *Proceedings of FIG Working Week 2004,* Athens, Greece, April 2004, 30.07.2011, Available from http://www.fig.net/pub/athens/papers/ts07/ts07_3_erol_celik.pdf

Erol, B. & Çelik, R.N. (2006). Modelling Local GPS/Levelling Geoid: Assessment of Inverse Distance Weighting and Geostatistical Kriging Methods. *Geoinformation Science Journal,* Vol.6, No.1, pp. 78-83

Erol, B., Erol, S. & Çelik, R.N. (2005). Precise Geoid Model Determination Using GPS Technique and Geodetic Applications, In: *Proceedings 2nd International Conference on Recent Advances in Space Technologies,* S. Kurnaz, F. Ince, S. Inbasioglu, S. Basturk (Eds.), pp.395-399, IEEE, Istanbul, Turkey, doi: 10.1109/RAST.2005.1512599

Erol, B., Erol, S. & Çelik, R.N. (2008). Height transformation using regional geoids and GPS/levelling in Turkey. *Survey Review,* Vol.40, No.307 pp. 2-18

Erol, B., Sideris, M.G. & Çelik, R.N. (2009). Comparison of Global Geopotential Models from the CHAMP and GRACE Missions for Regional Geoid Modeling in Turkey. *Studia Geopysica et Geodaetica,* Vol.53 pp.419-441.

Featherstone, W.E. (1998). Do we need a gravimetric geoid or a model of the base of the Australian Height Datum to transform GPS heights?. *The Australian Surveyor,* Vol.43, No.4 pp. 273-280

Featherstone, W.E. (2001). Absolute and relative testing of gravimetric geoid models using Global Positioning System and orthometric height data. *Computers & Geosciences,* Vol.27, No.7 pp. 807-814 doi:10.1016/S0098-3004(00)00169-2

Featherstone, W.E., Denith, M.C. & Kirby, J.F. (1998). Strategies for the accurate determination of orthometric heights from GPS. *Survey Review,* Vol.34, No.267, pp. 278-296

Forsberg, R. (1994). Terrain Effects in Geoid Computations, In: *Lecture Notes - International School for the Determination and Use of the Geoid,* 101–134, IGeS, DIIAR - Politecnico di Milano, Italy

Fotopoulos, G. (2003). *An Analysis on the Optimal Combination of Geoid, Orthometric and Ellipsoidal Height Data,* PhD Thesis, UCGE Report 20185, Geomatics Engineering Department, University of Calgary, Canada

Fotopoulos, G. (2005). Calibration of geoid error models via a combined adjustment of ellipsoidal, orthometric and gravimetric geoid height data. *Journal of Geodesy,* Vol.79, No.1-3, pp. 111-123

Fotopoulos, G., Kotsakis, C. & Sideris, M.G. (2001). How accurately can we determine orthometric height differences from GPS and geoid data?. *Journal of Surveying Engineering,* Vol.129, No.1, pp. 1-10

Förste, C., Bruinsma, S., Shako, R., Marty, J.C., Flechtner, F., Abrikosov, O., Dahle, C., Lemoine, J.M., Neumayer, H., Biancale, R., Barthelmes, F., König, R. & Balmino, G. (2011). EIGEN-6 - A new combined global gravity field model including GOCE data from the collaboration of GFZ-Potsdam and GRGS-Toulouse. *Geophysical Research Abstracts,* Vol.13

Förste, C., Stubenvoll, R., König, R., Raimondo, J.C., Flechtner, F., Barthelmes, F., Kusche, J., Dahle, C., Neumayer, H., Biancale, R., Lemoine, J.M. & Bruinsma, S. (2009). Evaluation of EGM2008 by comparison with other recent global gravity field models. *Newton's Bulletin (Special Issue)*, Vol.4, pp 26-37

GFZ (2006). The CHAMP Mission, 30.07.2011, Available from http://www.gfz-potsdam.de/pb1/op/champ/results/grav/010_eigen-champ03s.html

GGM02 (2004). GRACE Gravity Model 02, Center of Space Research (CSR) of the University of Texas at Austin, U.S., 30.07.2011, Available from http://www.csr.utexas.edu/grace/gravity/

GOCE (2009). European Space Agency GOCE (Gravity field and steady-state Ocean Circulation Explorer) Project Website, 30.07.2011, Available from http://earth.esa.int/GOCE/

Goiginger, H., Hoeck, E., Rieser, D., Mayer-Guerr, T., Maier, A., Krauss, S., Pail, R., Fecher, T., Gruber, T., Brockmann, J.M., Krasbutter, I., Schuh, W.D., Jaeggi, A., Prange, L., Hausleitner, W., Baur, O. & Kusche, J. (2011). The combined satellite-only global gravity field model GOCO02S, *Proceedings of 2011 General Assembly of the European Geosciences Union*, Vienna, Austria, April 2011

Gruber, T. (2004). Validation concepts for gravity field models from new satellite missions, 30.07.2011, Available from http://earth.esa.int/workshops/goce04/participants/

Haagmans, R., de Min, E. & van Gelderen, M. (1993). Fast evaluation of convolution integrals on the sphere using 1D FFT and a comparison with existing methods for Stokes' integral. *Manuscripta Geodaetica*, Vol.18, No.5 pp. 227-241

Heiskanen, W.A. & Moritz, H. (1967). *Physical Geodesy*, W.H. Freeman and Company, San Fransisco

Hirt, C. & Seeber, G. (2007). High-resolution local gravity field determination at the submilimeter level using a digital zenith camerasystem, In: *Dynamic Planet*, P. Tregoning & C. Rizos, (Eds.), 316-321, Springer Verlag, Berlin, Heidelberg

Hofmann-Wellenhof, B. & Moritz, H. (2006). *Physical Geodesy* (2nd Edition), Springer, 978-3211335444, New York

ICGEM (2011). Table of Available Models. *International Center for Global Earth Models, GeoForschungsZentrum Potsdam* (GFZ), Germany, 30.07.2011, Available from http://icgem.gfz-potsdam.de/ICGEM/ICGEM.html

Jang, J.S. (1993). ANFIS:adaptive-network based fuzzy inference system. *IEEE Transactions on Systems, Man, and Cybernetics*, Vol.23, No.3, pp. 665-685 doi: 10.1109/21.256541

Kavzaoğlu ,T. & Saka, M.H. (2005). Modelling local GPS/Levelling geoid undulations using artificial neural networks. *Journal of Geodesy* Vol.78, pp 520-527. doi: 10.1007/s00190-004-0420-3

Kiamehr, R. & Sjöberg, L.E. (2005). Comparison of the qualities of recent global and local gravimetric geoid models in Iran. *Studia Geophysica et Geodaetica*, Vol.49 pp. 289-304

Kılıçoğlu, A., Demir, C. & Fırat, O. (2005). Data and Methods Used in Computation of New Turkish Geoid 2003 (TG-03), Proceedings of Turkish National Geodesy Commission 2005 Year Annual Meeting: Geoid and Vertical, Trabzon, October 2005

Kılıçoğlu, A. & Fırat, O. (2003). Geoid Modelling with the Purpose of Determining Orthometric Heights using GPS for Large Scale Map Production and Case Studies, *Proceedings of Turkish National Geodesy Commission 2003 Year Scientific Conference-Invited Paper*, Konya, Turkey, September 2003, 30.07.2011, Available from http://www.harita.selcuk.edu.tr/arsiv/calistay2003/02ak_jeoid.pdf

Klokočník, J., Reigber, C., Schwintzer, P., Wagner, C.A. & Kostelecký, J. (2002) Evaluation of pre-CHAMP gravity field models GRIM5-S1 and GRIM5-C1 with satellite crossover altimetry. *Journal of Geodesy*, Vol.76, pp.189-198

Kotsakis, C. & Sideris, M..G. (1999). On the adjustment of combined GPS/levelling/geoid networks. *Journal of Geodesy*, Vol.73, No.8, pp. 412-421

Lambeck, K. & Coleman, R. (1983). The Earth's shape and gravity field: a report of progress from 1958 and 1982. *Geophysical Journal of the Royal Astronomical Society*, Vol.74 pp. 25–54

Lemoine, F.G., Kenyon, S.C., Factor, J.K., Trimmer R.G., Pavlis, N.K., Chinn, D.S., Cox, C.M., Klosko, S.M., Luthcke, S.B., Torrence, M.H., Wang, Y.M., Williamson, R.G., Pavlis, E.C., Rapp, R.H. & Olson, T.R. (1998). *The Development of the joint NASA GSFC and NIMA Geopotential Model EGM96*, Technical Paper, NASA/TP-1998-206861, National Aeronautics and Space Administration, Maryland, U:S. 575 pp.

LSMSDPR (2005) *Large Scale Map and Spatial Data Production Regulation*, Turkey

Merry, C.L. (2007). Evaluation of global geopotential models in determining the quasi-geoid for Southern Africa. *Survey Review*, Vol.39 pp. 180–192

Mikhail, E.M. & Ackermann, F. (1976). *Observations and Least Squares*, Harper & Row Publishers, ISBN 0-7002-2481-5, New York

Pavlis, N.K., Holmes S.A., Kenyon S.C. & Factor J.K. (2008). An Earth Gravitational Model to Degree 2160: EGM2008, *Proceedings of 2008 General Assembly of the European Geosciences Union*, Vienna, Austria, April 2008

Rangelova, E., Fotopoulos, G. & Sideris, M.G. (2010). Implementing a Dynamic Geoid as a Vertical Datum for Orthometric Heights in Canada, In: *Gravity, Geoid and Earth Observation*, S.P. Mertikas, (Ed.), 295-302, Springer Verlag, DOI 10.1007/978-3-642-10634-7_38, Berlin, Heidelberg

Rao, C.R. (1971). Estimation of Variance Components - MINQUE Theory. *Journal of Multivariate Statistics*, Vol.1, pp. 257-275

Rodriguez-Caderot, G., Lacy, M.C., Gil, A.J. & Blazquez, B. (2006). Comparing recent geopotential models in Andalusia (Southern Spain). *Studia Geophysica et Geodaetica*, Vol.50 pp 619–631

Roland, M. & Denker, H. (2003). Evaluation of Terrestrial Gravity Data by New Global Gravity Field, In: *Gravity and Geoid*, I.N. Tziavos, (Ed.), 256-261, Ziti Publishing, Greece

Rummel, R., Balmino, G., Johannessen, J., Visser, P. & Woodworth, P. (2002). Dedicated gravity field missions–principles and aims. *Journal of Geodynamics*, Vol.33 pp.3-20

Sadiq, M. & Ahmad, Z. (2009). On the selection of optimal global geopotential model for geoid modelling: A case study in Pakistan. *Advances in Space Research*, Vol.44 pp 627-639

Schwarz, K.P., Sideris, M.G. & Forsberg, R. (1987). Orthometric Heights Without Leveling. *Journal of Surveying Engineering*, Vol.113, pp. 28-40

Sen, A. & Srivastava, R.M. (1990). *Regression Analysis: Theory, Methods and Applications*, Springer Texts in Statistics, Springer, New York

Sideris, M.G. (1994). Geoid Determination by FFT Techniques, In: *Lecture Notes - International School for the Determination and Use of the Geoid*, 213–272, IGeS, DIIAR - Politecnico di Milano, Italy

Sjöberg, L. (1984). Non-negative Variance Component Estimation in the Gauss-Helmert Adjustment Model. *Manuscripta Geodaetica*, Vol.9, pp. 247-280

Stopar, B., Ambrožič, T., Kuhar, M. & Turk, G. (2006). GPS-Derived Geoid Using Artificial Neural Network and Least Squares Collocation. *Survey Review,* Vol.38, No.300, pp. 513-524

Takagi, T. & Sugeno, M. (1985). Fuzzy identification of systems and its application to modelling and control. *IEEE Transactions on Systems, Man and Cybernetics,* Vol.15, pp. 116-132

Tapley, B., Ries, J., Bettadpur, S., Chambers, M., Cheng, M., Condi, F., Gunter, B., Kang, Z., Nagel, P., Pastor, R., Pekker, T., Poole, S. & Wang F. (2005). GGM02 - an improved Earth gravity field model from GRACE. *Journal of Geodesy,* Vol.79, pp.467–478

Tapley, B., Ries, J., Bettadpur, S., Chambers, D., Cheng, M., Condi, F. & Poole, S. (2007). The GGM03 Mean Earth Gravity Model from GRACE. *EOS Transactions, AGU,* Vol.88, No.52 G42A-03

TNUGG (2003). Turkish National Union of Geodesy and Geophysics National Reports of Geodesy Commission of Turkey for 1999-2003, Presented In: *XXIII.General Assembly of the International Union of Geodesy and Geophysics,* 12.07.2011, Available from http:// www.iugg.org/members/nationalreports/turkey.pdf

TNUGG (2011). Turkish National Union of Geodesy and Geophysics National Reports of Geodesy Commission of Turkey for 2007-2011, 12.07.2011, Available from http://www.iag-aig.org/attach/5015ba0f03bf732e1543f4120f15ec9a/turkey.pdf

Torge, W. (1980). *Geodesy,* Walter de Gruyter, New York

Tscherning, C., Arabelos, D. & Strykowski, G. (2000). The 1-cm geoid after GOCE. In: *Gravity, Geoid and Geodynamics,* M.G. Sideris, (Ed.), 267-270, Springer Verlag, Berlin, Heidelberg

Tung, W.L. & Quek, C. (2009). A Mamdani-Takagi-Sugeno based Linguistic Neural-fuzzy Inference System for Improved Interpretability-Accuracy Representation, *Proceedings of IEEE International Conference on Fuzzy Systems,* August 2009, pp.367-372 doi: 10.1109/FUZZY.2009.5277194

USGS (1997). GTOPO30 information website, U.S. Geological Survey EROS data center, 27.December.2010, Available from http://eros.usgs.gov/#/Find_Data/Products_and_Data_Available/gtopo30_info

USGS (2010). Shuttle Radar Topography Mission (Mapping the World in 3 Dimensions) website, U.S. Geological Survey, 21.01.2011, Available from http://srtm.usgs.gov/index.php

Ustun, A & Abbak, R.A. (2010). On global and regional spectral evaluation of global geopotential models. *Journal of Geophysics and Engineering,* Vol.7, pp. 369-379 doi:10.1088/1742-2132/7/4/003

Vaniček, P. & Krakiwsky, E.J. (1986). *Geodesy: The Concepts* (2nd Edition), Elsevier Science, 978-0444877772, New York

Yanalak, M. & Baykal, O. (2001). Transformation of Ellipsoid Heights to Local Leveling Heights. *Journal of Surveying Engineering,* Vol.127, No.3, pp. 90-103

Yılmaz, M. (2010). Adaptive network based on fuzzy inference system estimates of geoid heights interpolation. *Scientific Research and Essays,* Vol.5, No.16 pp. 2148-2154

Yılmaz, M. & Arslan, E. (2008). Effect of the Type of Membership Function on Geoid Height Modeling with Fuzzy Logic. *Survey Review,* 40(310), pp. 379-391 doi: 10. 1179/003962608X325439

Yılmaz, N. & Karaali, C. (2010). Comparison of Global and Local Gravimetric Geoid Models in Turkey. *Scientific Research and Essays,* Vol.5, No.14 pp. 1829–1839

Zadeh, L.A. (1965). Fuzzy sets. *Information and Control,* Vol.8, pp. 338-353

6

Estimation of Satellite-User Ranges Through GNSS Code Phase Measurements

Marco Pini[1], Gianluca Falco[1] and Letizia Lo Presti[2]
[1]*Istituto Superiore Mario Boella*
[2]*Politecnico di Torino*
Italy

1. Introduction

A Global Navigation Satellite System (GNSS) receiver is able to compute the user position through a trilateration procedure, which includes the measure of the distance between the receiver and a set of satellites. Two different approaches are tipically used and implemented in commercial receivers. The former relies on code tracking, the latter leverages carrier phase measurements performed during carrier tracking.

This chapter focuses on the first approach and discusses the procedures that GNSS receivers perform to finely estimate satellite-user ranges. First, in section 2 we introduce the concept of pseudorange and in section 3 we give some fundamentals on primary signal processing blocks of every GNSS receiver: signal acquisition, tracking and data demodulation. In section 4, two common methods used to estimate the user-satellite range, on the basis of code phase measurements are presented. Finally, section 5 completes the chapter, providing an example of combined Position, Velocity and Time (PVT) computation for a GPS/Galileo receiver.

2. Theory and methods

Let us start with a simple example to introduce the concepts we will describe in the next sections.

John usually bikes to school following a straight path, keeping a constant speed. John wants to measure the distance between his house and the school and decides to compute such a distance by measuring the time it takes to go to school. He uses the following formula:

$$x = v \cdot t \tag{1}$$

where:

- x is the distance estimated by John;
- v is the average speed, read on the bike speedometer;
- t is the difference between the time instant when John arrives at school and the time instant when he leaves home. In both cases, John reads the time on his digital watch.

The following day, John repeats the experiment, but he measures t as the difference between the arrival time read on the school clock and the leaving time, read on his watch. John realizes that the estimated distance is significantly different from that estimated the previous day. Likely, his watch and the clock at school are not synchronized. In this case, the measured time interval can be written as follow:

$$\tilde{t} = t + \delta t \tag{2}$$

Equation (2) takes into account the bias δt between John's watch and the school clock. Considering this term, John understands that this reflects to an error δx on the estimated distance.

$$\tilde{x} = v \cdot \tilde{t} = v \cdot (t + \delta t) = x + \delta x \tag{3}$$

At this point, John wants to compare the result with one of his friends. He asks Alice to do the same measurement from her house, since John knows that his house is exactly *500 m* away from hers. Before the measurements, Alice and John synchronize their watches. Referring the measurements taken by John and Alice to the subscripts J and A, equation (3) becomes:

$$\begin{cases} x_J = \tilde{x}_J - \delta x_J = v_J(\tilde{t}_J - \delta t) \\ x_A = \tilde{x}_A - \delta x_A = v_A(\tilde{t}_A - \delta t) \end{cases} \tag{4}$$

where:

- $\tilde{x}_J$ and $\tilde{x}_A$ are the distances estimated by John and Alice, respectively;
- x_J and x_A are the unknown distances John and Alice want to measure;
- $\tilde{t}_J$ and $\tilde{t}_A$ are the time intervals measured by John and Alice;
- v_J and v_A are the average speeds of John and Alice read on their speedometers;
- δt is the unknown bias between Alice and John's watches and the school clock.

Recalling that Alice's house is *500 m* away from John's, the previous system of equations can be rewritten as:

$$\begin{cases} x_J = v_J(\tilde{t}_J - \delta t) \\ x_J + 500 = v_A(\tilde{t}_A - \delta t) \end{cases} \tag{5}$$

This new system has two equations and two unknowns: x_J and δt. In few steps, John can finally compute the distance between his house and the school, realizing that he obtains the same result of the first experiment. The conclusion of this simple example is that in 1 dimension, if the clocks used to measure time intervals are not synchronized, we need an additional equation to solve the problem.

Bringing the concept to a three-dimensional space, it is easy to understand that we need four equations to solve the problem and determine the unknown user position respect to a reference system. This is the case of Global Navigation Satellite System (GNSS) receivers.

Referring to the geometry sketched in Fig. 1, there are satellites in view broadcasting ranging signals, while a user on the Earth wants to estimate his unknown coordinates (x_u, y_u, z_u). The satellites continuously transmit their positions (i.e. (x_k, y_k, z_k) considering the *k-th* satellite coordinates), keeping their clocks synchronized to a common time scale. The user estimates the distances ρ_k with a set of satellites, measuring the travel time from the satellite to the receiving antenna.

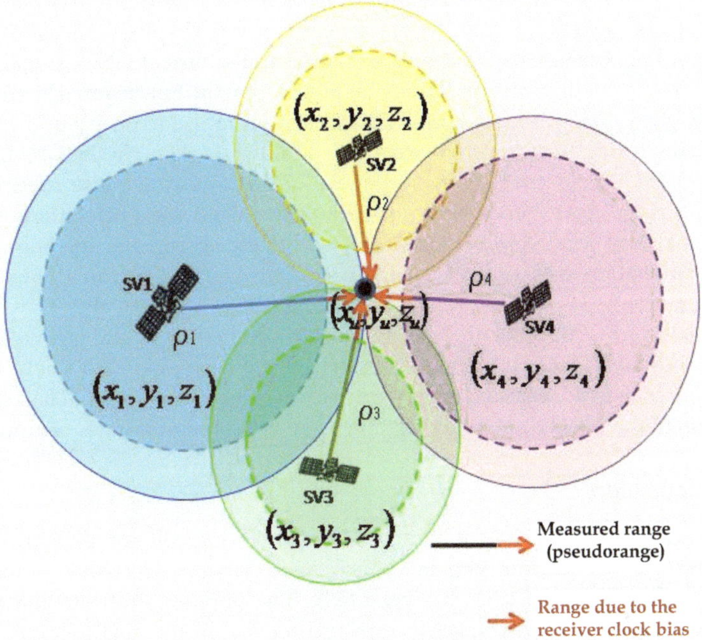

Fig. 1. Example of trilateration in case of clock biased receiver

The user needs at least 4 equations to be able to compute (x_k, y_k, z_k), because of the bias δt between his clock and the satellite time scale. Due to the presence of a common bias that affects all the measures of distance between the user and the satellites, we have to refer to such a distance as a **pseudorange** ρ_k instead of a range. From this moment on, the reader has to keep in mind this distinction.

$$\begin{cases} \rho_1 = \sqrt{(x_1 - x_u)^2 + (y_1 - y_u)^2 + (z_1 - z_u)^2} + \delta t \cdot c \\ \rho_2 = \sqrt{(x_2 - x_u)^2 + (y_2 - y_u)^2 + (z_2 - z_u)^2} + \delta t \cdot c \\ \rho_3 = \sqrt{(x_3 - x_u)^2 + (y_3 - y_u)^2 + (z_s - z_u)^2} + \delta t \cdot c \\ \rho_4 = \sqrt{(x_4 - x_u)^2 + (y_4 - y_u)^2 + (z_4 - z_u)^2} + \delta t \cdot c \end{cases} \tag{6}$$

The system (6) is the set of equations that every GNSS receiver has to solve. With the problem stated above and having in mind the task of the receiver, this chapter explains the operations performed to measure the user-satellite ranges. The focus will be mainly on measurements taken on the received spreading codes, while for carrier-phase measurements, interested readers can find comprehensive theory in (Misra & Enge, 2001; Jonge & Teunissen, 1996).

3. From the incoming signal to the pseudorange

When the GPS signal arrives at the receiver, it is very weak and the received power, proportional to the distance between the satellite and the user, is well below the noise floor. However, GPS receivers are able to compute their position with an accuracy that ranges from a couple of meters to centimeters in case of carrier-phase measurements. Such performance are possible thanks to the spread-spectrum nature of GNSS signals. It is useful to recall that each satellite utilizes Direct Sequence Spread Spectrum (DSSS) modulation (Kaplan & Hegarty, 2006), broadcasting the navigation message on pseudo random noise (PRN) spreading codes, over the same frequency. Taking as example the GPS L1 C/A code, each satellite uses a Gold code, quasi-orthogonal with respect to those used by the other satellites. Applying signal processing algorithms based on the correlations between the incoming signal and local replicas, the receiver can de-spread the incoming signal and retrieve the navigation message. Such algorithms are used to perform two fundamental processes, commonly known as *acquisition* and *tracking*, respectively. The first aims at roughly estimating the Doppler frequency and the code delay of the received signal. The tracking phase adjusts the parameters assessed by the acquisition, to finely measure the phase of each tracked GPS signal, keeping trace of changes in the future. The estimate of the code delay for all the tracked satellites is at the basis of the pseudoranges computation.

3.1 Signal acquisition

The first task of a GNSS receiver is to detect the presence of the satellites in view. This is performed by the acquisition system, which also provides a coarse estimate of two parameters of the received Signal In Space (SIS): the Doppler shift and the delay of the received spreading code with respect to the local replica. In the next sections, we will see that the precise alignment between the received and the local spreading codes, is fundamental for the measure of user-satellites ranges, that is necessary to fix the receiver position.

There are two mathematical disciplines which govern the operation performed by acquisition systems: the *Estimation theory* and the *Signal Detection theory*. These two extensive theories are described in various literature, whereas comprehensive analysis and applications can be found in many papers. For a complete mathematical background of the operation performed by GNSS signal acquisition, interested readers can refer to (Kay, 1993, 1998).

Keeping our description terse, real acquisition systems search for a satellite in view, correlating the received signal with a local replica of the spreading code and a local carrier. The search consists in finding the values of code delay and carrier frequency of the local signals that maximize the correlation. Exploiting the concepts and the methodology of the *Estimation theory*, it is possible to show that the Maximum Likelihood (ML) estimate of the vector $p=(\tau,f_d)$, whose elements are the two unknowns of the received signal $y_{IF}[n]$, can be obtained by maximizing the following function

$$\hat{p}_{ML} = \arg\max_{\bar{p}} \left| \frac{1}{L} \sum_{n=0}^{L-1} y_{IF}[n] \bar{r}_{IF}[n] \right|^2 = \arg\max_{\bar{p}} R(\bar{\tau},\bar{f}_d) \qquad (7)$$

where:

- $y_{IF}[n]$ represents the incoming signal, as stream of samples at the ADC output;
- L is the number of samples used to process a portion of the incoming signal;
- $\bar{p} = (\bar{\tau}, \bar{f}_d)$ is a vector of test variables: $\bar{\tau}$ represents the code delay and $\bar{f}_d$ the Doppler shift. $\bar{p}$ is defined in a proper support D_p containing all possible values that can be assumed by the elements of $p = (\tau, f_d)$. D_p is known as search space.
- $\bar{r}_{IF}[n]$ is the local signal sampled with a rate equal to the sampling frequency used by the ADC and can be expressed as follows:

$$\bar{r}_{IF}[n] = c(nT_s - \bar{\tau})e^{j2\pi\bar{f}_d nT_s} \tag{8}$$

where $c(nT_s - \bar{\tau})$ is the local spreading code with delay $\bar{\tau}$, $e^{j2\pi\bar{f}_d nT_s}$ represents the local carrier, in-phase and quadrature, T_s is the sampling interval.

Real acquisition systems find the values of $\bar{\tau}$ and $\bar{f}_d$ that maximize equation (7). As an example, Fig. 2 reports $R(\bar{\tau}, \bar{f}_d)$ over a predefined search space. A correlation peak corresponding to a defined pair of $\bar{\tau}$ and $\bar{f}_d$ clearly raises above the cross-correlation noise floor and indicates a first rough alignment between the incoming and the local signals.

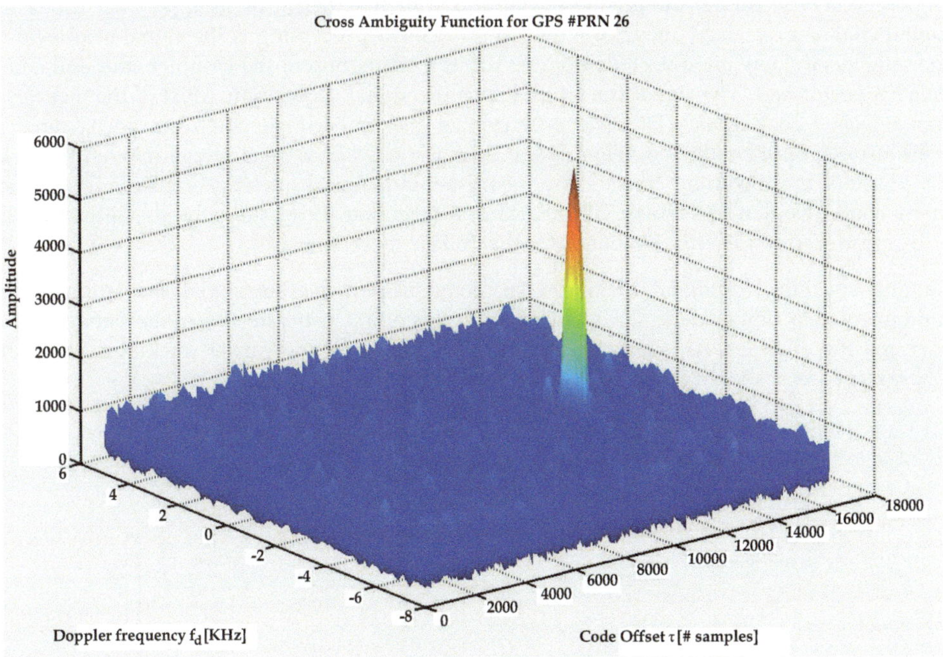

Fig. 2. Two-dimensional function evaluated by GNSS signal acquisition

Generally, the first estimate of τ and f_d that maximizes equation (7) is followed by a decision process. The maximum of $R(\bar{\tau}, \bar{f}_d)$ is taken as decision variable and compared against a threshold, that is often set according to the Neyman-Pearson (NP) theorem (Kay, 1998). If the maximum is higher than the threshold, the satellite is considered present, otherwise absent. Note that the performance of real acquisition algorithms are evaluated in terms of *Probability of Detection* and *Probability of False Alarm.*

It is important to highlight that for civilian GNSS signals (i.e. GPS L1 C/A, Galileo E1-B), the spreading code contained in $y_{IF}[n]$ is a periodic sequence with period equal to the code period T_p (i.e. 1 ms for the GPS L1 C/A code, 4 ms for the Galileo E1-B): therefore the delay τ can be estimated only in the range $(0,T_p)$. In practice only a portion of this infinite sequence enters into the summation in equation (7) (i.e. the samples of the portion of signal under test for $n = 0,\ldots,L-1$). This means that for a given value of $\bar{f}_d$, the correlation assumes the form of a circular correlation when the interval $(0,L-1)$ contains an integer number of code periods. This remark is quite important and helps to understand why real acquisition systems are based on Fast Fourier Transforms (FFTs). In fact, FFTs are used to implement fast circular correlations and scan the search space efficiently. Insights on the design of FFT-based signal acquisition system is out of scope for this chapter. However, one can find many algorithms proposed in recent literature and can refer to (Borre et al., 2006) for didactical examples.

3.2 Code and carrier tracking

Digital receivers sample the analog signal and split the stream of samples over different digital channels. As seen above, the first step in GNSS processing is the signal acquisition: the satellites in view are detected and a first rough estimation of the Doppler shift and code delay is performed. The signal tracking follows the signal acquisition. Most of the receivers use a Delay Lock Loop (DLL) to synchronize the spreading code from each satellite (Parkinson & Spilker, 1996), while a Phase Lock Loop (PLL) is generally employed to track the phase of the incoming carrier. The theory behind digital tracking loops is reported in many books (Kaplan & Hegarty, 2006; Parkinson & Spilker, 1996). Here the signal tracking is only introduced to give fundamentals for the following sections.

Roughly speaking, the signal tracking relies on the properties of the signal correlation and is fundamental to demodulate the navigation message and estimate the range between the user and the satellites. A generic block diagram of the code and carrier tracking system for GNSS receivers is shown in Fig. 3.

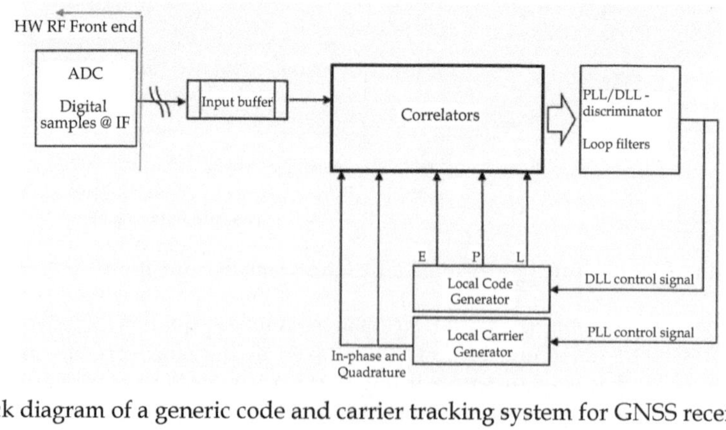

Fig. 3. Block diagram of a generic code and carrier tracking system for GNSS receivers

The stream of samples at the ADC output (i.e. $y_{IF}[n]$) is correlated with the local code and with two carriers, one in phase and one in quadrature, respectively. At the end of each integration period, the values of correlation are used to generate feedback control signals, one for the DLL and one for the PLL. *Early minus Late* DLLs use additional replicas of the local code, shifted of 0.5 chips earlier and later than the reference one, which is referred as *Prompt code*. The Early and Late correlations are combined to generate the DLL feedback on the basis of a proper discrimination function. Such a feedback is filtered to smooth the noise effect and is used to steer the code generator, that prepares the local code for the next loop iteration. In such a way the DLL continues to track the correlation peak in the time domain. The PLL works in a similar way. Generally, the in-phase and quadrature Prompt correlations are passed to a *Costas-PLL* (that is not sensitive to navigation bits transitions) (Kaplan & Hegarty, 2006; Misra & Enge, 2001) that generates the loop control signal. This is filtered and applied to the local carrier generator, that prepares the local carrier for the next iteration. This process repeats over time, making the receiver able to track the correlation peak in frequency domain.

When both the DLL and PLL are locked, the incoming signal is despread and converted to baseband. The navigation data bits appear at the output of the in-phase Prompt correlator and can be decoded. In addition, with the DLL locked, the local and the incoming codes are aligned. Referring to the local code, the receiver exactly knows when a new code period starts and is able to recognize navigation data bits and boundaries of the navigation message. The receivers stays synchronized to the tracked satellites, continuously counting the number of received chips, full code periods, navigation bits and message frames. These counters are fundamental to measure the misalignment over different channels, tracking different satellites, and are used to compute the pseudoranges. For sake of completeness, note that real receivers generally use architecture more complex than that reported in Fig. 2. For example, a Frequency Lock Loop (FLL) is employed to refine the rough estimate performed by the signal acquisition and ease the PLL lock, reducing the transient time between the signal acquisition and the steady-state carrier/code tracking. Recently new techniques based on digital signal processing have been developed in order to obtain higher precision and reduced computational load, improving the robustness against noise and interference. In this section, we have recalled only some fundamentals of code and carrier tracking, with the goal of providing the necessary background for the following part of the chapter.

3.3 Navigation message demodulation, frame and page synchronization

Once the tracking loops are locked (i.e. the local code keeps the alignment with the incoming code and the local carrier is exactly a replica of the received one), the navigation data bits appear at the output of the Prompt correlator, on the in-phase branch of the tracking loops. Considering the GPS L1 C/A code, using an integration time equal to the code period, we obtain a bit value every ms. However, due to the low signal power, real receivers usually set the integration time to 20 ms, which is the inverse of the navigation data rate (i.e. 50 Hz). Fig. 4.a shows 1 second of normalized navigation data bits at the Prompt correlator output, using an integration time of both 1 ms (blue) and 20 ms (red). The same example could be repeated considering the Galileo E1-B signal. In this case, a proper value of integration time is 4 ms, that corresponds to either the code period and the inverse of the navigation data rate. An example of navigation data bits, recovered processing the signal tranmitted by a simulated Galileo satellite, is shown in Fig. 4.b.

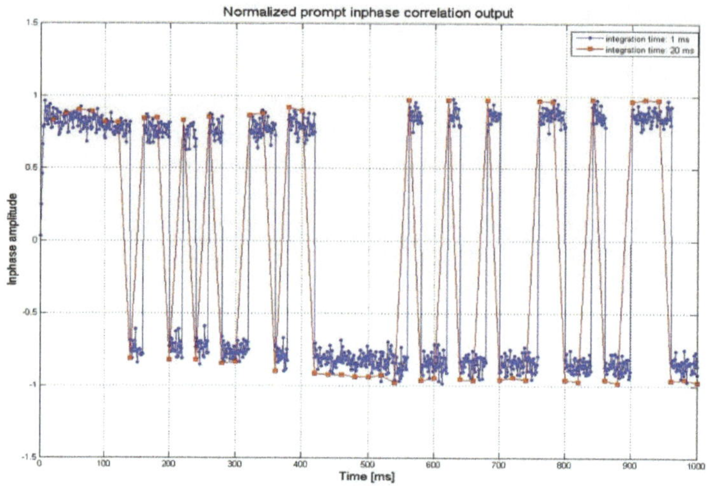

(a)

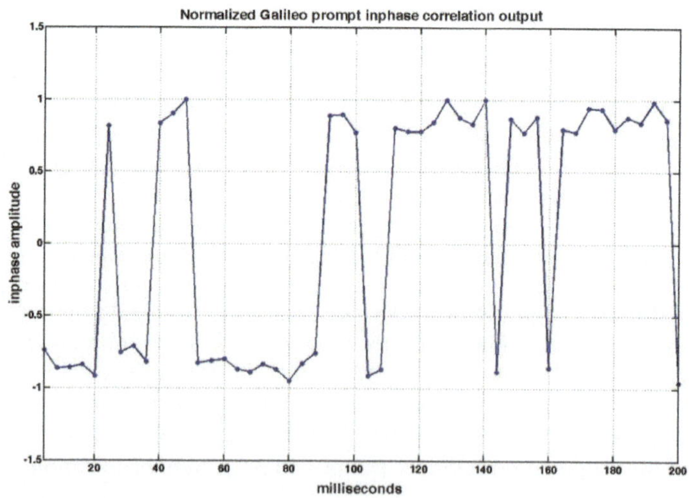

(b)

Fig. 4. Navigation data bits at the output of the inphase Prompt correlator both for GPS (a) and Galileo (b) signals

The stream of data bits must be decoded to recover the message broadcast by the satellite. The navigation data follow the scheme defined in the GPS Interface Control Document (Arinc Research Corporation, 1991) in case of GPS, while all the information regarding the navigation message of the Galileo Open Service can be found in (European Commission, 2010).

Since the navigation format is out of scope of this chapter we will give just an introduction to the argument by showing the general structure both of the GPS and Galileo message. In Fig. 5 the overall navigation data in case of GPS L1 C/A code is depicted.

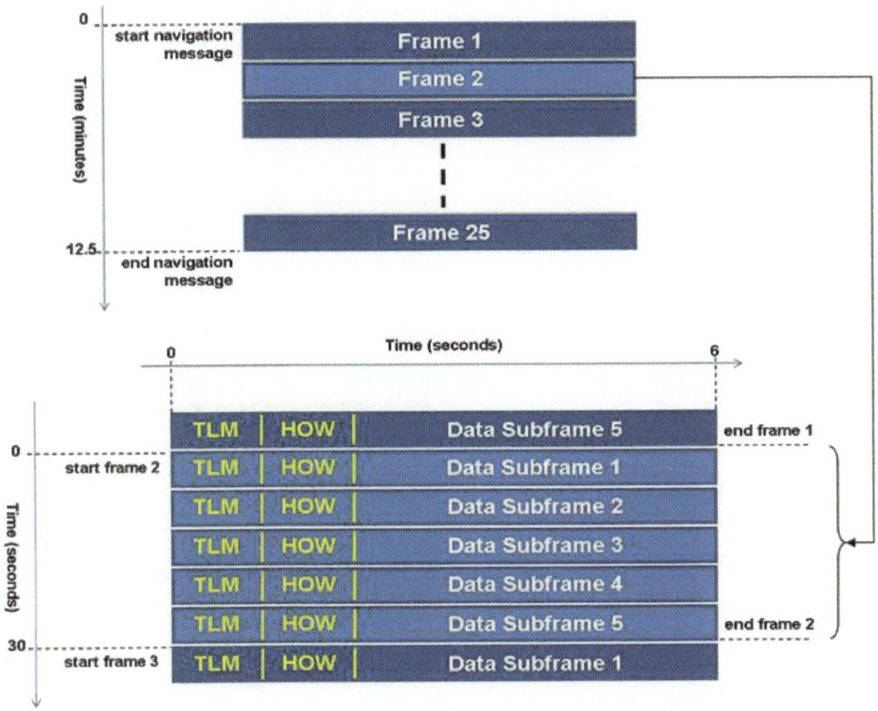

Fig. 5. Structure of the navigation message included in the GPS civil signal, transmitted on the L1 frequency

The rate of the navigation data bits is 50 bits per second. The whole message is 12.5 minutes long and is divided in 25 frames. Each frame lasts 30 seconds and is further divided in 5 subframes, six seconds long. Each subframe of the navigation message always starts with two special words, the Telemetry (TLM) and the Handover word (HOW).

In case of the Galileo E1 signal, the complete navigation message is transmitted on the data channel (E1-B) as a sequence of frames. A frame is composed of several sub-frames, and a sub-frame, in turn, is composed of several pages. The page is the basic structure to build the navigation message. Fig. 6 shows the structure of the Galileo data and an example of page for the E1-B message.

Prior to the navigation data decoding, the receives seeks for the preamble, a defined sequence of n bits, that marks the beginning of a subframe for the GPS L1 C/A, a page for the Galileo E1-B. A simple, but efficient, way to detect the preamble is to correlate the navigation data stream with a local binary sequence equal to the preamble. A maximum is detected when such a local sequence is aligned with the preamble. Naturally, the bit pattern used for the preamble can occur anywhere in the received data stream, thus an additional check must be carried out to authenticate the real preamble (e.g. in case of GPS, only when the maximum of correlation is found exactly every 6 seconds). When the beginning of the subframe is identified, the content of the subframe can be decoded. The receiver retrieves all the orbital parameters (i.e. ephemeris) necessary to compute the satellite position corresponding to the transmission of the subframe. Through the process used for navigation data decoding, the receiver is able to understand which subframe and word a certain bit belongs to. In this way, the receiver can have an exact, precise and real-time "comprehension" of each sample/bit broadcast by the satellite. This aspect will be fundamental in the computation of the pseudoranges as it will be explained in the next section.

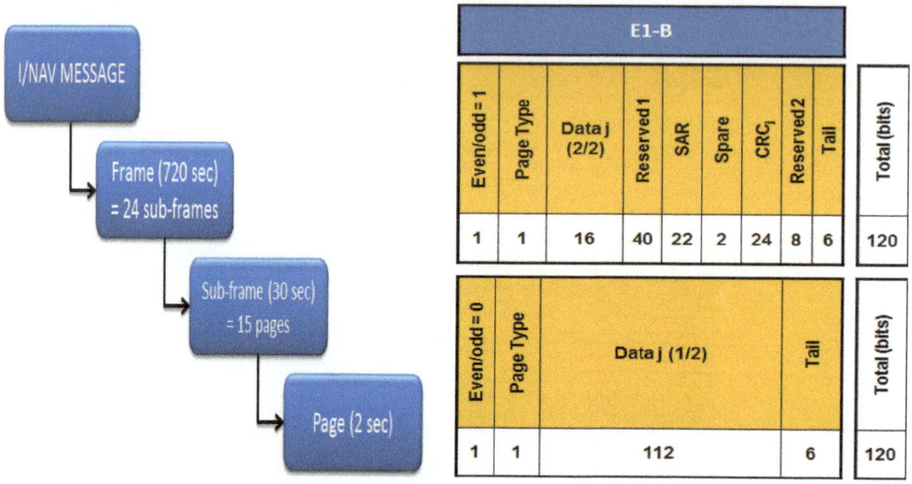

(a) (b)

Fig. 6. Galileo I/NAV navigation message structure (a) and I/NAV nominal page with bits allocation (b)

4. Performing range measurements using GNSS signals

In this section we focus on the measurements of the pseudorange, describing some methods commonly used to estimate the distance between the satellite and the user's receiver.

So far, we have explained how the detection of a preamble is an effective way to recognize the beginning of a subframe (a page in case of Galileo) and the starting point for decoding the navigation message. Here and in the following we want to introduce how GNSS receivers use the detection of a preamble to compute a valid pseudorange and estimate the user's position and velocity. According to (Borre et al., 2006), the pseudorange estimations can be divided into two sets of computations: the first is devoted to find the initial set of pseudoranges, the second keeps track of the pseudoranges after the first set is estimated.

4.1 Computation of the first set of pseudoranges

Before proceeding with the explanation of the pseudorange computation, it is useful to recall some hypothesis, that will be taken as true from now on.

All the clocks on-board of the satellites are assumed perfectly synchronized to a reference GNSS time-scale. In other words, we assume that the first chip of a definite subframe/page leaves the satellites at the same instant t_{tx}^{GNSS}. In addition:

- all the satellites belonging to the same system (i.e. GPS, GLONASS, Galileo) are synchronized each others but they are not with respect to different GNSSs;
- the receiver clock is not synchronized with the GNSS time-scale (as the school clock in the example of section 2 was not synchronized to Alice and John's watches). The actual time at the receiver can be written as $t^R = t^{GNSS} - \Delta b$, where t^{GNSS} is the actual time on the GNSS time-scale and Δb is the bias with respect to the clock on board of the satellite. For sake of simplicity, we assume that Δb remains constant over time. In the notation, the superscripts refer to the time-scale, while we use subscripts to identify definite time instants;
- all the examples and equations are given for the GPS satellites only but the explanation can be considered valid and easily extended for other GNSS systems too.

With these hypothesis in mind, once the preamble has been correctly detected, every navigation data for each satellite in view can be tagged with additional information such as the corresponding subframe, the number of bits read from the beginning of that subframe as well as the number of samples processed up to that time instant by acquisition and tracking stages. In this way, it will be easy to make comparisons among channels and calculate the time delay of the satellites. In fact, "*during the collection of the digitized data there is no absolute time reference and the only time reference is the sampling frequency. Moreover, the pseudorange can be measured only in a relative way because the clock bias of the receiver is an unknown quantity*" (Tsui, 2000). Therefore the pseudorange can be computed as the distance (or time) between two reference points. The way the reference points are chosen makes the main difference in the two methods that are commonly used in commercial receivers for the pseudorange computation and that we can call "**common transmission time**" and "**common reception time**", respectively.

4.1.1 Common transmission time

According to this approach, since all the satellites are synchronized, they broadcast the same preamble at the same moment, which is received by the user at different instants, due to different propagation delays. This approach follows what pragmatically happens in a real scenario where the satellites have different distances with respect to the user.

The left side of Fig. 7 represents the same subframe transmitted by the satellites at t_{tx}^{GPS}. On the right, Fig. 7 shows the local codes displacement at the receiver, assuming four tracked satellites. The blue rectangular is the TLM word of the subframe, which is received at different instants $t_{rx,i}^{GPS}$, because of the different traveling times τ_i. These can be written as :

$$\tau_i = t_{rx,i}^{GPS} - t_{tx}^{GPS} \tag{9}$$

where $t_{rx,i}^{GPS}$ corresponds to the time instants $t_{rx,i}^{R} = t_{rx,i}^{GPS} - \Delta b$ on the receiver time-scale.

The receiver recovers t_{tx}^{GPS} decoding the HOW of the previous subframe, which includes a truncated version of the absolute GPS time. The receiver reads $t_{rx,i}^{R}$, but it is not able to compute $t_{rx,i}^{GPS}$, since Δb is unknown. If the receiver was able to compute τ_i, the distances between the receiver and satellites would be simply obtained as:

$$\rho_i = \tau_i \cdot c \tag{10}$$

where c stands for the speed of light.

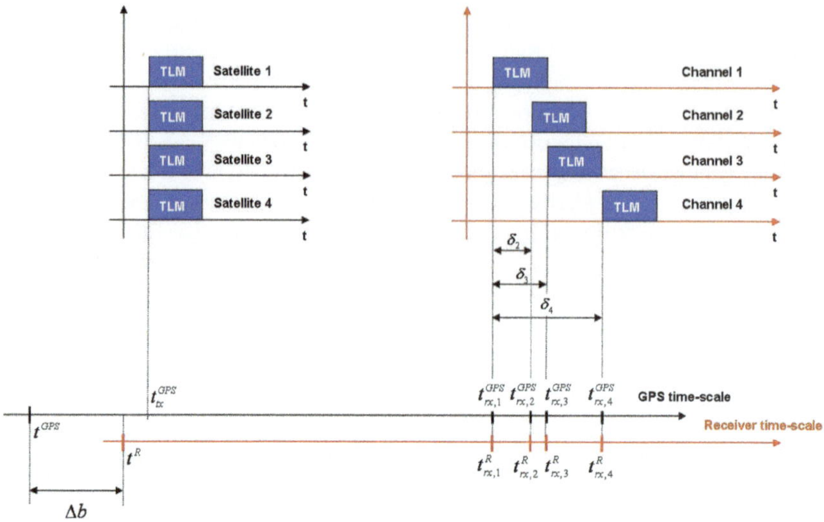

Fig. 7. Pseudorange computation based on "comon tranmission time", evaluating the beginning of a subframe for GPS system

Referring to Fig. 7, the satellite tracked on channel 1 is taken as reference, since the subframe transmitted at t_{tx}^{GPS} arrives first. In other words this means that the satellite tracked on channel 1 has the shortest distance respect to the receiver. Since the same subframe from the other satellites is received at different times, the receiver has to count, for each tracked satellite, the amount of time past from the reception of the subframe on the reference channel. Regarding to this concept, it is important to stress that the measurement of the delay between the reference satellite and the others in view has not necessarily to be performed on the beginning of a subframe, but it must be computed consistently (i.e. with respect to the same word or data bit belonging to the same subframe).

When the receiver is able to compute, for each tracked satellite, the relative time difference with respect to the reference channel, the relative pseudoranges can be evaluated.

In formula this difference can be written as:

$$\delta_i = t_{rx,i}^R - t_{rx,1}^R \quad \forall i = 1,..,4 \tag{11}$$

δ_i are measured through time counters, that are continuously updated by the tracking structures of each channel. With these time differences, the set of distances between the receiver and the satellites can be written as follows:

$$\rho_i = \rho_1 + c \cdot \Delta b + c \cdot \delta_i \quad \forall i = 1,..,4 \tag{12}$$

where:

- δ_i are the time differences between the beginning of the subframe received on channel i and the beginning of the subframe received on the reference channel. $\delta_i > 0 \forall i \neq 0$ and $\delta_1 = 0$;

- ρ_1 is the reference pseudorange, corresponds to the satellite closest to the user. Even if the receiver does not know the distance between this satellite and the user, a realistic value of ρ_1 can be set as an approximation. In fact, by considering that a typical travel time from the satellites to the Earth is on the order of 65-83 ms, an appropriated value could be $\tau_1 = 70$ ms (Borre et al., 2006), thus $\rho_1 = \tau_1 \cdot c = 20985.47 Km$. It is important to note that such approximated reference pseudorange, does not affect the computation of the user's position. Eventually, a the error due to on such approximation falls in the terms that takes into account the clock bias.;
- Δb has not been determined yet, but it will be solved computing the set of equations (6).

4.1.2 Common reception time

The second approach performs the pseudoranges estimation, setting a common receiving time t_u^R over all the channels. Also in this case, the reference channel is the one that receives the subframe transmitted at t_{tx}^{GPS} first. For all the tracked satellites (including the reference one), the receiver counts the elapsed time between the reception of subframe and t_u^R .This means that the receiver measure the delays as:

$$\delta_i = t_u^R - t_{rx,1}^R \quad \forall i = 1,..,4 \tag{13}$$

Fig. 8 depicts the method of fixing a unique time of reception for four GPS satellites in view

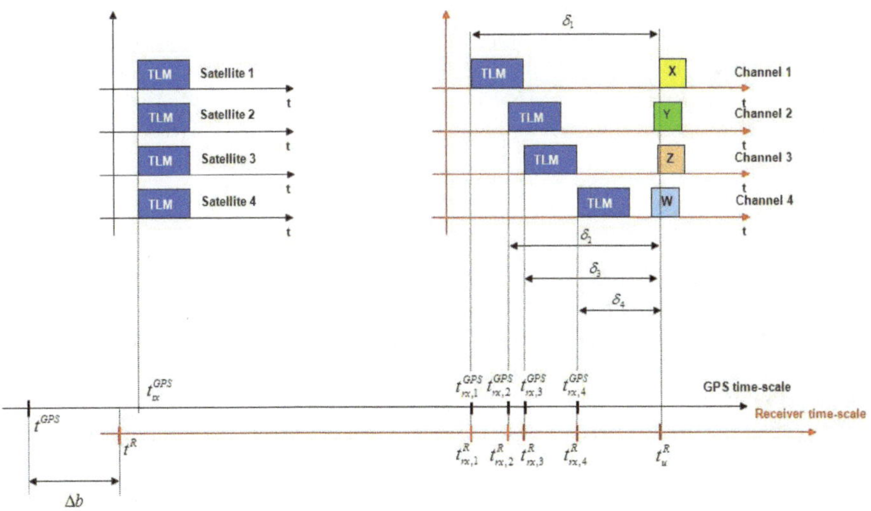

Fig. 8. Pseudorange computation based on "comon reception time", evaluating the start of the subframe for a GPS system

Once the δ_i has been computed, the receiver is able to calculate the pseudorange easily. This can be accomplished by evaluating the delta-difference (Δ_i) of delays with respect to the satellites and the reference one. The aforementioned relative difference Δ_i is stated as:

$$\Delta_i = \delta_i - \delta_1 \quad \forall i = 1,..,4 \tag{14}$$

and consequently, by modifying equation (12), the pseudoranges can be written as:

$$\rho_i = \rho_1 + c \cdot \Delta b + c \cdot \Delta_i \quad \forall i = 1,..,4 \tag{15}$$

where, as in the case of "common transmission time":

- ρ_1 is the reference pseudorange ;
- Δb represents the clock bias between the one on board of the satellite and the receiver's one.

This second method is usually employed in commercial GPS receivers. The main reason behind this choice is the relative simplicity and suitability of that approach in real-time implementations, since it does not require to wait until all the channels have received the same data bit (e.g. the beginning of the same subframe) to compute the pseudoranges. This concept gets more clear if we consider that, during the data demodulation and the tracking process, the receiver continously counts the number of samples processed on that channels, as well as the number of frames, subframes and data bits decoded. As a consequence, through a system of counters, it becomes easy to compute the time difference Δ_i among the channels at a certain t_u^R .

4.2 Computation of the subsequent sets of pseudoranges

Once the initial set ot pseudoranges has been computed, subsequent pseudoranges can be estimated. In this case, the computation of the reference pseudorange (i.e. ρ_1) can be refined with respect to the approximated value set during the first estimate (see section 4.1.1 for details). In fact, at this stage, the receiver has already computed the first estimate of its position and is able to accurately calculate the geometrical distance between the satellite and itself.

As far as the pseudoranges of the other satellites in view is concerned, let suppose that the receiver performs a new PVT every second.

According to the method based on common transmission time, the receiver has to measure a delay of 1s for the reference channel. By considering that a GPS navigation data lasts for 20 ms, it means that, after 1 second, 50 bits have been decoded for the reference satellite, starting from the beginning of the subframe. In order to estimate the time difference, the receiver must wait until each channel has demodulated 50 bits after the beginning of the subframe. Then, the pseudoranges can be computed as stated in equation (6) and the process repeats over time.

On the contrary, if the receiver follows the"common reception time", it moves ahead t_u^R of 1s, before measuring the time difference among the channels. Again, it is important to stress that this reception time is fixed by the receiver and it is independent from the number of navigation bits have been read for each tracked satellite.

The receiver can compute the user's position estimation at a rate much higher than 1 Hz. If we consider as the reference time the beginning of a new C/A code (i.e. every ms), the receiver can update the PVT at a 100 Hz rate.

5. Position, velocity and time (PVT) computation

This section completes the chapter and deals with the estimation of the user's PVT that comes after the measurement of a set of pseudoranges, for at least four satellites in view.

In order to have an accurate estimate of the user's position, the receiver has to consider additional error sources that typically affect the measured pseudorange and that have to be compensated. These sources include atmospheric effects (e.g. ionosphere and troposhere, that generate a delay in the signal broadcast by the satellite) and other kinds of noise related to the presence of multipath and interference.

A valid PVT can be estimated after the receiver retrives the satellites' positions, (i.e.: x_i, y_i, z_i as stated in equation (6)) from the navigation message. To compute the satellite position, the receiver needs the ephemeris and the time of transmission, which is usually referred to the beginning of the subframes. All the information the receiver needs is embedded in the navigation message. The time of transmission can be read every 6 seconds at the beginning of a subframe in a specific word that corresponds to the HOW. From the HOW the receiver retrieves a truncated version of the absolute GPS time (TOW). This number is referred to as Z-count. The Z-count is the number of seconds passed since the last GPS week rollover in units of 1.5s. The truncated Z-count in the HOW corresponds to the time of transmission of the next navigation data subframe. To get the time of transmission of the current subframe, the Z-count should be multiplied by 6 and 6s should be subtracted from the result (Borre et al. 2006).

Therefore, if we assume to perform the pseudorange estimation at the beginning of a new subframe, the time of transmission will be exactly equal to the value reported in the HOW of the previous subframe. Otherwise, if we implement the computation of the pseudorange in a different instant, we have to count the time elapsed between the beginning of the subframe and that instant. The way the time of transmission is computed represents the main difference between the two aforementioned methods (i.e. "common transmission time" and "common reception time"). According to the first method, all the satellites transmit the signal at the same time and, if we assume to calculate the pseudorange at the beginning of a subframe, this correspond to the TOW. On the contrary, if we consider the second approach, we have to keep in mind a different time of transmission for each satellite. Practically speaking, we have to sum up the TOW with the δ_i delay that elapsed from the starting point of the subframe and t_u^R. Since every satellite has a different distance with respect to the Earth, it follows that the δ_i delay will vary according to the satellite under consideration.

When four satellite have been correctly tracked, the full set of equations can be rewritten after having removed the satellite offset and atmospheric effects. According to the "common transmission time", the equations can be stated as in equation (6):

$$\begin{cases} \sqrt{(x_1-x_u)^2+(y_1-y_u)^2+(z_1-z_u)^2} = \rho_1 + c\cdot \Delta b \\ \sqrt{(x_2-x_u)^2+(y_2-y_u)^2+(z_2-z_u)^2} = \rho_1 + c\cdot \Delta b + c\cdot\delta_2 \\ \sqrt{(x_3-x_u)^2+(y_3-y_u)^2+(z_3-z_u)^2} = \rho_1 + c\cdot \Delta b + c\cdot\delta_3 \\ \sqrt{(x_4-x_u)^2+(y_4-y_u)^2+(z_4-z_u)^2} = \rho_1 + c\cdot \Delta b + c\cdot\delta_4 \end{cases} \tag{18}$$

where ρ_1 corresponds to the reference channel relative to the satellite with the shortest path to the user. In case we want to follow the second approach (i.e. "common reception time"), equation (18) keeps the same except for the time delay δ_i that has to be substituted by Δ_i.

In case two different GNSS systems are tracked and used for the PVT calculation, equation (18) has to be slightly modified. For example, let assume to have 4 GPS and 2 Galileo satellites in view, respectively. By following the "common transmission time" method, we can rewrite equation (18) as:

$$\begin{cases} \sqrt{(x_1^{GPS}-x_u)^2+(y_1^{GPS}-y_u)^2+(z_1^{GPS}-z_u)^2} = \rho_{1,GPS}+c\cdot\Delta b_{GPS} \\ \sqrt{(x_2^{GPS}-x_u)^2+(y_2^{GPS}-y_u)^2+(z_2^{GPS}-z_u)^2} = \rho_{1,GPS}+c\cdot\Delta b_{GPS}+c\cdot\delta_{2,GPS} \\ \sqrt{(x_3^{GPS}-x_u)^2+(y_3^{GPS}-y_u)^2+(z_3^{GPS}-z_u)^2} = \rho_{1,GPS}+c\cdot\Delta b_{GPS}+c\cdot\delta_{3,GPS} \\ \sqrt{(x_4^{GPS}-x_u)^2+(y_4^{GPS}-y_u)^2+(z_4^{GPS}-z_u)^2} = \rho_{1,GPS}+c\cdot\Delta b_{GPS}+c\cdot\delta_{4,GPS} \\ \sqrt{(x_1^{Gal}-x_u)^2+(y_1^{Gal}-y_u)^2+(z_1^{Gal}-z_u)^2} = \rho_{1,Gal}+c\cdot\Delta b_{GPS}+c\cdot\Delta b_{GPS/Gal} \\ \sqrt{(x_2^{Gal}-x_u)^2+(y_2^{Gal}-y_u)^2+(z_2^{Gal}-z_u)^2} = \rho_{1,Gal}+c\cdot\Delta b_{GPS}+c\cdot\Delta b_{GPS/Gal}+c\cdot\delta_{2,Gal} \end{cases} \tag{19}$$

When we work with more than one GNSS we have to keep in mind that different GNSSs are not synchronized among each others. This fact implies to introduce additional unknowns that take into account the time-bias between the GNSS systems. For example, if we consider a GPS/Galileo receiver as stated in equation (19), we need a variable that estimates the bias offset between the GPS and the Galileo time scales. Finally the receiver is able to compute a valid position and velocity. One of the most commonly used algorithm for the position estimation is based on the least-squares (LS) method. The description of this technique is out of scope in this Chapter and a lot of material can be found in the scientific literature (Bjork, 1990; Borre et al., 2006; Kaplan & Hegarty, 2006).

Another noteworthy technique that is used in most of the commercial receivers to improve the accuracy of the PVT computed by using the LS approach, is the so-called Kalman filter (Anderson & Moore, 1979; Brown & Hwang, 1997; Kalman, 1960). By combining a system model with the measurements, this algorithm is able to smooth the solution calculated by the LS as well as to provide estimation of the user's position even when less than four satellites are tracked (e.g. this can be done by using the modeled system only).

5.1 Examples using GPS and Galileo data

This section provides an example of the evaluation of user's position, presenting the results obtained with the LS algorithm. Most of these results have been taken from (Rao et al. 2011).

Taking as an example the GPS satellite with PRN 30, Fig. 9 shows the comparison of the pseudoranges estimation obtained implementing the common transmission time and the common reception time methods. The blue marks represents the pseudorange computed by considering the "common transmission time", while the reds correspond to observables calculated fixing a unique time of reception. Though these two methods are conceptually different, as expected, no significant differences can be noticed in the pseudoranges estimates, that are substantially similar, but shifted in time due to the different computation instant.

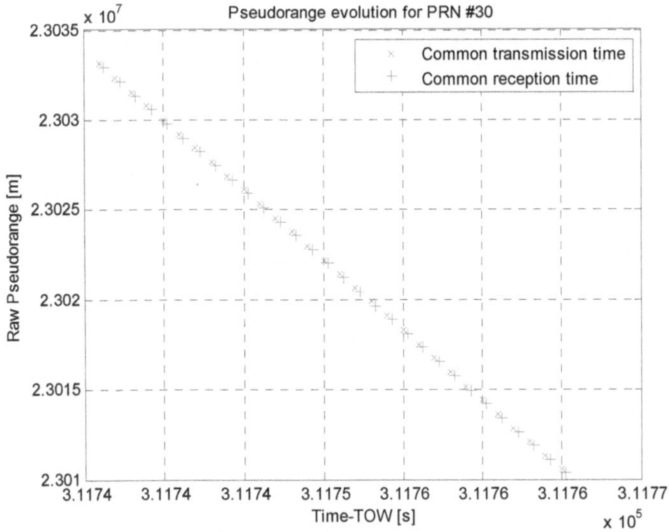

Fig. 9. Comparison between pseudoranges computed by using common reception time and common transmission time

If we suppose to start the PVT computation at the beginning of a subframe, updating the computation every second, using the common transmission time method, the first set of pseudorange corresponds to a time of transmission equal to TOW and is always an integer time of seconds for the subsequent sets. On the contrary, using the common reception time method, the pseudorange is computed at a transmission time that is not the same as TOW, but changes according to the reception time that has been fixed in the receiver. An example of real values of transmission time, following the two different approaches is reported in Table 1 for GPS satellite with PRN 30.

TIME OF TRANSMISSION [s] for GPS SV 30	
Common transmission time	Common reception time
311736	311736.277662467
311737	311737.277664239
311738	311738.27766595
311739	311739.277667661
311740	311740.277669371
311741	311741.277671082
311742	311742.277672854
311743	311743.277674503
311744	311744.277676275
311745	311745.277677986

Table 1. Different time of transmission according to the "common transmission" and "common reception" methods

An example of position estimation using the LS method is reported in Fig. 10. The LS algorithm have been run on the data sets of pseudoranges computed according to the two techniques and using both real GPS and simulated Galileo satellite signals.

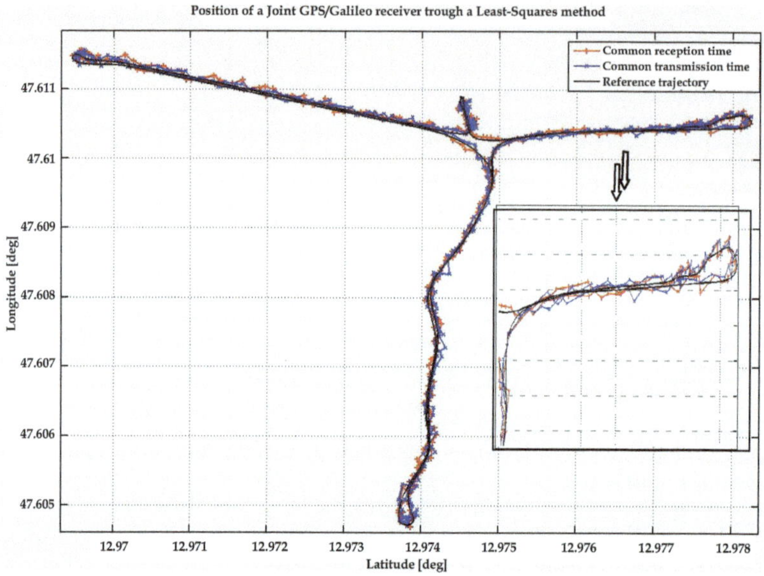

Fig. 10. PVT solution for a joint GPS/Galileo receiver by using "common reception time" and "common transmission time" for the pseudoranges computation and a LS-based receiver

As expected, the trajectory of the user estimated by the two methods does not significantly differ and the variance of the positioning error along the three axis X,Y,Z has the same magnitude in both cases. This fact proves once again the benefit of using a unique time of reception, which is particularly suitable without affecting the position accuracy.

6. Conclusion

In this chapter we have examined the GPS code-phase measurements in order to compute precise satellite-user ranges and to estimate the receiver's position accurately. Since the clocks on board of the satellites are not synchronized with the clock of the receiver, measures of code phase gives pseudoranges instead of ranges. Then, by limiting the discussion only at the pseudoranges computed through the code-phase estimation, two different methods have been presented. The former considers that all the satellites are synchronized and each navigation message is received by the user at different time instants. Therefore, measuring the time offset among all the channels and assigning a nominal travel time to the closest satellite, we are able to calculate the pseudoranges. On the other hand, the latter technique foresees the measure of time delays by fixing a common reception time over all the receiver channels. The first method is the most intuitive and didactic, while the second is more suitable for real-time implementations and is often employed in commercial

GPS receivers. In both the approaches, a fundamental role is played by the tracking stage whose aim is to continuously refine the misalignment between the incoming signal and the local replica in order to perform the code de-spreading and retrieve the navigation data. On the basis of the local code evolution, GNSS receivers measure code phase delays, implementing a set of time-counters that accumulate the number of processed frames, subframes, data bits, code periods and samples for all the tracked satellites. The presented theory is completed by a real data example of PVT, in case of a joint processing of GPS/Galileo signals, exploiting code-phase pseudorange measurements.

7. References

Anderson, O.D.B. & Moore, J.B. (1979). *Optimal Filtering,* Prentice Hall Inc., ISBN 0-486-43938-0, New Jersey, USA.

Arinc Research Corporation. (1991). *Interface control document. ICD-GPS-200,* Available from:< http://www.navcen.uscg.gov/pubs/gps/icd200/default.html >.

Bjork, A. (1990). Least Squares Methods, In: *Handbook of Numerical Analysis Vol. 1- Finite Difference Methods, Part1, Solution Equations in* R^N, Elsevier, pp. 466-647, North-Holland, ISBN: 0-4447-0366-7, Amsterdam, Holland.

Borre, K; Akos, D.M; Bertelsen, N.; Rinder, P. & Jensen, S.H. (2006). *A Software-defined GPS and Galileo Receiver: A Single-frequency approach,* Birkhäuser, ISBN 0-8176-4390-7, Boston, USA.

Brown, R.G. & Hwang, P.Y.C. (1997). *Introduction to random signals and applied Kalman filtering,* John Wiley & Sons, Inc., ISBN: 0-471-12839-2, New York, USA.

European Commission. (2010). *Open Service Signal-In-Space Interface Control Document. OS-SIS-GALILEO-ICD,* Avaiable from:
http://ec.europa.eu/enterprise/policies/satnav/galileo/open-service/index_en.htm

Jonge de, P.J & Teunissen, P.J.G. (1996). Computational aspects of the LAMBDA method for GPS ambiguity resolution. *Proceedings of ION GPS-96, 9th International Technical Meeting of the Satellite Division of the Institute of Navigation* , Kansas City, Missouri, Sept. 17-20.

Kalman, R.E. (1960). A new approach to linear filtering and prediction problems. *Transactions on ASME of Journal of Basic Engineering,* Vol.82, No. Series D, pp. 35-45.

Kaplan, E.D & Hegarty, C.J. (2006). *Understanding GPS: principles and applications,* House Inc., ISBN: 1-58053-894-0, Boston, USA.

Kay, S.M. (1993). *Fundamentals of Statistical Signal Processing, volume I: Estimation Theory,* Prentice Hall Inc., ISBN 0-13-345711-7, New Jersey, USA.

Kay, S.M. (1998). *Fundamentals of Statistical Signal Processing, volume II: Detection Theory,* Prentice Hall Inc., ISBN 0-13-504135-X, New Jersey, USA.

Misra, P. & Enge, P. (2001). *Global Positioning System: Signals, Measurements, and Performance,* Ganga-Jamuna Press, ISBN 0-9709544-1-7, Massachussets, USA.

Parkinson, W.B. & Spilker, J.J. (1996). *Global Positioning System: Theory and Applications, volume I and II,* American Institute of Aeronautics and Astronautics, Inc., Washington, USA.

Rao, M.; Falco, G. & Falletti, M. (2011). SDR Joint GPS and Galileo Receiver: from Theory to Practice. Submitted to *IET-Radar, Sonar and Navigation*.

Tsui, J.B. (2000). *Fundamentals of Global Positioning System Receivers*, John Wiley & Sons, Inc., ISBN 0-471-38154-3, New York, USA.

7

Achievable Positioning Accuracies in a Network of GNSS Reference Stations

Paolo Dabove, Mattia De Agostino and Ambrogio Manzino
Politecnico di Torino
Italy

1. Introduction

The Network Real Time Kinematic (NRTK) positioning is nowadays a very common practice not only in academia but also in the professional world. Since its appearance, over 10 years ago, a growing number of people use this type of positioning not only for topographic applications, but also for the control of vehicles fleets, precision agriculture, land monitoring, etc.. To support these users several networks of Continuous Operating Reference Stations (CORSs) were born. These networks offer real-time services for NRTK positioning, providing a centimetric positioning accuracy with an average distance of 25-35 kms between the reference stations.

What is the effective distance between reference stations that allows to achieve the precision required for real-time positioning, using both geodetic and GIS receivers? How the positional accuracy changes with increasing distances between CORS? Can a service of geostationary satellites, such as the European EGNOS, be an alternative to the network positioning for medium-low cost receivers? These are only some of the questions that this chapter try to answer.

First, the GNSS network positioning will be discussed, with particular attention to the differential GNSS corrections such as the Master Auxiliary Concept (MAC), Virtual Reference Station (VRS) and Flächen Korrektur Parameter (FKP).

After this short review, the results obtained during a national experiment designed to verify both the quality and the potential of existing real-time and post-processing positioning services will be presented, with particular attention to the variability of the same depending on the network geometry, the type of rover receiver and the duration of his survey, as well as the use of the different GNSS constellations currently available for our area.

This experiment was conducted using already existing CORSs. Three real-time networks, characterized by different distances between the stations (50, 100 and 150 kms), were designed. The real-time products were tested, for each network, by sessions during 24-hour on a centroid point, using both geodetic and GIS receivers provided by different companies. A so large time session is made to avoid, on final results, the constellation geometry based influence, making results fully comparable.

In addition to the real-time network corrections, a post-processing analysis will be conducted, using the raw data acquired from geodetic and GIS receivers and combining them to the RINEX from the nearest network station and to the RINEX of virtual stations, generated by the network software close to the measurement site.

The ultimate goal of this chapter is to quantify the accuracy achievable nowadays with geodetic and GIS receivers when they are used into a network of reference stations, as well as to verify (or deny) the possibility that, thanks to the continuous GNSS modernization program, the improvement of new satellite constellations and new algorithms for computing and positioning, networks that are characterized by large distances between reference stations can be used for high accuracy real-time positioning.

2. The network positioning concept

Between 1990 and 1995, the carrier-phase differential positioning has known an enormous evolution due to phase ambiguity fixing method named "On The Fly" Ambiguity Resolution (Landau & Euler, 1992). Using this technique, a cycle slip recovery, also for moving items, was not more problematic, but positioning problems when distances between master and rover exceed 10-15 kms were not solved. For this reasons, at the end of the 90's, the Network Real Time Kinematic (NRTK) or, more generally, Ground Based Augmentation Model (GBAS) was realized. (Vollath et al, 2000, Raquet & Lachapelle, 2001, Rizos, 2002).

First, to understand the network positioning concept it is necessary keep in mind some concepts about differential positioning. To do this, it is possible to write the carrier-phase equation in a metric form:

$$\phi_k^p(i) = \rho_k^p - cdT_k + cdt^p - \alpha_i I_k^p + T_k^p + Mi_k^p + E_k^p + \lambda_i Ni_k^p + \varepsilon_k^p \tag{1}$$

In this equation, the $\phi_k^p(i)$ term represents the carrier-phase measurement on the i-th frequency. On the right-hand side of the equation, in addition to the geometric range ρ_k^p between the satellite p and the receiver k, it is possible to find the biases related to receiver and satellite clocks multiply by the speed light (cdT_k and cdt^p), the ionospheric propagation delay $\alpha_i I_k^p$ (with a known coefficient $\alpha_i = f_1^2 / f_i^2$ that depend by the i-th frequency), the tropospheric propagation delay T_k^p, the multipath error Mi_k^p, the ephemeris error E_k^p, the carrier-phase ambiguity multiply by the frequency length $\lambda_i Ni_k^p$ and finally the random errors ε_k^p.

Single differences can be written considering two receivers (h and k). Neglecting multipath error, that depends only by the rover site and therefore can not be modelled, it is possible to write:

$$\phi_{hk}^p(i) = \phi_h^p(i) - \phi_k^p(i) = \rho_{hk}^p + \lambda_i Ni_{hk}^p - cdT_{hk} - \alpha_i I_{hk}^p + T_{hk}^p + E_{hk}^p + \varepsilon_{hk}^p \tag{2}$$

After that, double differences equations can be written considering two receivers (h and k) and two satellites (p and q). Subtracting the single difference calculated for the satellite q from those one calculated for the satellite p, it is possible to obtain the double differences equation, neglecting random errors contribution:

$$\phi_{hk}^{pq}(i) = \phi_{hk}^p(i) - \phi_{hk}^q(i) = \rho_{hk}^{pq} + \lambda_i Ni_{hk}^{pq} - \alpha_i I_{hk}^{pq} + T_{hk}^{pq} + E_{hk}^{pq} + \cancel{\varepsilon_{hk}^{pq}} \tag{3}$$

When the distance between the two receivers is lower than 10 kms, the atmospheric propagation delays and the ephemeris errors can be irrelevant, allowing to achieve a centimetrical accuracy. Over this distance, these errors grow up and can not be neglected. Otherwise, these errors are very spatially correlated and can be spatially modelled (Wübbena et al., 1996). However, to be able to predict and use in real-time these biases, three conditions must be satisfied: the knowledge with a centimetric accuracy of the masters positions, a control centre able to process in real-time data of all the stations, the continuous carrier-phase ambiguity fixing also when inter-station distances reach 80-100 kms. This concept is equal to bring to the left-hand side of (3), among the known terms, the first two terms on the right-hand side, i.e.:

$$\phi_{hk}^{pq}(i) - \rho_{hk}^{pq} - \lambda_i Ni_{hk}^{pq} = \alpha_i I_{hk}^{pq} + T_{hk}^{pq} + E_{hk}^{pq} \tag{4}$$

In this way, it is possible to model, not only between stations *h* and *k*, but also among all the reference stations of the network, the residual ionospheric and tropospheric biases and the ephemeris error. When these errors are modelled, they can be broadcasted to any rover receiver.

3. From the concept to the implementation

First, it is possible to note that the (4) was written using two satellites. Otherwise, if a network of GNSS reference stations is considered, the same satellites that are visible and usable for the two or more master stations can not be necessarily visible from the rover receiver. Therefore, it is better to move from ionospheric and tropospheric delays, which depend by a couple of satellites, to something which depends only by a single satellite.

The network biases can be calculated in real-time using double differences, single differences or non-differential equations. The achievement of a common value of ambiguities is required. A theoretical proof of the equivalence between the non-differential and differential methods, in particular, can be found in Schaffrin & Grafarend (1986). The use of a differential method has pros and cons. The best advantage of the differential method is that the unknown parameters are fewer. Otherwise, the main disadvantage is that there is a correlation problem.

Although the approaches are identical, in recent years the trend is to use a non-differential approach. The network state parameters are evaluated by the use of a Kalman filter.

This methodology is obviously more complex, since both dispersive and non-dispersive components of each station are considered as unknowns in the Kalman filter state vector. This increased complexity is balanced by many advantages. The use of a Kalman filter allows to increase the number of equations available at each epoch, including for example measurements related to satellites that are not tracked by all the stations, in order to make the network estimation more robust also when one or more permanent stations are not available (e.g. transmission problems).

3.1 The non-differential model

Starting from pseudorange and carrier-phase equations written for the L1 ($\overline{P}_k^p$ and $\overline{\phi}_k^p$) and L2 ($\overline{\overline{P}}_k^p$ and $\overline{\overline{\phi}}_k^p$) GPS frequencies considering only one receiver (*k*) and one satellite (*p*), it is possible to separate the unknowns that depend on the receiver and the satellite:

$$\begin{bmatrix} \bar{P}_k^p \\ \bar{\bar{P}}_k^p \\ \bar{\phi}_k^p \\ \bar{\bar{\phi}}_k^p \end{bmatrix} = \begin{bmatrix} 1 & 1 & 0 & 0 \\ 1 & \alpha & 0 & 0 \\ 1 & -1 & \lambda_1 & 0 \\ 1 & -\alpha & 0 & \lambda_2 \end{bmatrix} \begin{bmatrix} \rho_k^p + cdt^p - cdT_k + T_k^p \\ I_k^p \\ \bar{N}_k^p \\ \bar{\bar{N}}_k^p \end{bmatrix} \tag{5}$$

Although it stands to reason that the pseudorange measurements have a beneficial contribution to the unknowns estimation, now only the phase equations are considered. In addition, the geometric range ρ_k^p can be moved on the left-hand side of the equation.

$$\begin{bmatrix} \bar{\phi}_k^p - \rho_k^p \\ \bar{\bar{\phi}}_k^p - \rho_k^p \end{bmatrix} = \begin{bmatrix} 1 & -1 & \lambda_1 & 0 \\ 1 & -\alpha & 0 & \lambda_2 \end{bmatrix} \begin{bmatrix} cdt^p - cdT_k + T_k^p \\ I_k^p \\ \bar{N}_k^p \\ \bar{\bar{N}}_k^p \end{bmatrix} \tag{6}$$

After that, in the right-hand side, it is possible to separate the tropospheric bias from the clock errors:

$$\begin{bmatrix} \bar{\phi}_k^p - \rho_k^p \\ \bar{\bar{\phi}}_k^p - \rho_k^p \end{bmatrix} = \begin{bmatrix} 1 & 1 & -1 & \lambda_1 & 0 \\ 1 & 1 & -\alpha & 0 & \lambda_2 \end{bmatrix} \begin{bmatrix} cdt^p - cdT_k \\ T_k^p \\ I_k^p \\ \bar{N}_k^p \\ \bar{\bar{N}}_k^p \end{bmatrix} \tag{7}$$

Through (7), and after few mathematical processes that are not shown, it is possible to separate the tropospheric propagation delay and the clock errors, solving the network positioning in a non-differential way.

4. The biases interpolation

After the dispersive and not-dispersive biases estimation, three solutions can be followed:

- to consider data from the reference stations of the network and to interpolate these data on the rover position, generating a virtual reference station close to the rover (VRS positioning);
- to model with a plane the biases and to broadcast the model parameters to the rover (FKP positioning);
- to broadcast to the rover the estimated biases together with data from a master reference station of the network (MAC positioning).

4.1 The VRS positioning

When the previous biases are estimated, the easiest and oldest way to broadcast differential corrections is the VRS (Virtual Reference Stations).

As mentioned before, the idea is to create a synthetic correction generated as if the reference station is close to the rover. For this reason, the rover communicates its approximate position (e.g. through an NMEA GGA message). Using this position, it is possible to interpolate data following different strategies, e.g. using:

- plane triangles (Vollath et al., 2000, Landau et al., 2002);
- an Inverse Distance Weighting (IDW) estimation;
- least squares method for estimating polynomial coefficients;
- collocation (Raquet et al., 2001).

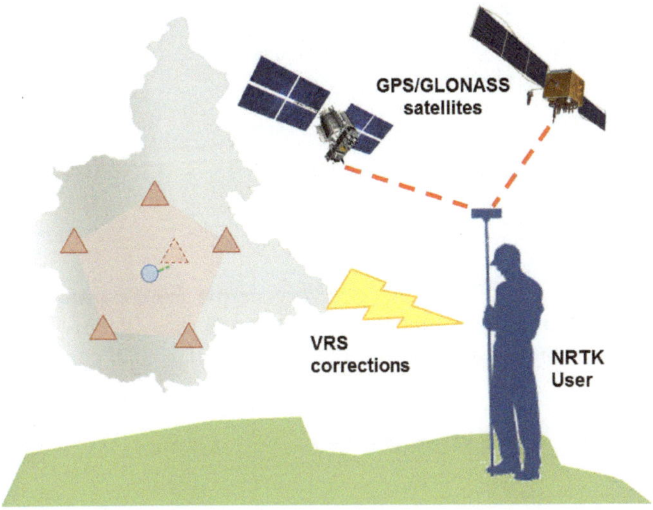

Fig. 1. The VRS positioning concept

This strategy can also be applied for post-processing positioning, by mean of a virtual data file (usually, in RINEX format). This file contains the observations that a virtual reference station may acquire in a well-known position selected by the network user.

As said above, the VRS positioning is the oldest network positioning strategy. Even if it has some advantages and is widely applied, there are also some disadvantages. The VRS method not allows, for example, a multi-base positioning (such as other methods) and it is not always well-regulated and repeatable.

4.2 The FKP positioning

Another differential network strategy is to calculate the interpolative area parameters and to broadcast them together with data from one reference station. This allows to have a one-way communication system and to maintain a relatively low transmission load.

The idea was first used by Wübbena et al. (1996, 2002), who uses a flat to interpolate the network biases in a given area. This positioning strategy was called FKP, which is the acronym of the German sentence "Flächen Korrektur Parameter" (flat correction parameters, in English).

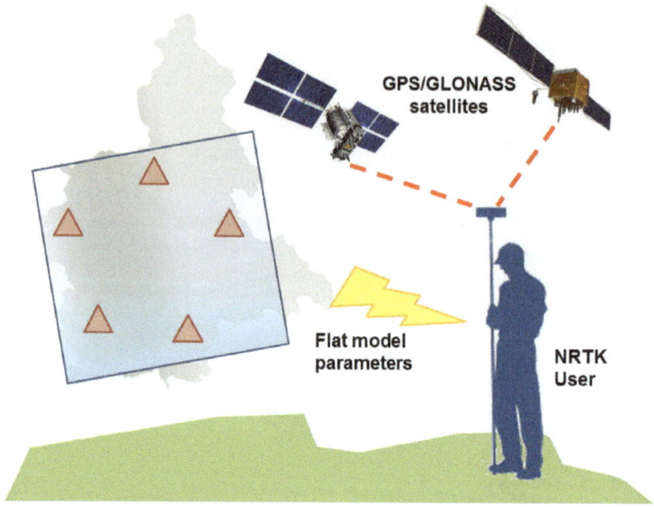

Fig. 2. The FKP positioning concept

The application of the positioning strategy is very simple. Four parameters, named E_0, N_0, E_1, N_1 can be computed considering the estimated values of geometric and ionospheric delays, using a given reference position (φ_R, λ_R). After that, it is possible to calculate the terms:

$$\begin{aligned} \delta r_0 &= 6.37\left(N_0\left(\varphi - \varphi_R\right) + E_0\left(\lambda - \lambda_R\right)\cos\varphi_R\right) \\ \delta r_I &= 6.37H\left(N_1\left(\varphi - \varphi_R\right) + E_1\left(\lambda - \lambda_R\right)\cos\varphi_R\right) \end{aligned} \quad (8)$$

where:

$$H = 1 + 16\left(0.53 - E/\pi\right)^3 \quad (9)$$

where E is the satellite elevation (in radians). Finally, the two carrier-phase corrections (in meters) are:

$$\begin{aligned} \delta r_{f1} &= \delta r_0 + \left(60/77\right)\delta r_I \\ \delta r_{f2} &= \delta r_0 + \left(77/60\right)\delta r_I \end{aligned} \quad (10)$$

4.3 The MAC positioning

In 2001, Euler et al. (2001) had proposed a new approach to the use and transmission of network corrections called Master Auxiliary Concept (MAC). The concept is the same as above: a common level of network ambiguity fixing is estimated and the corrections are transmitted to the rover separating dispersive and non-dispersive components.

In the MAC positioning, the coordinates and the biases of a single reference station (master station) are broadcasted to the rover in addition to the single differences (both corrections and coordinates) of the other stations in the network (auxiliary stations).

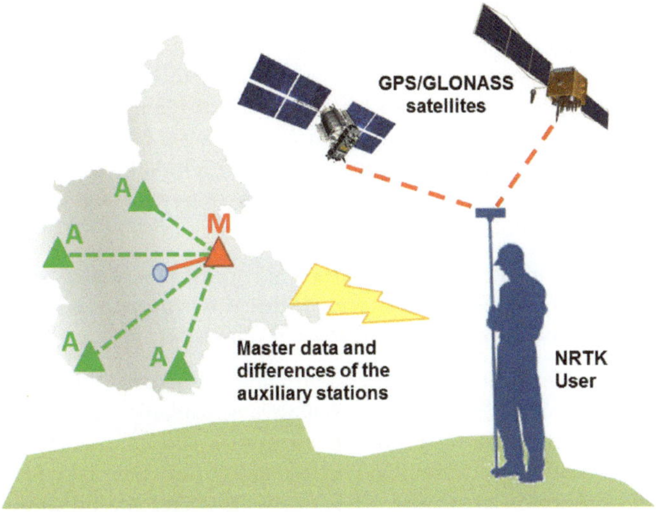

Fig. 3. The MAC positioning concept

These single differences are numerically small and have not a relevant transmission size, considering for example that the tropospheric corrections can be transmitted with a lower rate than the ionospheric ones (e.g. 2 - 5 seconds).

This new strategy implies that the network software should not perform any interpolation of the estimated biases. This interpolation, however, is only shifted to the rover, which has the possibility to choose different interpolative models or to apply a multi-base positioning.

Therefore, the rover receiver must have more computing power, so this positioning mode does not fit well to older receivers.

For very large networks, it is possible to transmit data from a subset of the network stations (sub-network or cell). Even in this case the positioning performed by the rover is accurate and fast. Even in this case, the result of the rover positioning is independent of the used cell.

5. NRTK developments and problems

It wonders if, due to the GPS and GLONASS modernization and the development of the Compass and Galileo constellations, the NRTK positioning will become obsolete. Over the last ten years, several authors (e.g. Chen et al., 2004) ask this question. In summary:

- After a large number of simulations, it is possible to conclude that, in the master-rover differential real-time positioning, the phase ambiguity solution will be almost instantaneous, making it unnecessary the use of a network of GNSS reference stations. However, in high ionospheric activity scenarios, the ambiguity fixing probability in the master-rover positioning will be very low. With a reference station network the ionosphere bias will be reduced (Stankov & Jakowskia, 2007).
- In differential positioning, the maximum distance between the master and the rover will be increased from 10 kms to 20 kms with the same reliability. Even the network

spacing will be increase with more frequencies (up to 80-100 kms), but the rover receiver will improve the reliability of fixing.

- The tropospheric biases will not be removed by using three or more frequencies. With a higher number of satellites tracked by a network of reference stations, instead, these errors will be estimated with a greater accuracy (Zhang & Lachapelle, 2001).
- Regardless of the GNSS future improvements, the multipath error in the rover will still be present. However, this error can be modelled on reference stations, giving the benefit of a more reliable estimation of the other biases.

To achieve a high real time positioning accuracy, in conclusion, a network solution will be required. The new GNSS constellations will decrease the time-to-fix the ambiguities, for both the network reference stations and the rover receiver.

A major technical problem of the network positioning is the correction signal broadcasting. A GPRS/GSM coverage in the survey site is not always available: in these cases, a radio link between the master and the rover receiver is a common solution. Otherwise, this solution does not involve the network positioning. A possible solution, especially for large networks, would be the integration between the NRTK correction and the SBAS architecture. This integration, in fact, does not require the use of additional antennas but only the payment of an access fee to the satellite band. Another solution might be to use digital subcarriers of TV channels. This solution fits very well, for example, with the MAC technique in the RTCM3 format.

On the other hand, the two-way internet communication could allow the network manager to offer additional services that are not usually provided. It may be possible, for example, to broadcast the number of satellite with a fixed network ambiguity, the maps on the survey site, the geoid undulation, etc... At the same time, the network user could transmit measurement data, updating in real time its survey, in addition to the quality of the data and of the fixed ambiguities, and to other parameters that can provide useful information for increase the reliability of GNSS positioning.

6. The experiments

In the previous sections, the network positioning concept and the different correction strategies were presented. But what is the effective distance between reference stations that allows to achieve the precision required for real-time positioning, using both geodetic and GIS receivers? And how the positional accuracy changes with increasing distances between Continuously Operating Reference Stations (CORSs)? These are only some of the questions that the experiments reported in the following try to answer.

The experiments were based on three different networks, with different inter-station distances: the first one (in the following, "red network" or "small network"), with distances of about 50 kms, is comparable with the existing GNSS networks in Italy. The second network ("green network" or "medium network") is characterized by distances of about 100 kms, which is the average spacing of the national geodetic network which materializes the Italian reference system (*Rete Dinamica Nazionale* - RDN). The last one ("blue network" or "large network") has inter-station distances of about 150 kms and it is used to verify the possibility of use not too thick networks.

The experiments were conducted using, as rover site, the reference one located on the roof of the headquarters of the Politecnico di Torino at Vercelli. For this reason, other reference stations have been chosen in the north-west side of Italy, so that the rover can be in a centroid point with respect to the three different GNSS networks (Fig. 4). The reference stations that are involved in this test (see the Table 1) belong to networks operated by public entities (such as administrative regions) and by private organizations (for example Surveyor Colleges or private companies).

	Station ID	*Station Name*	*Receiver type (IGS name)*	*Antenna type (IGS name)*
Red Network (50kms)	ALES	Alessandria	LEICA GRX1200+GNSS	LEIAR25.R3
	CRES	Crescentino	LEICA GRX1200+GNSS	LEIAR25.R3
	BIEL	Biella	LEICA GRX1200+GNSS	LEIAR25.R3
	LENT	Lenta	TPS NETG3	TPSCR.G3
	VIGE	Vigevano	TPS ODYSSEY_E	TPSCR3_GGD
Green Network (100 kms)	TORI	Torino	LEICA GRX1200+GNSS	LEIAR25.R3
	CHAT	Chatillon	TPS NETG3	TPSCR.G3
	LUIN	Luino	TPS NETG3	TPSG3_A1
	CREA	Crema	TPS ODYSSEY_E	TPSCR3_GGD
	SESC	Serravalle Scrivia	TPS NETG3	TPSCR.G3
Blue Network (150 kms)	CARP	Carpenedolo	LEICA GRX1200+GNSS	LEIAS10
	BUSL	Bussoleno	LEICA GRX1200+GNSS	LEIAR25.R3
	LOAN	Loano	TPS NETG3	TPSCR.G3
	TARO	Borgo val di Taro	TPS ODYSSEY_E	TPSCR3_GGD
	DOMO	Domodossola	TPS NETG3	TPSCR.G3

Table 1. Characteristics of permanent stations

The reference coordinates of all the stations were computed by processing 15 days of data with the Bernese GPS scientific software (version 5.0), linking the networks reference system with the ETRF2000 (2008.0), that is the Italian reference system materialized by the RDN. The different antennas used for the rover receivers were mounted on a pillar, as mentioned above, located on the roof of the Politecnico di Torino at Vercelli (Fig. 2).

The network software that was used is GNSMART (GNSS State Monitoring and Representation Technique), distributed by Geo++®. This software allows to quantify and to estimate tropospheric and ionospheric errors in addition to allows the modelling of the satellites ephemeris errors, of the multipath estimation and of the satellite and receiver clock errors. Even if it has no theoretical limitations to the minimum number of permanent stations, for a correct functioning, at least 5 stations are suggested.

In the experiments, double frequency, geodetic GNSS receivers of the main companies operating in Italy were used. The characteristics of the instruments are shown in the Table 2.

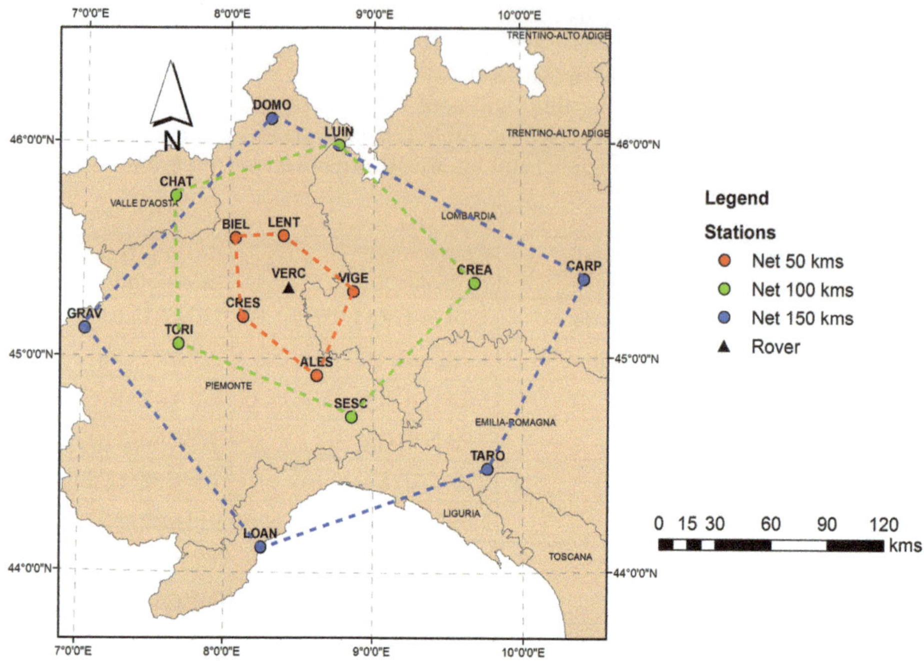

Fig. 4. Types of networks

	GX1230+GNSS (Leica Geosystems)	*GRS-1 (Topcon)*	*S9 GNSS (Stonex)*
Image		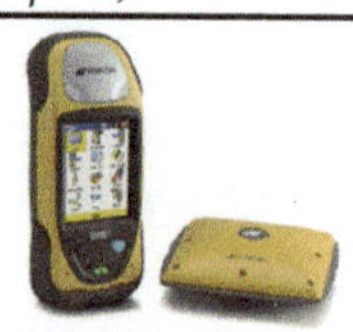	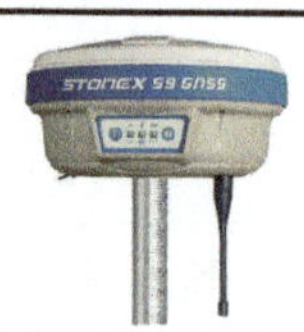
Antenna	LEIAX1203+GNSS	TPSPG_A1	TRM55970.00
Nr. of channels	120	72	220
Constellations	GPS+GLONASS	GPS+GLONASS	GPS+GLONASS
Position update rate	20 Hz	N/A	1 Hz
Type of protocols	RTCM 2.x RTCM 3.0 CMR / CMR+	RTCM 2.x RTCM 3.0 CMR / CMR+	RTCM 2.x RTCM 3.0 CMR / CMR+
Internal modem	Yes	Yes	Yes
Type of connection	GSM	GSM	GSM/GPRS

Table 2. Geodetic receivers

Fig. 5. Rover installation

In addition to the geodetic receivers, GIS L1 receivers with their external antennas were used in the experiments. As previously, the characteristics of these instruments are summarized in the Table 3.

	Zeno 10 (Leica Geosystems)	*GRS-1 (Topcon)*	*GeoXH (Trimble)*
Image		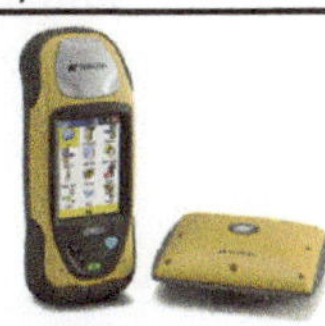	
Antenna	LEIAT502	TPSPG_A5	TRM53406.00
Nr. of channels	14	72	220
Constellations	GPS+GLONASS	GPS+GLONASS	GPS only
Position update rate	5 Hz	N/A	1 Hz
External antenna	Yes	Yes	Yes
Time to first position	35 ÷ 120 s	N/A	45 s
Type of protocols	RTCM 2.x RTCM 3.0 CMR / CMR+	RTCM 2.x RTCM 3.0 CMR / CMR+	RTCM 2.x RTCM 3.0 CMR / CMR+
Phase corrections	Yes	Yes	No
Internal modem	Yes	Yes	No
Type of connection	GSM/UMTS 3.5G	GSM	No

Table 3. GIS instruments

7. Real time positioning accuracies

The results reported below are average values, considered significant for the two sets of instruments used in the experiments. All data collected during the experiment were analysed using two types of charts:

- Cumulative Distribution Function (CDF), i.e. a curve that describes the probability that a variable X with a given probability distribution will be found at a value less than or equal to x%.
- Cumulative moving average, i.e. the arithmetic mean of a series of values over a period that increase with respect to time. Assuming equidistant measuring or sampling times, it can be computed as the sum of the values over a period divided by the number of values.

In the following, the planimetric and elevation positioning errors for both the instrument categories, for the different network size and products are analysed.

7.1 Geodetic receivers

The tests carried out using geodetic receivers have involved the use of the three types of NRTK corrections analysed in the previous paragraphs: VRS, MAC and FKP. For each receiver and each NRTK correction, 24 hours of real time positioning results have been stored. For this analysis, only the positions with both fixed ambiguities and a HDOP (Horizontal Diluition Of Precision) index lower than 4 have been considered.

Analysing the stored positions, it was possible to highlight the behaviour that each receiver has depending on the type of differential correction: the Fig. 6 shows the quality of the planimetric and height positioning that one of the receivers used has into the "red" network.

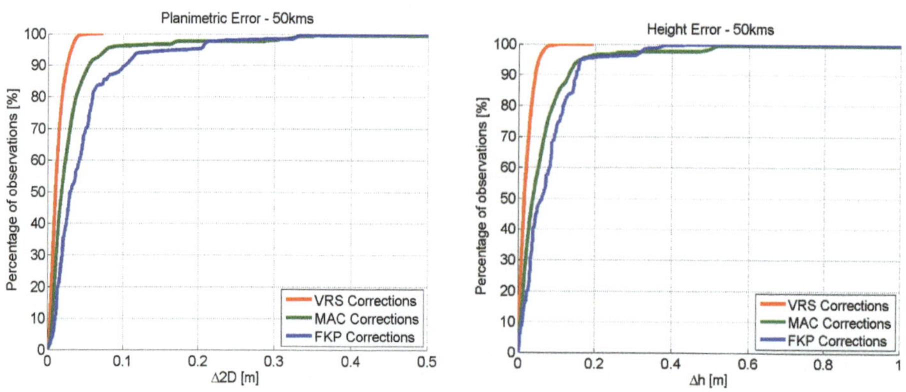

Fig. 6. CDF of planimetric (left) and elevation (right) errors of a geodetic receiver

In addition to the different behaviour that a receiver has with respect to differential correction, it is possible to consider also the position quality variation of different receivers. The Fig. 7 shows the planimetric and elevation error distributions using the three different geodetic receivers presented in Table 2, with a VRS correction broadcasted by the "green" network.

The analysis of the curves above allows to highlight a homogeneous behaviour among the receivers, which are separated only at the end of the distribution (around the 85% of probability). The planimetric accuracy, for example, changes from about 2 cms (95% of probability) for the "Receiver 2" to about 7 cms for "Receiver 1" and "Receiver 3".

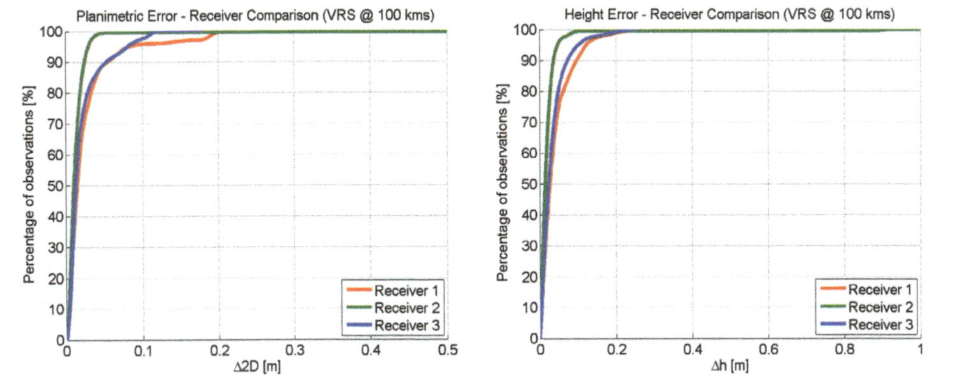

Fig. 7. CDF of planimetric (left) and elevation (right) errors of the three geodetic receivers

For this reason, it is acceptable to consider the behaviour of an "average" receiver, focusing on the variation in the positioning quality with respect to the size of the network.

The following sections show the results of these tests with the different NRTK correction used.

7.1.1 VRS positioning

The VRS is without doubt the most used differential correction in real-time positioning, as well as the easiest to manage for each receiver. As seen above, in fact, this type of differential correction is based on generating, starting from the data of network CORSs, a virtual reference station close to the measurement site.

It is therefore expected a deteriorating positioning quality with the increasing of the inter-station distances. If this variable is increasing, in fact, there are numerous inaccuracies that can be made by the interpolation step during the generation of the VRS correction.

The Fig. 8 shows the average behaviour of a geodetic receiver in the case of a VRS correction broadcasted by networks of different sizes.

The CDF analysis brings out an effective increase of the errors (both planimetric and altimetric) when the size of the GNSS network grows up. The planimetric error, for example, changes from values below 5 cms (95% of reliability) considering the "red" network up to 10 and 15 cm with the "green" and the "blue" one, respectively. A similar behaviour can be observed for the elevation error, with values from 6 cm ("red" network) to 10 cm ("green" network) and to about 25 cm ("blue" network).

Even with regard to the cumulative moving average, VRS positioning with a "red" network achieves a centimetre accuracy after few minutes, with a trend that remains constant over the time.

It is also interesting to analyse the trend of the cumulative moving average when "green" and "blue" networks are used: in both cases, there was a significant improvement of the position quality when the measurement period increases. This trend allows to reach a centimetre accuracy after a few hours-length measurement. This behaviour is evident for example when the "blue" network is used, where the effect of few outliers positions, not detected in real-time by the receiver, disappears only after 5 hours of measurements.

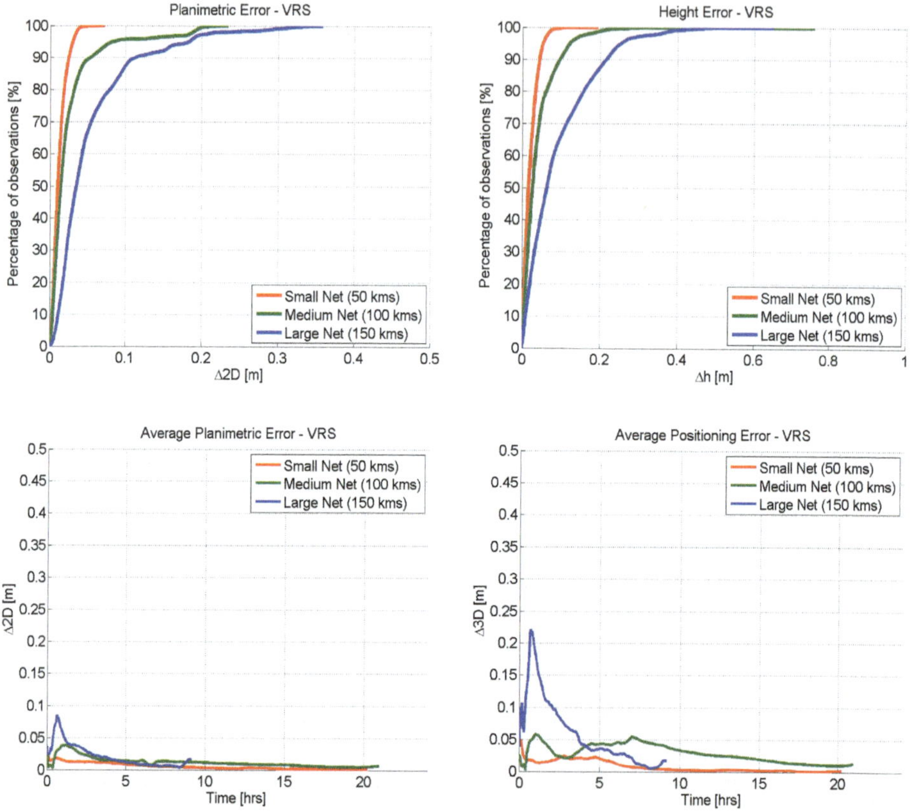

Fig. 8. Positioning quality of a geodetic receiver when a VRS correction is used: CDF of the planimetric error (top left) and of the elevation error (top right), cumulative moving average of the planimetric error (bottom left) and of the three-dimensional positioning error (bottom right)

7.1.2 MAC positioning

As said previously, the MAC correction is realized using observations from a single reference station (master) with additional information of other CORSs within a well-defined cell of the network (auxiliary stations). For this reason, this correction should be less sensitive to the variation of GNSS network sizes. As long as the distance between the master station and the rover is maintained below the permissible values for differential real-time positioning, the variation of the network size represents only a minor contribution to the differential correction (i.e., the contribution due to the auxiliary stations). The tests allow to extract an "average" behaviour of a geodetic dual frequency receiver when a MAC correction is used. This behaviour is summarized in the Fig. 9.

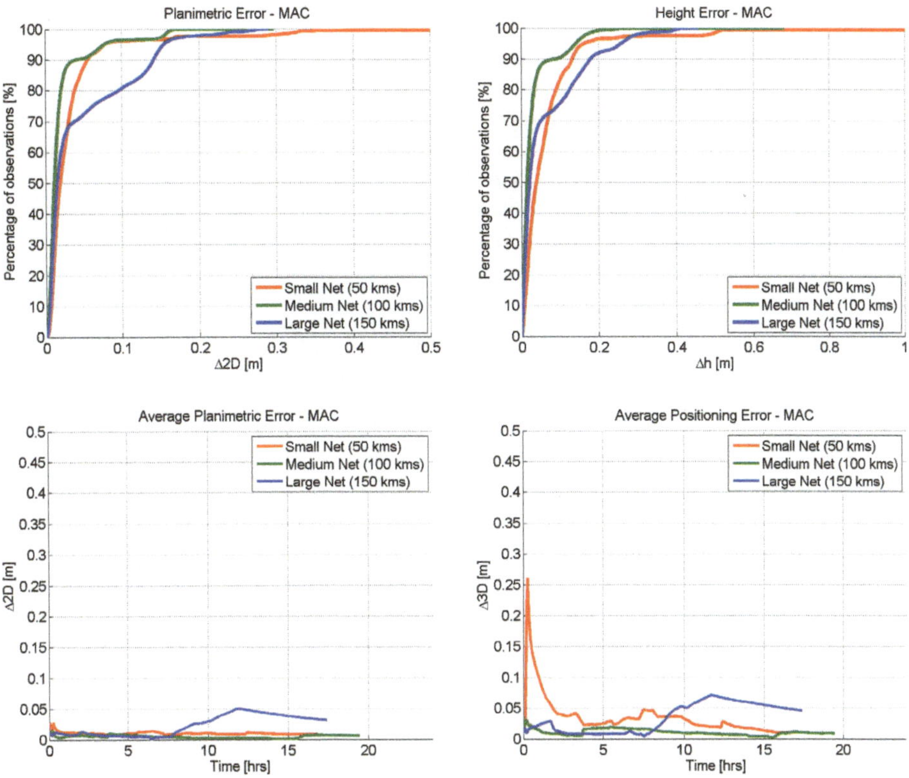

Fig. 9. Positioning quality of a geodetic receiver when a MAC correction is used: CDF of the planimetric error (top left) and of the elevation error (top right), cumulative moving average of the planimetric error (bottom left) and of the three-dimensional positioning error (bottom right)

The curves above confirms what was expected. Analysing, for example, the CDF curves of planimetric and elevation error, it is possible to see how positioning errors do not increase excessively when switching between the "red" and the "green" network. In these cases, the positioning quality is similar, and reaches about 5 cms (95% of observations) in planimetry and about 10 cms in altitude. A significant positioning deterioration occurs when differential MAC corrections broadcasted by the "blue" network are used. In this case, the master station is very far from the rover, causing problems on the quality of the positioning (15 cms in planimetry and 25 cms in elevation).

The trend of cumulative moving averages allows to highlight once again the similar behaviour of the MAC positioning performed with a "red" and a "green" network, as seen in bottom right in the Fig. 9. The cumulative moving average also shows how the MAC positioning with the "blue" network is not perfectly consistent over the time: as it is possible to see, after about 8 hours of measurement there is a worsening of the three-dimensional positioning quality, due to measurement error variations that are not well modelled by a so wide network.

7.1.3 FKP positioning

The FKP positioning, as seen in previous sections, consists of broadcasting to the rover the bias flat model parameters estimated by the GNSS network software. The hypothesis that spatial delay variations can be arranged along a plane is certainly reliable for small networks, but become trivial when the inter-station distances become too high. In that case, in fact, local atmospheric phenomena, which can cause considerable disturbances in the GNSS observations, are not taken into account. The positioning results obtained by the use of a geodetic receiver corrected by a FKP model are shown in the Fig. 10.

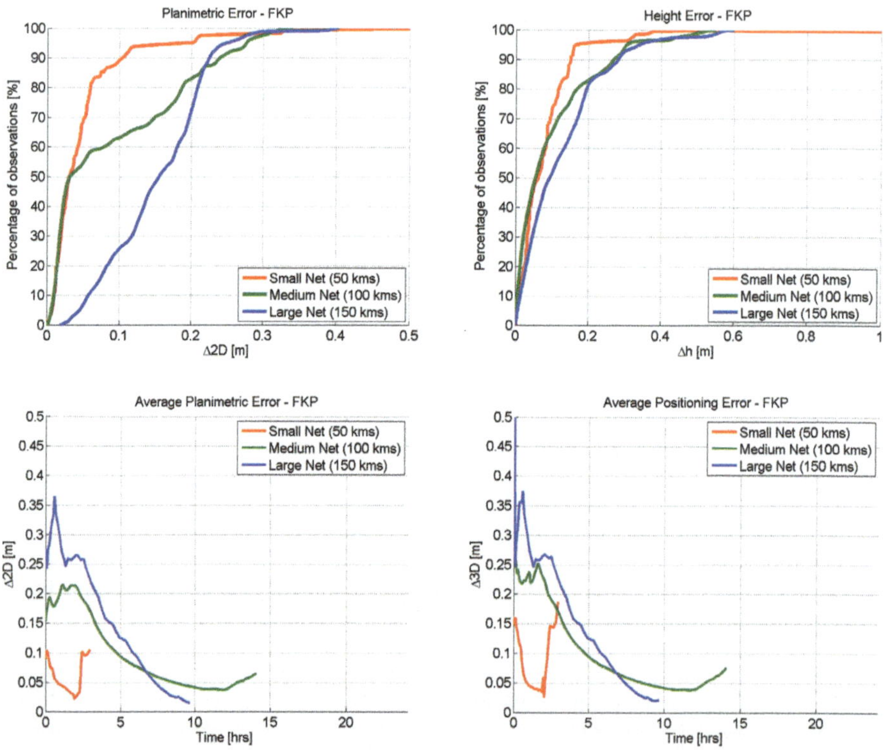

Fig. 10. Positioning quality of a geodetic receiver when a FKP correction is used: CDF of the planimetric error (top left) and of the elevation error (top right), cumulative moving average of the planimetric error (bottom left) and of the three-dimensional positioning error (bottom right)

As it is possible to see in the previous figures, a flat interpolation model allows to achieve a positioning error equal to or slightly greater than 10 cms (95% of reliability) only when small networks (e.g. the "red" one) are used. When medium-sized and large-sized networks (the "green" and "blue" ones, respectively) are used, the planimetric average error exceeds 20 cms. A very similar trend is found also for the elevation error, which increases from 15 cms ("red" network) to over 30 cms ("green" and "blue" networks). There are, in this case, no significant differences between the two wider networks.

With regard to the performance of the cumulative moving average, it is possible to see that a positioning error always lower than 5 cm can be achieved only by averaging several hours of data. The analysis of the average length of the lines shows the small number of epochs with a fix ambiguity values (almost always less than 50% of the total measured times). This is not due to NRTK corrections transmission problems, but to the use of FKP corrections by the receiver (the flat model does not fit well with the rover measurement errors).

7.2 GIS receivers

The tests carried out on the three GIS receivers shown in Table 3 were designed to study their accuracy within GNSS networks with different inter-station distances. The corrections from a VRS (used by all receivers in this class) and from the nearest reference station (NRT) were tested. Given the receiver category and the metric accuracies expectations, the EGNOS[1] corrections were also used, in order to assess whether could be, for GIS receivers, the benefits of a network of GNSS reference stations compared with the area corrections broadcasted by a geostationary satellites constellation.

In the following, the results obtained using VRS corrections are discussed. After that, the comparison between these results and those obtained using corrections from the NRT station and from the EGNOS satellites are presented.

7.2.1 VRS positioning

First, planimetric and elevation accuracies achievable with a GIS receiver into networks with different inter-station distances are analysed. As said above, 24 hours of measurements (to be independent of satellites geometry) and only positions with a HDOP index lower than or equal to 4 (to exclude outliers) were considered.

The Fig. 11, in the next page, shows the results obtained considering an "average" receiver. From the pictures analysis, it may notice that the positioning accuracy changes when the inter-station distance increases. However, it is possible to see that, unlike the geodetic receivers, in this case there is not a significant positioning deterioration with the increasing network size (from the "red" network to the "blue" one). The planimetric error at the 95% of reliability, for example, goes from 80 cms ("red" network) to 60 cms ("green" network) and to about 1 m ("blue" network). The improvement obtained by considering a medium-sized ("green") network is not surprising, but it must be analysed considering the quality of GIS receivers. This behaviour shows a substantial stability of the positioning accuracy, which remains always around metric values. This trend is more evident when the elevation error is analysed. Cumulative moving average lines achieve a sub-decimetric accuracy only after about 5 hours, showing no particular differences between the three different networks.

7.2.2 NRT and EGNOS positioning

The analysis carried out considering the positioning quality with VRS corrections were compared with these obtained using the corrections from both the nearest reference station and the European geostationary satellites constellation EGNOS. In order to highlight the benefits of differential corrections, the stand-alone positioning results are also reported in the figures.

---

[1] http://www.esa.int/esaNA/egnos.html

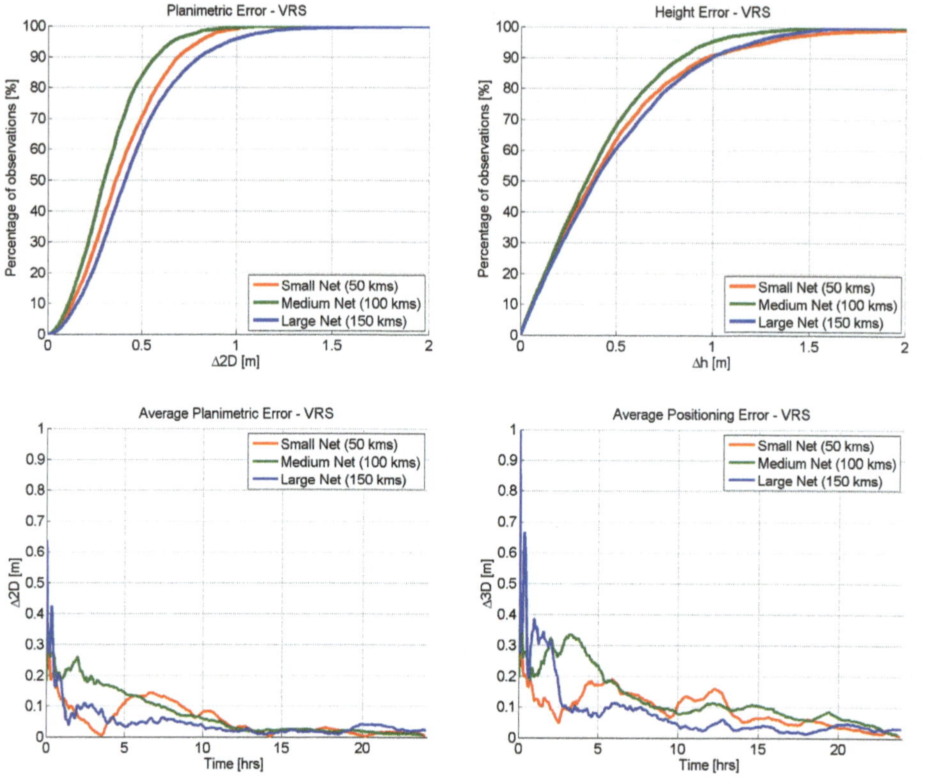

Fig. 11. Positioning quality of a GIS receiver when a VRS correction is used: CDF of the planimetric error (top left) and of the elevation error (top right), cumulative moving average of the planimetric error (bottom left) and of the three-dimensional positioning error (bottom right)

The Fig. 12 shows the comparison between the stand-alone positioning error and this one obtained using the two corrections said above. The figure shows the results obtained considering the network with inter-station distances of about 100 kms ("green" network), which in previous tests gave better results. As shown, both the NRT and EGNOS corrections allow to obtain a positioning quality that is fully comparable to that one achievable using VRS corrections. This result, although it may seem in contrast with the virtual stations and with the GNSS network positioning concepts, must not surprise. Common GIS receivers, in fact, are not able to well use carrier-phase corrections that difficultly can be modelled when the reference stations are too far from the measurement site.

The analysis of figures above allows also to highlight benefits due to the use of differential corrections with respect to the stand-alone positioning. The planimetric error (at the 95% of reliability), for example, decreases from values close to 1.7 m for stand-alone positioning up to about 70 cms when differential corrections are used. This improvement is even more evident observing the height accuracy trend (which decreases from about 4.5 ms to 1 m) and when cumulative moving average is considered (Fig. 13).

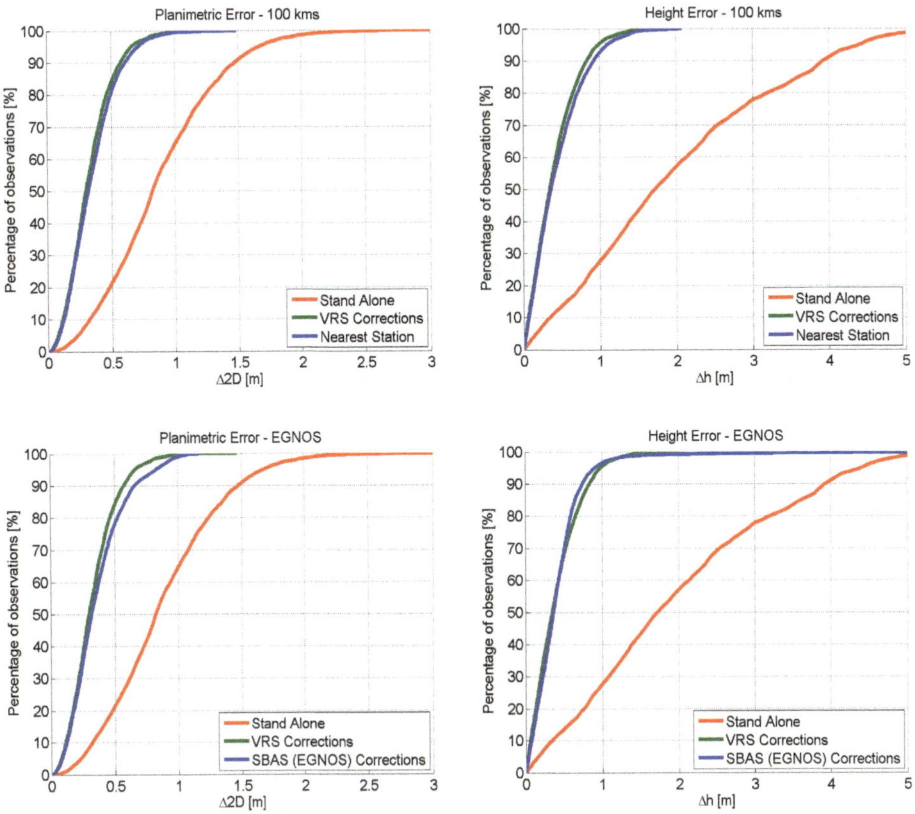

Fig. 12. Positioning quality comparison of a GIS receiver: CDF of planimetric error (left) and of elevation error (right) when a NRT correction is used (top) and when an EGNOS area correction is involved (bottom)

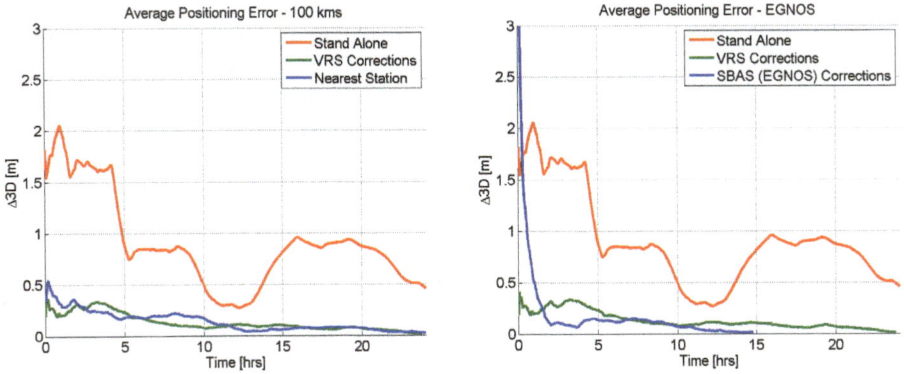

Fig. 13. Positioning quality comparison of a GIS receiver: cumulative moving average of the positioning error when NRT (left) and EGNOS (right) corrections are used

8. Post-processing positioning accuracies

Raw data files in a RINEX format were stored in order to estimate the accuracies achievable in post-processing and the performance when the average inter-station distance increases.

These files, with a length of 24 hours, were split in many shorter files with different duration, in order to statistically evaluate the planimetric and altimetric accuracy.

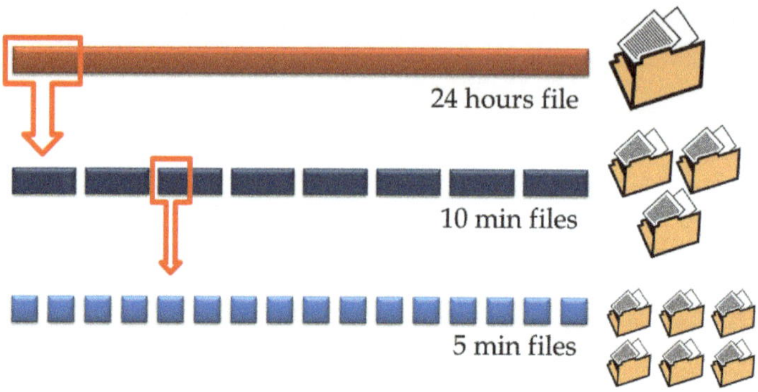

Fig. 14. Data files split schema

The data files were processed by a commercial software (Leica Geomatics Office™ v.8.0) based on the double differences approach, using as master station the nearest permanent station to each considered network and a VRS generated by the network software close to the measurement site.

The post-processing results show a no significant difference among the three geodetic receivers, due to the goodness raw data quality. The same behaviour was observed also when the three GIS receivers were used. For this reason, an "average" instrument for each class of receivers is considered in the following analysis.

8.1 Geodetic receivers

Raw data files of a geodetic receiver were split in many files of 5 and 10 minutes long, and they were post-processed as said above. The CDF of the planimetric and altimetric error (calculated using the "true position" evaluated from the network adjustment previously described) were computed for each time session.

The Fig. 15, in the next page, shows the results obtained using the nearest station for the three considered networks. A low deterioration of the positioning accuracy can be observed when different reference stations (at different distances from the rover) were used as master.

This can be seen, for instance, considering the planimetric accuracy obtained by the post-processing of the 5 minutes long data. In this case, for the "green" and the "blue" networks, a significant degradation can be observed only in the last 10% of the distribution.

Considering the 10 minutes long files, no significant improvements are observed, as expected, in the "red" network, while a better accuracy can be seen when "green" and "blue" networks are used. Also the percentage of epochs with fixed ambiguities are similar between the "red" and the "green" networks (98-99% of epochs using 5 minutes files and 99-100% using 10 minutes files). For the "blue" one, this percentage decreases to 92% and 97% respectively.

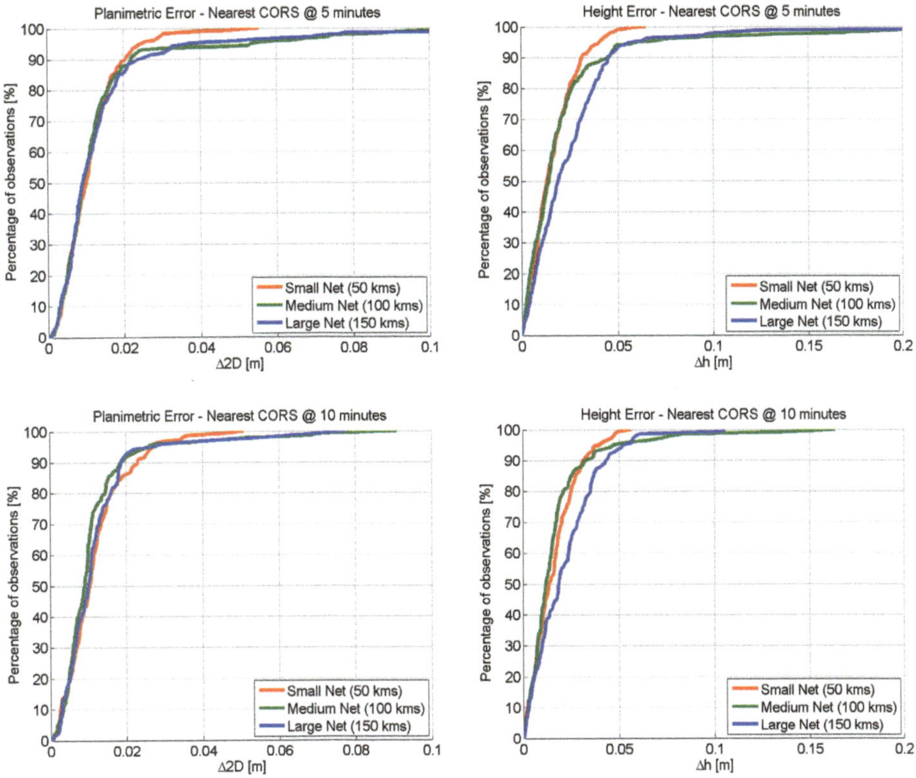

Fig. 15. Positioning quality of a geodetic receiver after the post-processing with the nearest reference station. CDF of the planimetric (left) and altimetric (right) error using static time sessions of 5 (top) and 10 (bottom) minutes

These results clearly show what is the limit (about 2 cms) of the post-processing approach when a master station is located farther than 25-30 kms from the rover. This distance is until today comparable with the actual inter-station distances of GNSS networks. To improve the 2 cms limit with a reasonable reliability, two strategies can be adopted:

- increase the static measurement length, resulting in a lack of the productivity;
- use a post-processing network product, i.e. a virtual RINEX file generated from the error models estimated by the GNSS network.

In the last case, the main advantage for the user consists in having a raw data file located close to the rover.

In this way, the rover has a higher probability to fix the phase ambiguities. Otherwise, this product shows the problems already discussed for the real-time VRS positioning. The VRS

RINEX files, in fact, are generated interpolating the error model estimated by the network software. When the inter-station distances grows up, a positioning quality deterioration is expected, due to the approximations made in the interpolation process of a wider area. The Fig. 16, that shows the results obtained using a VRS RINEX file as master, confirms what was expected.

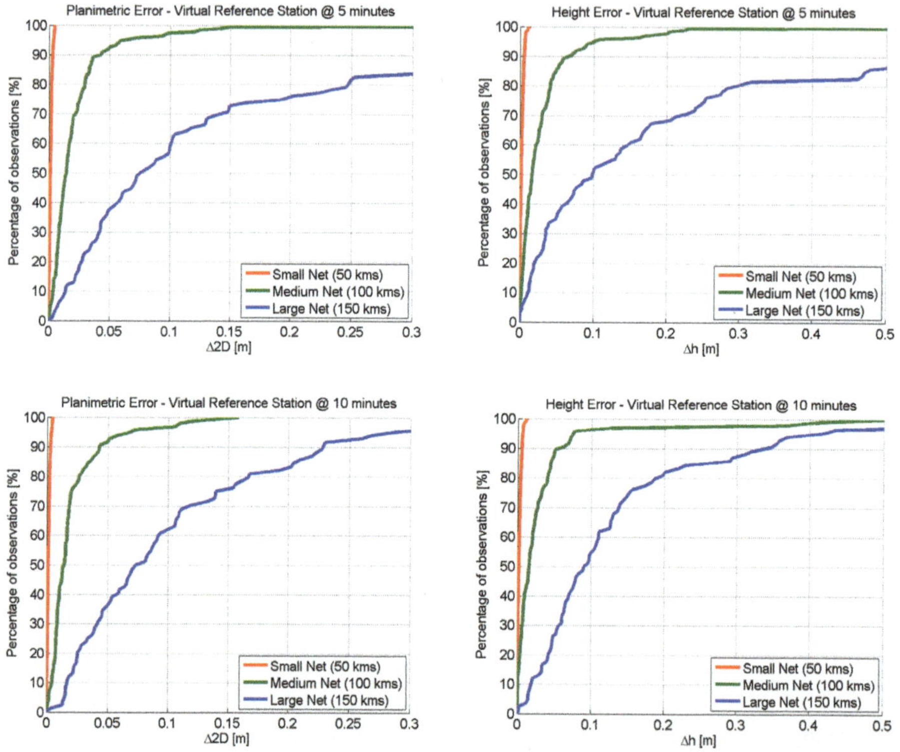

Fig. 16. Positioning quality of a geodetic receiver after the post-processing with the VRS. CDF of the planimetric (left) and altimetric (right) error using static time sessions of 5 (top) and 10 (bottom) minutes

It is easy to note that the VRS post-processing positioning improves the planimetric and altimetric accuracy only in the case of small- ("red") and medium-sized ("green") networks (about 1 cm and 4 cms respectively, considering the planimetric error).

These results confirm the goodness of this product when GNSS permanent stations, far each other about 50-100 kms, are used. When these distances exceed 100 kms, the positioning quality is comparable, or even worse, with the one obtained using the nearest reference station as master.

The percentage of fixed ambiguities is about 100% of the epochs considering 5 minutes or 10 minutes long files in "red" network, while it is very low for "green" (50%) and "blue" (12%) networks.

8.2 GIS receivers

As in the previous section, an "average" GIS receiver was considered, and the same processing methods were adopted. However, it was a priori decided to increase the static processing length session, splitting the raw data in 10 or 20 minutes long files. This is the average time that an operator could wait to achieve a sub-decimeter positioning accuracy using a low-cost receiver.

The Fig. 17 shows the post-processing results of raw data files, obtained using the nearest reference station.

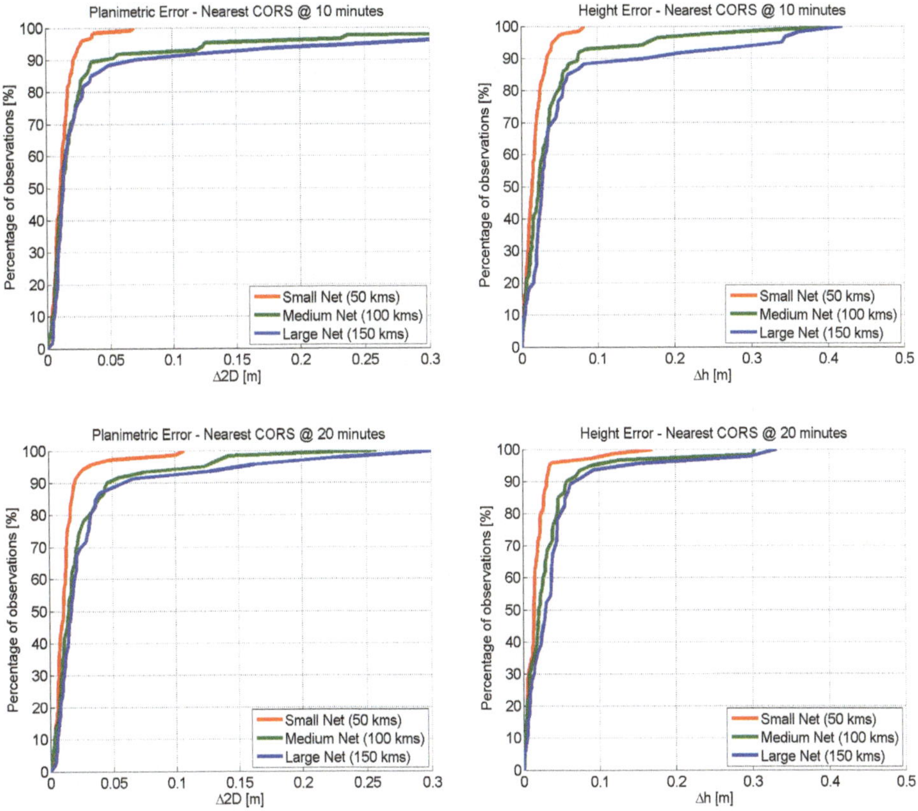

Fig. 17. Positioning quality of a GIS receiver after the post-processing with the nearest reference station. CDF of the planimetric (left) and altimetric (right) error using static time sessions of 10 (top) and 20 (bottom) minutes

A low deterioration both in planimetric and in altimetric accuracy can be observed, when the master is farther than 30 kms from the rover. This deterioration is not due to the low quality of raw data, but, instead, because GIS receivers are not able to track the L2 frequency. This frequency, in fact, allows to linearly combine measurements to reduce some of the biases (e.g. iono-free combination).

It should also be noted that increasing the measurement time from 10 to 20 minutes does not entail a real improvement in the positioning accuracy, that reaches values from about 2-3 cms ("red" network) to 7 cms ("green" network) at the 90% of reliability.

As before, a better accuracy can be obtained using VRS RINEX files generated by the network software. The analysis of the positioning accuracy, shown in Fig. 18, confirms the expected behaviour, already seen for geodetic receivers.

A reduction of the maximum planimetric and altimetric error for "red" and "green" networks is observed (few centimetres at 90% of reliability).

The percentage of measurement sessions with fixed ambiguities, goes from 68% ("red" network) to 48% ("green" network) and only to 31% ("blue" network). These percentages do not appreciably changes when 20 minutes long files are considered.

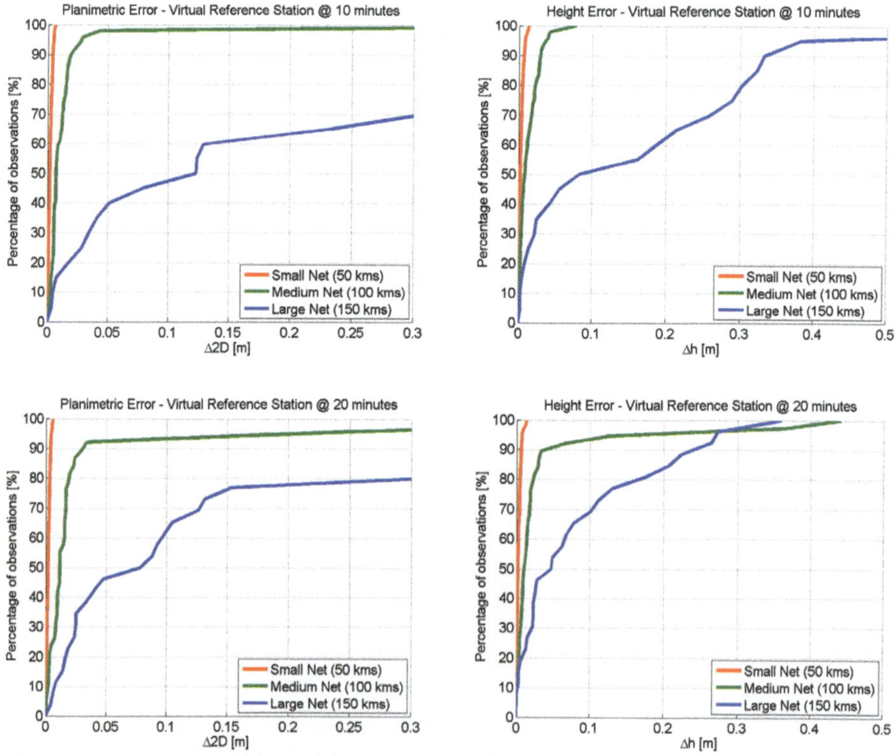

Fig. 18. Positioning quality of a GIS receiver after the post-processing with the VRS file. CDF of the planimetric (left) and altimetric (right) error using static time sessions of 10 (top) and 20 (bottom) minutes

The combined use of single frequency instruments and virtual data is very useful only for the "red" network, while the virtual data generated by the "green" and the "blue" networks significantly increase the errors, as clearly shown in Fig. 18.

It is representative the planimetric positioning achieved with the VRS generated by the wider ("blue") network, which has a percentage of about 20-30% of the data that lies outside the maximum axis value (30 cms). In this case, the percentage of measurement sessions with fixed ambiguities is 100% ("red" network) and collapse to 40% ("green" network) and only to 10% ("blue" network).

9. Conclusions

In this chapter, the accuracy of geodetic and GIS receivers in small-, medium- and large-sized networks of GNSS reference stations were analysed, comparing the results obtained with different network products. The accuracies achieved with a 95% of reliability, referring to a well-known rover position and using 24 hours of measurement, were considered.

Geodetic receivers can benefit from the VRS corrections transmitted by networks with inter-station distances up to 100 kms, allowing it to achieve planimetric accuracies from 2 to 8 cms and from 5 to 12 cms in elevation. A similar behaviour can be found when MAC corrections are used. This network product, in fact, provides comparable results for the small- and medium-sized networks (about 5 cm in planimetry and 10 cm in elevation).

If large networks are considered, the NRTK positioning is often inefficient and unreliable. Due to their lower accuracy to model biases of large areas, FKP corrections are not suitable for positioning even in medium-sized networks.

The performance of GIS receivers in real-time is poorly influenced by the size of the network. Planimetric error achieves accuracies from 65 to 85 cms in the three considered networks, and elevation error is always about 1 m. This improvement is noticeable when it is compared to the stand-alone position, with planimetric accuracies of 1.7 m and 4.5 m in altitude. Even with the EGNOS corrections it is possible to reach the same altitude accuracy (1 m at 95%) and a planimetric accuracy of about 75 cms. Using the network differential corrections, a planimetric accuracy of 50 cm can be achieved by averaging few minutes of real-time positions.

Regarding the post-processing positioning, no substantial differences were noted in the accuracy considering static session of 5 and 10 minutes long for geodetic receivers, and of 10 and 20 minutes long for GIS receivers.

For geodetic instruments, it is found that the positioning using a VRS RINEX file allows an improvement only when small-sized networks are involved. For wider networks, the best accuracies are always obtained using the RINEX file from the nearest reference station, although the number of ambiguity fixes may drop up to about 30% of the epochs.

Considering GIS receivers, the best performance is obtained when the nearest station data are used in a small-sized network (inter-station distances of about 50 kms), with a planimetric error of 2 cms and an elevation error of 3 cms. A VRS RINEX file generated by a large network does not improve the position accuracy with respect to the results obtained from the nearest station, while some advantages can be found when a medium-sized network is involved. The planimetric accuracy, in fact, goes from 10 cms, when data from the nearest station are used, to about 4 cms considering virtual data generated by a GNSS network. A similar behaviour can be also found when elevation accuracy is considered (from 15 cms to 8 cms).

The experiments were funded as part of the National Research Project PRIN 2008 "The new Italian Geodetic Reference System: continuous monitoring and application of control management of the territory", financed by the Italian Ministry of Education, University and Research in 2008.

10. References

Chen, X.; Vollath, U. & Landau, H. (2004). Will GALILEO/modernized GPS obsolete Network RTK, *Proceedings of ENC-GNSS 2004*, Rotterdam, The Netherlands, May 2004.

Euler, H.J.; Keenan, C.R.; Zebhauser, B.E. & Wübbena G. (2001). Study of a simplified approach in utilizing information from permanent reference station arrays, *Proceedings of the 14th International Technical Meeting of the Satellite Division of The Institute of Navigation (ION GPS 2001)*, Salt Lake City (UT-USA), September 2001.

Landau, H. & Euler, H.J. (1992). On-the-Fly ambiguity resolution for precise differential positioning, *Proceedings of the 5th International Technical Meeting of the Satellite Division of The Institute of Navigation (ION GPS 1992)*, Albuquerque (NM-USA), September 1992.

Landau, H., Vollath, U. & Chen, X. (2002). Virtual Reference Station Systems. *Journal of Global Positioning Systems*, Vol. 1(2), pp. 137-143.

Raquet, J. & Lachapelle, G. (2001). Multiple Reference RTK Positioning. *GPS World*, Vol.12(4), pp. 48-53.

Raquet, J.; Lachapelle, G. & Fortes, L. (2001). Use of a covariance analysis technique for predicting performance of regional area differential code and carrier-phase networks. *Navigation*, Vol. 48(1), pp. 25-34.

Rizos, C. (2002). Network RTK research and implementation - a geodetic perspective. *Journal of Global Positioning Systems*, Vol. 1(2), pp. 144-150.

Schaffrin, B. & Grafarend, E. (1986). Generating classes of equivalent linear models by nuisance parameter elimination. *Manuscripta Geodaetica*, Vol.11, pp. 262–271.

Stankov, S.M. & Jakowskia, N. (2007). Ionospheric effects on GNSS reference network integrity. *Journal of Atmospheric and Solar-Terrestrial Physics*, Vol. 69 (4-5), pp. 485-499.

Vollath, U.; Buecherl, A.; Landau, H.; Pagels, C. & Wagner, B. (2000). Multi-Base RTK using Virtual Reference Stations, *Proceedings of the 13th International Technical Meeting of the Satellite Division of The Institute of Navigation (ION GPS 2000)*, Salt Lake City (UT-USA), September 2000.

Wübbena, G. & Bagge, A. (2002). RTCM Message Type 59–FKP for transmission of FKP. *Geo++ White paper*, N.2002.01.

Wübbena, G.; Bagge, A.; Seeber, G.; Böder, V. & Hankemeier, P. (1996). Reducing distance dependent errors for real-time precise DGPS applications by establishing reference station networks, *Proceedings of the 9th International Technical Meeting of the Satellite Division of The Institute of Navigation (ION GPS 1996)*, Kansas City (KS-USA), September 1996.

Zhang, J. & Lachapelle, G. (2001). Precise estimation of residual tropospheric delays using a regional GPS network for real-time kinematic applications. *Journal of Geodesy*, Vol.75 (5-6), pp. 255-266.

8

Precise Real-Time Positioning Using Network RTK

Ahmed El-Mowafy
Curtin University
Australia

1. Introduction

In the classic RTK method using a single reference station the rover needs to work within a short range from the reference station due to the spatial decorrelation of distance-dependent errors induced by the ionosphere, troposphere and orbital errors. The operating range of RTK positioning is thus dependent on the existing atmospheric conditions and is usually limited to a distance of up to 10-20 km. In addition, no redundancy of the reference stations is usually available if the reference station experiences any malfunctioning. The constraint of the limited reference-to-rover range in RTK can be removed by using a method known as Network RTK (NRTK), whereby a network of reference stations with ranges usually less than 100 km is used. The network stations continuously collect satellite observations and send them to a central processing facility, at which the station observations are processed in a common network adjustment and observation errors and their corrections are computed. The observation corrections obtained from the network are sent to the user, operating within the coverage area of the network, to mitigate his observation errors.

In this chapter, the principles of Network RTK are first discussed and the advantages and disadvantages of the method are given. Next, the network design parameters are discussed, which include network baseline lengths and configuration, the communication method between the computing centre and the user, and the amount of calculations required by the network processing centre and by the user. Description of possible network processing techniques, their basic models, and a comparison between their advantages and disadvantages are given. Finally, some important NRTK applications are discussed including the use of NRTK in engineering surveying, machine automation and in the airborne mapping and navigation. Results from real-time testing are discussed.

2. Principles of the network RTK

The aim of network RTK is to minimise the influence of the distance dependant errors on the computed position of a rover within the bounds of the network. NRTK provides redundancy of reference stations in the solution, such that if observations from one reference station are not available, a solution is still possible since the observations are gathered and processed in a common network adjustment. Figure 1 illustrates a simple demonstration of the concept of NRTK through representation of the relationship between the modelled distance-dependent errors and their actual values. The error planes at the three shown reference stations are at different levels. The NRTK provides an error surface formed from the errors at the three reference stations (a plane in this case). The actual change of error between the reference stations is shown in red. If a user is close to any of the stations, assuming having the same level of error of that reference station will give reasonable accuracy and results in small positioning errors at the rover. As the user moves away from the reference station, the magnitude of the differential error between the actual and the reference station error level increases. On the other hand, the differential error between the actual error and the NRTK estimated error, interpolated on the NRTK error surface at the location of the rover when used, is significantly minimised.

In principle, the RTK network approach consists of four basic segments: data collection at the reference stations; manipulation of the data and generation of corrections at the network processing centre; broadcasting the corrections, and finally positioning at the rover utilizing information from the NRTK. In the first segment, multiple reference stations simultaneously collect GNSS satellite observations and send them to the control centre, where a main computer directly controls all the reference stations, mostly via the Internet. All reference stations should use geodetic-grade multi-frequency GNSS receivers. The incoming GNSS observation data from all operating reference stations are screened for blunders and next their ambiguities are fixed. The control computer uses these data in processing a networking solution, and the data are archived for post-processing use. The network information are then broadcast to users. The network information depends on the processing algorithm and may include any of the following: observations from one reference station (physical or virtual), coefficients for interpolation of corrections within the coverage area, and observation corrections at a group of reference stations. To increase reliability, it is recommended to let a second computer work in real time as a backup to the main computer in the event of any malfunctioning.

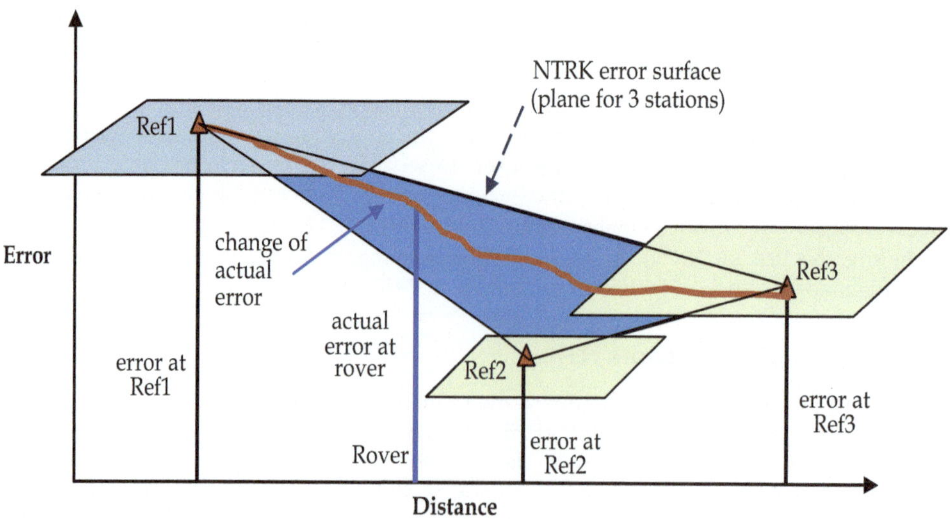

Fig. 1. Relationship between errors in a small NRTK coverage area

NRTK usually requires a minimum of three reference stations to generate corrections for the network area. In general there is no restriction concerning the network size, it can be regional, national, or even international. However, reference station separation is usually restricted to less than 100 km to allow for quick and reliable ambiguity resolution. As the number of stations increases, redundancy increases, and better corrections can be estimated. If one or two reference stations fail at the same time, their contribution can be eliminated from the solution and the remaining reference stations can still provide the user with corrections and give reliable results (El-Mowafy *et al.*, 2003, Hu *et al.*, 2003). Typically, a NRTK server system would consist of the following components (e.g. Leica Geo systems, 2011):

- A site server connected to each reference station receiver,
- A network server that acquires the data from the site servers and sends it to the processing centre,
- A cluster server that hosts the network processing software. The software performs several tasks including: quality check of data, apply antenna phase centre corrections, ambiguity fixing, modelling and estimation of systematic errors, interpolation of errors (corrections) in some techniques (e.g. VRS, PRS) and generation of virtual observations, model coefficients in others (FKP), or Mac data.
- A firewall is usually established to protect the above servers from being accessed by a user.
- RTK proxy server to deal with requests from the users and send back network information.
- The user interface to send/receive data from the NRTK centre.

The main advantages of the Network RTK can be summarised as follows:

- Cost and labour reduction, as there is no need to set up a base reference station for each user.
- Accuracy of the computed rover positions are more homogeneous and consistent as error mitigation refers to one processing software, which uses the same functional and stochastic modelling and assumptions, and use the same datum.
- Accuracy is maintained over larger distances between the reference stations and the rover.
- The same area can be covered with fewer reference stations compared to the number of permanent reference stations required using single reference RTK. The separation distances between network stations are tens of kilometres, usually kept less than 100 km.
- NRTK provides higher reliability and availability of RTK corrections with improved redundancy, such that if one station suffers from malfunctioning a solution can still be obtained from the rest of the reference stations.
- Network RTK is capable of supporting multiple users and applications.

Network RTK has though some disadvantages, which are:

- The cost of subscription with a NRTK provider.
- The cost of wireless communication with the network (typically via a wireless mobile using for instance GPRS technology).
- The dependence on an external source to provide essential information.

3. Network design parameters

Establishing a network RTK usually starts after a thorough cost/benefit analysis. At the design stage, the following main factors should be considered:

1. Baseline lengths (distances between the reference stations) station locations and network configuration (number and geometric distribution of reference stations).
2. The communication method between the computing centre and the user.
3. Calculations required by the network control and by the user (network algorithm).

These factors are discussed in the following sections.

3.1 Distance between the reference stations and network configuration

The main advantage of the network approach is that it improves modelling of the distance-dependent errors over long distances (El-Mowafy *et al.*, 2003 and Euler *et al.*, 2004). The observation corrections (computed as the same value of the errors but with opposite sign) can be generated after removing cycle slips and determining double differenced phase ambiguities between the reference stations. A major technical challenge in NRTK is ambiguity resolution within a reasonable short period over such large distances between reference stations. In order to achieve a fast and reliable ambiguity resolution, the distance between the reference stations is better chosen not to exceed 100 km (Wübbena and Willgalis, 2001). Typically, baseline lengths in NRTK range between 20 km and 100 km (70 km on average).

In principle, a minimum of three stations is required to generate RTK network corrections, but in practice this number should not be less than five. The increased redundancy of reference stations improves positioning accuracy and ambiguity resolution and helps to sustain network availability and reliability in the case of a temporary failure of any reference station. However, the degree of redundancy should be evaluated by means of a cost/benefit analysis, balancing the need to improve the economical aspects of establishing and running the network and keeping the required degree of redundancy. Hence, selection of baseline lengths by the network designers should satisfy the following conditions:

- covering of the whole area of interest with reliable corrections;
- maintaining sufficient station redundancy;
- achieving a reliable ambiguity resolution with acceptable confidence level at every location within the network area as long as a minimum of five satellites are observed;
- Ensuring reliable communications between the reference stations and the network centre (mostly via the internet through land lines, and in remote areas through satellite communication);
- choosing sites free from multipath and radio frequency interference. It is also preferable to have the references stations situated at a similar altitude.

In network configuration, the following can be taken into consideration:

- For a limited number of reference stations, it is recommended to shape the network as a polygon with one or more central stations.
- A compact shape of the network is preferable (i.e. a circular network is better than a rectangular network).

- Some geographic regions (i.e. equatorial or high latitude) will require a denser network than in the mid latitudes due to poorer-satellite geometry, satellite availability, and ionosphere disturbance etc.

In practice, a main factor affecting the choice of station distances and network configuration is finding suitable sites for the reference stations. The main considerations are availability of communication infrastructure and obtaining approval of site owners (whether a government or a private sector).

It is also possible to integrate observation corrections estimated from networks of different sizes. For instance, errors of regional or even global nature, such as satellite orbital errors, clocks and the regional behavior of the ionosphere, which slowly change, can be estimated from regional networks. The local ionospheric and tropospheric errors on the other hand can be estimated from local networks. Thus, RTK networks can be configured such that areas of heavy usage can be covered by a close-meshed reference station network for highest accuracy and reliability in positioning, whereas less important areas are covered by a wide-meshed network of regional or national extension (Wübbena *et al.*, 2001).

3.2 Communication method between the processing centre and the user

Real-time applications require a communication link between a service provider and the user. Currently, there are two main modes of communication that can be used in network RTK; either a duplex (bi-directional) communication or one-direction communication. Each method has its advantages and disadvantages. In choosing which communication method to use, the designer has to consider economical factors such as: operational cost by the user, cost of maintenance of existing infrastructure and/or building new one, and the amount of computations needed by the rover and the processing centre. The technical aspects that need to be addressed include:

- expected signal strength at different locations,
- number of users,
- range and coverage (Wu, 2009) ,
- transmission bandwidth,
- protocol,
- reliability and error correction,
- latency (one second and shorter data transmission latencies are required for cm level positioning accuracy).

At present, the duplex communication mode is the mostly used method. In this mode, a cellular modem such as a General Packet Radio Service (GPRS) or Global System for Mobile Communications (GSM) are used. GPRS is usually preferred as it is more economical than GSM since the user only pays for the data packets received, not for the entire call duration when using GSM. GPRS can provide a stable and reliable connection with latencies less than one second (Hu et al., 2002). The duplex approach has a restriction on the number of users, as this number is limited by the ability of the NRTK processing centre to simultaneously perform calculations for all users. This may also result in extended latency in receiving the network information. For a limited number of users this latency is usually less than three seconds. On the other hand, the one-direction communication method mainly employs VHF or UHF broadcasting or encodes the RTK corrections into a broadcast TV audio sub-carrier signal (Petrovski *et al.*, 2001). For VHF broadcasting, allocation of suitable broadcast radio frequency and obtaining its license is an important issue in the early development of a network RTK. The main advantage of this method is that there is no restriction on the number of users concurrently using the NRTK service. However, the main disadvantage of the method is the high cost of the infrastructure needed to build radio signal repeaters, if needed, to cover the whole area. In addition, some problems can be experienced due to the possibility of receiving signals of varying strength in different locations, and possible frequency jamming. A mix of both communication methods is however possible (Cruddace *et al.*, (2002).

The data transmission from the reference stations to the control centre server and from the control centre server to the user for RTK corrections is mostly carried out via the Network Transport of RTCM via Internet Protocal (Ntrip), BKG, 2011. Ntrip is an open source and can be downloaded from the internet (LENZ, 2004). Ntrip was built over the TCP/IP foundation and is an application - level protocol for streaming GNSS data over the internet. It was first developed by the German Federal Agency for Cartography and Geodesy (BKG). Ntrip uses HTTP and has three components: Ntrip Client, Server and Ntrip Caster. Ntrip is designed for disseminating differential correction data (e.g. in the RTCM-SC104 format) or other kind of GNSS streaming data to stationary or mobile users over the internet. It allows simultaneous PC, Laptop, PDA, or receiver connections to a broadcasting host. Ntrip supports wireless internet access through mobile IP networks like GSM, GPRS, EDGE, or UMTS (BKG, 2011).

To reduce latency, the amount of data transmitted to the rover should be minimised. One possible solution is to change (optimise) the update rates for the different parameters to follow their physical behaviour. Distance dependent errors can thus be separated into a dispersive component, consisting mainly of the ionospheric refraction, and a non-dispersive component consisting of the tropospheric refraction and orbit errors. Different proposals for optimising the update rates have been made. An update rate of 15 seconds seems reasonable for non-dispersive correction differences, while an update rate of only 10 seconds may be sufficient for the dispersive contribution (Euler *et al.*, 2004). However, the impact of these rates on the Time-To-First-Fix (TTF) of carrier phase ambiguities should be carefully studied, as it lies at the top of the user interests (El-Mowafy, 2005).

The type of communications used also affects the network algorithm and the amount of calculations required at the processing centre and by the user. For instance, if a bi-directional communication is used, the processing centre can individualise the network information for a user based on his/her approximate location. Thus, the computations made at the user receiver are minimised. On the other hand, if the data link is one-directional, the user has to make the necessary interpolation of errors at his location and has to identify a suitable reference station to use.

3.3 NRTK solution methods

Currently several solution methods can be applied in Network RTK, including the Virtual Reference Station (VRS), Pseudo-Reference Station (PRS), individualised Master-Auxiliary corrections (iMAX), Area-Parameter Corrections (Flächenkorrekturparameter -FKP- in its German origin), and the Master-Auxiliary (MAC) method. In VRS, PRS and iMAX referencing is made to a non-physical reference station located in the vicinity of the approximate position of the rover and virtual observations are generated to refer to this non-physical reference station. The user typically has no information about the size of errors and their behaviour. In contrast to the non-physical network approach, FKP and MAC broadcast raw reference station observations and network information separately. The network information is represented by dispersive and non-dispersive corrections and the rover software decides how the network information is applied. A summary of these methods is given in section 5.

Once the network errors are computed at the reference station, distance-dependent errors need to be interpolated at the location of the user receiver. Several methods can be used for such interpolation process including: the use of linear interpolation, using a linear combination model, applying an inverse-distance linear interpolation or a low-order surface model (used for example in the FKP technique), utilisation of the least-squares collocation approach, or using Kriging techniques (see for instance Fototpoulos, 2000, Dai *et al.*, 2001, Wu, 2009 and Al-Shaery *et al.*, 2010).

4. Estimation of the dispersive and non-dispersive errors at the network reference stations

The mathematical equation of the code and phase observations for the receiver (j) and the satellite (s) at time (t) can be written as:

$$C_j^s = \left|\vec{R}_j^s\right| + \frac{\vec{R}_j^s}{\left|\vec{R}_j^s\right|}\delta\vec{r}^s + c(\delta t_j - \delta t^s) + T_j^s(t) + \delta T_j^s(t) - I_j^s(t) - \delta I_j^s(t) + p_j^s + \varepsilon_{Pj}^s \quad (1)$$

$$\phi_j^s = \left|\vec{R}_j^s\right| + \frac{\vec{R}_j^s}{\left|\vec{R}_j^s\right|}\delta\vec{r}^s + c\,(\delta t_j - \delta t^s) + T_j^s(t) + \delta T_j^s(t) - I_j^s(t) - \delta I_j^s(t) + \frac{c}{f}N_j^s + p_i^s + \varepsilon_{\phi j}^s \quad (2)$$

Where:

C_j^s, ϕ_j^s	code and phase observations, respectively;
$\left\|\vec{R}_j^s\right\|$	geometric range between the user's antenna and the satellite;
$\delta\vec{r}^s$	orbit error;
c	speed of light;
δt_j, δt^s	receiver and satellite clock errors, respectively;
$T_j^s(t)$	modelled tropospheric refraction delay (mainly the hydrostatic, dry component of the troposphere);
$\delta T_j^s(t)$	residual tropospheric refraction delay (mainly the unmodelled wet troposphere);
$I_j^s(t)$	modelled ionospheric refraction delay if applied (frequency dependent);
$\delta I_j^s(t)$	residual ionospheric refraction (frequency dependent);
f	signal frequency;
N_j^s	integer phase ambiguity;
p_j^s	total site dependent errors (antenna phase centre and multipath δM_j^s);
ε_{Pj}^s, $\varepsilon_{\phi j}^s$	code and phase remaining random noise, respectively.

From the above equations, one can see that positioning accuracy from GNSS phase observations is limited by two types of errors: the distance dependent errors, which include orbit, ionosphere and troposphere errors, and station dependent errors, which include multipath, antenna phase centre variation, and receiver hardware biases. The network estimation methodology uses the known information of the antennae and site to reduce station related errors and focuses on estimating the distance-dependent errors.

For the station-dependent errors, multipath can be minimised using choke rings and modelling the site specific multipath pattern taking advantage of the fixed reflector to antenna geometry at reference stations and of the daily repeatability of multipath. This can be done utilizing techniques such as the Hilbert Huang transformation to decompose the time-shifted post-fit GPS phase signal residuals (Hsieh and Wu, 2008). Another approach is to include multipath error in the network estimation process, which will average out the uncorrelated multipath errors. To minimise the antenna phase centre variation, the definition of the network reference stations antennae has always to be consistent. This can be done by using the same antenna model type for all reference stations and unifying antenna orientation. To eliminate the phase centre variation, an absolute calibration of each antenna is recommended. However, most current networks only apply relative calibration of the antennae, which is a standard calibration process that can be applied for the type of antenna used, determined relative to a reference antenna (typically a Dome Margolin Model T with choke ring).

The distance dependent errors can be separated into a dispersive component (i.e. frequency dependent), which is the error induced by the ionosphere, and a non-dispersive component, that include orbital and tropospheric errors. Estimation of the dispersive and non-dispersive errors at the network reference stations can be performed in several ways. In one approach the state of individual GPS errors in real time can be estimated by processing all stations of the network simultaneously using un-differenced observables (Wübbena and Willgalis, 2001, Zebhauser et al., 2002, Wübbena et al., 2005). Then, the state vector ($\vec{X}$) at station j reads:

$$\vec{X} = \left(N_j^s,\ \delta t_j,\ \delta t^s,\ \delta\vec{r}^s,\ \delta T_j^s,\ \delta r_{j\,I}^s,\ \delta M_j^s\right)^T \tag{3}$$

The orbital and tropospheric errors are combined to form the geometric (non-dispersive) error $\delta r_{j\,o}^s$, the ionospheric dispersive error $\delta r_{j\,I}^s$ (replacing the terms I_j^s and δI_j^s and in Equations 1 and 2). The state space approach has some advantages; the main one is its ability to constrain each bias by specific models (Wübbena and Willgalis, 2001). Also, a change in the network configuration caused by the breakdown of one of the reference stations can be compensated without much effort. Moreover, in the case of irregular conditions of one of the state parameters, warnings can be issued to the users.

Another popular method for estimation of network errors is using single difference linear combination of observations. The dispersive and non-dispersive components are determined for satellite *s* and between the reference stations *j* and *k*, using dual-frequency receivers of L1 and L2, as follows:

$$\delta\Delta r_{jk_{I,1}}^s = \left(\frac{f_2^2}{f_2^2 - f_1^2}\,\delta\Delta r_{jk}^s\right)_{L1} - \left(\frac{f_2^2}{f_2^2 - f_1^2}\,\delta\Delta r_{jk}^s\right)_{L2} \tag{4}$$

$$\delta\Delta r^{s}_{jk_{o,1}} = \left(\frac{f_1^2}{f_1^2-f_2^2}\,\delta\Delta r^{s}_{jk}\right)_{L1} - \left(\frac{f_2^2}{f_1^2-f_2^2}\,\delta\Delta r^{s}_{jk}\right)_{L2} \quad (5)$$

5. Summary of network RTK processing techniques

In this section, the most common network RTK techniques used at present are discussed, namely: the VRS, FKP, and Mac.

5.1 The Virtual Reference Station (VRS) method

The VRS technique is currently the most popular NRTK method due to the fact that it does not require changes in the user software, i.e. it is compatible with existing software. The rover applies the standard differential positioning of its observations with observations from a 'virtual' reference station. The distance-dependent errors are computed for each pair of satellites, and for each master-to-another-reference station. The VRS method requires bi-directional communication. The rover sends its approximate position via a wireless communication link (typically a cellular modem in NMEA format) to the network processing centre where computations are carried out for each user (Vollath *et al.*, 2000, Hu *et al.*, 2003). Some network providers use only the nearest three-to-five reference stations to compute the measurement errors for a specific user, see Figure 2. The estimated network measurement distance-dependent errors are interpolated for a virtual reference station (VRS). The VRS location is typically selected at the initial approximate position of the rover. For a kinematic user, this VRS location is kept to preserve the ambiguity values determined from its solution until the range between the VRS and the actual position of the rover becomes too long for precise differential positioning. Then, a new VRS at the most recent position of the user is established.

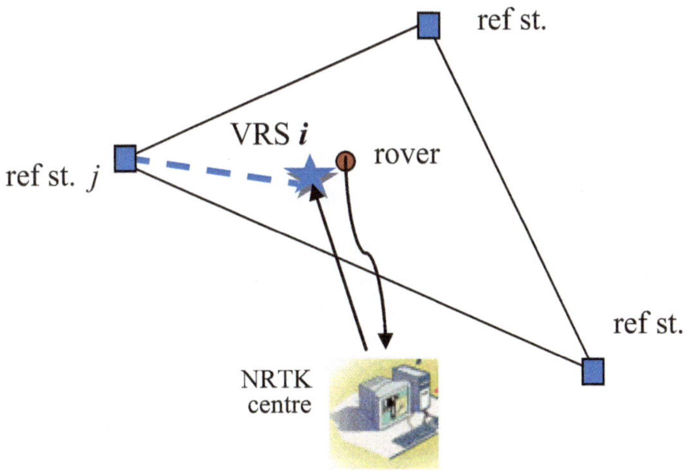

Fig. 2. VRS concept

To construct the observations at the VRS, the VRS and the satellite known positions are firstly used to compute the range between the satellite and the VRS. Similarly, the range between the satellite and the master station is computed, where the master station is usually selected as the nearest continuously operating reference station to the user. The range difference between the VRS (point *i)* and master station (point *j)* with respect to satellite (*s*) reads:

$$\Delta R_{ij}^{s} = R_{j}^{s} - R_{i}^{s} \quad (6)$$

Where:

R_{i}^{s} satellite to VRS range
R_{j}^{s} satellite to the master station range
ΔR_{ij}^{s} range difference

The interpolated distance-dependent errors, dispersive $\delta\Delta r_{ij_I}^{s}$ and non-dispersive $\delta\Delta r_{ij_O}^{s}$ or their total, at the VRS are added to the master station's observations to generate the VRS observations on a satellite-by-satellite and epoch-by-epoch basis for L1 and L2 frequencies, such that:

$$\phi_{ij}^{s} = \phi_{j}^{s} + (\Delta R_{ij}^{s} - \delta\Delta r_{ij_I}^{s} + \delta\Delta r_{ij_O}^{s} + \Delta T_{ij}^{s})/\lambda \quad (7)$$

$$P_{ij}^{s} = P_{j}^{s} + \Delta R_{ij}^{s} + \delta\Delta r_{ij_I}^{s} + \delta\Delta r_{ij_O}^{s} + \Delta T_{ij}^{s} \quad (8)$$

where ΔT_{ji} represents the difference in the modelled part of the troposphere, which is usually subtracted before computation of the network errors, and thus need to be reconsidered. Previous testing of the VRS systems for kinematic ground surveying showed that system positioning accuracy was typically 1-2 cm in plane coordinates and 3-5 cm in height (El-Mowafy *et al.*, 2003). The raw measurements or their corrections are sent to the rover in RTCM version 2.1 format using message types 18/19 or 20/21, respectively. If the former types are used, the constructed VRS observations are computed by the network centre whereas if the latter message types are used, the user has to compute the VRS observations.

A variation to the VRS concept is the Pseudo Reference Station approach (PRS), where the virtual reference station is taken at a pre-selected grid point instead of the approximate position of the user. The virtual observations in this case will also refer to a non-physical reference station. At start of the survey, the baseline length is typically a few metres for the VRS approach and could be several kilometres for the PRS method.

The basic advantages of the VRS mode are that it does not need software changes in the user receiver, and no special formats and conventions are needed. However, a main drawback of the method is the presence of a restriction on the number of users according to the capacity of the network processing centre due to the fact that VRS observations are customised for each user. For kinematic applications, re-determination of VRS may be needed according to the distance between the user and the VRS. In addition, if RTCM message types 18/19 are used, the user will have no information about error sizes, which always helps in interpretation and analysis of positioning results.

5.2 The Area-Parameter Corrections-Flächenkorrekturparameter (FKP) method

The Flächenkorrekturparameter (FKP) or "Area-Parameter Corrections" method represents the network information using coefficients of a surface centred at the location of a physical reference station (Wübbena and Bagge, 2006). Raw reference station observations and network information are broadcast separately. The method requires advanced software in the rover receiver to do interpolation of corrections. However, unlike the VRS approach, the user has information about error sizes, which helps in quality control and analysis of results. The rover software decides how the network information is applied. For instance, the user can apply the corrections at his location to mitigate observation errors and do differential positioning with the broadcast master station data. Alternatively, the user can utilise the network information to construct a VRS at a nearby location, or the user may apply the Precise Point Positioning (PPP) in what is known as PPP-RTK (Teunissen et al., 2010). FKP can apply an open message with one-directional communication (from centre to users) to cover a certain area. In this case, no restrictions would exist on the number of users. A bi-directional communication can also be employed.

In FKP, the residuals at the network reference stations are assumed to define a surface which is "parallel" to the WGS-84 ellipsoid in the height of the reference station. For baselines less than 100 km, the spatial variations of the residuals can be approximated by a low-order surface model, e.g. a plane, using a bilinear polynomial in the form:

$$\delta r(t) = a(t)\,(\phi - \phi_R) + b(t)\,(\lambda - \lambda_R) + c(t) \tag{9}$$

where:

a, b, c coefficients defining the plane at time (t). a and b model the trend of change of error within the area, and c is used for modelling the station specific errors of the master station if undifferenced observations are used or the averaged value of the station specific errors of all the stations if double difference observations are used (Wu et al., 2009).

ϕ, λ geographic coordinates of the interpolated point (in radians)

ϕ_R, λ_R geographic coordinates of the reference point (in radians)

The coefficients are estimated from a weighted least squares solution from the computed residuals at each reference station using Equations 3 or 4 and 5. For instance, for *n* number of reference stations we have:

$$\begin{bmatrix} \delta r_{R-1} \\ \delta r_{R-2} \\ \vdots \\ \delta r_{R-n} \end{bmatrix} = \begin{bmatrix} \Delta\lambda_{R-1} & \Delta\phi_{R-1} & 1 \\ \Delta\lambda_{R-2} & \Delta\phi_{R-2} & 1 \\ \vdots & \vdots & \vdots \\ \Delta\lambda_{R-n} & \Delta\phi_{R-n} & 1 \end{bmatrix} \begin{bmatrix} a \\ b \\ c \end{bmatrix} \tag{10}$$

where $\Delta\lambda_{R\text{-}j}$ and $\Delta\phi_{R\text{-}j}$ are the difference in latitude and longitude between the reference station R and station j, respectively.

The least-squares estimates for the coefficients can be obtained by (Wu, 2009):

$$\begin{bmatrix} \hat{a} \\ \hat{b} \\ \hat{c} \end{bmatrix} = (A^T A)^{-1} A^T \,\widetilde{\delta r} \tag{11}$$

Where:

$$A = \begin{bmatrix} \Delta\lambda_{R-1} & \Delta\phi_{R-1} & 1 \\ \Delta\lambda_{R-2} & \Delta\phi_{R-2} & 1 \\ \vdots & \vdots & \vdots \\ \Delta\lambda_{R-n} & \Delta\phi_{R-n} & 1 \end{bmatrix} \text{ and } \widetilde{\delta r} = \begin{bmatrix} \delta r_{R-1} \\ \delta r_{R-2} \\ \vdots \\ \delta r_{R-n} \end{bmatrix} \tag{12}$$

Typically in FKP, two planes are computed for each satellite, centred at each reference station, one plane for the dispersive and another for the non-dispersive corrections. The corrections at the rover are determined through interpolation using the inclination parameters of the correction planes. Results of Euler *et al.*, 2002, showed that the plane surface model gave good results when modelling the regional trends of the correction differences. Although low-order (linear-plane) surface models are usually utilised, longer baselines between reference stations may require polynomials of a higher order.

The surface area model was discussed in several publications, e.g. Varner, 2000, Fotopoulos and Cannon, 2001 and Wübbena and Bagge, 2002. An example is given by the latter study for generation of FKP, where the errors are computed as follows:

$$\delta r_o(t) = 6.37\,(FKP_{N_o}\,(\phi - \phi_R) + FKP_{E_o}\,(\lambda - \lambda_R)\cos(\phi_R))\,(t) \quad (13)$$

$$\delta r_I(t) = 6.37\,\alpha\,(FKP_{N_I}\,(\phi - \phi_R) + FKP_{E_I}\,(\lambda - \lambda_R)\cos(\phi_R))\,(t) \quad (14)$$

$$\alpha = 1 + 16\,(0.53 - \theta/\pi)^3 \quad (15)$$

Where:

δr_o estimated non-dispersive geometric (orbital and tropospheric) error
δr_I estimated dispersive ionospheric error
FKP_{N_I} The FKP parameter in north-south direction for the ionospheric signal "narrow lane" in ppm
FKP_{E_I} The FKP parameter in east-west direction for the ionospheric signal "narrow lane" in ppm
FKP_{N_o} The FKP parameter in north-south direction for the geometric signal "ionosphere-free" in ppm
FKP_{E_o} The FKP parameter in east-west direction for the geometric signal "ionosphere-free" in ppm
θ the satellite elevation angle in radians

After interpolating the dispersive and non-dispersive errors, they are combined to generate the range residuals for L1 and L2 frequency observations, which read:

$$\delta r^s_{i\,1} = \delta r_o + \frac{f_2}{f_1}\,\delta r_I \quad (16)$$

$$\delta r^s_{i\,2} = \delta r_o + \frac{f_1}{f_2}\,\delta r_I \quad (17)$$

Where:

$\delta r^s_{i\,1}, \delta r^s_{i\,2}$ total measurement errors for the frequencies f_1 and f_2.
f_2, f_1 frequencies of L_1 and L_2 signals.

Finally, the range R, derived from the carrier-phase measurements are corrected as follows:

$$R^s_{i\,1corrected} = R^s_{i\,1} - \delta\, r^s_{i\,1} \quad (18)$$

$$R^s_{i\,2corrected} = R^s_{i\,2} - \delta\, r^s_{i\,2} \quad (19)$$

The drawbacks of the FKP method include the need of the rover to perform interpolation of measurement corrections, possible inconsistency at the edge of two adjacent planes due to the use of the linear plane surfaces, and large data formats are needed. In Radio Technical Commission for Maritime services (RTCM) format version 3.1, FKP corrections can be sent via message types 1034 and 1035 for GPS and GLONASS observations, respectively.

5.3 The Master-Auxiliary (MAC) method

In the Mac approach, the rover sends its approximate position via NMEA format to the network processing centre. The centre determines for this specific user the appropriate master station, which is usuall selected the closest reference station to its position, and identifies the auxiliary reference stations. These stations are chosen within a catch circle of a predefined radius (e.g. 70 km) around the rover, and with a pre-set number (e.g. from 3 to 14). Figure 3 illustrates the Mac concept. In one Mac approach, a network RTK of large number of reference stations can be subdivided into clusters (Leica Geo systems, 2011). The processing centre defines the appropriate cluster to a user and defines the appropriate network corrections applicable to that user.

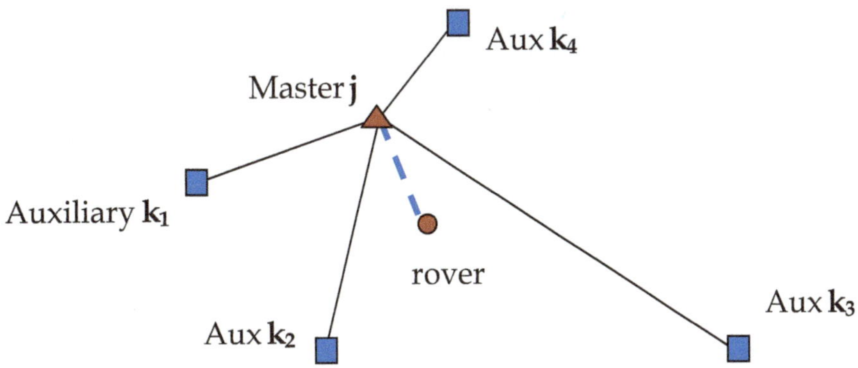

Fig. 3. The Master-Auxiliary NRTK

The rover can receive different types of information according to the strategy used by the Mac processing centre which may include:

- The coordinates and raw measurements of the Master station.
- Measurement corrections at the Master station.
- Correction differences between the Master and Auxiliary stations. These differences when being added to the corrections of the Master will give the corrections at the Auxiliary stations.

The latter Mac corrections can be received via RTCM v3.1 message types 1014-1018, 1030-1031 and 1034-1035 etc.

In Equations (7 and 8), the single observation differences between the Master station *j* and an Auxiliary station *k* for satellite *s* reads (Takac and Lienhart, 2008):

$$\phi_{jk}^{s} = \phi_{j}^{s} + (\Delta R_{jk}^{s} - \delta\Delta r_{jk_I}^{s} + \delta\Delta r_{jk_o}^{s} + \Delta T_{jk}^{s})/\lambda \tag{20}$$

$$P^{s}_{jk} = P^{s}_{j} + \Delta R^{s}_{jk} + \delta\Delta r^{s}_{jk_{I}} + \delta\Delta r^{s}_{jk_{o}} + \Delta T^{s}_{jk} \tag{21}$$

One characteristic of the Mac approach is that its data are sent to the user at the same ambiguity level. This can be explained as follows. In the Mac method the carrier phase ambiguities are determined with respect to fixed single difference ambiguity values. However, ambiguity fixing is more reliably performed using a double difference approach. Thus, the ambiguities from the satellite *s* can be determined from that of the reference satellite *q* and their double difference as follows:

$$N^{s}_{kj} = N^{q}_{kj} + N^{q,s}_{kj} \tag{22}$$

Therefore, the ambiguity bias, which is the difference between the true ambiguity and the estimated ambiguity for the reference satellite, usually known as the ambiguity level, should be estimated. It is common to all estimated ambiguities of satellites observed from one baseline and cancels out in double differencing.

After receiving the Mac information, the rover software is free to decide the method of interpolating the corrections at its location. The processing centre can do the interpolation if needed (individualised I-Max). The rover software is also free on how the Mac information be used to determine its position. For instance, the rover can apply double differencing with the Master reference station as the base. It can also do that after removing the errors from both the Master reference station and its position.

6. PPP- RTK

A more recent direction of NRTK implementation is its integration with the precise point positioning (PPP) technique, Wübbena *et al.*, 2005. In a standalone PPP mode, undifferenced observations are used and the satellite related errors are mitigated by using satellite clock corrections and utilising precise orbits to avoid the orbital errors associated with the use of broadcast ephemeris. These satellite products are typically provided from a processing centre analysing global data such as the International GNSS Service (IGS). Since only one receiver is used in PPP, the ambiguities are solved as part of the unknowns with real numbers and not fixed. As a result, several minutes of data are needed when processing to achieve a reliable convergence of the solution. As the ambiguities are solved as real numbers, only accuracy at the sub-decimetre, at best, is achievable from PPP. However, it is possible to integrate PPP and NRTK into a seamless positioning service, which can provide an accuracy of a few centimetres (Li et al., 2011). The concept of PPP-RTK is to augment PPP estimation with precise un-differenced atmospheric corrections and satellite clock corrections from a reference network, so that instantaneous ambiguity fixing is achievable for users within the network coverage.

A few techniques have been yet proposed for PPP-RTK. In the method presented in Teunissen et al., 2010, un-differenced observation equations for the network stations are used, and thus the design matrix of the network will show a rank defect. This rank defect is eliminated through an appropriate reparametrization (i.e. reduction and redefinition of the unknown parameters). This results in redefined satellite clocks and ambiguities. The tropospheric delay is lumped with the phase and pseudo range satellite clock errors and the ambiguity becomes a between receiver single-differenced ambiguity. Eventually, a full-rank system of observations can be obtained.

In PPP-RTK, the function of the network is to provide the user with satellite clocks and interpolated ionospheric delays. When these precise estimates are passed on to the user, the above given definition of these clocks ensures that the ambiguities of the user are also integer and ambiguity resolution is available at the user side. Satellite clocks for each epoch are added as pseudo-observations, with appropriate variance matrix. The precise IGS orbits are used. For the network processing in Teunissen et al., 2010, a Kalman filter is used, assuming the ambiguities are time-invariant, while for the user, an epoch-by-epoch least-squares processing is used, thus providing truly instantaneous single-epoch solutions. The integer ambiguity resolution of both network and user is based on the LAMBDA method (Teunissen, 1995), with the Fixed Failure Ratio Test (Teunissen and Verhagen, 2009).

7. Network RTK applications

In this section important applications that can benefit from the few cm-level positioning precision and accuracy achievable by using a single GNSS receiver with NRTK are presented.

7.1 NRTK in engineering surveying

Surveying works in construction sites are usually dependent on determination of accurate coordinates and heights. The 3D positioning versatility and accuracy achievable from NRTK encourages the use of this technique for construction surveying works, particularly for large sites when a rapid survey is needed. The method helps in reducing field expenses and time due to reduction of the size of surveying crew, elimination of the need for frequent setups of the surveying instruments, and the reduction of the need for accurate local traverses or multiple control stations within the site. Studies showed that RTK GPS and the traditional techniques employing total stations gave statistically compatible results (El-Mowafy, 2000).

With a typical accuracy of 1-5 cm, the NRTK GPS technique can be utilised for medium accuracy construction survey works such as:

- grading,
- staking out of marks with medium accuracy, such as roads, footings, pipelines, utilities, landscaping, fences etc.,
- cadastral surveys,
- mapping,
- checking of the as-built structures,
- site exploration for new projects.

A NRTK GNSS system can be integrated with the total station for instantaneous determination of the total station location by mounting the GNSS antenna directly on top of the total station alidade in open sites. Thus, the need for establishing permanent horizontal control stations onsite can be minimised. For orientation determination, the total station can be sighted at a back station, where its coordinates can be instantaneously determined using the NRTK-GNSS technique. This process improves the economics of surveying work, and reduces the overall surveying time, including the time required for the initialisation of the total station at each setup. However, one should note that performance of surveying with RTK GNSS in construction sites are affected by satellite availability, multipath errors resulting from working near buildings, and latency of the reference data. The influence of the number of satellites in view, dilution of precision (DOP) and age of corrections over the accuracy and stability of the NRTK GPS solution was discussed in some studies, e.g. Aponte et al., 2009.

The performance of surveying with the network RTK approach in construction sites was evaluated by two tests. The first test was executed for checking performance in determination of planimetric coordinates and the second test for evaluation of performance in height determination. The first test was carried out during construction of a large building in Dubai for checking of positions of surveying marks of the footings, landscaping and the access road of the building. 48 points were used for checking purposes including 18 points on the boundary of the road, 19 points for the footings, and 11 points for the landscaping. These points were set out using a calibrated total station of 1 second precision. Next, point coordinates were computed from the working drawings and uploaded to a GPS controller, where a single GPS dual-frequency receiver was independently used for positioning of the test marks by utilizing data from the Dubai NRTK, known as Dubai Virtual Reference System (DVRS). The network consists of five continuously operating reference stations with baseline lengths ranging from 23.4 km to 90.8 km and uses the VRS algorithm. The position of each test point was determined after 10 seconds of data collection, which were recorded at 2 seconds interval. The shifts between the positions determined from the two methods, namely: calibrated total station and NRTK using observations from GPS, were measured and compared to represent the precision of the latter method if used in construction sites instead of the former method. Figure 4 illustrates the differences between the two methods. The statistics of coordinate differences between the two methods in easting and northing are given in Table 1. The average norm of the spatial differences ($\sqrt{(diff_E)^2 + (diff_N)^2}$) was generally less than 1.45 cm whereas the maximum difference was 3.45 cm. The small errors can be explained by the presence of one of the network reference stations within a few kilometres, which would be picked up by the system as the master reference station. Thus, most orbital and atmospheric errors were cancelled and the remaining errors would mainly be due to data noise and multipath. These results show that the GPS-RTK network approach can be used in the setting out of medium-accuracy surveying marks.

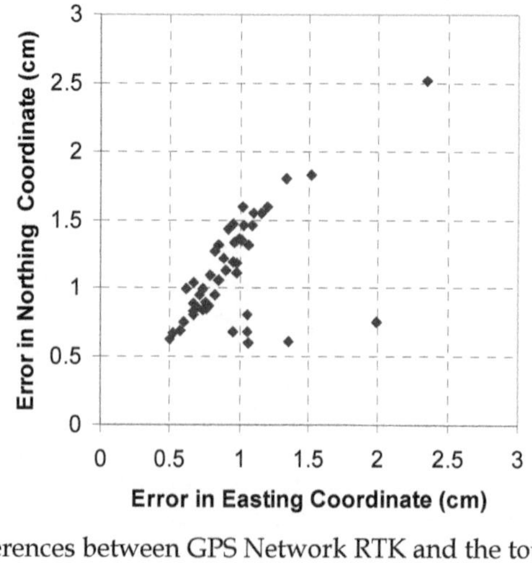

Fig. 4. Positioning differences between GPS Network RTK and the total station

	Average (cm)	Maximum (cm)	RMS (cm)
E	0.95	2.35	1.00
N	1.12	2.52	1.18

Table 1. Statistics of positioning differences between GPS RTK-Network with the total station

To check repeatability of results (internal accuracy), another independent survey was carried out after two days and 4 hours to re-determine the coordinates of the same marked points previously determined by GPS. The average and maximum values of the differences between the results of the two surveys are given in Table 2. As the table demonstrates, repeatability testing showed that the average value for differences in the total planimeteric coordinate estimation was at the cm level, while for ellipsoidal height determination it was 1.56 cm. These differences can be attributed to changes in the quality of the measurements used, which mainly resulted from differences in the number of the observed satellites and their geometric distribution. These parameters have affected the quality of the network computations of the measurement corrections and the quality of coordinate estimation at the rover.

	Average (cm)	Maximum (cm)
E	0.85	1.23
N	1.15	1.88
h	1.56	3.22

Table 2. Statistics of coordinate discrepancies between different observing sessions

Unlike traditional levelling, GPS derived heights are referenced to an ellipsoidal datum (WGS 84) and do not depend on local gravity variations, whereas in most levelling works and mapping orthometric heights are used. Orthometric heights reflect changes in topography as well as local variations in gravity. They are referenced to the geoid, which is an equi-potential level surface of the Earth that is closely associated with the mean sea level on a global basis. To convert ellipsoidal heights from GPS (h_{GPS}) into orthometric heights (H), geoid heights are needed, such that:

$$H = h_{GPS} - N \tag{23}$$

where (N) is the geoid height. Thus, with the use of one receiver and employing NRTK to determine ellipsoidal heights, orthometric heights can be determined if a good geoid model is available.

To assess accuracy of orthometric height determination by using NRTK, the second test was performed in Dubai, using the DVRS network. The Dubai gravimetric geoid model was used, which was developed by integrating a comprehensive set of gravity measurements with GPS, levelling and digital elevation data. The computed geoid fits GPS/levelling at the 3-4 cm level RMS (Forsberg *et al.* 2001). The test was performed on a network consisting of 41 benchmarks of the second order levelling network. Orthometric heights at these benchmarks were first estimated by combining ellipsoidal heights determined by using the Dubai NRTK with the local gravimetric geoid model data and were next compared to known orthometric heights of the benchmarks. The test area spanned approximately 22.7 km x 7.8 km in the Easting and Northing directions respectively, representing the area acquiring the most demanding survey works in the Emirate of Dubai. The height difference between the highest and lowest points in this test was approximately 34.5 meters. Each test point was occupied for a period of a few seconds, representing an ordinary working environment. The standard deviations of the ellipsoidal height determination for the occupied points of the test network ranged between 1.05 cm and 5.47 cm (El-Mowafy *et al.*, 2005).

Figure 5 shows the differences in orthometric heights between using the "NRTK GPS + geoid heights" and the known orthometric heights of the benchmarks. On average, differences were within ±5 cm, with a maximum value of 7.04 cm. The statistical results of the differences are presented in Table 3. The average value of the absolute differences was 2.4 cm with 3.05 cm standard deviation. The differences towards the north-east were greater than those at the south-west region of the test. This can mainly be attributed to accuracy of the geoid model used in the test area. Figure 6 illustrates the surface plot of the height difference between the two methods. These results show that no significant systematic errors were present. The achieved accuracy is considered precise enough for third order levelling, which represents the majority of levelling works being carried out.

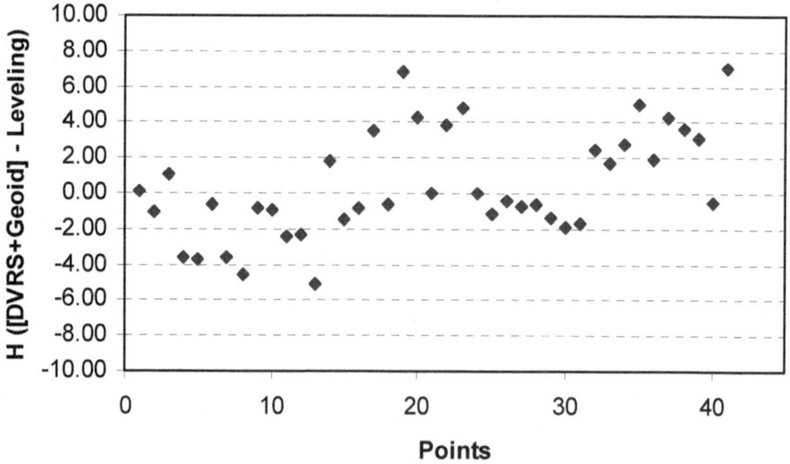

Fig. 5. Height differences between GPS Network RTK geoid heights and precise levelling (cm)

Average of absolute values	Max. difference	σ
2.40	7.04	3.05

Table 3. Statistics of height differences between the GPS network RTK + geoid and precise levelling (cm)

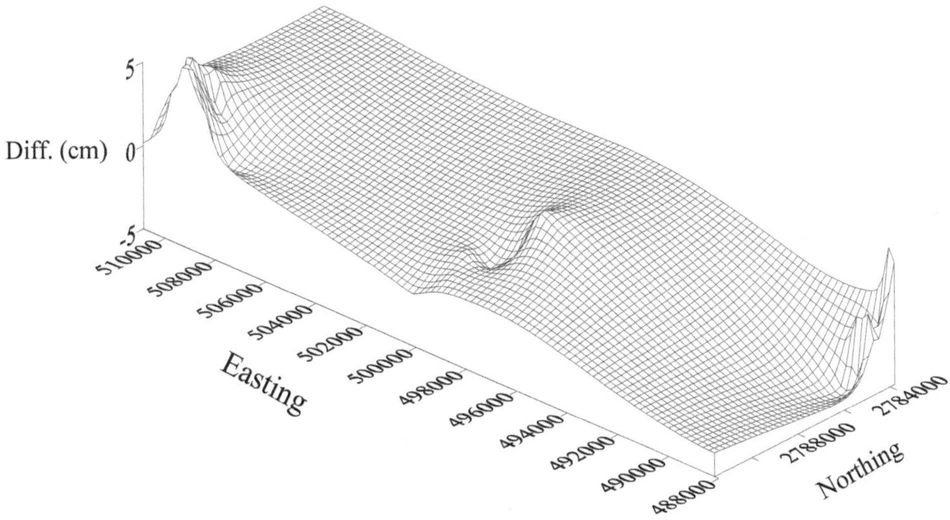

Fig. 6. Surface plot of the height differences

7.2 Using RTK GPS for remotely monitoring and controlling machine automation

The use of Real-time GNSS positioning for machine automation can enhance its productivity and functionality. Real-time GNSS positioning can provide cm accuracy, facilitating high performance and output for machines that require positioning data. The Supervisory Control and Data Acquisition (SCADA) is a good example for supporting a field automated system. Having real-time accurate GNSS positioning information as an input to the SCADA system gives a lot of opportunities to develop many solutions for planning, design, construction and monitoring field operations that require precise positioning.

An automated machine such as a field tractor, excavator, or a driller can be automatically operated, unmanned and fully remotely monitored and controlled. The machine can be controlled by the analogue and digital outputs from a Remote Terminal Unit (RTU). The RTU can have preloaded Programmable Logic Control (PLC) software that activates the outputs according to the field inputs from the machine primary sensors as well as the real-time accurate coordinates fed from a GNSS unit receiving measurement corrections from a NRTK centre. The field operations, events, logs and alarms can be fully remotely monitored through a SCADA system. For the SDADA system programming software, standards such as IEC1131-3 can be used, which is the international standard for controller programming languages. It specifies the syntax, semantics and display for the PLC programming languages. An open standard communication protocol such as Distributed Network Protocol (DNP) can be used. DNP is a set of standards and interoperable communication protocol used between companies in processing automation systems. These open source standards and interoperable platforms will allow the system design to be implemented in any of the industrial proven commercial SCADA systems.

The functionality and process automation of the system can be described in two scenarios. In the first scenario, the machine can be operated in a semi-automated mode with online telemetric control. The real-time GNSS positioning is computed and a Geographic Information System (GIS) is utilised at a control centre that operates the machine in a telemetric mode. In the second scenario, the system can be fully automated based on pre-set instructions and automated integration with GIS planning, database and geo-coded maps. In this case, the GIS system is uploaded in the field machine.

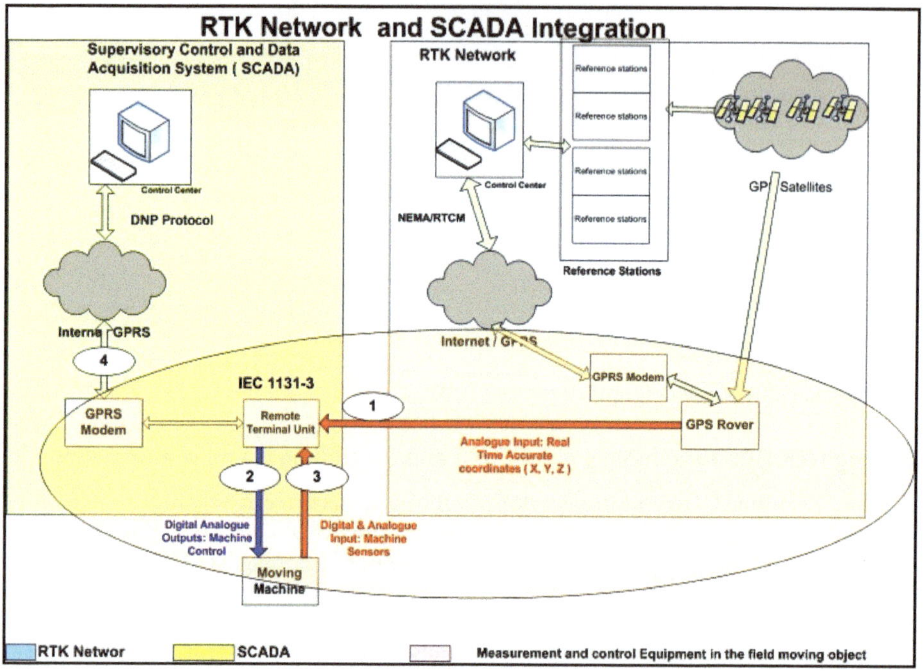

Fig. 7. Proposed network RTK GNSS and SCADA integration

Figure 7 illustrates a developed architecture for the first scenario, where the machine is remotely controlled from a SCADA centre. In this case, SCADA monitoring and controlling hardware and software can be automated in the field with a built-in computer that has all required software and is placed inside the machine. The SCADA system can be integrated with the GIS through an interface where the mimics are truly geo-referenced maps representing the reality of the field in a full 3D mode. The GNSS and the primary sensors are sent through the RTU to the SCADA system to update the GIS map with real-time information. The following procedure can be applied (El-Mowafy and Al-Musawa, 2009):

1. The roving GNSS will be mounted on the field machine and sends its observations to the network centre. Positioning information is computed at the centre for telemetric control of the machine.
2. The control centre operates the SCADA monitoring and controlling hardware and software. The field automated process is fully monitored and controlled in a remote mode. Report, alarms, trends and historical records can be retrieved at the centre.
3. SCADA system chooses the planned field works based on the 3D accurate coordinates of the field points and using GIS.

4. The centre sends primary information to the field RTU unit that controls the machine. These commands can be either full machine commands or only the main commands that transfer between phases of field operations if detailed information is pre-fed to the RTU.
5. The RTU based on a preloaded automation process program will produce the controlling outputs based on a set of variable inputs from the primary sensors as well as the input of the position information received from the centre.
6. Real-time data from the GNSS and the primary sensors are sent remotely from the field through the RTU to the SCADA system to update it with the actual work progress in a feedback loop.

The logical sequence of the soft PLC process program of the RTU is illustrated in Figure 8 as a flow chart. As the figure depicts, the program starts with initialising and testing the system availability and health status. It reads the current positioning information of the machine as well as the status of the inputs from primary sensors. It executes the RTU process program while monitoring any alarms or malfunction signals for an emergency stop. The RTU events, logs and alarms are sent to the remote SCADA centre.

The machine automation in this scenario can be fully automated in the field. In this case, the GNSS unit mounted on the field machine receives the corrections from the NRTK and continuously feeds the RTU with the real-time positioning information representing the exact location of the machine in the field. The automation system is uploaded on a computer fitted in the machine and all control is pre-programmed and work is executed online in a feedback loop to keep up with the pre-set design. The real-time 3D accurate coordinates of the field points are to be input to a GIS system on board, where its output is fused with the PLC program. Work orders can be downloaded remotely to the field RTU at any stage if a change of plans is required. The RTU can also be connected through a modem to a remote monitoring centre with the ability to control and adjust the field process based on any new input. The field automated process can thus be fully monitored and controlled in a remote mode. Report, alarms, trends and historical records can be archived at the centre.

To investigate positioning precision that can be obtained from network RTK for machine automation, a test was performed in Abu Dhabi. The Abu Dhabi network RTK was used in this test. The network consists of 20 reference stations with separating distances between stations ranging between 60 km to 209 km. Several types of NRTK techniques can be implemented as per user choice, including the VRS method, the MAC approach, the FKP, and standard RTK using a single nearby reference station. The proposed approach was tested in the marine mode for one hour where a Leica 1200 GPS system dual-frequency receiver was mounted on a dredger working in a small island close to Abu Dhabi main island. The NRTK positioning system was operating at a sampling rate of 1 Hz. In addition, the data were internally stored for post-mission processing to act as a reference for comparison with network RTK results to assess its accuracy for the application being tested. The rover data were referenced in this case to station ADCC of the Abu Dhabi continuously operating reference network. The distance between this station and the test trajectory was 6 km on average, giving stable ambiguity fixing with precise positioning output. When comparing the two sets of positioning results (Network RTK and post-mission processing) the differences were at the cm range. The average precision of the determined positions in NRTK mode was 2.85 cm for the horizontal components and 4.1 cm for the height. Statistics of the positioning differences between the two methods are given in Table 4. During testing, availability of NRTK was higher than 95%. These results show that the NRTK can be successfully used for positioning of field machines.

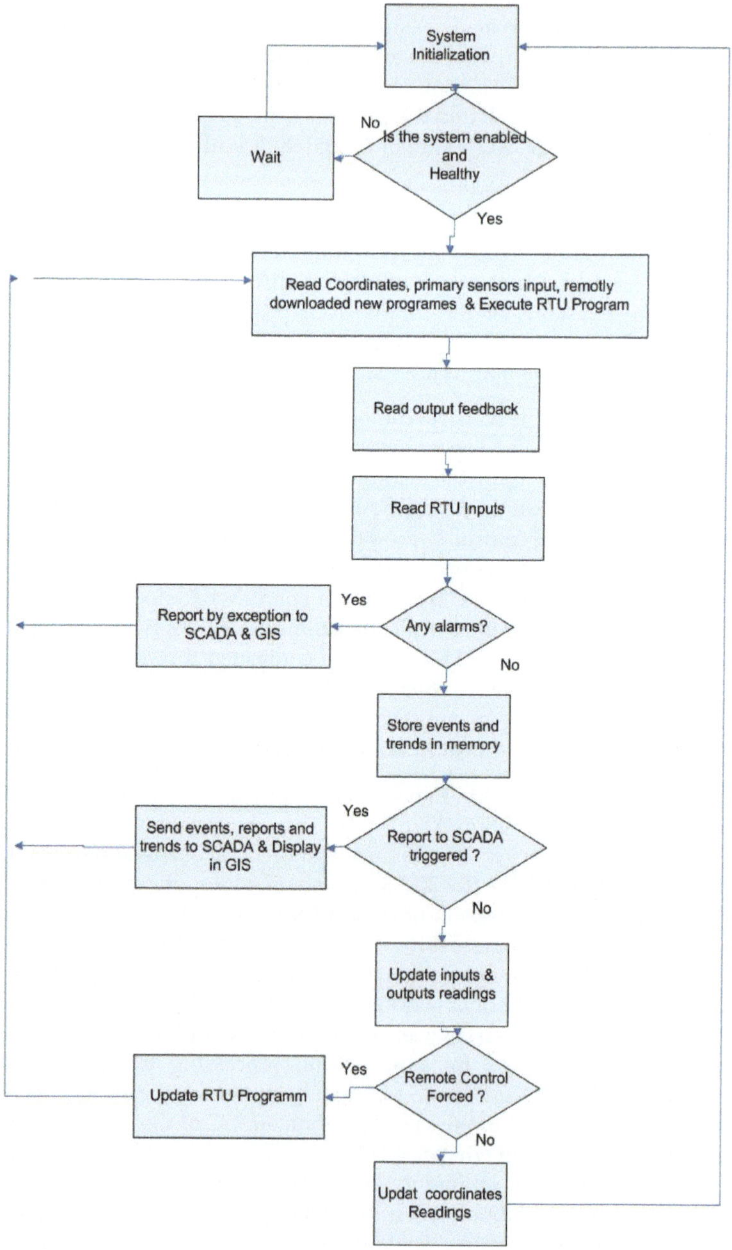

Fig. 8. Flowchart of the RTK GNSS & SCADA/GIS Logic control

S.D	Maximum (cm)	Average (cm)
σ_E	3.60	1.92
σ_N	3.93	2.11
σ_h	9.03	4.10

Table 4. Difference between network RTK and post-mission positioning in machine automation testing

7.3 Using network RTK in the airborne mode

The network RTK approach is mostly used in static or kinematic ground applications. In this section, the use of the NRTK approach in the airborne mode is discussed. At present, positioning by GNSS is a widely used technique in the airborne mode for geo-referencing of aerial mapping data and surveillance by Unmanned Aerial Vehicles (UAV). In aviation, it is estimated that from 2015, most new commercial aircraft will be fitted with GNSS to enhance precise navigation and make it safer (Pedreira, 2009). However, at the moment, GPS is the only approved system as a stand-alone aid for non-precision approaches (Radišić, 2010), e.g. as a supplementary navigation system and for positioning in non safety-of-life applications. This is mainly due to the need to achieve high level of performance in terms of integrity, availability and reliability in the airborne navigation, which GPS on its own cannot reach due to the limited number of satellites available in one site at any particular instance. This situation is expected to improve with the addition of the new systems such as Galileo and Compass. When using network RTK in the airborne navigation, additional concerns have to be addressed, which include:

- Due to the high dynamics involved in the airborne navigation, a high update rate of sending the corrections is needed compared with the rate implemented for land applications. This rate has a direct impact on the Time-To-First-Fix of phase ambiguities, and thus on the overall positioning feasibility and accuracy (El-Mowafy, 2004).
- The format of GPS measurement corrections should be standardised to ensure that the system is independent of any single receiver manufacturer. The use of the RTCM Version 3.1 standards is thus recommended.

The main advantages of using network RTK for precise airborne positioning can be summarised as follows:

- No dedicated ground reference stations are needed for post-mission or real-time applications.
- Unlike standard differential positioning, the distance between the aircraft receiver and the nearest reference station does not present a concern as long as the aircraft flies within the network RTK area of coverage.
- In navigation, due to the fact that networks RTK usually have an area of coverage that extends to several hundreds of kilometres, each network can cover more than one airport, including small airports, unlike the current Local Area Augmentation systems

(LAAS) implemented only in some major airports. NRTK systems can also be used in search and rescue operations, emergency landing, road traffic monitoring from the air, as well as emergency response.
- Compared to LAAS, no significant additional infrastructure cost is involved as the hardware and software of the GNSS-NRTK are available in most developed countries and the establishment of new networks is currently underway or planned in different regions worldwide.
- Network RTK provides cm to decimetre positioning accuracy even in the case of malfunctioning of some reference stations, particularly for dense networks.
- Network RTK can give better runway utilisation by improving airport surface navigation. It can also enhance air traffic management by increasing dynamic flight planning.

The use of the VRS technique in the airborne mode is not generally recommended since in this high velocity environment continuously updated approximate coordinates have to be used for the VRS computation. This is similar to having a moving reference station. A system reset should thus be frequently performed when the VRS coordinates are changing, which will result in frequent initialisation of the carrier-phase ambiguities. Therefore, it is preferable to keep the VRS location for the longest possible range. An alternative approach would be to apply the PRS technique, where the PRS points are chosen along the path of the final approach and close to and at the airport. Furthermore, the duplex communication mode used in the VRS technique is limited by the ability of the processing centre to simultaneously perform calculations for all users. As this number grows, extended latency in receiving the corrections may result. Additionally, the possibility of signal breaks in the duplex communication mode is more than the case of using a one-direction communication. Thus, the use of a one-directional communication method, e.g. applying the FKP method, would be more appropriate for the airborne mode. The PRS and Mac techniques can also be implemented in the one-directional mode, whereby the PRS or the Master-Auxiliary stations are selected to cover a specific area, such as the airport. The establishment of ground transmitters at the airport can improve availability of the corrections.

The feasibility of using real-time reference networks for positioning in the airborne mode was examined using the DVRS NRTK over the city of Dubai. Flight tests using a helicopter and a small fixed-wing airplane were carried out. The trajectory of the fixed-wing aircraft test is illustrated in Figure 9. The main parameters under investigation were the achievable accuracy and availability of VRS measurements. In these tests, aircraft positions were determined using a dual-frequency GPS receiver (Leica SR530). The data were processed in real time at one-second intervals. The DVRS reference stations collect and process data at five-second intervals. Thus, the NRTK data were interpolated in time for the rover receiver to compute positions at the one-second interval.

To assess performance of NRTK approach for this test, the results were compared with positions determined from a standard double-difference technique whereby the observations of the aircraft receiver were stored and processed in a post-mission mode. The aircraft data in this case were referenced to one of the DVRS network stations located within a range of a few kilometres from the flight route. Precise IGS orbits were used in the post-mission processing. The differences between the two methods (NRTK and post-mission processing) are given in Figure 10. For the test at hand, the DVRS data were lost for some periods, which ranged from a few seconds to three minutes. The periods when the DVRS data were available are shown in the dashed areas in Figure 10. The temporary break in reception of the NRTK data can be attributed to the use of GSM signals as the means of communication between the DVRS centre and the aircraft at the time of the test. However, the GSM signals were only used for testing purposes. In practice, the problem of breaks in receiving the network corrections can be significantly alleviated by using more robust means of delivering NRTK service to the aircraft.

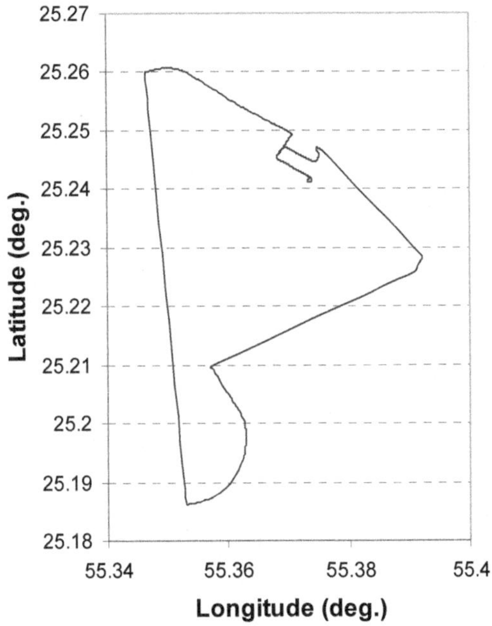

Fig. 9. Trajectory of a fixed-wing aircraft test

When comparing the results of positioning obtained by the DVRS NRTK with the post-mission double-difference positioning for the periods where NRTK data were received and phase ambiguities were fixed, the average 2-D and height positioning discrepancies between the two methods were at a few cm level, as they were 1.6 cm and 2.8 cm respectively. The differences can mainly be attributed to the model assumptions and procedure in the two techniques. During the period when phase ambiguities were only solved in a float solution, the differences were 26.3 cm and 52.5 cm. However, when the DVRS NRTK data were lost, positioning accuracy deteriorated to the metre level. In this case, supporting methods, such as good prediction algorithms and integration with other sensors, e.g. a geodetic-grade inertial system, are needed to cover the short periods when breaks in reception of the measurement corrections take place. Several methods for prediction of NRTK observation corrections as a time series were investigated in El-Mowafy, 2008. Different time-series prediction methods were investigated for different types of errors. The double exponential smoothing prediction approach performed best in most of the cases when studying the satellite clock error corrections. Winters' method and the Autoregressive Integrated Moving Average (ARIMA) model were the best methods for predicting the orbital and wet tropospheric errors, respectively.

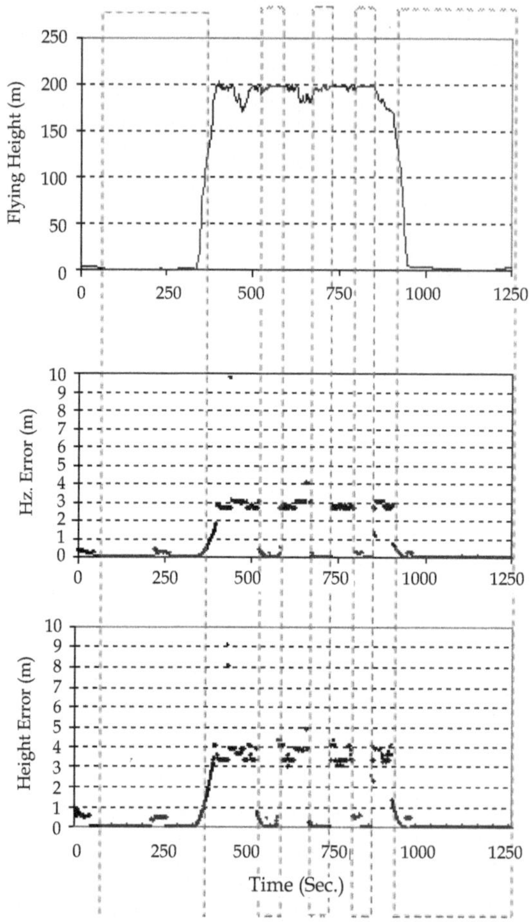

Fig. 10. Fixed-wing aircraft test results

8. References

Al-Shaery, A.M., Lim, S., & Rizos, C. (2010). Functional models of ordinary kriging for medium range real-time kinematic positioning based on the Virtual Reference Station technique, *Proceedings of 23rd Int. Tech. Meeting of the Satellite Division of the U.S. Inst. of Navigation*, , pp. 2513-2521, Portland, Oregan, USA, September 21-24, 2010.

Aponte, J., Meng, X., Hill, C., Moore, T., Burbidge M. & Dodson A. (2009). Quality assessment of a network-based RTK GPS service in the UK. *Journal of Applied Geodesy*, Vol. 3, pp. 25 – 34.

BKG (2011). Networked Transport of RTCM via Internet Protocol. 21.08.2011, Available: http://igs.bkg.bund.de/ntrip.

Cruddace, P., Wilson, I., Greaves, M., Euler, H-J., Keenan, R. & Wüebbena, G. (2002). The Long Road to Establishing A National Network RTK Solution, *Proceedings FIG XXII Int. Congress*, Washington, D.C., April 19-26, 2002.

Dai, L., Han, S., Wang, J. & Rizos, C. (2001). A study on GPS/GLONASS multiple reference station techniques for precise real-time carrier phase positioning, *Proceedings 14th Int. Tech. Meeting of the Satellite Division of the U.S. Inst. of Navigation*, pp. 392-403, Salt Lake City, UT, September 11-14, 2001.

El-Mowafy, A. (2000). Performance Analysis of the RTK Technique in an Urban Environment, *the Australian Surveyor*, Vol. 45, No. 1, pp. 47-54.

El-Mowafy, A., Fashir, H., Al Marzooqi, Y., Al Habbai, A. & Babiker, T. (2003). Testing of the DVRS National GPS-RTK Network, *Proceedings of the 8th ISU International Symposium*, Strasbourg, France, May 25-28, 2003.

El-Mowafy, A. (2004). Using Multiple Reference Station GPS Networks for Aircraft Precision Approach and Airport Surface Navigation, *Proceedings of GNSS 2004, The 2004 International Symposium on GNSS/GPS*, Sydney, Australia, December 6–8, 2004.

El-Mowafy, A. (2005). Analysis of the Design Parameters of Multi-Reference Station RTK GPS Networks, *Journal of Satellite and Land Information Science (SaLIS)*, Vol. 65, No. 1, pp. 17-26.

El-Mowafy, A. (2008). Improving the Performance of RTK-GPS Reference Networks for Precise Airborne Navigation, *Navigation*, Journal of the Institute Of Navigation (ION), Vol. 57, No. 3, pp. 215-223.

El-Mowafy, A. & Al-Musawa, M. (2009). Utilization of GIS and RTK GPS Reference Networks for Machine Automation, *Proceeding of the 6th International Symposium on Mechatronics and its Applications*, Sharjah, UAE, March 24-26, 2009.

Euler, H.J., Townsend, B.R. & Wübbena, G. (2002). Comparison of Different Proposals for Reference Station Network Information Distribution Formats, *Proceedings of the International Technical Meeting, ION GPS-02*, pp. 2334 - 2341, Portland, Oregon. September *2002*.

Euler, H-J., Seeger, S., Zelzer, O., Takac, F. & Zebhauser, B. E. (2004). Improvement of Positioning Performance Using Standardized Network RTK Messages, Proceedings of ION NTM, San Diego, CA, January 26-28, 2004.

Fotopoulos, G. (2000). Parameterization of DGPS Carrier Phase Errors Over a Regional Network of Reference Stations. Master thesis. Department of Geomatics Engineering, University of Calgary, Calgary, Canada.

Fotopoulos, G. & Cannon, M. E. (2001). An Overview of Multi-Reference Station Methods for cm-Level Positioning. GPS Solutions, Vol. 4, No. 3, pp. 1–10.

Forsberg, R., Strykowski, G. & Tscherning, C. C. (2001). Geoid Model for Dubai Emirate, Report No. SP296, Dubai Municipality.

Hsieha, C.H. & Wu, J. (2008). Multipath Reduction on Repetition in Time Series from the Permanent GPS Phase Residuals, *The International Archives of the Photogrammetry, Remote Sensing and Spatial Information Sciences*, Vol. XXXVII, Part B4, Beijing, pp. 911-916.

Hu, G. R., Khoo, V. H. S., Goh, P. C. and Law, C. L. (2002). Internet-based GPS VRS RTK positioning with a multiple reference station network. *Journal of Global Positioning Systems*. Vol. 1, No. 2, pp. 113 - 120.

Hu, G.R., Khoo, H.S., Goh, P.C. & Law, C.L. (2003). Development and Assessment of GPS Virtual Reference Stations for RTK Positioning, *Journal of Geodesy*, Vol. 77, No. 5, pp. 292–302.

Leica Geo. Systems (2011). Using Network RTK, 15.07.2011, Available from: http://smartnet.leica-geosystems.eu/spiderweb/2fNetworkRTK.html.

LENZ, E. (2004). Networked Transport of RTCM via Internet Protocol (NTRIP) - Application and Benefit in Modern Surveying Systems. FIG Working Week 2004, Athens, Greece. May 22-27.

Li, X., Zhang, Z. & M. Ge (2011). Regional reference network augmented precise point positioning for instantaneous ambiguity resolution. *Journal of Geodesy*. Vol. 85, No3, pp. 151-158.

Pedreira, P. (2009). Optimistic Outlook for Galileo, *GIM International*, pp. 6-13.

Radišić, T., Novak, D. & Bucak, T. (2010). The Effect of Terrain Mask on RAIM Availability, *Journal of Navigation*, Vol. 63, No. 1, pp. 105-117.

Petrovski, I., Kawaguchi, S., Torimoto, H., Fujii, K., Sasano, K., Cannon, M.E. & Lachapelle, G. (2001). Practical Issues of Virtual Reference Station Implementation for Nationwide RTK Network, *Proceedings of GNSS 2001, The 5th GNSS International Symposium*, Seville, Spain, 8-11 May, 2001.

Takac, F. & Lienhart, W. (2008). SmartRTK: A Novel Method of Processing Standardised RTCM Network RTK Information for High Precision Positioning, Proceedings of ENC GNSS 2008, Toulouse, France, April 22-25, 2008.

Teunissen, P. J. G. (1995). The least squares ambiguity decorrelation adjustment: a method for fast GPS integer ambiguity estimation, Journal of Geodesy, Vol. 70, No. 1-2, pp. 65-82.

Teunissen, P. J. G. & Verhagen, S., (2009). The GNSS ambiguity ratio-test revisited, Survey Review, Vol. 41, No. 312, pp. 138-151.

Teunissen, P., Odijk, D.J.G & Zhang, B. (2010). Results of CORS Network Based PPP with Integer Ambiguity Resolution, Journal of Aeronautics, Astronautics and Aviation, Series A, Vol. 42, No. 4, pp. 223-230.

Varner, C. (2000). DGPS carrier phase networks and partial derivative algorithms. Ph.D Thesis, Dept. of Geomatics Engineering, University of Calgary, Calgary, Canada.

Vollath, U., Buecherl, A., Landau, H., Pagels, C. & Wager, B. (2000). Multi-base RTK positioning using virtual reference station. *Proceedings 13th Int Tech Meeting Satellite Division, US ION*, Salt Lake City, UT, 19–22 September, 2000.

Wu, S., Zhang, K., & Silcock D. (2009). Differences in Accuracies and Fitting Surface Planes of Two Error Models for NRTK in GPSnet. *Journal of Global Positioning Systems*, Vol.8, No.2, pp.154-163.

Wu, S., (2009). Performance of Regional Atmospheric Error Models for NRTK in GPSnet and the Implementation of NRTK System, Ph.D Thesis, School of Mathematical and Geospatial Sciences, RMIT University, Melbourne, Australia.

Wübbena, G., Bagge, A. & Schmitz, M. (2001). Network-Based Techniques for RTK Applications. *Proceedings the GPS JIN 2001 Symposium, GPS Society, Japan Institute of Navigation*, Tokyo, Japan, November 14-16, 2001.

Wübbena, G. & Willgalis, S. (2001). State Space Approach for Precise Real Time Positioning in GPS Reference Networks. *Proceedings of International Symposium on Kinematic Systems in Geodesy, Geomatics and Navigation*, KIS-01, Banff, Canada , June 5-8, 2001.

Wübbena, G., Schmitz, M. & Bagge, A. (2005). PPP-RTK: Precise Point Positioning Using State-Space Representation in RTK Networks, *Proceedings of the 18th International Technical Meeting of the Satellite Division of The Institute of Navigation ION GNSS 2005*, Long Beach, California, September 13-16, 2005, pp. 2584-2594.

Wübbena, G. & Bagge, A. (2006). RTCM Message Type 59 - FKP for transmission of FKP, Version 1.1, Geo++ GmbH White Paper Nr. 2006.01, 17.7.2009, Available: http://www.geopp.de/download/geopprtcmfkp59- 1.1.pdf .

Zebhauser, B.E, Euler, H-J, Keenan, C.R & Wübbena, G. (2002). A Novel Approach for the Use of Information from Reference Station Networks Conforming to RTCM V2.3 and Future V3.0. *Proceedings of PLANS 2002*, Palm Springs, California, April 15-18, 2002.

9

A Decision-Rule Topological Map-Matching Algorithm with Multiple Spatial Data

Carola A. Blazquez
Universidad Andres Bello
Department of Engineering Science
Chile

1. Introduction

Intelligent Transportation System (ITS) applications such as congestion and traffic management employ Global Positioning Systems (GPS) technology to collect positioning data in two or three dimensions of events, incidents, or vehicles. This information is integrated with Geographic Information Systems (GIS) to determine the roadway upon which events and incidents occur, point features such as traffic signs are located, or vehicles are traveling.

Vehicle trajectories displayed on a digital map are not situated on top of the roadway centerlines, which represent the real world. Therefore, when both GPS measurements and roadway centerline maps are very accurate, a GPS data point is associated with the nearest roadway by calculating the minimum perpendicular distance between each roadway representation and the GPS data point. This process is called "snapping". Unfortunately, a spatial mismatch occurs when a GPS data point is snapped to an incorrect roadway centerline due to roadway network complexities, inadequate GPS data collection procedures, and lack of accuracy in the digital roadway map and the GPS measurements, or combinations of them (Chen et al., 2005). Figure 1 shows an example where errors in the location of the measured GPS data point cause an incorrect snap to the nearest road 2 instead of snapping to road 1.

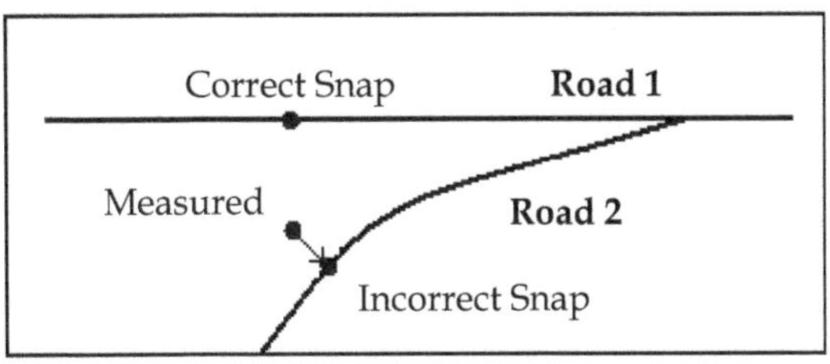

Fig. 1. Measured GPS Data Point with Error Snapped to the Wrong Roadway Centerline

Generally, spatial mismatches or map-matching problems occur at overpasses and underpasses, converging and diverging roadways such as ramps and divided highways, or when roads are close together. Figure 2 presents GPS measurements of a vehicle traveling at a major highway interchange containing ramps, overpasses, and underpasses. This example indicates that multiple spatial mismatches may occur at interchanges.

As a consequence of the map-matching problem, any subsequent usage, visualization, computation, evaluation, analysis, planning, and decision-making may be impacted negatively and produce erroneous perceptions. For example, the calculated cumulative distance traveled by a vehicle along a roadway network is incorrect and, therefore, calculated values for performance measures such as fuel consumption or decision management tools that depend upon cumulative distance are wrong. Additionally, any non-spatial data collected from vehicle sensors such as speed data or emission levels are associated with incorrect roadway centerlines. Furthermore, GPS data points might be incorrectly assigned to roadways along which no measurements were ever taken affecting transportation applications such as road use charging based on the total mileage driven by vehicle (Cozzens, 2009; Sheridan, 2011). The need to overcome spatial mismatches in ITS applications is a major motivation for implementing map-matching algorithms.

Fig. 2. GPS Data Points Collected by a Vehicle While Traveling at a Highway Interchange

Section 2 presents a literature review of map-matching algorithms developed to solve spatial ambiguities. Section 3 describes the proposed topological decision-rule map-matching algorithm and an example of its implementation. Results of the performance analysis with real spatial data are presented in section 4. Finally, section 5 presents a summary and the main conclusions of this chapter, and further research topics to be addressed.

2. Map-matching methods

The problem of resolving spatial ambiguities has been widely studied over the years. The following map-matching algorithms are described in the literature with different levels of complexity ranging from simple geometric techniques to complex, advanced approaches.

2.1 Semi-deterministic map-matching

The earliest map-matching algorithm, before GPS was developed in the 1970's, followed a semi-deterministic model (French, 1989). This model assumes that the vehicle has an initial location on a roadway and a given direction of travel. Conditional tests are applied to determine whether the vehicle is traveling on the known road by comparing turns from the vehicle location to a segment of the digital road map. A correction is performed whenever the heading of the vehicle changes (Morisue & Ikeda, 1989). However, for this technique to work, the vehicle is generally assumed to follow a predetermined road. There is considerable uncertainty when the vehicle travels off-road because there is no longer any way to correct for errors (Zhao, 1997; Czerniak, 2002).

2.2 Probabilistic map-matching

The probabilistic approach, described later, has the advantage of not assuming that the vehicle is always on a road. Vehicle heading error is calculated with an elliptical or rectangular confidence region and error models are developed within which the true vehicle location can be determined. If the vehicle position within the region contains one intersection or road segment, a match is made and the coordinates on the road are used in the next position calculation. If more than one road or intersection lies within the region, connectivity checks are made to determine the most probable location of the vehicle given earlier vehicle positions. As a result, the algorithm yields the best match segment along with the most probable matching point on the segment (Zhao, 1997; Czerniak, 2002).

2.3 Fuzzy logic map-matching

Fuzzy logic is an effective way to deal with tasks that involve qualitative terms and concepts, vagueness, and human intervention. Expert knowledge and experiences employed by a fuzzy logic based map-matching algorithm are represented as a set of rules to determine vehicle location (e.g., if the difference between the orientation of the roadway segment and the heading of the vehicle is small, then resemblance between the vehicle travel path and the candidate route is high).

S. Kim and J. H. Kim (2001) propose an adaptive fuzzy-network-based C-measure algorithm that identifies the roadway on which a vehicle is traveling by comparing C-measures associated with each candidate roadway. These measures are membership functions that represent the certainty of the existence of a vehicle on a specific roadway. After the roadway is identified, the algorithm determines the vehicle position on the roadway by orthogonal projection. The algorithm requires the distance between the vehicle's GPS coordinates and its projected position on the roadway to be small. Furthermore, the shape of the roadway must be similar to the trajectory of the vehicle.

Jagadeesh et al. (2004) developed a map-matching algorithm based on the inferences and a simple fuzzy rule set. This algorithm evaluates the likelihood of candidate roads to be the actual traveled road. Three fuzzy rules are employed for this purpose, which include heading comparison, road resemblance, and verification of off road vehicles. Test results with simulated data indicate that the algorithm is capable of achieving high accuracy.

Quddus et al. (2006) describe a map-matching algorithm based on fuzzy logic theory. The proposed algorithm employs an integrated navigation system and digital map data to identify the correct link and determine the vehicle location on the selected link. Although the algorithm was tested successfully in different road networks, the authors consider that future evaluation of the algorithm is required under urban conditions.

Yet another map-matching algorithm based on fuzzy theory is proposed by Guo and Luo (2009). First, the algorithm compares the similarity degree between the trajectory curve of the road and all candidate roads to identify the road on which a vehicle is traveling. Subsequently, fuzzy preference relations are adopted to perform a multi-criteria decision and a look-ahead technique is employed to improve the matching accuracy. The algorithm requires testing and analysis with GPS data in addition to cell phone positions.

2.4 Kalman filter approach

There has been abundant research on application of Kalman filters in combination with GPS and dead-reckoning signals to solve spatial mismatches. This integrated technology improves positioning accuracy by estimating white noise and error in the GPS and then correcting the vehicle's position (Jo et al., 1996; W. Kim et al., 2000; Zhao et al., 2003). For example, Quddus et al. (2003) present a general map-matching algorithm that integrates GPS and dead-reckoning sensor data (position, velocity, and time) through an extended Kalman filter and uses them as input to improve performance of the algorithm. The physical location of the vehicle on a roadway link is determined empirically from the weighted averages of two state determinations of the vehicle position based on topological information and external sensors.

Yang et al. (2003) present an improved map-matching algorithm that employs Kalman filtering to filter unreasonable GPS data and the Dempster-Shafer (D-S) theory to correctly snap GPS vehicle coordinates to the digital roadway map. The D-S theory allows explicit representation of ignorance and combination of evidence and operates with a smaller set of uncertainties. Although the authors report satisfying results, they suggest additional research to verify the accurate performance of the algorithm.

Nassreddine et al. (2009) describe a map-matching method based on D-S theory and interval analysis to compute accurate vehicle positions from an initial estimated position on a digital road network. The authors state that the proposed technique proves to be successful at junctions and parallel roads. However, real world data needs to be examined in addition to simulated data.

2.5 Particle filtering and map-matching

Particle filtering, based on a stochastic process, is another approach to the map-matching problem. Particle filters are recursive implementations of Monte Carlo-based statistical signal processing (Crisan & Doucet, 2002). Gustafsson et al. (2002) evaluate in real time a map-matching particle filter used to match a vehicle's horizontal driven path to a digital roadway map. They conclude that the particle filter converged relatively rapid after a few iterations of the algorithm. The challenge of this map-matching technique is to find nonlinear relations and non-Gaussian sensor models that provide the most information about the vehicle's position. The authors assert that research is still needed to seek a reliable way to detect divergence and to restart the filter.

Toledo-Moreo et al. (2009) present a multiple-hypothesis particle-filter based algorithm to solve the map-matching problem with integrity provision at the lane level. The proposed system joins measurements from a GPS receiver, an odometer, and a gyroscope along with road information in digital maps. A set of six experiments were conducted with real data for a period of 30 minutes proving the feasibility of the approach for lane-level applications. The authors mention that outlier removal, multipath effect mitigation, and additional method validation are tasks that need to be addressed in the future.

2.6 Personal navigation assistants and map-matching

White et al. (2000) discuss solutions to the map-matching problem for personal navigation assistants (PNA). Four different map-matching algorithms were implemented and tested: 1) use of minimum distance (point-to-curve), 2) comparison of heading information with arc and trajectory, 3) use of topology to select roads that are reachable from the current road, and 4) construction of piece-wise linear curves from different paths, followed by comparison of them to centerline curves using points (curve-to-curve matching). The authors conclude that these algorithms performed better when the distance between the GPS point and the closest road is small and that correct matches tend to occur at greater speeds on straight roadways.

Freitas et al. (2009) explain the necessity of map-matching algorithms to correctly locate GPS positions on a map when using PNA, particularly for dynamic route guidance systems. The authors describe an approach to update digital maps through the use of GPS points, in order to identify map incongruence. The proposed system was designed as a prototype and lacks of extensive testing, however, it correctly processes and implements methods for map-matching and detecting discrepancies between the real network and digital maps.

2.7 Topological network-based algorithms

Taylor et al. (2001) describe an algorithm called "Road Reduction Filter (RRF)" that uses differential corrections and height aids. RRF identifies all possible roadway candidates while systematically removing incorrect ones. RRF is improved by using shortest path network analysis and drive restriction information. A shortest path network routine calculates the distance through the roadway network from a vehicle's previous position to each potential present position offered by the algorithm. The drive restriction information routine selects roadways using direction and access information.

Greenfeld (2002) presents a map-matching procedure that consists of two algorithms. One algorithm assesses similarity between characteristics of the roadway network and the positioning pattern of the vehicle. The second algorithm performs topological analysis and applies a weighting scheme to match each GPS data point to the roadway network. The highest weighted score determines the most likely candidate for a correct match. The author indicates that further research is needed to determine the correct position of the vehicle along a roadway segment and to verify the accuracy performance of the algorithms.

Doherty et al. (2000) studied an algorithm that automatically matches GPS data points to roadway segments along a network. First, the algorithm joins GPS points to create a linear object forming the vehicle's track. Subsequently, it creates a buffer zone around the linear object, and then identifies all the roadways that are totally included within the buffer to select the correct one.

Marchal et al. (2005) presents an innovative map-matching algorithm that relies on GPS measurements and network topology. The algorithm consists of maintaining a set of candidate paths as GPS data are processed and computing matching scores for each path. The path with the best score represents the correct vehicle route. According to the authors further research is needed to improve the robustness of the algorithm.

Yet another topological map-matching algorithm is proposed by Wang and Yang (2009). The algorithm presents high accuracy and solves spatial ambiguities in complex roadway networks, specifically near intersections and parallel roads. Nevertheless, the topological algorithm was tested on only four road intersections with a 2-second sampling interval of GPS measurements.

Velaga et al (2009) describe an enhanced weight-based topological map-matching algorithm for ITS. The algorithm was tested with real data under different operational environments. However, the optimal algorithmic weights for different factors such as heading, proximity, connectivity, and turn-restriction still need to be estimated with a range of real-world field data from different road environments.

Blazquez and Vonderohe (2005) propose a topological map-matching algorithm that resolves spatial ambiguities that occur with intelligent winter maintenance vehicle data collected in Wisconsin. The algorithm computes shortest paths between snapped GPS data points using network topology and turn restrictions. If similarity exists between calculated and recorded vehicle speed values, then the path is feasible and snapped GPS locations are correct. If the path is not viable, then GPS data points are snapped to alternative roadway centerlines, shortest paths are recalculated, and speeds are again compared. The authors studied this problem further and published the effects of controlling parameters on the performance of the map-matching algorithm (Blazquez & Vonderohe, 2009). The current chapter discusses and describes in more detail the performance analysis of this map-matching algorithm.

2.8 Other map-matching algorithms

According to Zhao (1997), many pattern recognition methods (e.g., neural network) could be used for map-matching. Neural networks are dynamic systems that consist of many interconnected layered nodes (neurons). These networks need to be trained to arrange the layers and interconnections to model real-world applications. Other pattern recognition methods can be used to work with positioning sensors such as GPS. The underlying principle of these methods is that the digital map is used to filter out vehicle sensor errors and to determine the best position.

Schlingelhof et al. (2008) present a two-dimension map-matching algorithm based on a lane-level model. The output of this algorithm is the road segment identification number, the relative vehicle position along this segment, and the relative transversal vehicle position with respect to one of the border lines. The road selection algorithm consists of extracting candidate segments, computing positioning solution residuals, and selecting the most likely segment. The authors state that the first results obtained with real measurements are encouraging. However, these should be generalized to enhanced maps.

Li et al. (2005) present a novel map-matching method using least-squares position estimation, and digital mapping and height data to augment the vehicle position calculation. Experiment results indicate that combining the algorithm with height aiding improves the vehicle position accuracy when the number of visible satellites is reduced.

3. Decision-rule topological map-matching algorithm

3.1 Description

The decision-rule topological map-matching algorithm determines the correct roadway centerline for vehicle travel by obtaining feasible shortest paths between snapped GPS data points in post-processing mode. The algorithm selects all roadways within a buffer around a GPS data point and snaps the point to the closest roadway by obtaining the minimum perpendicular distance from the data point to each roadway. Figure 3 illustrates that GPS data points 1 and 2 (shown as circles) are snapped to ramp 2 because it is the closest roadway contained with the buffers around the points. Subsequently, the shortest path (displayed with a bold arrow) is obtained between the two snapped GPS data points S1 and S2 (shown as squares). Only paths that follow allowable traffic directions and allowable turns are employed. The travel speed between these two snapped GPS points is determined by the length of the shortest path and the difference in time stamps for the points. The computed speed is compared to the average of the speeds at the data points collected by the vehicle while traveling. If the computed speed is within a specified tolerance of the average recorded speed, then the obtained shortest path is viable and the snapped locations for points 1 and 2 are accepted as correct.

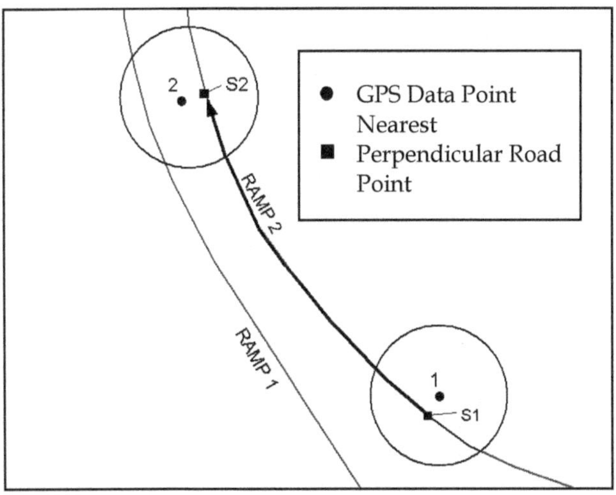

Fig. 3. Example of Snapping to the Correct Roadway for Two GPS Data Points Using the Map-Matching Algorithm

The map-matching algorithm advances to GPS data point 3, snaps this point to the closest roadway centerline within its buffer, and calculates the shortest path between snapped point S2 and the newly-snapped GPS data point S3. If the path between S2 and S3 is not feasible because the speed comparison yields a large disparity, then the algorithm determines if feasible routes exist between the preceding and subsequent points bounding the GPS data points of concern, as illustrated in the example of Figure 4. This example shows that there is no feasible path between snapped points S2 and S3 when network topology and turn restrictions are employed. Therefore, the map-matching algorithm looks ahead by snapping point 4 to the nearest roadway centerline within its buffer, and determines if the shortest path between snapped points S3 and S4 is possible. Since the tested path is not feasible, the algorithm snaps point 3 to the next nearest roadway centerline within its buffer obtaining point alt3, shown as a triangle.

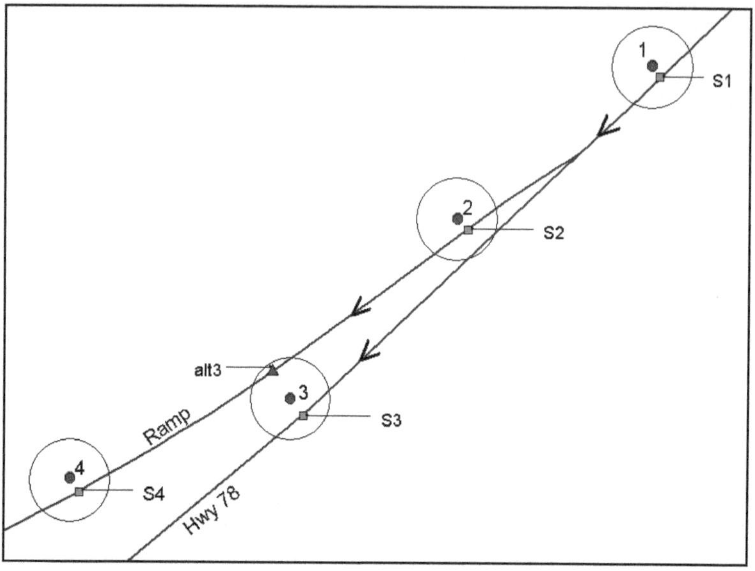

Fig. 4. Example of an Alternative Roadway Centerline Snapping

Subsequently, the upper part of the algorithm (shown in Figure 5) for alternative roadway centerline search and feasibility path check is initiated. This algorithm verifies if a path is feasible between the alternative snapped location for point 3 (where Ki = 3), and former and succeeding neighboring snapped points 2 and 4 (where Ki-1 = 2 and Kj = 4). If the shortest paths between these three points are not feasible because the speed comparison fails, then the algorithm searches for other roadway centerlines within the buffer around point 3 that have not already been used in a feasibility path check. When finding a new candidate, point 3 is then snapped to it and the feasibility of shortest paths between snapped points 2, 3, and 4 (Ki-1, Ki, Kj) is checked again. If these paths are feasible, then the spatial ambiguity is resolved, and the algorithm terminates. If no alternative roadway centerline exists within the buffer for GPS data point 3, then the algorithm continues by snapping data point 4 to alternative roadway candidates contained within its buffer, and the upper part of the algorithm is executed again. If no other roadway centerlines exist within the buffer of GPS data point 4 or no feasible paths are obtained, then the lower part of the algorithm is executed and feasible paths between preceding and subsequent data points are examined. If none of the consecutive data points aid in solving the spatial mismatch between the snapped points for 2 and 3, then it is likely that no roadway centerlines within their buffers yield a feasible path and larger buffers and/or more consecutive data points need to be utilized by the algorithm. Once a feasible path is obtained, the intermediate points not employed during the map-matching process are snapped to the roadway along that feasible path.

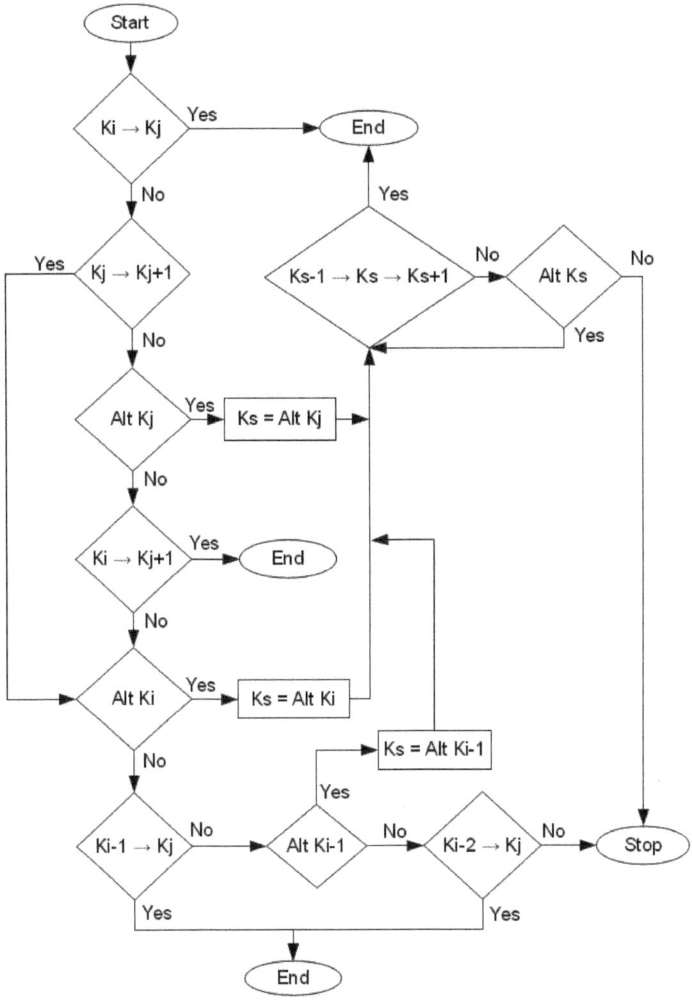

Fig. 5. Flow Diagram with the Step Sequence of the Map-Matching Algorithm

3.2 Example of an implementation of the algorithm

The example illustrated in Figure 6 includes a set of Differential GPS (DGPS) data points collected every five seconds by a winter maintenance vehicle during the 2002-2003 winter season in Columbia County, Wisconsin. The spatial mismatch, occurring at the diverging roadways in this figure, is resolved by implementing the decision-rule map-matching algorithm. Points 0, 2, 3, and 4 are snapped to the nearest roadway within their 35-foot buffers, resulting in points S0, S2, S3, and S4 (shown as rectangles). Points S0, S3, and S4 are on the Interstate 39 centerline, while point S2 is situated on the ramp centerline. Note that no roadways are contained within the buffer for GPS data point 1, thus, this point is not used in determining the feasible path.

The shortest path between points S0 and S2 is computed using network topology and allowable turns. Consequently, the speed comparison shown in Table 1 is performed to determine if this path is feasible. In this case, the obtained path is feasible since the difference between the average calculated and recorded speeds (26.8 and 31.5 mi/h, respectively) is within tolerance (25 mi/h). Therefore, the current snapped positions for points 0 and 2 are initially assumed to be correct. The main algorithm continues by finding the shortest path between the next pair of snapped points, S2 and S3. This path is not feasible when using network topology because if the vehicle was located at S2, it would have to exit the ramp and travel approximately 5,125.9 feet in 5 seconds at an average speed of 699 mi/h to reach snapped point S3. Hence, either point S2 or S3 or both were snapped to an incorrect roadway centerline. The map-matching algorithm now obtains the shortest path between points S3 and S4 and determines that the difference between calculated and average recorded speeds with values of 29 mi/h and 35 mi/h, respectively, is within tolerance. Therefore, an alternative roadway centerline is sought within the buffer around point 2. Interstate 39 is found to be the next nearest roadway, resulting in alternative point alt2, shown as a triangle in Figure 6. Consequently, feasibility is checked for paths between the preceding points S0 and alt2, and between alt2 and its successor, snapped point S3. As indicated in Table 1, both computed shortest paths are feasible. The calculated speeds along these paths are within 25 mi/h of their respective average recorded speeds for the vehicle. Therefore, the spatial ambiguity at the diverging roadway is resolved and the correct roadway for point 2 is Interstate 39. Data point S1 is then obtained by snapping point 1 to the Interstate 39 centerline.

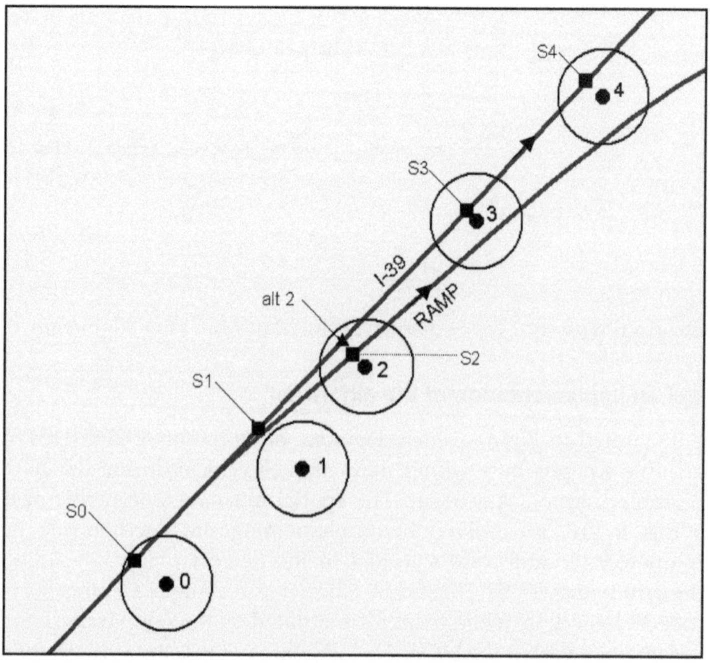

Fig. 6. Example of Map-Matching Algorithm at Diverging Roadways

Data Points	Shortest Path Distance (ft)	Calculated Speed (mi/h)	Average Recorded Speed (mi/h)	Is Path Feasible?
S0 → S2	392.6	26.8	31.5	YES
S2 → S3	5125.9	699	33	NO
S3 → S4	213	29	35	YES
S0 → alt2	392.8	26.8	31.5	YES
alt2 → S3	215.7	29.4	33	YES

Table 1. Speed Comparison for Determining Feasibility of Shortest Paths

4. Performance analysis of the decision-rule map-matching algorithm

Success in solving spatial ambiguities depends on the values assigned to each variable of the map-matching algorithm. The analysis in this chapter examines the performance of the map-matching algorithm as values of the following parameters vary: 1) buffer size, 2) speed range, 3) number of consecutive data points, 4) temporal resolution, and 5) DGPS positional error.

4.1 Spatial data description

The data employed in this study were collected by winter maintenance vehicles in Columbia and Portage Counties, Wisconsin, and Polk County, Iowa. These counties have different accuracy roadway centerline maps with 1:2,400, 1:12,000, and 1:100,000 nominal scales, respectively, and employ different AVL/DGPS systems for data collection. Selected data sets with sampling intervals of 2 and 10 seconds were collected for different storm events and vehicle operators driving through various routes over the 2000-2001, 2001-2002, and 2002-2003 winter seasons. These routes include federal, state, and interstate highways, and local roads. Figures 7, 8, and 9 display examples of data collected in Columbia, Portage, and Polk counties every 2, 10, and 10 seconds, respectively. Notice that none of the counties employed an integrated dead reckoning system and heading information was not available during the data collection process.

Fig. 7. DGPS Data Points Collected in Portage County Every 10 seconds

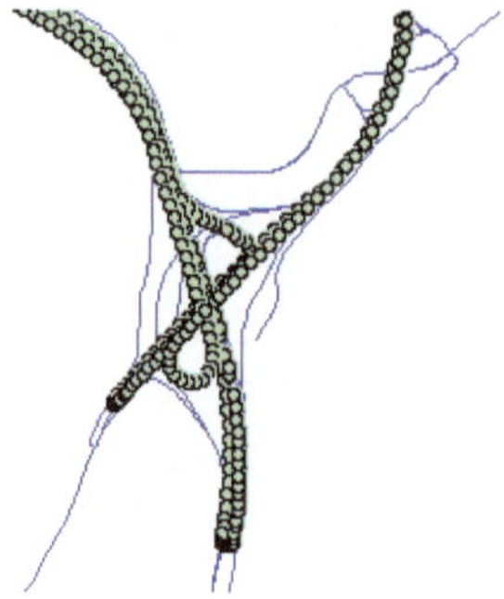

Fig. 8. DGPS Data Points Collected in Columbia County Every 2 seconds

Fig. 9. DGPS Data Points Collected in Polk County Every 10 seconds

4.2 DGPS data point classification

This section identifies different cases (i.e., false negatives, false positives, no solution, incorrect and correct snap, and solved spatial ambiguities) obtained from comparing snapping results to the true roadway centerline on which a vehicle is traveling. The true vehicle path was obtained by performing a visual examination of the collected data. Data points are classified in these cases before and after applying the map-matching algorithm.

4.2.1 False negatives and false positives

False Negatives (FN) occur when data points fail to snap to any roadway centerline when they should have snapped to one. False Positives (FP) are data points that snapped to some roadway centerline when they should have not snapped to any centerline. Figure 10 shows an example of three successive GPS data points (1, 2, and 3) considered as FN. They should have snapped to Interstate 39 east bound direction, however, their buffers with radius r are too small to include any roadway centerline.

4.2.2 Solved / not solved cases

If roadway centerlines exist within the buffer of a data point, then a correct snap occurs when this point snaps along the true route of the vehicle. Conversely, an incorrect snap is obtained when a data point snaps to a roadway that is not on the true route of the vehicle.

Correct and incorrect snaps are computed before and after applying the map-matching algorithm.

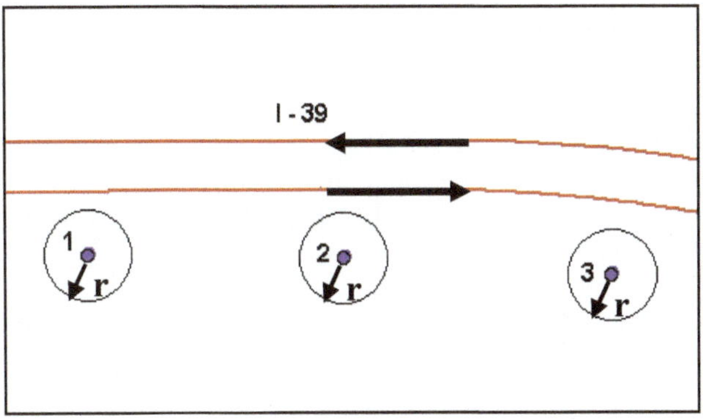

Fig. 10. Example of Three Consecutive GPS Data Points Considered False Negatives

Figure 11 presents the cases of snapped and not snapped data points before and after applying the map-matching algorithm. The group of data points that does not snap to any roadway contains either FN or points that have no solution. Data points that have roadway centerlines within their buffers are either snapped correctly or incorrectly, or are FP. A data point that snaps incorrectly before applying the algorithm and snaps correctly afterwards is

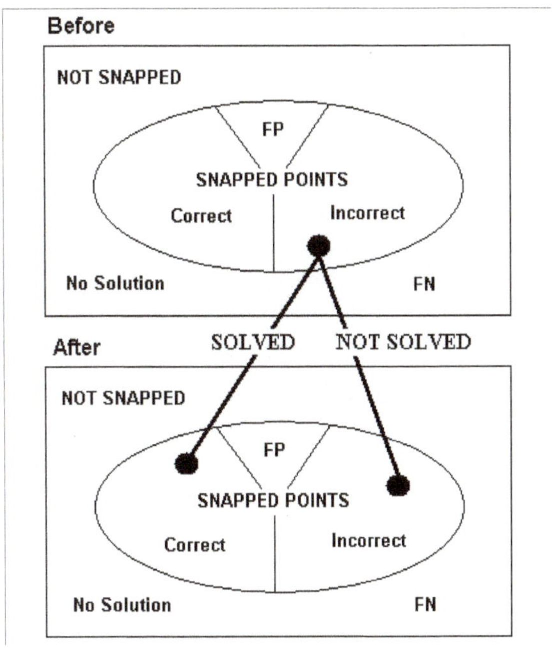

Fig. 11. Cases for Data Points Snapped and Not Snapped Before and After Applying the Algorithm

regarded as a solved case. If a data point is snapped incorrectly before applying the algorithm and it is snapped incorrectly after applying the algorithm, then the spatial mismatch is not solved. If this occurs, then some neighboring data points may be left incorrectly snapped. Note that FN, FP, and no solution are not included in the solved and not solved case analysis.

In the following section of this chapter, FN data points are minimized and solved spatial mismatches are maximized after applying the algorithm. Although FP and no solution cases occur due to spatial database incompleteness, they amount to less than 0.5% of the total number of data points examined in this study. Therefore, these two cases were not taken into account in the analysis.

4.3 Analysis of the impact of variables on the performance of the map-matching algorithm

This section examines each algorithmic variable independently to determine its effect on the performance of the map-matching algorithm. These variables are classified into two groups. One group consists of parameters controlled by the user (i.e., buffer size, speed range, number of consecutive data points) and the other group comprises parameters controlled through the data (i.e., temporal resolution and DGPS error).

4.3.1 Buffer size

The appropriate buffer size employed during the snapping process when solving spatial ambiguities depends on the quality and geometry of the spatial data. This proximity parameter used to select roadway centerlines around data points is critical for solving the map-matching problem and, therefore, for the success of the algorithm. Buffers that are overly small in size might not include any roadways. While extremely large buffers make the algorithm less efficient since it needs to examine more roadways, many of which will not be correct.

Roadways are typically represented by centerlines that do not account for lane widths. Therefore, data points will almost always appear offset some distance from roadway centerlines in addition to being affected by errors in the DGPS measurements and digital roadway maps (Wolf & Ghilani, 1997). Hence, the buffer size parameter was tested at 10-ft increments from 20 ft to 60 ft for data collected in Columbia and Portage Counties, and at 20-ft increments from 20 to 100 ft for data collected in Polk County. The latter is due to the smaller scale of the Polk County roadway centerline map. These buffer size values were predetermined through the computation of average distance percentages between the data points and roadway centerlines. As different buffer sizes were analyzed and tested against the map-matching algorithm, the speed range tolerance and number of consecutive data points were maintained constant with values 25 mi/h and 5, respectively.

Figure 12 shows a chart with the average percentages of FN before and after applying the algorithm, as the buffer size varies for Columbia, Portage, and Polk County. This figure indicates that lower FN percentages are obtained after applying the algorithm for all three counties. Portage and Polk counties present the largest decrease of FN percentages with an average difference of 20% before and after executing the algorithm. Overall, average percentages of FN data points diminish as the buffer size increases since more data points are snapped to roadway centerlines.

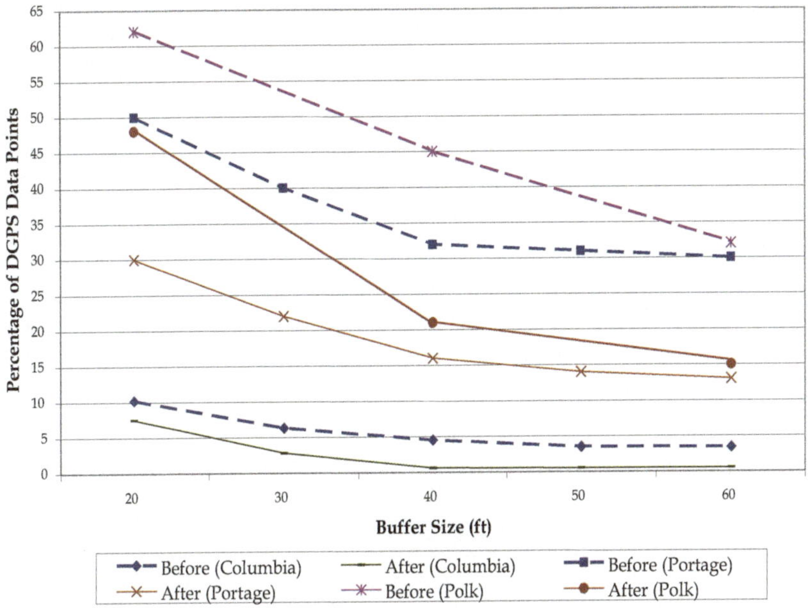

Fig. 12. FN Percentages Before and After Applying Algorithm by Buffer Size for Columbia, Portage, and Polk Counties

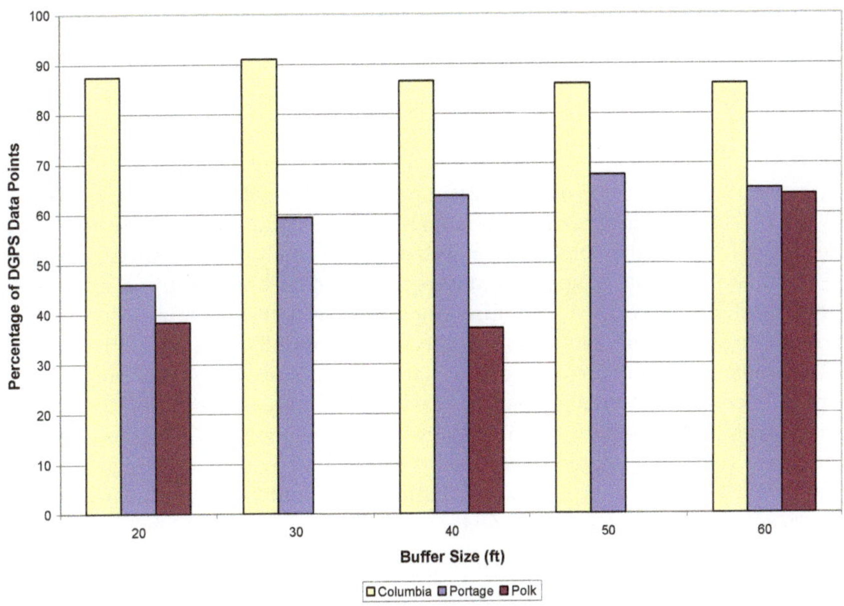

Fig. 13. Percentages of Solved Cases After Applying Algorithm by Buffer Size for Columbia, Portage, and Polk Counties

Figure 13 presents the percentage of solved spatial ambiguities after applying the map-matching algorithm for Columbia, Portage, and Polk counties. This chart indicates that over 90% of incorrectly snapped data points collected in Columbia County were solved by the algorithm when employing a 30-foot buffer size. Whereas, solved cases reached their maximum values (68% and 64%) for Portage and Polk counties with 50 and 60-foot buffers, respectively. As mentioned earlier, Polk County data was tested for buffer sizes every 20 feet, thus, there is no data for buffer sizes equal to 30 and 50 feet.

4.3.2 Speed

The map-matching algorithm determines the correct roadway centerline on which a vehicle is traveling by computing feasible shortest paths between snapped data points. This feasibility is sensitive to the allowable range utilized when comparing computed and recorded speeds. The analysis of this variable examines the effect that it has on the performance of the map-matching algorithm.

The average recorded speed (v) is computed using the recorded speeds (v_1 and v_2), as shown in Equation 1. Equation 2 presents the computed speed calculation (s) given the shortest distance traveled (D) and timestamps (t_1 and t_2) between a pair of snapped data points. Subsequently, the algorithm accepts a tested path as feasible if the average recorded speed is within the equally distributed speed range shown in Equation 3.

$$v = \frac{v_1 + v_2}{2} \tag{1}$$

$$s = \frac{D}{(t_2 - t_1)} \tag{2}$$

$$v \in s \pm \frac{\text{SpeedRange}}{2} \tag{3}$$

FN curves were computed for various buffer sizes and different speed range tolerances from 5 to 35 mi/h with increments of 5 mi/h for the three counties. Analysis results for this variable show that feasible paths are rejected when small speed ranges are employed leaving FN data points not snapped to any roadway centerline. On the contrary, as speed range increases, FN percentages diminish since feasible paths are found during the speed comparison process. Figure 14 shows FN curves for Columbia County with data collected every 2 seconds. These curves are approximately parallel as the speed range varies, and stabilize for speed ranges greater than 15 mi/h. Speed ranges equal to or greater than 25 mi/h are needed to minimize FN percentages in Portage and Polk counties. Further speed range increase does not improve the results because all feasible paths are accepted. In general, FN curves are steeper for small buffer sizes, and approach near-zero slope for buffer sizes equal to or greater than 40 feet.

Analysis results for this variable indicate that the percentage of solved cases increases as speed range also increases. The percentage of solved cases has the highest value of approximately 90% when the algorithm employs speed ranges equal to or greater than 20 mi/hr and a 30-foot buffer for Columbia County data. Conversely, there is no considerable increase in the percentage of solved cases remaining at 68% for Portage County data when speed range values equal to or greater than 25 mi/h are employed. The percentage of solved cases for Polk County remained constant at 50% for speed ranges equal to or greater than 15 mi/h, independent of buffer size. Thus, the map-matching algorithm is sensitive to speed range values, particularly when small speed ranges are employed since feasible paths are rejected.

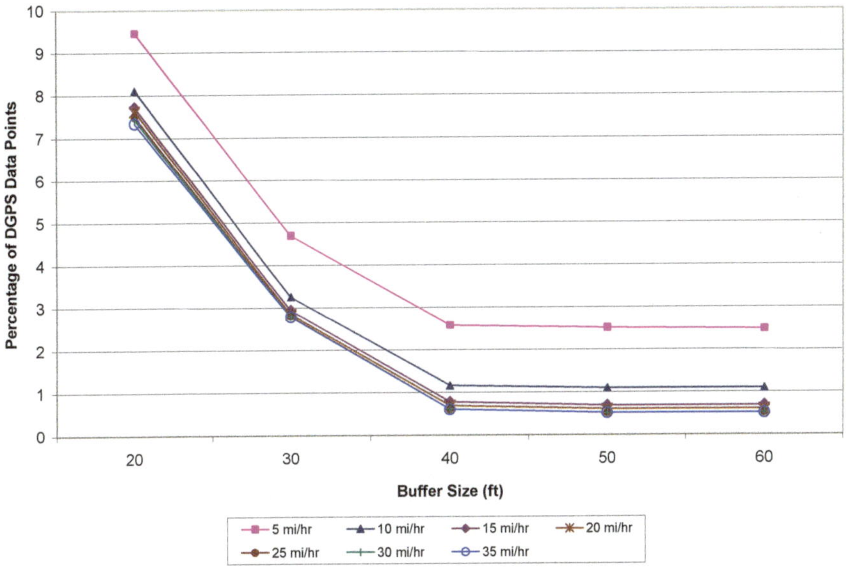

Fig. 14. FN Percentages After Applying Algorithm for Different Speed Ranges by Buffer Size for Columbia County

4.3.3 Number of consecutive GPS data points

If no feasible paths are obtained between a pair of snapped data points, then the algorithm tests for viable routes between preceding and subsequent data points, as described in Figure 5. If a small buffer size is utilized, several successive data points do not snap to any roadway centerline generating FN data points. Thus, the number of consecutive data points used by the algorithm needs to be incremented to consider adjacent data points that are correctly snapped and minimize FN percentages.

Although the map-matching algorithm may employ any number of consecutive data points, the performance of the map-matching algorithm was analyzed with a number of consecutive data points between three and eight. A previous test determined that this range of consecutive data points is suffice for solving spatial ambiguities with the spatial and temporal data employed in this study.

Similar to the FN curve behavior due to speed range variations, FN curves for different number of consecutive data points are parallel for the three counties and converge to constant values as the buffer size increases. Figure 15 shows the percentage of FN data points as the number of consecutive data points varies by county with a 40-foot buffer size. No significant improvements are identified in the percentage of FN for the three counties as the number of consecutive data points varies, except for Portage County data that presents a decrease in the amount of FN when increasing the number of consecutive data points from three to four.

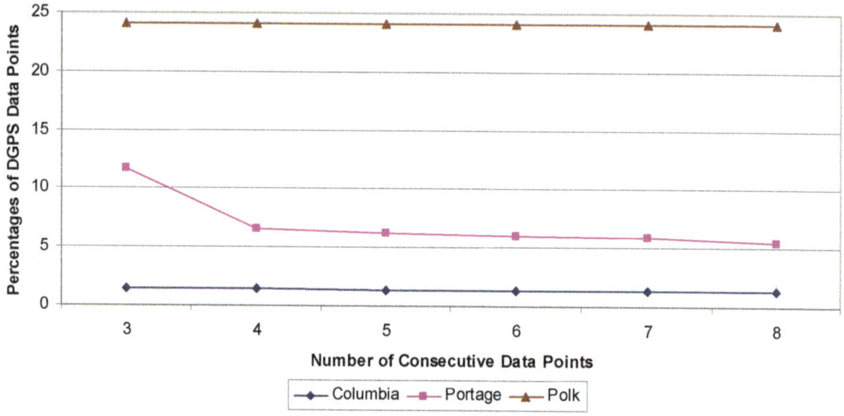

Fig. 15. Average Percentages of FN After Applying Algorithm by Number of Consecutive Data Points and County

The percentage of solved spatial mismatches increased as the number of consecutive data points increased. Eight consecutive data points solve almost 100% of initial incorrect snaps for Columbia County data when employing a 40-foot buffer. The largest percentage of solved mismatches (over 70%) after applying the algorithm occurs with a 50-foot buffer for Portage County. While the percentage of solved cases in Polk County remained constant at 50%, as the buffer size and number of consecutive data points increased. The results of this analysis show that increasing the number of consecutive data points solves a larger number of spatial ambiguities. By increasing this number, the algorithm resolves ambiguities that arise when alternative roadway centerlines are equally viable.

4.3.4 Temporal resolution

The outcome of the map-matching technique is not only affected by spatial inaccuracies, it is also influenced by the collection frequency of the data points. As temporal resolution increases, the tracking of the vehicle becomes more accurate. On the other hand, the sampling interval impacts the sizes of the data sets. Processing of large data sets takes significant CPU time, and increases storage requirements. Hence, there is a tradeoff between decreasing the sampling interval and quality of collected speed data.

Data sets collected in Columbia County with an original 2-second time interval were processed to generate data files with lower temporal resolutions varying from 2 to 30 with increments of 4 seconds. Similarly, data collected every 10 seconds in Portage and Polk counties were processed to create data files with temporal resolutions equal to 10, 20, and 30 seconds, respectively. The speed range and number of consecutive points remained constant with values 25 mi/h and 5, respectively.

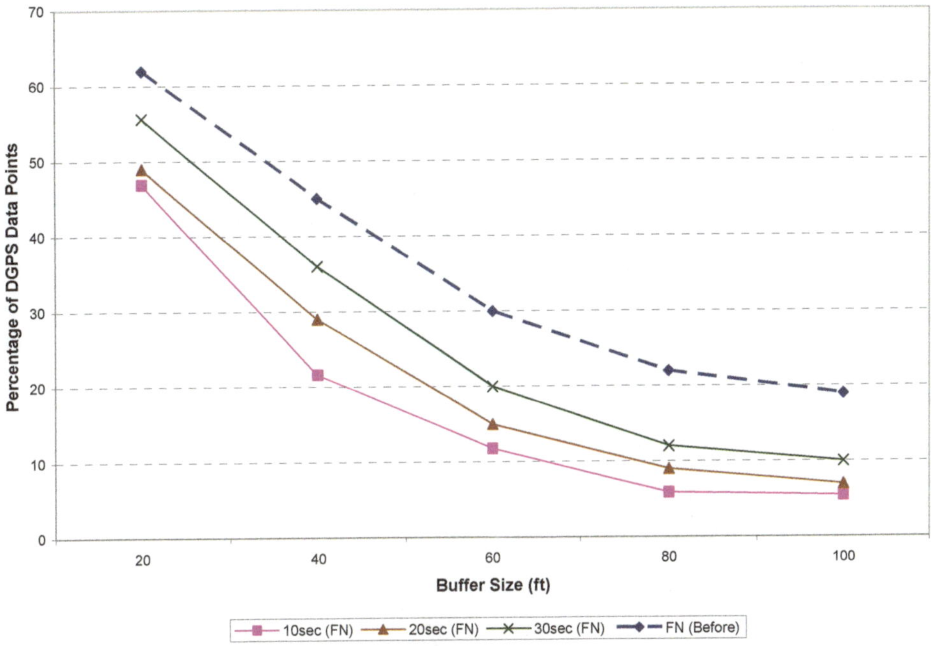

Fig. 16. FN Percentages Before and After Applying Algorithm for Different Temporal Resolutions by Buffer Size for Polk County

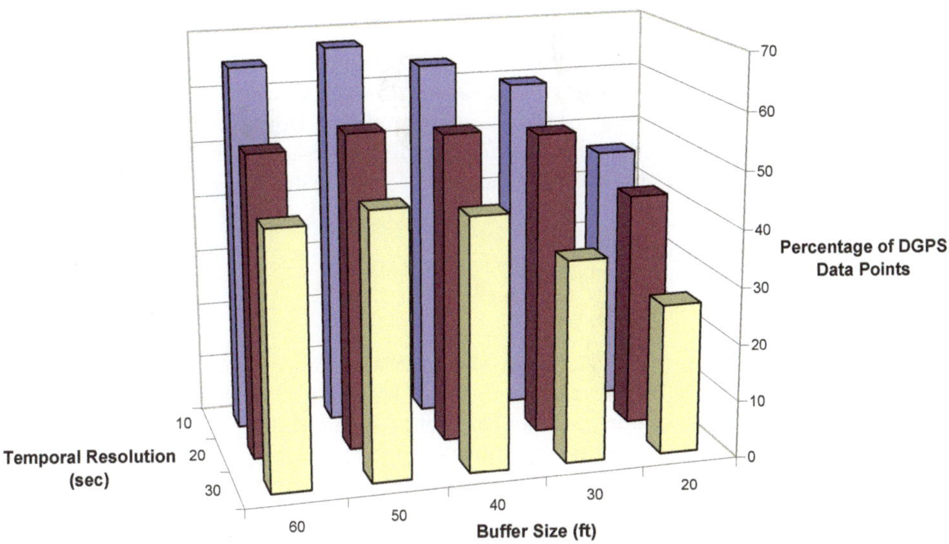

Fig. 17. Percentages of Solved Cases for Different Temporal Resolutions by Buffer Size for Portage County

Figure 16 illustrates FN curves before and after applying the algorithm for different temporal resolutions with data originally collected every 10 seconds in Polk County. The graph presents relatively parallel FN curves for all data collection frequencies. These curves show that as temporal resolution increases, the percentages of FN data points decreases. FN curves after applying the algorithm show lower percentages of FN data points compared to before executing the algorithm. FN curves for Columbia and Portage counties behave similarly with different sampling intervals. All county cases illustrate that larger amount of FN points occur when using smaller buffers independent of data collection frequency. Figure 17 shows the variation of solved cases as temporal resolution increases in Portage County. Percentages of solved spatial ambiguities increase as data is collected at higher frequencies, being the largest at a 50-foot buffer with 68%. This percentage decreases in average for Columbia County data from approximately 80% to 20% as sampling intervals increase from 5 to 30 second for all buffer sizes. The same behavior is apparent for solved case percentages in Polk County as data is collected more frequently.

4.3.5 GPS error

GPS measurements are affected by both systematic and random errors. Their combined magnitudes will affect the accuracy of the positioning results. Systematic errors obey physical or mathematical law, and can be computed and applied to measurements to eliminate their effects (Ghilani & Wolf, 2006). Random errors occur because of stochastic noise in the measurement process producing different coordinates each time a measurement is achieved, even during short intervals. This type of error is assumed to be Gaussian affecting both latitude and longitude or X, Y coordinates. DGPS is a method that increases the accuracy of CA code measurements by canceling some of the inherent systematic errors. Any potentially remaining systematic errors were not modeled in this study, and only the effects of random errors were examined.

Random errors were simulated by using a normal distribution random number generator (Box & Muller, 1958) for known means and different standard deviations. If U1 and U2 are a pair of independent uniformly-distributed random numbers from the rectangular density function on the interval (0, 1), then a pair of independent random numbers (X_1 and X_2) from a normal distribution with mean zero and standard deviation σ are generated using Equations 4 and 5.

$$X_1 = (-2\ \log U_1)^{1/2} \cos(2\pi U_2) \tag{4}$$

$$X_2 = (-2\ \log U_1)^{1/2} \sin(2\pi U_2) \tag{5}$$

Experiments conducted by the Wisconsin Winter Maintenance Concept Vehicle project concluded that random DGPS errors were on the order of 2 to 5 meters, root-mean-square (Vonderohe et al., 2001). Therefore, a mean value of zero and standard deviations of ±2 and ±5 meters were employed in this analysis. Speed range and number of consecutive points values were held fixed as 2- and 5-meter standard deviation errors were introduced in the DGPS data points.

Percentages for FN and solved cases were computed to compare the performance of the algorithm for original and perturbed DGPS data points. Figure 18 presents variations in the percentage of FN data points for original and perturbed data by county for a 40-foot buffer

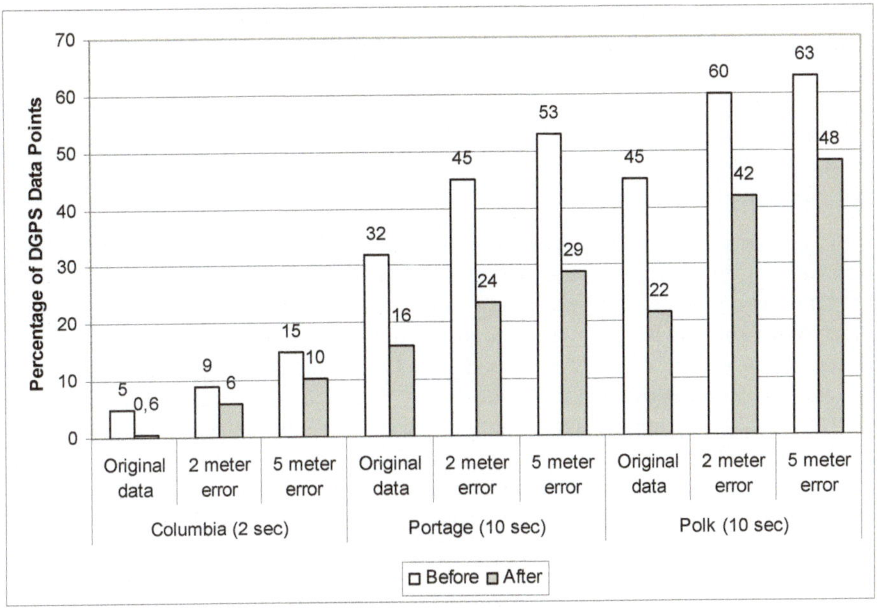

Fig. 18. FN Percentages Before and After Applying the Algorithm for Original Data, 2 m, and 5 m Error with a 40-foot Buffer by County

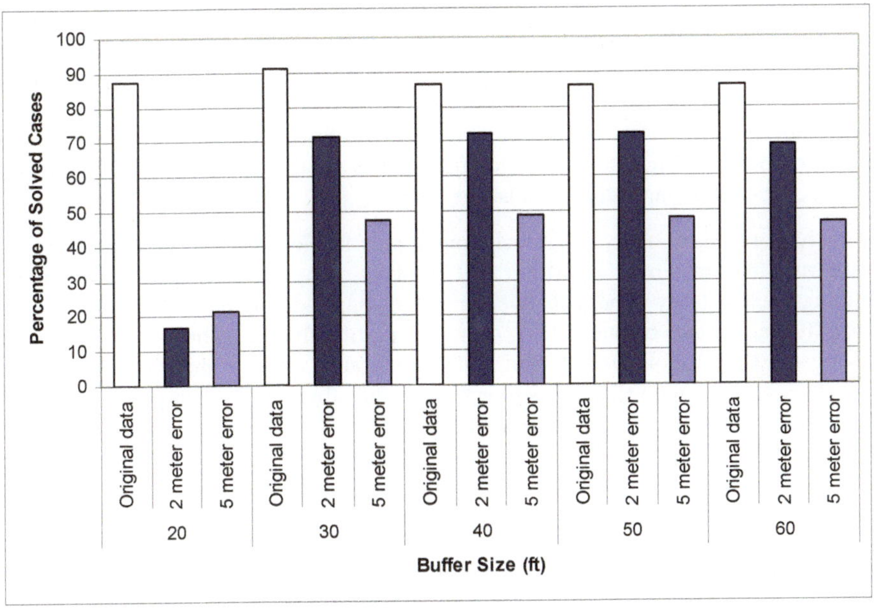

Fig. 19. Percentage of Solved Cases After Applying the Algorithm for Original Data, 2 m, and 5 m Error by Buffer Size in Columbia County

before and after applying the algorithm. All FN percentages decrease after executing the algorithm, independent from the spatial data quality. Average FN percentages computed with original data present smaller values than data perturbed with 2- and 5-m error before and after applying the algorithm. For example, FN percentages increase from 22% to 48% for Polk County after executing the algorithm when introducing a 5-m error. In general, the percentage of data points that should snap to a roadway centerline increases when there is larger error in the DGPS data points.

Figure 19 presents the percentages of solved spatial ambiguities by the algorithm before and after perturbing the DGPS data points with original and simulated random errors (2-and 5-meter standard deviation) for Columbia County. This figure shows that the percentage of incorrect snaps solved after applying the algorithm for original Columbia County data are larger than those computed with perturbed data. On average, the percentage of solved cases decreases approximately 20% and 40% for data with 2- and 5-meter error for all buffer sizes, except for the 20-foot buffer. This small buffer is not able to accommodate the spatial ambiguities that arise with simulated data. Similarly, Portage and Polk counties present a drop in the percentages of solved data points from approximately 68% and 50% for original data to approximately 10% and 15%, respectively, for 5 m perturbed data.

5. Summary and conclusions

Transportation applications employ AVL/DGPS technology to collect vehicle positions and other sensor data. Normally, DGPS data points are associated with roadways by snapping to the nearest centerline in a GIS environment. The map-matching problem or spatial ambiguities arise during this association due to errors in DGPS measurements and digital cartography. Such ambiguities are common at underpasses and converging or diverging roadways. These can result in DGPS data points being snapped to incorrect roadway centerlines affecting the calculation of cumulative distance traveled by the vehicles along a roadway network, or the allocation of non-spatial data collected from vehicle sensors to incorrect roadways. Thus, this problem propagates to the computation of performance measures or decision management tools.

This study contributes with the development and implementation of a post-processing decision-rule map-matching algorithm that resolves many of these spatial ambiguities by examining the feasibility of paths between pairs of snapped data points. A viable path is the shortest-distance path between two snapped points that a vehicle can travel, while following network topology and turn restrictions, at a speed comparable to its average recorded speed. If a given shortest path is not feasible, then DGPS data points are related to other roadway centerlines within their buffers and new shortest paths are calculated; or adjacent DGPS data points are used to determine feasible paths. Examples were presented to describe the step-by-step process of the map-matching algorithm. Five variables were studied independently to analyze the performance of the map-matching algorithm. These variables are buffer size, speed range tolerance, number of consecutive points, temporal resolution, and positional error in the DGPS data points. Data collection frequency and DGPS error are variables controlled externally through the data, while buffer size, speed range, and number of consecutive data points are algorithm parameters are controlled by the user.

The results of this study indicate that the success of the map-matching algorithm in solving spatial ambiguities depends on not only by the variables employed by the algorithm, but also by the sampling interval and the quality of the spatial measurements and roadway map scale. If lower spatial data qualities and less frequent sampling intervals are used, then the algorithm requires larger buffers and speed ranges to obtain best results. On the other hand, if GPS data points collected more frequently are snapped to higher accuracy maps, such as the Columbia County case, then larger percentages of incorrect snaps are solved and smaller buffer sizes are adequate. By increasing the number of consecutive data points, a larger number of spatial ambiguities are solved, particularly when alternative roadway centerlines are equally viable, and FN percentages are reduced since more combinations are examined between pairs of snapped DGPS data points. However, no significant variations in the solved results for Polk County are apparent as the number of consecutive data points increases since lower spatial data accuracies were used in this county. Table 2 presents the best and worst variable values encountered when solving incorrect snaps after applying the map-matching algorithm by county. This table indicates that larger speed range values, and numbers of consecutive points provide better results in maximizing solved cases. Stable percentage values are reached as both speed range and number of consecutive points reach certain values. While small speed ranges tend to reject tested paths, larger speed ranges accept most of these paths without improving the performance of the algorithm. Similarly, larger percentages of solved cases are obtained as the number of consecutive points increases since additional combinations between pairs of snapped data points are examined. Overall, higher parameter values yield better results as data are collected less frequently and snapped to lower quality roadway maps.

County	Buffer Size (ft)		Speed Range (mi/hr)		Number of Consecutive Points	
	Best	Worst	Best	Worst	Best	Worst
Columbia	30	≥50	35	5	8	3
Portage	50	20	≥25	5	8	3
Polk	40	20	≥15	5	≥3	≥3

Table 2. Best and Worst Variable Values for Solved Cases by County

Introducing positional error in the DGPS data points decreases the percentage of solved incorrect snaps and total number of snapped data points obtained before and after applying the algorithm. As the positional error increments from 2 to 5 meters in standard deviation, the percentage of solved cases decrease and FN percentages increase for all counties. Thus, larger buffer sizes and speed ranges are needed for lower quality data. Future research is required to explore these parameter values against additional spatial data qualities derived from multiple ITS applications. Further research may involve online implementation of the map-matching algorithm, in which spatial ambiguities are solved as GPS measurements are collected in real-time.

6. References

Blazquez, C., & Vonderohe, A. (2005). Simple Map-Matching Algorithm Applied to Intelligent Winter Maintenance Vehicle Data. *Journal of Transportation Research Board,* Vol. 1935, pp. 68-76.

Blazquez, C., & Vonderohe, A. (2009). Effects of Controlling Parameters on Performance of a Decision-Rule Map-Matching Algorithm. *ASCE Journal of Transportation Engineering*, Vol. 135, No. 12, (December, 2009), pp. 966-973.

Box, G. & Muller, M. (1958) A Note on the Generation of Random Normal Deviates. *Annals of Mathematical Statistics*, Vol. 29, No. 2, pp. 610-611.

Chen, W., Li, Z., Yu, M., & Chen, Y. (2005). Effects of sensor errors on the performance of map-matching. *Journal of Navigation*, Vol. 58, pp. 273-282.

Cozzens, T. (2009) Mileage-Based Road Tax Gets Pumped. *GPS World*, Vol. 20, No. 5.

Crisan, D., & Doucet A. (2002). A Survey of Convergence Results on Particle Filtering Methods for Practitioners. *IEEE Transactions on Signal Processing*, Vol. 50, No. 3, pp. 736-746.

Czerniak R. (2002). Collecting, Processing, and Integrating GPS Data into GIS. In *Transportation Research Board: NCHRP Synthesis of Highway Practice 301*, Transportation Research Board, National Research Council, Washington, D.C.

Doherty, S.T., Noel, N., Gosselin, M.L., Sirois, C., and Ueno, M. (2000). Moving Beyond Observed Outcomes: Integrating Global Positioning Systems and Interactive Computer-Based Travel Behavior Surveys. *Transportation Research Circular E-C026, Personal Travel: The Long and Short of It*, Transportation Research Board, National Research Council, Washington, D.C.

Freitas, T., Coelho, A., & Rossetti, R. (2009) Improving digital maps through GPS data processing. *Proceedings of the 12th International IEEE Conference on Intelligent Transportation Systems, ITSC*, St. Louis, Missouri., October 2009.

French, R. (1989). Map Matching Origins, Approaches and Applications. *Proceedings of the Second International Symposium on Land Vehicle Navigation*, Munster, Germany, July 1989.

Ghilani, C., & Wolf, P. (2006). *Adjustment Computations-Spatial Data Analysis*, John Wiley & Sons, Inc., ISBN 9780471697282, Hoboken, New Jersey.

Greenfeld, J.S. (2002). Matching GPS Observations to Location on a Digital Map. *Proceedings of Transportation Research Board 81st Annual Meeting*, Transportation Research Board, Washington, D.C., January 2002.

Guo, L. & Luo, D. Y. (2009) Development of an Integrated Map Matching Algorithm Based on Fuzzy Theory. *Proceedings of Second International Conference on Intelligent Computation Technology and Automation*, Zhangjiajie, China, October 2009.

Gustafsson, F., Gunnarsson, F., Bergman, N., Forssell, U., Jansson, J., Karlsson, R., & Nordlund, P. (2002). Particle Filters for Positioning, Navigation, and Tracking. *IEEE Transactions on Signal Processing*, Vol. 50, No. 2, pp. 425-437.

Jagadeesh, G., Srikanthan, T., & Zhang, X. (2004) A Map Matching Method for GPS Based Real-Time Vehicle Location. *The Journal of Navigation*, Vol. 54, pp. 429-440

Jo, T., Haseyama, M., & Kitajima, H. (1996). A Map Matching Method with the Innovation of the Kalman Filter. *IEICE Transactions on Fundamentals of Electronics, Communications and Computer Sciences*, Vol. 79-A, No. 11.

Kim, S., & Kim, J.H. (2001). Adaptive Fuzzy-Network-Based C-Measure Map-Matching Algorithm for Car Navigation System. *IEEE Transactions on Industrial Electronics*, Vol. 48, No. 2, pp.432-441.

Kim, W., Lee, G.I., & Lee, J.G. (2000). Efficient Use of Digital Road Map in Various Positioning for ITS. *Proceedings of IEEE PLANS, Position Location and Navigation Symposium*, San Diego, California, March 2000.

Li, J., Taylor, G., & Kidner, D. (2005) Accuracy and Reliability of Map-Matched GPS Coordinates: The Dependence on Terrain Model Resolution and Interpolation Algorithm. *Computers & Geosciences*, Vol. 31, pp. 241-251.

Marchal, F., Hackney, J., & Axhausen, K.W. (2005). Efficient Map-Matching of Large GPS Data Set - Tests on a Speed Monitoring Experiment in Zurich. *Proceeding of 84th Annual Meeting Transportation Research Board*, Washington, DC., January 2005.

Morisue, F., & Ikeda, K. (1989). Evaluation of Map-Matching Techniques. *Proceedings of Vehicle Navigation and Information Systems Conference Record*, Toronto, Canada, September 1989.

Nassreddine, G., Abdallah, F., & Denoeux, T. (2009) Map Matching Algorithm Using Interval Analysis and Dempster-Shafer Theory. *Proceedings of IEEE Intelligent Vehicles Symposium (IV'09)*, Xi'an ,China, June 2009.

Quddus, M., Ochieng, W., Zhao, L., & Noland, R. (2003). A General Map Matching Algorithm for Transportation Telematics Applications. *GPS Solutions*, Vol. 7, No. 3, pp. 157-167.

Quddus, M., Noland, R., & Ochieng, W. (2006). A High Accuracy Fuzzy Logic Based Map Matching Algorithm for Road Transport. *Journal of Intelligent Transportation Systems*, Vol. 10, No. 3, pp. 103-115.

Schlingelhof, M., Betaille, D., Bonnifait, P., and Demaseure, K. (2008) Advanced Positioning Technologies for Co-operative Systems. *IET Intelligent Transport Systems*, Vol. 2, No. 2, pp. 81-91.

Sheridan, K., T. Dyjas, C. Botteron, J. Leclere, F. Dominic, and G. Marucco (2011) Demands of Roads. An Assisted-GNSS Solution Uses the EGNOS Data Access Service. *GPS World*, Vol. 22, No. 3.

Taylor, G., Blewitt, G., Steup, D., Corbett, S., & Car, A. (2001). Road Reduction Filtering for GPS-GIS Navigation. *Transactions in GIS*, Volume 5, No. 3, pp. 193-207.

Toledo-Moreo, R., Betaille, D., & Peyret, F. (2010) Lane-Level Integrity Provision For Navigation And Map Matching With GNSS, Dead Reckoning, And Enhanced Maps. *IEEE Transactions on Intelligent Transportation Systems*. Vol. 11, No. 1, pp. 100-113.

Velaga, N., Quddus, M., & Bristow, A. (2009). Developing an Enhanced Weight-Based Topological Map-Matching Algorithm for Intelligent Transport Systems. *Transportation Research Part C: Emerging Technologies*, Vol. 17, No. 6, pp. 672-683.

Vonderohe, A., Malhotra, A., Sheth, V., Mezera, D., & Adams, A. (2001). *Report Wisconsin Winter Maintenance Concept Vehicle: Data Management Year 1.* Department of Civil and Environmental Engineering, University of Wisconsin-Madison.

Wang, Z., & Yang, Z. (2009) Research on the Map Matching of Typical Region Based on the Topological Analysis. *Proceedings of 2nd International Conference on Intelligent Computation Technology and Automation*, Zhangjiajie, China October 2009.

White, C., Bernstein, D., & KornHauser, A. (2000). Some Map Matching Algorithms for Personal Navigation Assistants. *Transportation Research Part C: Emerging Technologies*, Vol. 8, No. 1, pp. 91-108.

Wolf, P. & Ghilani, C. (1997). *Adjustment Computations-Statistics and Least Squares in Surveying and GIS*, John Wiley & Sons, ISBN 0-471-16833-5, New York.

Yang, D., Cai, B., & Yuan, Y. (2003). An Improved Map-Matching Algorithm Used in Vehicle Navigation System. *Proceedings of IEEE International Conference on Intelligent Transportation Systems*, Vol. 2, Shanghai, China, October 2003.

Zhao, Y. (1997). *Vehicle location and navigation systems*, Artech House, Inc., ISBN 0-89006-861-5, Norwood, MA.

Zhao, L., Ochieng, W., Quddus, M., & Noland, R. (2003). An Extended Kalman Filter Algorithm for Integrating GPS and Low Cost Dead Reckoning System Data for Vehicle Performance and Emissions Monitoring. *Journal of Navigation*, Vol. 56, pp. 257-275.

10

Improved Inertial/Odometry/GPS Positioning of Wheeled Robots Even in GPS-Denied Environments

Eric North[1], Jacques Georgy[2], Umar Iqbal[3], Mohammed Tarbochi[4] and Aboelmagd Noureldin[5]

[1]*Canadian Forces Aerospace and Telecommunications Engineering Support Squadron*
[2]*Trusted Positioning Inc.*
[3]*Electrical and Computer Engineering Department, Queen's University*
[4]*Electrical and Computer Engineering Department, Royal Military College*
[5]*Electrical and Computer Engineering Department, Royal Military College/Queen's University*
Canada

1. Introduction

As described by Pacis et al (Pacis et al., 2004) the control strategy from a navigational viewpoint used in a mobile platform ranges from tele-operated to autonomous. A tele-operated platform is a platform having no on-board intelligence and whose navigation is guided in real-time by a remote human operator. An autonomous platform is one that takes its own decisions using onboard sensors and processor. According to Pacis et al (Pacis et al., 2005) for autonomous mobile robot navigation the problems that must be dealt with are localization, path planning, obstacle avoidance and map building. The focus of this work is in the localization problem.

Localization is the problem of estimating robot's pose relative to its environment from sensor observations. Localization is a necessity for successful mobile robot systems, it has been referred to as "the most fundamental problem to providing a mobile robot with autonomous capabilities" (Cox, 1991). Furthermore, as confirmed in (Pacis et al., 2004) to achieve autonomous navigation the robot must maintain an accurate knowledge of its position and orientation. Successful achievement of all other navigation tasks depends on the robot's ability to know its position and orientation accurately. According to a review by Borenstein et al (Borenstein et al., 1997) of mobile robot positioning technologies, positioning systems are divided into seven categories falling in two groups. They classified the positioning techniques as: relative position measurement and absolute position measurement. The former includes odometry and inertial navigation while the latter includes magnetic compass, active beacons, global positioning system (GPS), landmark navigation and map-based positioning.

An unprecedented surge of developments in mobile robot outdoor navigation was witnessed after the US government removed selective availability (SA) of GPS. Examples of applications for these robots are autonomous lawnmowers and motorized wheelchairs. These devices are low-cost and are used on terrain that is not flat. GPS can be used to provide three-dimensional knowledge of the mobile robot's position. Unfortunately, GPS suffers from outages when line-of-sight is blocked between the robot and GPS satellites. These outages are caused by operating the robot in and around buildings, dense foliage and other obstructions. An inertial measurement unit (IMU), with three accelerometers and three gyroscopes, is a good choice in lieu of GPS during outages for providing a 3-D positioning solution. Since a low-cost solution is needed for certain mobile robots, a low-cost IMU based on a micro electro-mechanical system (MEMS) has to be used. However, MEMS-based inertial sensors suffer from several complex errors such as biases; moreover these errors have influential stochastic parts. Since inertial navigation systems (INS) involve integration operations using sensor readings, the subsequent errors will accumulate and cause a rapid degrade in the quality of position estimate. Odometry using wheel encoders is another type of dead reckoning that provides limited localization information, mostly two-dimensional (2-D). This information is not subject to the same magnitude of errors as the IMU, provided that the vehicle does not encounter excessive skidding or slipping. But these 2-D solutions will not be adequate if the robot often moves outside the horizontal plane.

While 2-D and 3-D solutions using sensors in a full-sized vehicle have been done in the work to-date, further research is needed in the area of 3-D localization of small wheeled mobile robots operating in large 3-D terrain. The majority of the previous work using small mobile robots shows that the terrain is flat and the paths of the robots are small (for example (Ohno et al., 2003)(Ollero et al., 1999)(Chong & Kleeman, 1997)). This work attempts to bridge the gap between full-sized vehicle navigation in 3-D and navigation of small wheeled mobile robots over large paths in uneven terrain. Furthermore, this work will provide a 3-D solution for a small wheeled mobile robot that is required to travel distances in excess of 1 km over hilly terrain.

This work aims at combining the advantages of inertial sensors and odometry while mitigating their disadvantages to provide enhanced low-cost mobile robot 3-D localization capabilities during GPS outages. This will be achieved through the use of a Kalman Filter (KF) that integrates odometry from wheel encoders, low cost MEMS-based inertial sensors and GPS in a loosely-coupled scheme. To enhance the performance and lower the cost further, the proposed technique uses a reduced inertial sensor system (RISS). To further enhance the solution during GPS outages, velocity updates computed from wheel speeds are used to reduce the drift of the estimated solution. Moreover, this work proposes the development of a predictive error model used in a KF for estimating the errors in positions, velocities and azimuth angle from RISS mechanization. The experimental results will show that this error model when combined in a KF with 3-D measurement updates of velocities using forward speed from encoders together with pitch and azimuth estimates is a good technique for greatly reducing localization errors.

The structure of the rest of this chapter is as follows: Section 2 presents the methodology used. It describes the equations used to implement the RISS mechanization and KF error-models. Section 3 is a description of the mobile robot and the setup used in the experiments. Section 4 presents the results and discussion of this work. Finally, Section 5 is the conclusion and discussion of future work.

2. Methodology

2.1 Reduced inertial sensor system

In addition to MEMS-based sensors, the concept of RISS is used in a navigation scheme for a full-sized vehicle (Iqbal et al., 2008) in order to further lower the cost of the positioning solution. The RISS used in (Iqbal et al., 2008) involves a single-axis gyroscope and two-axis accelerometer together with a built-in vehicle speed sensor to provide a 2-D navigation solution in denied GPS environments. With the assumption that the vehicle remains mostly in the horizontal plane, the vehicleï£¡s speed sensor is used with heading information obtained from the vertically-aligned gyroscope to determine the velocities along the East and North directions. Consequently, the vehicles longitude and latitude are determined. If pitch and roll angles are needed the two accelerometers pointing towards the forward and transverse directions are used together with odometer-derived speed and a reliable gravity model to determine these angles independently of the integration filter. In (Iqbal et al., 2009), 2-D RISS/GPS integration were presented using Kalman filter (KF) for a full-sized vehicle.

In this work a low-cost navigation system using a KF to integrate MEMS-based RISS with GPS in a loosely-coupled scheme is described. The RISS used herein is 3-D: it includes a three-axis accelerometer and a single-axis gyroscope aligned with the vertical axis of the body frame of the robot together with two wheel encoders. Here accelerometers are used to calculate 3-D velocity and position while the vertical gyroscope is used to calculate the azimuth angle (i.e. the heading of the robot). Pitch and roll are calculated based on the idea presented in (Noureldin et al., 2002)(Noureldin et al., 2004) using the two horizontal accelerometers and the forward velocity obtained from wheel encoders. This constitutes the RISS mechanization.

The benefits of eliminating the other two gyroscopes in this RISS mechanization scheme are as follows: (1) further decreases in system costs and (2) improvements in positioning accuracy by employing fewer inertial sensors and thus less contribution of sensor errors towards positional errors. Of particular importance is the reduction in error in pitch and roll calculations. Whereas full mechanization of pitch and roll from gyroscopes involves integration, their calculation in RISS mechanization using accelerometers does not. This last fact decreases the portion of positional error originating from pitch and roll errors.

2.2 Coordinate transformation from local level Iram (LLF) to body frame (B-F)

The local level frame (LLF) serves to represent mobile robot attitude and velocity for operation on or near the surface of the earth and is defined as an origin, x, y and z-axis. The origin coincides with the center of the sensor frame (origin of inertial sensor triad). The y-axis points to true north. The x-axis points to east. Finally, the z-axis completes the right-handed coordinate systems pointing up, perpendicular to the reference ellipsoid.

One of the important direction cosine matrices for specifying rotation between one coordinate frame to another is $\mathbf{R}_b^l$ which transforms a vector from b-frame to LLF, a requirement during the mechanization process. $\mathbf{R}_b^l$ is expressed in terms of yaw, pitch and roll Euler angles is defined as:

$$\mathbf{R}_b^l = \begin{bmatrix} \cos\psi\cos\theta - \sin\psi\sin\rho\sin\theta & -\sin\psi\cos\rho & \cos\psi\sin\theta + \sin\psi\sin\rho\sin\theta \\ \sin\psi\cos\theta + \cos\psi\sin\rho\sin\theta & \cos\psi\cos\rho & \sin\psi\sin\theta - \cos\psi\sin\rho\cos\theta \\ -\cos\rho\sin\theta & \sin\rho & \cos\rho\cos\theta \end{bmatrix} \quad (1)$$

2.3 Mobile robot odometry equation

The conventions and notation presented in (Chong & Kleeman, 1997) are used to create a kinematic model for the mobile robot. In this work, a simple model for mobile robot kinematics is considered. The wheels must be as thin as possible (one rolling point-of-contact between the terrain and each wheel). Also, there must not be any slipping along the longitudinal direction. There must not be any sliding along the transverse direction.

Define the instantaneous center of curvature (ICC) as a means of describing the curvilinear motion that the mobile robot makes on a plane. In a two-dimensional environment the plane that the robot travels on remains fixed for all possible positions and orientations of the mobile robot (some authors refer to the reference frame enclosed in this plane as the "global reference frame"(Chong & Kleeman, 1997). In this work, motion of the mobile robot on possibly distinct planes for each time interval is considered. The mobile robot travels on a plane that is fixed from time $k-1 \leq t \leq k+1$.

$$V_{T_k} = \frac{r_R\omega_{R_k} + r_L\omega_{L_k}}{2} \quad (2)$$

where:

V_{T_k} is the velocity of the robot measured from its center and tangent to the circular path contained on a plane from time $k-1 \leq t \leq k+1$;

r_R is the radius of the right drive wheel;

r_L is the radius of the left drive wheel;

ω_{R_k} is the angular velocity of the right drive wheel from time $k-1 \leq t \leq k+1$;

ω_{L_k} is the angular velocity of the left drive wheel from time $k-1 \leq t \leq k+1$;

k represents discrete time epochs; and

Δt is the sampling time.

Rotational speeds of the left and right drive wheels are measured using encoders which are used to calculate the forward velocity of the robot. The forward velocity is transformed into velocities in the local frame using the equations below. Equation 2 is expressed in the mobile robot frame. In order for us to use the velocities of the robot's wheels as a measurement update we must transform these quantities to the local frame. Using the following transformation we have:

$$V_{e_k}^{odo} = V_{T_k}\cos(\rho_k)\sin(\alpha_k) \quad (3)$$

$$V_{n_k}^{odo} = V_{T_k}\cos(\rho_k)\cos(\alpha_k) \quad (4)$$

$$V_{u_k}^{odo} = V_{T_k}\sin(\rho_k) \quad (5)$$

2.4 RISS mechanization

2.4.1 Attitude equations

The equations for calculating pitch and roll from accelerometers are based on the idea presented in (Noureldin et al., 2002)(Noureldin et al., 2004). The robot acceleration derived from wheel encoder measurements is removed from the forward accelerometer measurement before computing the pitch angle. The equation for pitch ρ and neglecting acceleration in the forward direction since the robot travels at very low speeds is as follows:

$$\rho = \sin^{-1}\left(\frac{f_y - a_f}{g}\right) \approx \sin^{-1}\left(\frac{f_y}{g}\right) \tag{6}$$

Where:

f_y is the forward accelerometer reading;

g is the acceleration due to gravity; and

a_f is the forward acceleration and is derived from the forward velocity of the robot calculated from the average velocity measured by the wheel encoders from Equation 2.

The transverse accelerometer has to be compensated for the normal acceleration of the vehicle and then it is used to calculate the roll angle. The equation for roll θ:

$$\theta = -\sin^{-1}\left(\frac{f_x + V_f \omega_z}{g\cos\rho}\right) \tag{7}$$

Where:

f_x is the transversal accelerometer reading; and

ω_z is the vertical gyroscope reading.

The equation for the time-rate-of-change of yaw according to (Iqbal et al., 2009) using the previous value for V_e from RISS mechanization:

$$\dot{\psi} = \omega_z - \omega_e \sin\phi - V_e \frac{\tan\phi}{R+h} \tag{8}$$

Integrating in discrete time gives us:

$$\psi(k) = \psi(k-1) + \dot{\psi}(k)\,\Delta t \tag{9}$$

2.4.2 Velocity equations

There are three accelerometers that can be used to measure acceleration in the body frame of the mobile robot. Use these acceleration values to compute a velocity increment in the current time-step in order to compute an estimate for velocities. Use roll, pitch and yaw to calculate a rotation matrix $\mathbf{R}_b^l$ from the body frame to the local frame in Equation 1. Calculate a skew-symmetric matrix ω_{ie}^l for the earth's rotation rate since the last velocity calculation. In addition, calculate the skew-symmetric matrix ω_{el}^l for the LLF change-of-orientation since last calculation.

Using $\mathbf{R}_b^l$ from Equation 1, calculate the skew-symmetric matrix for earth rotation rate ω_{ie}^l since the last velocity calculation:

$$\omega_{ie}^l = \begin{bmatrix} 0 & -\omega_e \sin\phi & \omega_e \cos\phi \\ \omega_e \sin\phi & 0 & 0 \\ -\omega_e \cos\phi & 0 & 0 \end{bmatrix} \tag{10}$$

In addition, calculate the skew-symmetric matrix for the L-frame change of orientation ω_{el}^l since last calculation:

$$\omega_{el}^l = \begin{bmatrix} 0 & -\frac{V_e \tan\phi}{N+h} & \frac{V_e}{N+h} \\ \frac{V_e \tan\phi}{N+h} & 0 & \frac{V_n}{M+h} \\ -\frac{V_e}{N+h} & -\frac{V_n}{M+h} & 0 \end{bmatrix} \tag{11}$$

Use the following equation to provide velocity increments of the mobile robot in the body frame:

$$\Delta\vec{V}^b = \begin{bmatrix} f_x\Delta t \\ f_y\Delta t \\ f_z\Delta t \end{bmatrix} \tag{12}$$

With the three matrices $\mathbf{R}_b^l$, ω_{ie}^l and ω_{el}^l, the effect of gravity in the local frame in addition to the body-frame velocity increments $\Delta\vec{V}^b$ the new velocities are calculated by determining the rate-of-change for velocity increments $\Delta\vec{V}^l$ in the local frame as follows:

$$\Delta\vec{V}^l = \mathbf{R}_b^l\Delta\vec{V}^b - \left(2\omega_{ie}^l + \omega_{el}^l\right)\vec{V}^l\Delta t + \vec{g}^l\Delta t \tag{13}$$

Where $\vec{g}^l = [0 \quad 0 \quad -g]^T$. Integration is performed using the previous values for velocities $\vec{V}^l$ at time $k-1$ to get $\vec{V}^l$ at time k using $\Delta\vec{V}^l$ as follows:

$$\vec{V}^l(k) = \vec{V}^l(k-1) + 0.5\left[\Delta\vec{V}^l(k) + \Delta\vec{V}^l(k-1)\right] \tag{14}$$

2.4.3 Position equations

The equations for altitude h, latitude ϕ and longitude λ are as follows:

$$h(k) = h(k-1) + 0.5\left[V_u(k) + V_u(k-1)\right]\Delta t \tag{15}$$

$$\phi(k) = \phi(k-1) + 0.5\left[V_n(k) + V_n(k-1)\right]\frac{\Delta t}{R+h} \tag{16}$$

$$\lambda(k) = \lambda(k-1) + 0.5\left[V_e(k) + V_e(k-1)\right]\frac{\Delta t}{(R+h)\cos\phi} \tag{17}$$

It should be noted that any uncompensated bias or drift error in the accelerometer data will lead to growing errors when integrating acceleration to get velocity and again when integrating to get position. Furthermore, any uncompensated bias or drift error in the vertical gyroscope reading will lead to error growth when integrating to get yaw and again (together with velocity) to get position.

2.5 Kalman filtering

KF is the most commonly used technique for INS/GPS integration (Farrell & Barth, 1998)(Grewal et al., 2007). fig 1 shows a top-level view of the KF-based system used in this chapter for outdoor mobile robot localization. As mentioned previously, a loosely-coupled integration scheme is adopted in this chapter.

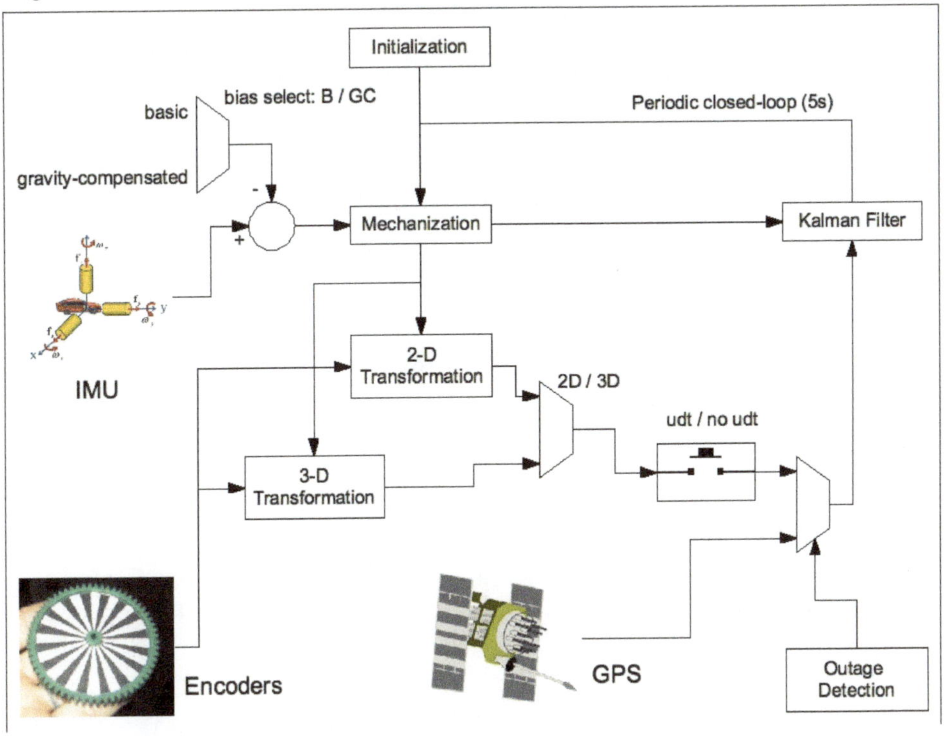

Fig. 1. An overview of the KF-based system used for outdoor mobile robot localization.

2.5.1 Error state vector

Since KF requires linearized models it estimates the error state not the navigation states themselves. The errors in the navigation states estimated by the filter are then used to correct the mechanization output and provide corrected navigation states. Leveraging the benefits of wheel encoders during GPS outages the KF presented in this section uses an error vector containing eleven states.

The linearized error-state system model used by the KF in this work is in the form:

$$\delta\dot{\vec{x}} = \mathbf{F}\delta\vec{x} + \mathbf{G}W(t) \tag{18}$$

The error state vector in Equation 18 consists of position errors (for latitude, longitude and altitude), velocity errors (along East, North and vertical Up), yaw error and inertial sensor stochastic drift for the single gyroscope and three accelerometers:

$$\delta\dot{\vec{x}} = \begin{bmatrix} \delta\dot{\phi} & \delta\dot{\lambda} & \delta\dot{h} & \delta\dot{V}_e & \delta\dot{V}_n & \delta\dot{V}_u & \delta\dot{\psi} & \delta\dot{\omega}_z & \delta\dot{f}_x & \delta\dot{f}_y & \delta\dot{f}_z \end{bmatrix}^T \tag{19}$$

The following sections contain derivations of the equations for each error state in the model. These equations use first order terms from the Taylor series expansion of the mechanization equations.

2.5.2 Position errors

From (Noureldin et al., 2009) the position components of the mechanization equations are linearized, yielding three error equations for latitude, longitude and altitude. Neglecting higher-order terms of the Taylor Series and writing in matrix form gives:

$$\delta\dot{\vec{r}}^l = \begin{bmatrix} \delta\dot{\phi} \\ \delta\dot{\lambda} \\ \delta\dot{h} \end{bmatrix} = \begin{bmatrix} 0 & 0 & -\frac{V_n}{(M+h)^2} \\ \frac{V_e \tan\phi}{(N+h)\cos\phi} & 0 & -\frac{V_e}{(N+h)^2\cos\phi} \\ 0 & 0 & 0 \end{bmatrix} \begin{bmatrix} \delta\phi \\ \delta\lambda \\ \delta h \end{bmatrix} + \begin{bmatrix} 0 & \frac{1}{M+h} & 0 \\ \frac{1}{(N+h)\cos\phi} & 0 & 0 \\ 0 & 0 & 1 \end{bmatrix} \begin{bmatrix} \delta V_e \\ \delta V_n \\ \delta V_u \end{bmatrix} \tag{20}$$

Where M and N are the respective Meridian and normal radii of the curvature of the Earth.

2.5.3 Velocity errors

The velocity components from the mechanization equations are linearized to provide velocity error equations; these equations are presented in (Noureldin et al., 2009). The velocity errors are function of errors in position, velocity, and attitude as well as accelerometer stochastic drift errors.

$$\delta\dot{\vec{V}}^l = \begin{bmatrix} \delta\dot{V}_e \\ \delta\dot{V}_n \\ \delta\dot{V}_u \end{bmatrix} = \begin{bmatrix} 2\omega_e V_u \sin\phi + 2\omega_e V_n \cos\phi + \frac{V_n V_e}{(N+h)\cos^2\phi} & 0 & 0 \\ -2\omega_e V_e \cos\phi - \frac{V_e^2}{(N+h)\cos^2\phi} & 0 & 0 \\ -2\omega_e V_e \sin\phi & 0 & \frac{2g}{M+h} \end{bmatrix} \begin{bmatrix} \delta\phi \\ \delta\lambda \\ \delta h \end{bmatrix}$$
$$+ \begin{bmatrix} -\frac{V_u}{N+h} + \frac{V_n \tan\phi}{N+h} & 2\omega_e \sin\phi + \frac{V_e \tan\phi}{N+h} & -2\left(\omega_e \cos\phi + \frac{V_e}{N+h}\right) \\ -2\left(\omega_e \sin\phi + \frac{V_e \tan\phi}{N+h}\right) & -\frac{V_u}{M+h} & -\frac{V_n}{M+h} \\ 2\left(\omega_e \cos\phi + \frac{V_e}{N+h}\right) & 2\frac{V_n}{M+h} & 0 \end{bmatrix} \begin{bmatrix} \delta V_e \\ \delta V_n \\ \delta V_u \end{bmatrix}$$
$$+ \begin{bmatrix} 0 & f_u & -f_n \\ -f_u & 0 & f_e \\ f_n & -f_e & 0 \end{bmatrix} \begin{bmatrix} \delta\rho \\ \delta\theta \\ \delta\psi \end{bmatrix} + \mathbf{R}_b^l \begin{bmatrix} \delta f_x \\ \delta f_y \\ \delta f_z \end{bmatrix} \tag{21}$$

In this work the errors in pitch $\delta\rho$ and $\delta\theta$ roll are not modelled inside the KF. This is because they don't suffer from error growth due to lack of integration operations. Therefore there are no dynamic error states for pitch and roll errors. Instead, expressions for $\delta\rho$ and $\delta\theta$ in $\delta\dot{\vec{V}}^l$ composed of other error states and pitch and roll equations in RISS mechanization are derived. The equation for velocity error is then re-arranged to accommodate the error terms belonging to $\delta\rho$ and $\delta\theta$.

The following is a derivation for the expression for $\delta\rho$ in Equation 21 using Equation 6. Take the derivative of the error component of ρ to give $\delta\rho$:

$$\delta\rho = \left[\frac{d}{df_y}\rho\right]\delta f_y = \left[\frac{d}{df_y}\sin^{-1}\left(\frac{f_y}{g}\right)\right]\delta f_y = \frac{1}{g\sqrt{1-\left(\frac{f_y}{g}\right)^2}}\delta f_y \tag{22}$$

A similar operation is performed for the error in roll, keeping in mind that the aim is to find an alternate expression for $\delta\theta$ that contains error terms other than $\delta\theta$ and $\delta\rho$. Using Equation 7, the partial derivatives of each component of θ in Equation 21 are used to give:

$$\delta\theta = \left[\frac{\partial}{\partial\boldsymbol{\theta}}\theta\right]\delta\boldsymbol{\theta} = -\left[\frac{\partial}{\partial\boldsymbol{\theta}}\sin^{-1}\left(\frac{f_x + V_f\omega_z}{g\cos\rho}\right)\right]\delta\boldsymbol{\theta} \tag{23}$$

Where $\delta\boldsymbol{\theta} = \begin{bmatrix}\delta V_f & \delta\rho & \delta\omega_z & \delta f_x\end{bmatrix}^T$. Taking partial derivatives results in:

$$\delta\theta = -\frac{1}{\sqrt{1-\left(\frac{f_x+V_f\omega_z}{g\cos\rho}\right)^2}}\left(\frac{\omega_z}{g\cos\rho}\delta V_f + \frac{f_x+V_f\omega_z}{g}\left(\frac{\sin\rho}{\cos^2\rho}\right)\delta\rho + \frac{V_f}{g\cos\rho}\delta\omega_z + \frac{1}{g\cos\rho}\delta f_x\right) \tag{24}$$

Express $\delta\rho$ and its associated terms in Equation 24 using the components from Equation 22 as follows:

$$\begin{aligned}\frac{f_x+V_f\omega_z}{g}\left(\frac{\sin\rho}{\cos^2\rho}\right)\delta\rho &= \frac{f_x+V_f\omega_z}{g}\left(\frac{\sin\rho}{\cos^2\rho}\right)\left(\frac{1}{g\sqrt{1-\left(\frac{f_y}{g}\right)^2}}\delta f_y\right)\\ &= \left(\frac{\left(f_x+V_f\omega_z\right)\sin\rho}{g^2\cos^2\rho\sqrt{1-\left(\frac{f_y}{g}\right)^2}}\right)\delta f_y\end{aligned} \tag{25}$$

Express δV_f contained in $\delta\theta$ from Equation 24 in terms of the three velocities along the east, north and up channels:

$$\begin{aligned}\delta V_f &= \left[\frac{\partial}{\partial\vec{V}_f}V_f\right]\delta\vec{V}_f = \left[\frac{\partial}{\partial\vec{V}_f}\sqrt{V_e^2+V_n^2+V_u^2}\right]\delta\vec{V}_f\\ &= \frac{1}{\not{2}\sqrt{V_e^2+V_n^2+V_u^2}}\left[\not{2}V_e\ \not{2}V_n\ \not{2}V_u\right]\begin{bmatrix}\delta V_e\\ \delta V_n\\ \delta V_u\end{bmatrix} = \frac{1}{V_f}\left[V_e\ V_n\ V_u\right]\begin{bmatrix}\delta V_e\\ \delta V_n\\ \delta V_u\end{bmatrix}\end{aligned} \tag{26}$$

Using Equations 25 and 26, Equation 24 can be rewritten as:

$$\delta\theta = -\left(\frac{\omega_z}{V_f\left(g\cos\rho\right)\sqrt{1-\left(\frac{f_x+V_f\omega_z}{g\cos\rho}\right)^2}}\right)\left[\,V_e\;V_n\;V_u\,\right]\begin{bmatrix}\delta V_e\\ \delta V_n\\ \delta V_u\end{bmatrix}$$
$$-\frac{V_f}{g\cos\rho\sqrt{1-\left(\frac{f_x+V_f\omega_z}{g\cos\rho}\right)^2}}\delta\omega_z$$
$$-\frac{1}{\sqrt{1-\left(\frac{f_x+V_f\omega_z}{g\cos\rho}\right)^2}}\begin{bmatrix}\frac{1}{g\cos\rho} & \frac{\left(f_x+V_f\omega_z\right)\sin\rho}{g^2\cos^2\rho\sqrt{1-\left(\frac{f_y}{g}\right)^2}}\end{bmatrix}\begin{bmatrix}\delta f_x\\ \delta f_y\end{bmatrix} \quad (27)$$

Using Equations 22 and 27 the equation for the attitude components of the velocity error can be re-arranged to accommodate the error terms belonging to ρ and θ. Similar terms will be grouped to produce a set of equations that describe the attitude errors within $\delta\dot{\vec{V}}^l$. The term $\delta\dot{\vec{V}}^l_{\text{att}}$ will be used to describe the attitude-portion of the velocity errors states. Once the equation for the components of velocity errors due to attitude errors is described it can be easily combined with the other terms in the velocity error states.

$$\delta\dot{\vec{V}}^l_{\text{att}} = \begin{bmatrix}0 & f_u & -f_n\\ -f_u & 0 & f_e\\ f_n & -f_e & 0\end{bmatrix}\begin{bmatrix}\delta\rho\\ \delta\theta\\ \delta\psi\end{bmatrix} = \begin{bmatrix}-f_n\\ f_e\\ 0\end{bmatrix}\delta\psi + \begin{bmatrix}\frac{-f_uV_f}{g\cos\rho\sqrt{1-\left(\frac{f_x+V_f\omega_z}{g\cos\rho}\right)^2}}\\ 0\\ \frac{f_eV_f}{g\cos\rho\sqrt{1-\left(\frac{f_x+V_f\omega_z}{g\cos\rho}\right)^2}}\end{bmatrix}\delta\omega_z$$
$$+\begin{bmatrix}\frac{-f_u}{g\cos\rho\sqrt{1-\left(\frac{f_x+V_f\omega_z}{g\cos\rho}\right)^2}} & \frac{-f_u\left(f_x+V_f\omega_z\right)\sin\rho}{g^2\cos^2\rho\sqrt{1-\left(\frac{f_y}{g}\right)^2}\sqrt{1-\left(\frac{f_x+V_f\omega_z}{g\cos\rho}\right)^2}}\\ 0 & \frac{-f_u}{g\sqrt{1-\left(\frac{f_y}{g}\right)^2}}\\ \frac{f_e}{g\cos\rho\sqrt{1-\left(\frac{f_x+V_f\omega_z}{g\cos\rho}\right)^2}} & \frac{f_e\left(f_x+V_f\omega_z\right)\sin\rho}{g^2\cos^2\rho\sqrt{1-\left(\frac{f_y}{g}\right)^2}\sqrt{1-\left(\frac{f_x+V_f\omega_z}{g\cos\rho}\right)^2}}+\frac{f_n}{g\sqrt{1-\left(\frac{f_y}{g}\right)^2}}\end{bmatrix}\begin{bmatrix}\delta f_x\\ \delta f_y\end{bmatrix}$$
$$+\begin{bmatrix}\sqrt{\sigma^2_{V_f}}\\ 0\\ \sqrt{\sigma^2_{V_f}}\end{bmatrix}W(t) \quad (28)$$

Experimental data shows that the forward speed originating from the encoders does not suffer from stochastic errors such as stochastic scale factor. This is due to the fact that the robot's wheels are essentially rigid. Therefore the error in forward speed δV_f contained in $\delta\boldsymbol{\theta}^T$ is expressed as a white noise term using the standard deviation of δV_f. The standard deviation is added to the process noise coupling state vector **G**.

2.5.4 Heading error

As mentioned earlier, pitch and roll do not have an error model in this work. This is because errors in pitch and roll do not accumulate as for the other measurements due to a lack of integration operations. To obtain an expression for heading error, use the yaw expression from mechanization for yaw ψ and linearize it by taking the first order terms in the Taylor series expansion. An expression for the error in yaw (and consequently azimuth) is determined:

$$\delta\dot{\psi} = \begin{bmatrix} -\left(\omega_e \cos\phi + \frac{V_e \csc\phi}{R+h}\right) & \frac{V_e \tan\phi}{(R+h)^2} & \frac{\tan\phi}{R+h} & 1 \end{bmatrix} \begin{bmatrix} \delta\phi \\ \delta h \\ \delta V_e \\ \delta\omega_z \end{bmatrix} \tag{29}$$

2.5.5 Measurement model

Section 2.5.1 described the error-state system model for the KF. The KF also needs a measurement model to be used in the update stage. There are two measurement update models used in this work. The first is when GPS is available and the second is used when there is a GPS outage. During GPS availability both GPS position and velocity are used and the differences between the RISS mechanization position and velocity and those of GPS are used as a measurement. The measurement model is as follows:

$$\vec{z}_k^{GPS} = \mathbf{H}_k^{GPS} \vec{x}_k + \vec{v}_k^{GPS} \tag{30}$$

Where the measurement state vector $\vec{z}_k^{GPS}$ is defined as:

$$\vec{z}_k^{GPS} = \begin{bmatrix} \phi_k^{INS} - \phi_k^{GPS} \\ \lambda_k^{INS} - \lambda_k^{GPS} \\ h_k^{INS} - h_k^{GPS} \\ V_{e_k}^{INS} - V_{e_k}^{GPS} \\ V_{n_k}^{INS} - V_{n_k}^{GPS} \\ V_{u_k}^{INS} - V_{u_k}^{GPS} \\ \omega_{z_k}^{INS} - \omega_{z_k}^{GPS} \end{bmatrix} \tag{31}$$

The design matrix **H** is:

$$
\mathbf{H}_k^{GPS} = \begin{bmatrix} 1\,0\,0\,0\,0\,0\,0\,0\,0\,0\,0\,0\,0\,0\,0 \\ 0\,1\,0\,0\,0\,0\,0\,0\,0\,0\,0\,0\,0\,0\,0 \\ 0\,0\,1\,0\,0\,0\,0\,0\,0\,0\,0\,0\,0\,0\,0 \\ 0\,0\,0\,1\,0\,0\,0\,0\,0\,0\,0\,0\,0\,0\,0 \\ 0\,0\,0\,0\,1\,0\,0\,0\,0\,0\,0\,0\,0\,0\,0 \\ 0\,0\,0\,0\,0\,1\,0\,0\,0\,0\,0\,0\,0\,0\,0 \end{bmatrix} \tag{32}
$$

And $\vec{v}_k^{GPS}$ is a white noise process with zero mean and unity variance. When GPS is not available, forward velocity derived from the wheel encoders together with pitch and azimuth estimates are used as a measurement update. The measurement model is as follows:

$$
\vec{z}_k^{odo} = \mathbf{H}_k^{odo} \vec{x}_k + \vec{v}_k^{odo} \tag{33}
$$

Where the measurement state vector $\vec{z}_k^{odo}$ is defined as:

$$
\vec{z}_k^{odo} = \begin{bmatrix} V_{e_k}^{INS} - V_{e_k}^{odo} \\ V_{n_k}^{INS} - V_{n_k}^{odo} \\ V_{u_k}^{INS} - V_{u_k}^{odo} \end{bmatrix} \tag{34}
$$

The design matrix **H** is:

$$
\mathbf{H}_k^{odo} = \begin{bmatrix} 0\,0\,0\,1\,0\,0\,0\,0\,0\,0\,0\,0\,0\,0\,0 \\ 0\,0\,0\,0\,1\,0\,0\,0\,0\,0\,0\,0\,0\,0\,0 \\ 0\,0\,0\,0\,0\,1\,0\,0\,0\,0\,0\,0\,0\,0\,0 \end{bmatrix} \tag{35}
$$

And $\vec{v}_k^{odo}$ is a white noise process having zero mean and unity variance.

3. Experimental setup

3.1 Wheeled mobile robot

Outdoor trajectory tests are conducted to assess the performance of the developed navigation solution; the tests are conducted using a mobile robot shown in fig 2. The robot was developed by a member of NavINST and members of the Electrical and Computer Engineering Department (ECE) at Royal Military College (RMC) of Canada. The mobile robot is three-wheeled and differentially-driven with a quadrature optical encoder coupled to the drive outputs of each motor. Appropriate scaling in the navigation scheme is used to provide angular velocity estimates of each wheel from these encoders. Power on-board the mobile robot comes from three sources, namely (1) two 12V batteries connected to the drive amplifiers and motors, (2) two 12V batteries connected to the sensors and encoder processor board and (3) a battery internal to the laptop mounted on the top level of the robot.

Fig. 2. The mobile robot used for the experiments in this work, custom-built by the author and members of the Electrical and Computer Engineering Department at Royal Military College of Canada.

3.2 Equipment

The inertial sensors used in this work include a MEMS-grade IMU made by Crossbow, model IMU300CC-100. Specifications of this IMU are in table 1 and detailed specifications can be found in (*IMU300CC - 6DOF Inertial Measurement Unit*, 2009). Velocity updates are provided by the forward speed of the robot, derived from encoders coupled to the drive output of each motor. The results of the presented navigation solution are evaluated with respect to a reference solution made by NovAtel where a Honeywell HG1700 high-end tactical grade IMU is integrated with a NovAtel OEM4 GPS receiver. The IMU and GPS receiver are integrated with a G2 Pro-Pack SPAN unit which is an off-the-shelf system developed by NovAtel. The details of this system are described in (*SPAN Technology System User Manual OM-20000062*, 2005). Biases and scale factors for the HG1700 IMU are in table 2 and detailed specification can be found in (*HG1700 Inertial Measurement Unit*, 2009). The high-cost NovAtel SPAN system provides a reference solution to validate the proposed method which uses the low-cost MEMS-based sensors. The SPAN system is also used to examine the overall performance during some GPS outages intentionally introduced in post-processing. A basic block diagram of the sensor electronics on-board the mobile robot appears in fig 3.

4. Results and discussion

Trajectories are carried out using the mobile robot described in Section 3 and sensor data is collected to test the developed solution in post-processing. Four navigation solutions are compared in order to show the benefit of using RISS instead of a full IMU and the benefit of

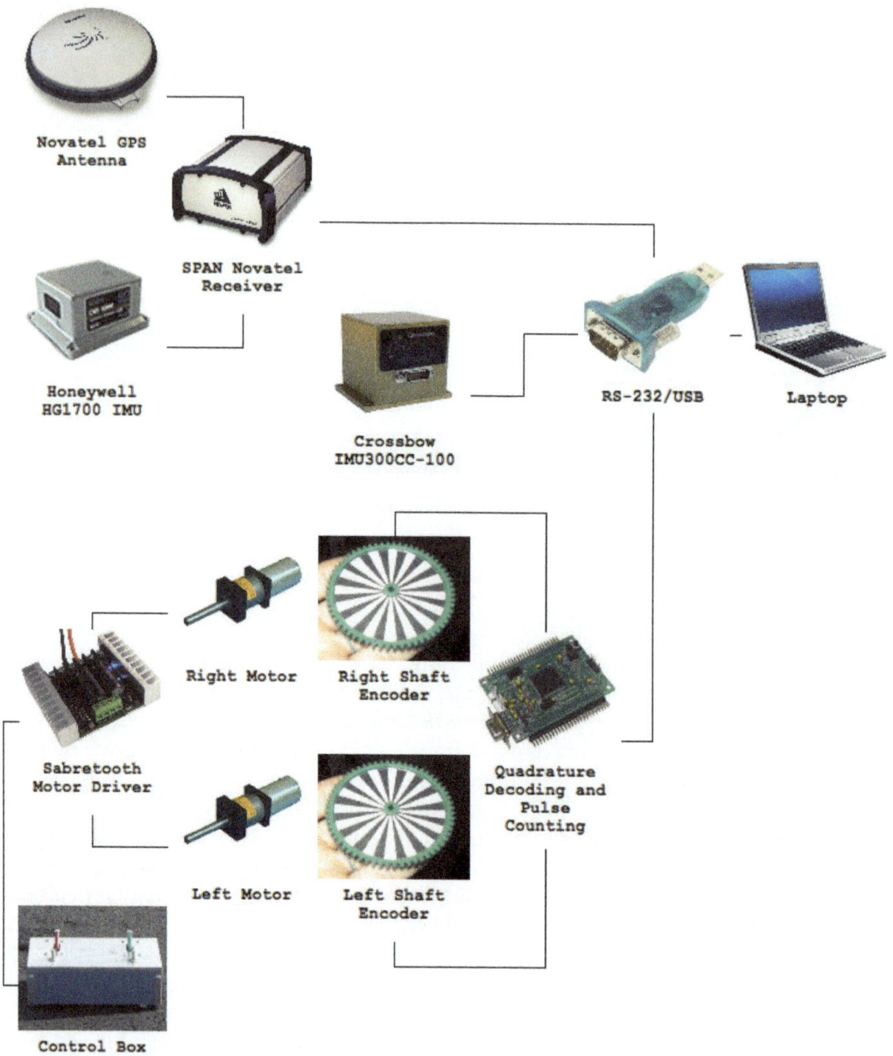

Fig. 3. Block diagram for the electronics on-board the mobile robot used in the experiments.

Crossbow IMU (IMU300CC)			
Gyroscopes		Accelerometers	
Bias, °/sec	$< \pm 2.000$	Bias, mg	$< \pm 30.000$
Scale Factor, %	< 1.000	Scale Factor, %	1.000
Random Walk, $°/\sqrt{hr}$	2.250	Random Walk, $m/(s\sqrt{hr})$	0.150

Table 1. Bias, scale factor error and random walk for the Crossbow IMU300CC IMU. Adapted from (*IMU300CC - 6DOF Inertial Measurement Unit*, 2009).

Honeywell IMU (HG1700)			
Gyroscopes		Accelerometers	
Bias, °/hr	1.000	Bias, mg	1.000
Scale Factor, ppm	150.000	Scale Factor, ppm	300.000
Random Walk, $°/\sqrt{hr}$	0.125		

Table 2. Bias, scale factor error and random walk for the Honeywell HG1700 IMU found in Novatel GPS/INS. Adapted from *SPAN Technology System User Manual OM-20000062* (2005) and *HG1700 Inertial Measurement Unit* (2009).

using velocity updates from wheel encoders during GPS outages. Each of the four navigation solutions are described as follows:

KF using RISS and velocity updates during GPS outages;

KF using RISS without updates during outages;

KF using full IMU with velocity updates during outages; and

KF using full IMU without updates during outages.

The errors in all the estimated solutions are calculated with respect to the NovAtel reference solution. Results for two trajectories are shown in this work. The first trajectory is shown in fig 4 and is located on-campus at RMC. It forms a loop with start and end at the same position and contains two different sections with hills both at an incline and decline to the robot's trajectory.

The second trajectory for this experiment is shown in fig 5 and is also located on-campus at RMC. This trajectory forms a loop with start and end at the same position and is much longer than the trajectory in fig 4. It contains several different sections which include hills both at an incline and decline to the robot's trajectory.

4.1 Trajectory 1

The ultimate check for the proposed systemï£¡s accuracy is during GPS signal blockage which can be intentionally introduced in post-processing. Since the presented solution is loosely coupled the outages represent complete blockages of GPS updates. Seven GPS outages are simulated with durations of 60 seconds each. The simulated outages are chosen such that they encompass straight portions, turns, and slopes.

Table 7 shows the root mean square (RMS) error in both the estimated 2-D horizontal position and the estimated altitude during seven GPS outages for the four compared solutions. The

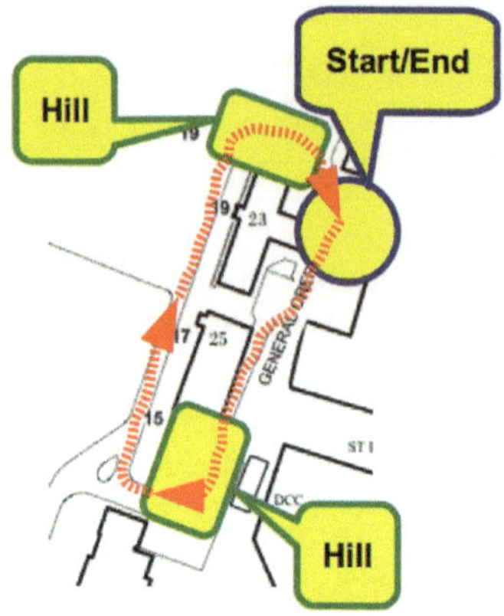

Fig. 4. The first trajectory for assessing each navigation solution.

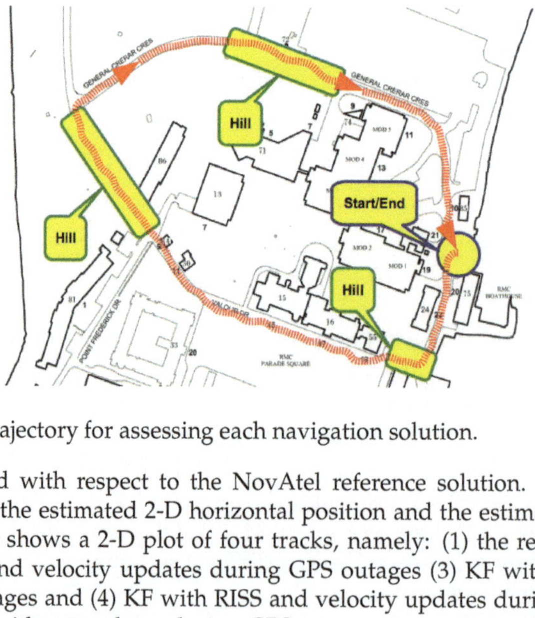

Fig. 5. The second trajectory for assessing each navigation solution.

errors are calculated with respect to the NovAtel reference solution. Table 8 shows the maximum errors in the estimated 2-D horizontal position and the estimated altitude during these outages. fig 6 shows a 2-D plot of four tracks, namely: (1) the reference solution, (2) KF with full IMU and velocity updates during GPS outages (3) KF with RISS and without updates during outages and (4) KF with RISS and velocity updates during outages. The KF with full IMU and without updates during GPS outages is not shown because the position errors would dramatically change the scale of the plot and make comparison of the other solutions very difficult.

The results in table 8 and fig 6 clearly show the advantage of RISS over a full IMU. There is a big difference in 2-D positional errors when one compares the results of KF with full IMU without updates with the results of KF with RISS without updates during GPS outages. While the former has an average of the maximum positional error for the seven GPS outages equal to 139.3 meters, the latter shows an error of 18.67 meters. The reason for this difference is the use of accelerometers vice the two gyroscopes used to get pitch and roll from the RISS.

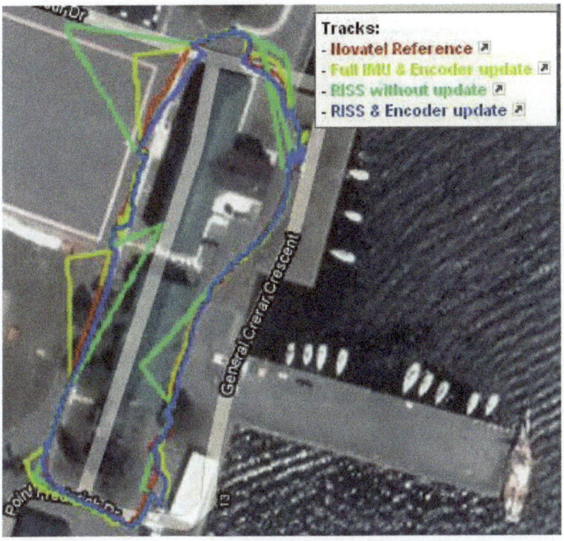

Fig. 6. Three solutions and reference for first trajectory: Red for reference, Yellow for KF using full IMU with velocity updates, Green for KF using RISS without updates, Blue for KF using RISS with velocity updates.

RMS Errors during GPS outages								
Outage No.	1	2	3	4	5	6	7	**Average**
Duration (s)	60	60	60	60	60	60	60	**60**
KF full IMU without updates								
h (m)	6.63	25.56	16.30	7.73	17.86	2.26	11.42	**12.54**
Pos 2D (m)	30.01	26.55	82.14	79.83	37.54	172.52	21.63	**64.32**
KF full IMU with velocity updates								
h (m)	1.39	13.91	17.22	18.07	1.46	5.38	6.11	**9.08**
Pos 2D (m)	5.64	1.61	3.32	3.54	7.28	9.10	4.53	**5.00**
KF RISS without velocity updates								
h (m)	1.20	4.96	9.21	0.27	1.02	0.66	3.86	**3.03**
Pos 2D (m)	16.43	6.32	2.47	4.07	10.75	21.49	7.21	**9.82**
KF RISS with velocity updates								
h (m)	1.43	0.68	3.61	0.44	0.43	1.38	1.86	**1.40**
Pos 2D (m)	2.48	1.21	2.96	0.76	2.39	3.05	1.60	**2.06**

Fig. 7. RMS errors for altitude and 2D position for the seven outages in first trajectory.

The benefit of using wheel encoders to provide velocity updates during GPS outages can be seen by two comparisons. The first comparison uses KF with full IMU and considers two sets of results, one with and the second without velocity updates. With velocity updates the solution has an average of the maximum positional error for the seven GPS outages equal to 8.77 meters while the case without updates has 139.3 meters of error. The second comparison uses KF with RISS and considers two sets of results, one with and the second without velocity updates. With velocity updates the solution has an average of the maximum positional error for the seven GPS outages equal to 3.35 meters while the case without updates has 18.67 meters of error.

Max Errors during GPS outages								
Outage No.	1	2	3	4	5	6	7	**Average**
Duration (s)	60	60	60	60	60	60	60	**60**
KF full IMU without updates								
h (m)	9.76	46.81	27.02	11.91	34.49	5.02	21.24	**22.32**
Pos 2D (m)	64.22	55.19	183.07	163.18	96.39	366.10	46.85	**139.29**
KF full IMU with velocity updates								
h (m)	4.96	24.31	29.16	28.15	1.69	8.97	9.46	**15.24**
Pos 2D (m)	8.98	2.00	5.62	7.44	13.23	17.00	7.15	**8.77**
KF RISS without velocity updates								
h (m)	4.97	9.46	13.36	0.61	2.17	1.84	4.72	**5.30**
Pos 2D (m)	34.42	12.40	4.83	9.26	17.88	38.39	13.51	**18.67**
KF RISS with velocity updates								
h (m)	3.51	2.83	3.75	0.58	0.82	1.71	2.04	**2.18**
Pos 2D (m)	3.36	1.99	5.91	1.23	3.21	4.79	2.97	**3.35**

Fig. 8. Maximum errors for altitude and 2D position for the seven outages in first trajectory.

When comparing KF with RISS and velocity updates to KF with full IMU and velocity updates the advantage of RISS can be clearly noticed. The former has an average of the maximum positional error for the seven GPS outages equal to 3.35 meters while the latter shows an error of 8.77 meters.

These results together with the trajectory plots in fig 6 demonstrate that the proposed 3-D localization solution using KF for RISS/GPS integration and employing velocity updates using wheel encoders outperforms all the other compared solutions. Furthermore this solution provides very good results when compared to the MEMS-based INS/GPS integration results in the literature.

4.2 Trajectory 2

The second trajectory is much longer than the first one and enables the examination of long GPS outages. Seven outages are simulated with the duration of each outage equal to 150 seconds. This duration is chosen to test the performance of the proposed navigation solution in long GPS outages. The simulated outages are also chosen such that they encompass straight portions, turns and slopes.

Table 10 shows the root mean square (RMS) error in both the estimated 2-D horizontal position and the estimated altitude during the seven GPS outages for the four compared solutions. Table 11 shows the maximum errors in the estimated 2-D horizontal position and the estimated altitude during each outage. fig 9 shows a 2-D plot of four tracks, namely: (1) reference solution, (2) KF with full IMU and velocity updates during GPS outages, (3) KF with RISS and without updates during outages and (4) KF with RISS and velocity updates during outages. As mentioned earlier the KF with full IMU and without updates during GPS outages is not shown because the position errors would dramatically change the scale of the plot and make comparison of the other solutions very difficult.

The results in table 10 and table 11 confirm the results of the first trajectory and they further demonstrate the advantage of RISS over a full IMU. One can see a great difference in 2-D positional errors when comparing the results of KF with full IMU without updates with the results of KF with RISS without updates during during GPS outages. While the former technique has an average of the maximum positional error for the seven GPS outages equal to

789.2 meters the latter shows an error of 68.29 meters. The reason for this huge enhancement of performance is the use of accelerometers vice the two gyroscopes used to get pitch and roll from the RISS.

As seen in the first trajectory the benefit of using velocity updates during GPS outages derived from the wheel encoders is seen by two comparisons. The first comparison uses KF with full IMU and considers two sets of results, one with and the second without velocity updates. With velocity updates the solution has an average of the maximum positional error for the seven GPS outages equal to 11.44 meters while the case without updates has 789.2 meters of error.

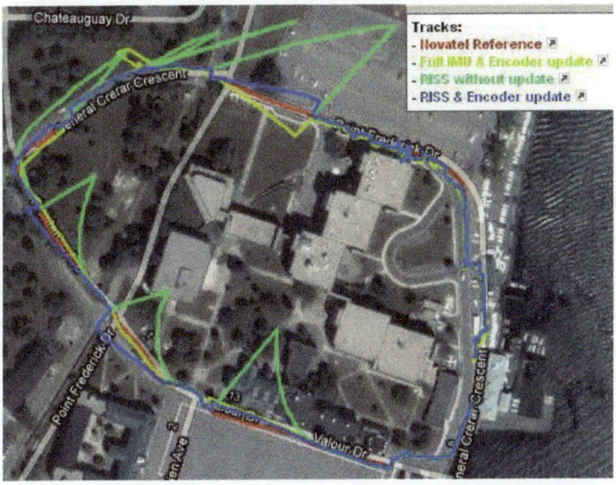

Fig. 9. Three solutions and reference for second trajectory: Red for reference, Yellow for KF using full IMU with velocity updates, Green for KF using RISS without updates, Blue for KF using RISS with velocity updates.

RMS Errors during GPS outages								
Outage No.	1	2	3	4	5	6	7	**Average**
Duration (s)	150	150	150	150	150	150	150	**150**
KF full IMU without updates								
h (m)	118.33	7.4394	70.7598	33.7392	49.571	138.992	393.058	**115.98**
Pos 2D (m)	224.16	395.08	474.40	155.13	209.99	654.03	308.26	**345.86**
KF full IMU with velocity updates								
h (m)	4.52	4.24	6.78	8.05	15.85	11.48	29.22	**11.45**
Pos 2D (m)	6.8248	3.8805	7.6824	3.4269	2.5439	13.9607	9.0778	**6.771**
KF RISS without velocity updates								
h (m)	32.44	19.20	20.55	33.36	50.89	24.52	50.80	**33.11**
Pos 2D (m)	39.03	22.72	48.03	55.68	57.21	27.28	52.12	**43.15**
KF RISS with velocity updates								
h (m)	2.86	2.39	2.33	3.18	2.09	4.52	3.86	**3.03**
Pos 2D (m)	4.35	9.58	5.71	4.26	0.85	1.04	7.41	**4.74**

Fig. 10. RMS errors for altitude and 2D position for the seven outages in second trajectory.

The second comparison uses KF with RISS and considers two sets of results, one with and the second without velocity updates. With velocity updates the solution has an average of

Max Errors during GPS outages								
Outage No.	1	2	3	4	5	6	7	**Average**
Duration (s)	150	150	150	150	150	150	150	**150**
KF full IMU without updates								
h (m)	261.20	10.21	145.67	60.44	106.94	314.92	897.83	**256.75**
Pos 2D (m)	528.2	874.4	1069.2	294.7	466.8	1601.6	689.5	**789.2**
KF full IMU with velocity updates								
h (m)	6.40	6.93	11.50	16.71	22.81	21.63	50.73	**19.53**
Pos 2D (m)	10.11	6.46	10.92	6.37	5.19	22.61	18.43	**11.44**
KF RISS without velocity updates								
h (m)	47.20	31.74	89.46	83.49	53.21	31.73	18.87	**50.81**
Pos 2D (m)	72.67	41.55	45.74	70.76	112.95	49.31	85.07	**68.29**
KF RISS with velocity updates								
h (m)	3.33	2.78	2.75	3.47	4.16	5.06	4.2	**3.68**
Pos 2D (m)	6.92	16.32	9.52	6.68	1.24	1.31	11.52	**7.64**

Fig. 11. Maximum errors for altitude and 2D position for the seven outages in second trajectory.

the maximum positional error for the seven GPS outages equal to 7.64 meters while the case without updates has 68.29 meters of error.

When comparing KF with RISS and velocity updates to KF with full IMU and velocity updates the advantage of RISS can be seen especially in the altitude component. The former has an average of the maximum positional error for the seven GPS outages equal to 7.64 meters while the latter shows an error of 11.44 meters. The former has an average of the maximum altitude error for the seven GPS outages equal to 3.68 meters while the latter shows an error of 19.53 meters.

These results together demonstrate that the 3-D localization solution using KF for RISS/GPS integration and employing velocity updates using wheel encoders outperforms all the other compared solutions. Furthermore this solution provides very good results when compared to the MEMS-based INS/GPS integration results in the literature.

5. Conclusion and future work

This chapter presented an outdoor 3-D localization solution for mobile robots using low-cost MEMS-based sensors, wheel encoders and GPS. A reduced inertial sensor system was used for both decreasing the cost and improving the performance. The integration was achieved using a loosely-coupled KF. In this work, a predictive error model for KF was developed for estimating the errors in positions, velocities and attitude provided by RISS mechanization. Using this error model inside the KF gives good results during GPS outages that outperformed the full IMU results. Furthermore, when this KF is used with measurement updates using forward velocity from encoders together with pitch and azimuth estimates (during GPS outages) it provides better results and outperforms all the compared solutions.

The positioning solutions in this work were tested with two real trajectories with seven simulated GPS outages whose duration was 60 seconds in the first trajectory and 150 seconds in the second trajectory. The proposed solutions were discussed and compared with each solution also compared against a reference solution. Considering the maximum error in horizontal positioning in the first trajectory, the KF with RISS and velocity updates during GPS outages achieved an average improvement of approximately 97.6% over KF with full IMU without any updates during outages, of approximately 61.8% over KF with full IMU with velocity updates during outages, and of approximately 82.0% over KF with RISS without any updates during outages. Considering the maximum error in horizontal positioning in the first trajectory, the KF with RISS and velocity updates during GPS outages achieved an average improvement of approximately 99.0% over KF with full IMU without any updates during outages, of approximately 33.2% over KF with full IMU with velocity updates during outages, and of approximately 88.8% over KF with RISS without any updates during outages. These results show the superiority of the proposed localization solution.

One problem unique to small wheeled robots with strap-down navigation systems is that there is a great deal of chassis rigidity that passes along any disturbances felt at the wheels of the robot. Small, low-cost robots do not have suspension systems found on full-size vehicles which prevent many disturbances from being measured by the accelerometers of a strap-down IMU. A future investigation is required regarding low-cost measures for dampening some of the vibrations caused by small obstacles and imperfections on the road surface. Prospective researchers should make a careful selection of tires for their small mobile robot that allow moderate deformation to small obstacles while preserving sufficient shape to maintain reliable estimates for velocities measured by the wheel encoders.

Kalman filtering is a good technique for reducing the stochastic error of a system since it requires little processing time compared to other algorithms. It is a suitable choice for deployment in low-cost, low-power, low-form-factor systems such as those found on small mobile robots. Further study is required to determine the performance of the techniques outlined in this work in the context of an embedded system operating in real-time.

6. Acknowledgements

To our family and friends for their love, support and commitment. This chapter wouldn't have been possible without them.

7. References

Borenstein, J., Everett, H., Feng, L. & Wehe, D. (1997). Mobile robot positioning: Sensors and techniques, *Journal of Robotic Systems* 14(4): 231–249.

Chong, K. S. & Kleeman, L. (1997). Accurate odometry and error modelling for a mobile robot, *Proceedings of the 1997 IEEE International Conference on Robotics and Automation*, Vol. 4, Albuquerque, NM, pp. 2783–2788.

Cox, I. J. (1991). Blanche - an experiment in guidance and navigation of an autonomous robot vehicle, *IEEE Transactions on Robotics and Automation* 7(2): 193–204.

Farrell, J. A. & Barth, M. (1998). *The Global Positioning System & Inertial Navigation*, McGraw-Hill.

Grewal, M. S., Weill, L. R. & Andrews, A. P. (2007). *Global Positioning Systems, Inertial Navigation, and Integration*, John Wiley and Sons.

HG1700 Inertial Measurement Unit (2009).
URL: *http://www51.honeywell.com/aero/common/documents/myaerospacecatalog-documents /Missiles-Munitions/HG1700_Inertial_Measurement_Unit.pdf*

IMU300CC - 6DOF Inertial Measurement Unit (2009).
URL: *www.xbow.com/Products/Product_pdf_files/Inertial_pdf/IMU300CC_Datasheet.pdf*

Iqbal, U., Karamat, T. B., Okou, A. F. & Noureldin, A. (2009). Experimental results on an integrated gps and multi sensor system for land vehicle positioning, *International Journal of Navigation and Observation* 2009.

Iqbal, U., Okou, A. F. & Noureldin, A. (2008). An integrated reduced inertial sensor system - riss/gps for land vehicle, *IEEE/ION Position, Location and Navigation Symposium (PLANS) 2008*, Monterey, California, USA, pp. 912–922.

Noureldin, A., Irvine-Halliday, D. & Mintchev, M. P. (2002). Accuracy limitations of fog-based continuous measurement-while-drilling surveying instruments for horizontal wells, *IEEE Transactions on Instrumentation and Measurement* 51(6): 1177–1191.

Noureldin, A., Irvine-Halliday, D. & Mintchev, M. P. (2004). Measurement-while-drilling surveying of highly-inclined and horizontal well sections utilizing single-axis gyro sensing system, *Measurement Science and Technology* 15(12): 2426–2434.

Noureldin, A., Karamat, T., Eberts, M. D. & El-Shafie, A. (2009). Performance enhancement of mems based ins/gps integration for low cost navigation applications, *IEEE Transactions on Vehicular Technology* 58(3): 1077–1096.

Ohno, K., Tsubouchi, T., Shigematsu, B., Maeyama, S. & Yuta, S. (2003). Outdoor navigation of a mobile robot between buildings based on DGPS and odometry data fusion, *Proceedings of the IEEE International Conference on Robotics and Automation*, Vol. 2, IEEE Press, pp. 1978–1984.

Ollero, A., Arrue, B. C., Ferruz, J., Heredia, G., Cuesta, F., López-Pichaco, F. & Nogales, C. (1999). Control and perception components for autonomous vehicle guidance. application to the romeo vehicles, *Control Engineering Practice* 7(10): 1291–1299.

Pacis, E., Everett, H., Farrington, N. & Bruemmer, D. (2004). Enhancing functionality and autonomy in man-portable robots, *SPIE Proc. 5804: Unmanned Ground Vehicle Technology VII*, Orlando, FL, USA.

Pacis, E., Everett, H., Farrington, N., Kogut, G., Sights, B., Kramer, T., Thompson, M., Bruemmer, D. & Few, D. (2005). Transitioning unmanned ground vehicle research technologies, *SPIE Proceedings 5804: Unmanned Ground Vehicle Technology VII*, Orlando, FL, USA.

SPAN Technology System User Manual OM-20000062 (2005).
URL: *www.novatel.com/Documents/Manuals/om-20000062.pdf*

11

Beyond Trilateration: GPS Positioning Geometry and Analytical Accuracy

Mohammed Ziaur Rahman
University of Malaya
Malaysia

1. Introduction

Trilateration/multilateration is the fundamental basis for most GPS positioning algorithms. It begins by finding range estimates to known satellite positions which provides a spherical Locus of Position (LOP) for the receiver. Ideally four such spherical LOPs can be solved to precisely determine the receiver position. Thus, it is an analytical approach that finds receiver position by solving required number of linear/quadratic equations. This method can determine the receiver position precisely when the equations are perfectly formulated. However, determining the exact range is nearly impossible in real-life due to many external factors such as noise interference, signal fading, multi-path propagation, weather condition, clock synchronization problem etc (Strang & Borre, 1997). Hence, trilateration fails to achieve sufficient accuracy under real world conditions. It is also argued that GPS algorithms are not at all tri/multi-lateration rather they are difference of measurement (time-difference or second order difference of two ranges) based hyperbolic formulations (Chaffee & Abel, 1994). However, there are widely used useful range-based algorithms such as Bancroft (1985) method. Therefore, trilateration is still predominantly associated with positioning (Bajaj et al., 2002).

In this chapter, we first discuss about the analytical accuracy of trilateration based positioning algorithms. Subsequently, we show how noise can impact positioning accuracy in real world. In Section 3, we present existing analytical algorithms for GPS along with two new analytical approaches using Paired Measurement Localization (PML) of (Rahman & Kleeman, 2009). PML approaches can cope up with practical improper range based equations and are computationally efficient for implementation by conventional and resource constraint GPS receivers. Section 4 draws some conclusions for this chapter.

2. Trilateration: its problems and alternative approaches

As alluded before, analytical approaches of positioning are based on accurate distance measurement from geo-stationary satellites. Trilateration is the basis of these techniques where the range measurements from $n+1$ satellites are used for an n-dimensional position estimation (Caffery, 2000).

In the ideal scenario when we can measure the precise range estimates of the GPS receiver, we can formulate a spherical locus of position for the receiver. The fundamental positioning geometry using three satellites placed in a hypothetical 2-Dimensional space is shown in Fig. 1(a).

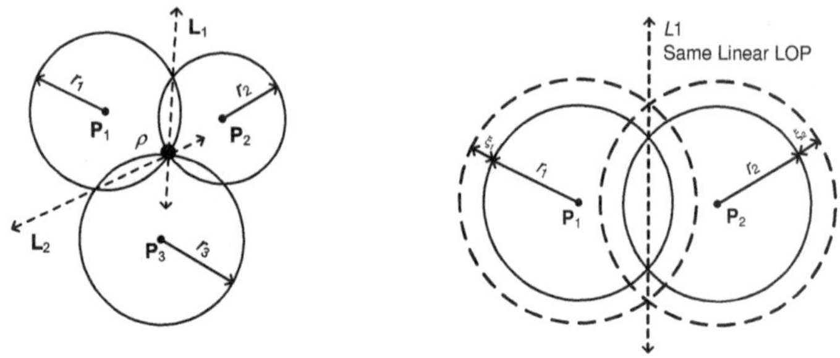

(a) Ideal 2-D trilateration scenario where linear form LOPs are found from the corresponding two circular LOPs (Caffery, 2000).

(b) LOP from equidistant satellites in presense of equal noise.

Fig. 1. Depiction of observation 1.

The circles surrounding the satellites with known positions $\mathbf{p}_1(x_1, y_1), \mathbf{p}_2(x_2, y_2)$ and $\mathbf{p}_3(x_3, y_3)$, denote the LOPs obtained from the individual range measurements for each satellite. Ideally, the LOPs surrounding satellite i is given by,

$$r_i^2 = \|\mathbf{p}_i - \rho\|^2 = \left((x - x_i)^2 + (y - y_i)^2\right) \tag{1}$$

In 2-D, it is feasible to calculate the exact receiver position using only three range measurements. Two range measurements can result in two solutions corresponding to the intersection of two circular LOPs. The third measurement resolves this ambiguity.

However, equating two circular LOPs will result in a straight line equation (in case of 3-D, it will be planar equation) passing through two intersecting points of the circular LOPs. This line does not represent the actual locus of the receiver position as it will be clarified later. However, following (Caffery, 2000) this line is referred as *Linear Form LOP* in the subsequent discussions. In Fig. 1, L_1 and L_2 are determined from the circular LOPs corresponding to satellite pairs ($\mathbf{p}_1$, $\mathbf{p}_2$) and ($\mathbf{p}_1$, $\mathbf{p}_3$) respectively, with the intersection point (x, y) of L_1 and L_2 denoting the actual position of the receiver.

As shown in Fig. 1,the positioning geometry works correctly for ideal case of exact range estimates being measured by the positioning devices. However, in reality it is quite difficult to measure the exact range both for external noise impact and internal errors such as receiver clock bias and satellite clock skews. However, we also showed the fact that accurate positioning can be obtained if the noise effect is exactly the same for two satellites. However, in case of variable noise presence for two satellite range estimates usual linear form LOP obtained from circular LOPs deviates significantly from the true position of the receiver and leads to a bad positioning geometry. This is further explained as follows.

As it is clarified before that the range equations are mostly not accurate in practical scenario. Though trilateration is a mathematical approach and ideally it can find the exact receiver position, however it cannot find the position very well when the range estimates are perturbed by noise. In this section we will specifically identify the problems of trilateration for inaccurate range equations. For the ease of understanding we still limit this discussion for 2-dimensions only.

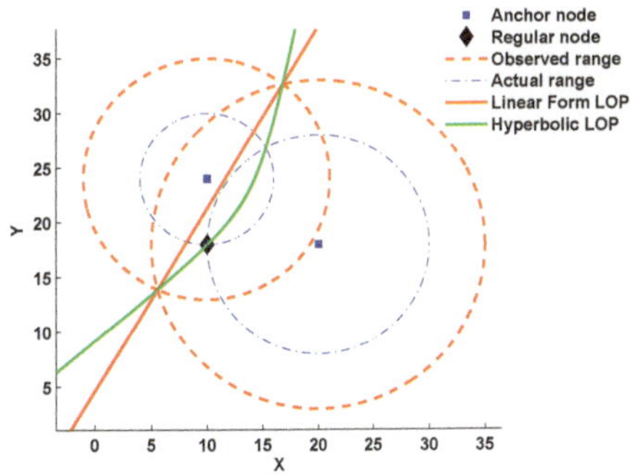

(a)

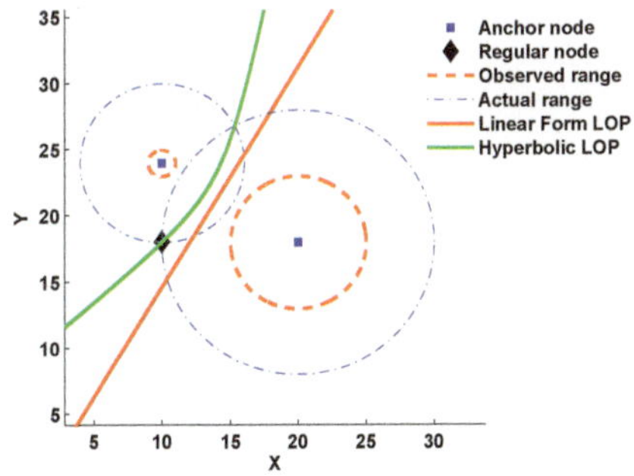

(b)

Fig. 2. The hyperbolic and linear form LOP of a receiver from range estimates by a pair of satellites under equal noise assumption. (a) The general case when two observed circular LOPs physically intersect. (b) The case when circular LOPs do not intersect due to noise and underestimation of the ranges.

At first, we present the following observation that identifies the case when the conventional trilateration works in consideration of noise.

Observation 1. *Assuming a receiver uses range estimates from two satellites that are located at the same distance from the receiver and have equal noise components, it is shown below that the locus of positions for that receiver (as the error components vary) is a straight line whose equation is independent of range estimates.*

Assume that due to noise, the range measurements for $\mathbf{p_1}\,(x_1, y_1)$, $\mathbf{p_2}\,(x_2, y_2)$ and $\mathbf{p_3}\,(x_3, y_3)$ are corrupted to give respective LOPs of radii $\tilde{r}_1 = r_1 + \xi_1$, $\tilde{r}_2 = r_2 + \xi_2$ and $\tilde{r}_3 = r_3 + \xi_3$, where $\tilde{r}_i$, r_i represent the observed and actual distance (pseudorange and actual range respectively) between the ith satellite and receiver respectively and ξ_i is the measurement noise at the receiver corresponding to the measurement. The circular LOP can then be expressed as:

$$(r_i + \xi_i)^2 = ||\mathbf{p}_i - \rho||^2 \tag{2}$$

where $\rho = (x, y)$ is the receiver position to be determined.

Equating the circular LOPs for $\mathbf{p}_1$ and $\mathbf{p}_2$ using (2), L_1 becomes:

$$\begin{aligned}&(x_2 - x_1)\,x + (y_2 - y_1)\,y = \\ &\tfrac{1}{2}\left(||\mathbf{p}_2||^2 - ||\mathbf{p}_1||^2 + (r_1 + \xi_1)^2 - (r_2 + \xi_2)^2\right)\end{aligned} \tag{3}$$

where the right hand side becomes independent of range parameters, *i.e.*, measurement values $\tilde{r}_1$ and $\tilde{r}_2$ whenever $\tilde{r}_1 = \tilde{r}_2 \Rightarrow r_1 + \xi_1 = r_2 + \xi_2$. One particular case is equidistant satellites and equal noise presence when the above condition is fulfilled. ■

The importance of this observation lies in the fact that it eliminates the signal propagation dependent parameters and receiver clock bias under assumed conditions completely. GPS measurements are mostly susceptible to these errors which are both device and environmentally dependent.

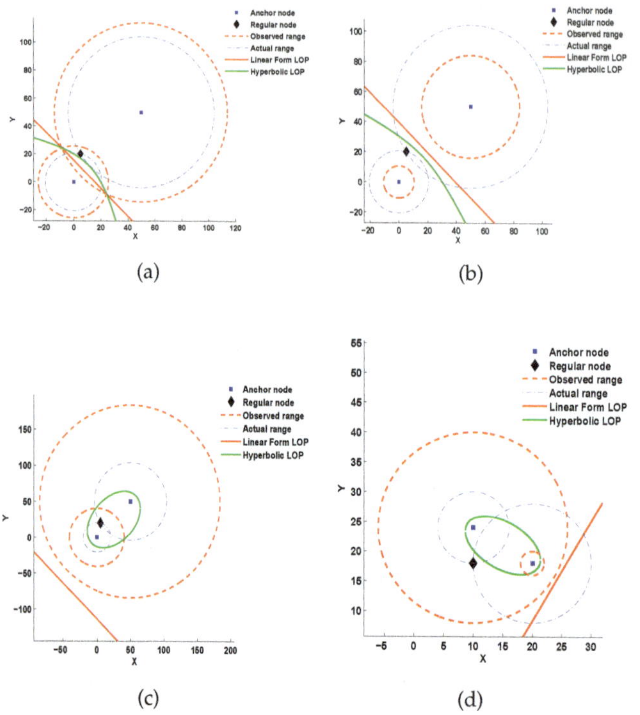

Fig. 3. The hyperbolic and linear form LOPs for unequal noise presence. (a) The general case when two observed circular LOPs physically intersect. (b) The case when observed circular LOPs do not intersect due to underestimation of the ranges. (c) The case when observed circular LOPs do not intersect but overlap completely due to overestimation of the ranges.(d) The case when ranging errors are of opposite signs.

Assuming equal noise presence, it is useful to explore paired measurements rather than individual ranges to mitigate the effect of noise. As the difference of the range estimates equate to actual difference for equal noise presence (*e.g.*, $\tilde{r_2} - \tilde{r_1} = r_2 - r_1$), the LOP for the receiver position is found by the locus of positions maintaining constant difference from the pair of satellites. Hence, the hyperbolic LOP of the receiver can be found independent of the noise parameters as shown in Fig. 2 and formulated below:

$$\sqrt{(x-x_2)^2+(y-y_2)^2} \quad - \quad \sqrt{(x-x_1)^2+(y-y_1)^2} \quad = \quad (\tilde{r_2}-\tilde{r_1}) \tag{4}$$

After algebraic manipulations, it takes the general hyperbolic form as follows for $\mathbf{p}_1 = (0,0), \mathbf{p}_2 = (a,0)$, and $\tilde{r_1} - \tilde{r_2} = c$.

$$\left(x-\frac{a}{2}\right)^2 - \frac{y^2}{\left(\frac{a^2}{c^2}-1\right)} = \frac{c^2}{4} \tag{5}$$

The hyperbolic LOP represents the actual LOP for a pair of satellites under the equal noise assumption. The linear form LOP does not truly represent the locus of the receiver in presence of noise unless both ranges to the satellites are equal as clarified in Fig. 2. Two possible cases could arise due to equal noise presence: *a*) the circular ranges have a physical intersection and *b*) the circular ranges do not have any physical intersection. In both cases, the hyperbolic LOP is able to represent the original receiver position whereas linear form LOP deviates from receiver position significantly. As establishing the LOP is the first step in positioning, any error present at this step could aggravate the result significantly and hence finding a LOP closer to the original receiver position is fundamental to achieving high accuracy positioning.

It is also crucial to compare the hyperbolic and linear form LOPs for unequal noise components in individual measurements as in reality this assumption can be void. In these general situations three possible cases could arise. *a*) the observed circular ranges have a physical intersection; *b*) the observed circular ranges do not have any common intersection region; and *c*) One of the observed circular ranges overlap completely within the other circular region.

These three cases are shown in Fig. 3 where Fig. 3(a), (b) shows the hyperbolic and linear form LOPs for noise ratio (ξ_1/ξ_2) of 2 while Fig. 3(c) shows the LOPs for noise ratio of 4. Fig. 3(c) also shows that for completely overlapped ranges the hyperbolic formulation turns into elliptic formulation. This is the case when coefficient of y^2 in (5) changes sign as the range difference becomes greater than distance between the satellites $(c > a)$. The noise presence generally attenuates the signal more than that of ideal propagation scenario causing overestimation of the range. However, it is theoretically possible to imagine the case where range is underestimated due to noise. The simultaneous overestimation and underestimation of ranges is supposed to be the most detrimental for LOP estimation and hence this case is shown in Fig. 3(d). It is evident from the figures that for all the three cases of unequal noise presence as well as for noise having different signs, hyperbolic formulation is better suited than linear form and the impact of noise is less detrimental on hyperbolic LOPs than it is on linear form LOPs.

3. Analytical approaches for global positioning

We have discussed about the mathematical basis for positioning and presented the problems of regular trilateration from the viewpoint of noisy measurements. The positioning algorithms for GPS need greater care for noise and often augmented by filtering process to mitigate the effect of noise. However, they still largely depend on basic analytical positioning both for initial estimation and for error correcting/filtering phase. In this chapter, we present the different analytical algorithms for GPS.

We begin with the 3-D analogous formula for equation 2 which represents a sphere.

$$\tilde{r}_i^2 = (r_i + \xi_i)^2 = \|\mathbf{p}_i - \rho\|^2 = \left((x - x_i)^2 + (y - y_i)^2 + (z - z_i)^2\right) \tag{6}$$

A generally acceptable modeling of the ranging error ξ_i is described by the following equation (Strang & Borre, 1997).

$$\xi_i = I_i + T_i + c\,(dt_i(t - \tau_i) - dt(t)) - e_i \tag{7}$$

where I_i is the ionospheric error, T_i is the tropospheric error, c is the speed of light, dt_i is the satellite clock offset, dt is the receiver clock offset, t is the receiver time and τ_i is the signal propagation time and e_i represents all other unmodelled error terms.

The equation 6 can be iteratively solved using Newton's method. However, the iterative approach will be computationally expensive. Moreover, the positioning accuracy will be poor as there is no proper formalism to identify and mitigate the error components.

3.1 Ordinary trilateration for positioning

Let $\mathbb{P}_i = \left(\mathbf{p}_1^i, \mathbf{p}_2^i\right)$ be an arbitrary satellite pair, where $\mathbf{p}_1^i = \left(x_1^i, y_1^i, z_1^i\right)$ and $\mathbf{p}_2^i = \left(x_2^i, y_2^i, z_2^i\right)$ represent satellite positions of the ith pair. Analogous to 2-D linear form LOP of equation 3, a 3-D planar form LOP is found as follows.

$$\begin{aligned}&\left(x_2^i - x_1^i\right)x + \left(y_2^i - y_1^i\right)y + \left(z_2^i - z_1^i\right)z = \\ &\tfrac{1}{2}\left(\|\mathbf{p}_2^i\|^2 - \|\mathbf{p}_1^i\|^2 + \left(r_1^i\right)^2 - \left(r_2^i\right)^2 + 2\xi\left(r_1^i - r_2^i\right)\right)\end{aligned} \tag{8}$$

Where it is assumed that the noise are equal and constant for a particular satellite pair *i.e.*, $\xi_1 = \xi_2 = \xi$.

The equation becomes linear in terms of x, y, z and ξ if the noise is represented by a single parameter ξ for all pairs. In that case there are four unknowns in this equation and therefore four equations will be required to solve them. In practicality, the assumption is susceptible for large positioning error and hence iterative refinement approach of the following is rather adopted for real implementations.

3.2 Iterative least squares estimate

The iterative approach works by having a preliminary estimate of the receiver position ($\rho^0 = [x^0 y^0 z^0]^T$). Let the rotation rate of the earth be ω. The position vectors in the earth centered earth fixed (ECEF) system of the receiver be donated by $\rho(t)_{ECEF}$ and geo-stationary position vector for satellite i be denoted by $\mathbf{p}_i(t)_{geo}$ where the argument t denotes the dependence on time. The range equation can be written as:

$$r_i = \|R_3(\omega\tau_i)\mathbf{p}_i(t - \tau_i)_{geo} - \rho(t)_{ECEF}\| \tag{9}$$

Where R_3 is the earth's rotation matrix as defined below.

$$R_3(\omega\tau_i) = \begin{bmatrix} \cos(\omega\tau_i) & \sin(\omega\tau_i) & 0 \\ -\sin(\omega\tau_i) & \cos(\omega\tau_i) & 0 \\ 0 & 0 & 1 \end{bmatrix}$$

Let

$$\begin{bmatrix} x_i \\ y_i \\ z_i \end{bmatrix} = R_3(\omega\tau_i)\mathbf{p}_i(t - \tau_i)_{geo} \ and \begin{bmatrix} x \\ y \\ z \end{bmatrix} = \rho(t)_{ECEF} \tag{10}$$

Now, omitting the refraction terms I_i and T_i and linearizing the equation 6, we get

$$-\frac{x_i - x^0}{(\tilde{r}_i)^0}\delta x - \frac{y_i - y^0}{(\tilde{r}_i)^0}\delta y - \frac{z_k - z^0}{(\tilde{r}_i)^0}\delta z + (c\,dt) = \tilde{r}_i - (\tilde{r}_i)^0 - \epsilon_i = b_i - \epsilon_i \tag{11}$$

where b_i denotes the correction to the preliminary range estimate.

When more than four observations are available we can compute the correction values $\langle \delta x, \delta y, \delta z \rangle$ for the preliminary estimate. The least squares formulation can be concisely written as follows.

$$\mathbf{Ax} = \begin{bmatrix} -\frac{x_1-x^0}{r_1} & -\frac{y_1-y^0}{r_1} & -\frac{z_1-z^0}{r_1} & 1 \\ -\frac{x_2-x^0}{r_2} & -\frac{y_2-y^0}{r_2} & -\frac{z_2-z^0}{r_2} & 1 \\ \vdots & \vdots & \vdots & \vdots \\ -\frac{x_m-x^0}{r_m} & -\frac{y_m-y^0}{r_m} & -\frac{z_m-z^0}{r_m} & 1 \end{bmatrix} \begin{bmatrix} \delta x \\ \delta y \\ \delta z \\ \delta c\, dt \end{bmatrix} = \mathbf{b} - \epsilon \tag{12}$$

The least squares solution is

$$\begin{bmatrix} \delta x \\ \delta y \\ \delta z \\ \delta c\, dt \end{bmatrix} = (A^T \Sigma^{-1} A)^{-1} A^T \Sigma^{-1} \mathbf{b} \tag{13}$$

If the code observations are independent and assumed to have equal variance, then the above can be simplified to

$$\begin{bmatrix} \delta x \\ \delta y \\ \delta z \\ \delta c\, dt \end{bmatrix} = (A^T A)^{-1} A^T \mathbf{b} \tag{14}$$

The final position vector can be estimated by $\rho = \left[x^0 + \delta x\; y^0 + \delta y\; z^0 + \delta z\right]^T$.

3.3 Bancroft's method (least squares solution)

We want to turn positioning into a linear algebra problem. Here is a clever method due to Bancroft (1985) that does some algbraic manipulations to reduce the equations to a least-squares problem. Multiplying things out in equation 6 and using the receiver clock bias $-b = \xi_i' s$ as the only noise parameter, we get

$$x_i^2 - 2x_i x + x^2 + y_i^2 - 2y_i y + y^2 + z_i^2 - 2z_i z + z^2 = r_i^2 - 2r_i b + b^2 \tag{15}$$

Rearranging,

$$\left(x_i^2 + y_i^2 + z_i^2 - r_i^2\right) - 2\left(x_i x + y_i y + z_i z - r_i b\right) + \left(x^2 + y^2 + z^2 - r^2\right) = 0 \tag{16}$$

Let $\rho = [x\, y\, z\, r]^T$ denote the receiver position vector and $\mathbf{p}_i = [x_i\, y_i\, z_i\, r_i]^T$ denote the ith satellite position and range vectors.

Using *Lorentz inner product* for 4-space defined by:

$$\langle \vec{u}, \vec{v} \rangle = u_1 v_1 + u_2 v_2 + u_3 v_3 - u_4 v_4$$

Equation 16 can be rewritten as:

$$\frac{1}{2}\langle \mathbf{p}_i, \mathbf{p}_i \rangle - \langle \mathbf{p}_i, \rho \rangle + \frac{1}{2}\langle \rho, \rho \rangle = 0; \tag{17}$$

In order to apply least squares estimation the equations for each satellite are organized as follows:

$$\mathbf{B} = \begin{bmatrix} x_1 & y_1 & z_1 & -r_1 \\ x_2 & y_2 & z_2 & -r_2 \\ \vdots & \vdots & \vdots & \vdots \\ x_m & y_m & z_m & -r_m \end{bmatrix},$$

$$\mathbf{a} = \frac{1}{2}\begin{bmatrix} \langle \mathbf{p}_1, \mathbf{p}_1 \rangle \\ \langle \mathbf{p}_2, \mathbf{p}_2 \rangle \\ \vdots \\ \langle \mathbf{p}_m, \mathbf{p}_m \rangle \end{bmatrix}, \mathbf{e} = \begin{bmatrix} 1 \\ 1 \\ \vdots \\ 1 \end{bmatrix}, \text{and } \wedge = \frac{1}{2}\langle \rho, \rho \rangle$$

We can now rewrite equation 17 as:

$$\begin{aligned} &\mathbf{a} - \mathbf{B}\rho + \wedge \mathbf{e} = 0 \\ &\Rightarrow \mathbf{B}\rho = \mathbf{a} + \wedge \mathbf{e} \end{aligned} \tag{18}$$

For more than 4 satellites, we can have closed form least squares solution as follows:

$$\rho = \mathbf{B}^{+}\mathbf{a} + \wedge \mathbf{e} \tag{19}$$

where $\mathbf{B}^{+} = (\mathbf{B}^T\mathbf{B})^{-1}\mathbf{B}^T$ is the pseudoinverse of Matrix **B**.

However, the solution ρ involves $\wedge$ which is defined in terms of unknown ρ. This problem is avoided by substituting ρ into the definition of the scalar $\wedge$ and using the linearity of the Lorentz inner product as follows:

$$\wedge = \frac{1}{2}\Big\langle \mathbf{B}^{+}(\mathbf{a} + \wedge \mathbf{e}), \mathbf{B}^{+}(\mathbf{a} + \wedge \mathbf{e}) \Big\rangle$$

After rearranging,

$$\wedge^2 \langle \mathbf{B}^{+}\mathbf{e}, \mathbf{B}^{+}\mathbf{e} \rangle + 2\wedge \Big(\langle \mathbf{B}^{+}\vec{\mathbf{e}}, \mathbf{B}^{+}\mathbf{a} \rangle - 1 \Big) + \langle \mathbf{B}^{+}\mathbf{a}, \mathbf{B}^{+}\mathbf{a} \rangle = 0 \tag{20}$$

This is a quadratic equation in $\wedge$ with coefficients $\langle \mathbf{B}^{+}\mathbf{e}, \mathbf{B}^{+}\mathbf{e} \rangle$, $2\left(\langle \mathbf{B}^{+}\mathbf{e}, \mathbf{B}^{+}\mathbf{a} \rangle - 1\right)$, and $\langle \mathbf{B}^{+}\mathbf{a}, \mathbf{B}^{+}\mathbf{a} \rangle$. All these three values can be computed and we can solve for two possible values of $\wedge$ using the quadratic equation. If we get the two solutions to this equation $\wedge_1$ and $\wedge_2$, then we can solve for two possible solutions ρ_1 and ρ_2 in equation 19. One of these solutions will make sense, it will be on the surface of the earth (which has a radius of approximately 6371 km), and one will not.

The major advantage of the Bancroft's method is to have a closed form least squares solution for GPS equations. It has the same advantage of least squares approach of using all the available satellites for location estimation. On the contrary, it uses the fundamental equation of spherical ranging that in the course of solution leads to planar form LOPs which are than hyperboloid LOPs. Therefore as discussed before, this method cannot be used for high-accuracy positioning in presence of noise.

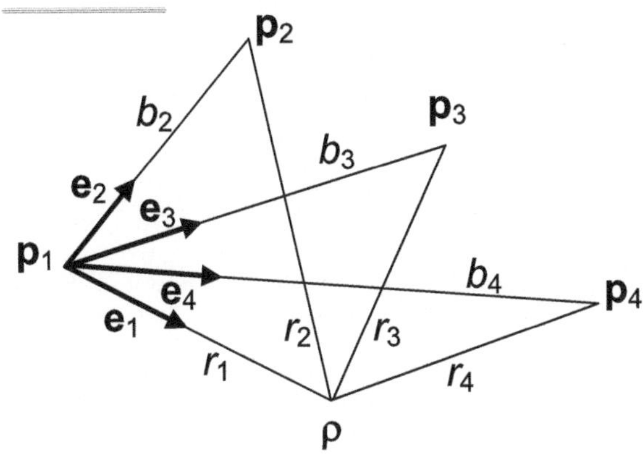

Fig. 4. Three dimensional vector representation for a receiver and four satellites.

3.4 Kleusberg's algorithm

Kleusberg (1994) provided a vector algebraic solution for GPS. The geometry of the 3-D positioning is shown in figure 4. It begins with the fundamental equation 6 for range estimates. It also uses difference equation given below analogous to equation 4 between two satellite measurements.

$$\sqrt{(x-x_i)^2+(y-y_i)^2+(z-z_i)^2}-\sqrt{(x-x_1)^2+(y-y_1)^2+(z-z_1)^2}=(\tilde{r}_i-\tilde{r}_1)=d_i \quad (21)$$

This represents a sheet of hyperboloid. We can find three such hyperboloids for $i = 2, 3$ and 4 that can be solved for determining the receiver position. Mathematically, there will be two solutions though one of which can be discarded from the knowledge of the earth's proximity.

Let b_2, b_3, b_4 be the known distances from satellite 1 to satellites 2, 3, 4 along unit vectors $\mathbf{e}_2, \mathbf{e}_3, \mathbf{e}_4$. From the cosine law for triangle $1-i-\rho$,

$$\tilde{r}_i^2 = b_i^2 + \tilde{r}_1^2 - 2b_i\tilde{r}_1\mathbf{e}_1\cdot\mathbf{e}_i \quad (22)$$

Squaring equation 21 and equating with $\tilde{r}_i^2$ of equation 22, we get

$$2\tilde{r}_1 = \frac{b_i^2-d_i^2}{d_i+b_i\mathbf{e}_1\cdot\mathbf{e}_i} \quad (23)$$

Using satellite pairs $(1,2), (1,3)$ and $(1,4)$; we can get three equations for $\tilde{r}_1$ as follows:

$$\frac{b_2^2-d_2^2}{d_2+b_2\mathbf{e}_1\cdot\mathbf{e}_2} = \frac{b_3^2-d_3^2}{d_3+b_3\mathbf{e}_1\cdot\mathbf{e}_3} = \frac{b_4^2-d_4^2}{d_4+b_4\mathbf{e}_1\cdot\mathbf{e}_4} \quad (24)$$

The only unknown in the above equation is the unit vector $\mathbf{e}_1$.

Some rewritings result in the two scalar equations as follows:

$$\begin{aligned}\mathbf{e}_1\cdot\mathbf{f}_2 &= u_2 \quad \text{and}\\ \mathbf{e}_1\cdot\mathbf{f}_3 &= u_3\end{aligned} \quad (25)$$

Where for $m = 2, 3$;

$$\mathbf{F}_m = \frac{b_m}{b_m^2 - d_m^2}\mathbf{e}_m - \frac{b_{m+1}}{b_{m+1}^2 - d_{m+1}^2}\mathbf{e}_{m+1}$$
$$\mathbf{f}_m = \frac{\mathbf{F}_m}{||\mathbf{F}_m||}$$
$$u_m = \frac{1}{||\mathbf{F}_m||}\left(\frac{d_{m+1}}{b_{m+1}^2 - d_{m+1}^2} - \frac{d_m}{b_m^2 - d_m^2}\right)$$

The unit vector $\mathbf{f}_2$ lies in the plane through satellites 1, 2 and 3. This plane is spanned by $\mathbf{e}_2$ and $\mathbf{e}_3$. Similarly $\mathbf{f}_3$ is in the plane determined by satellites 1, 3 and 4.

Equation 25 determines the cosine of the two unit vectors $\mathbf{f}_2$ and $\mathbf{f}_3$ with the desired unit vector $\mathbf{e}_1$. It will have two solutions, one above and one below the plane spanned by $\mathbf{f}_2$ and $\mathbf{f}_3$. In case these vectors are parallel their inner product is zero and there are infinitely many solutions and hence the position cannot be determined.

The algebraic solution to equation 25 can be derived using vector triple product identity,

$$\mathbf{e}_1 \times (\mathbf{f}_1 \times \mathbf{f}_2) = \mathbf{f}_1(\mathbf{e}_1 \cdot \mathbf{f}_2) - \mathbf{f}_2(\mathbf{e}_1 \cdot \mathbf{f}_1)$$

All the terms in the right hand of the above equation is readily computed using $\mathbf{u}_2, \mathbf{u}_3$. Substituting $\mathbf{h}$ for the right hand side and $\mathbf{g}$ for $\mathbf{f}_1 \times \mathbf{f}_2$, we get

$$\mathbf{e}_1 \times \mathbf{g} = \mathbf{h} \tag{26}$$

Multiplying both sides of the equation by $\mathbf{g}$ and applying the vector triple product identity,

$$\mathbf{e}_1(\mathbf{g} \cdot \mathbf{g}) - \mathbf{g}(\mathbf{g} \cdot \mathbf{e}_1) = \mathbf{g} \times \mathbf{h} \tag{27}$$

The scalar product in the second term of the left-hand side can be written in terms of the angle θ between unit vector $\mathbf{e}_1$ and $\mathbf{g}$ as follows

$$\mathbf{g} \cdot \mathbf{e}_1 = [\mathbf{g} \cdot \mathbf{g}]^{\frac{1}{2}} \cos\theta$$

The sine value of the angle can be found from equation 26 as follows:

$$[\mathbf{h} \cdot \mathbf{h}]^{\frac{1}{2}} = [(\mathbf{e}_1 \times \mathbf{g}) \cdot (\mathbf{e}_1 \times \mathbf{g})]^{\frac{1}{2}} = [\mathbf{g} \cdot \mathbf{g}]^{\frac{1}{2}} \sin\theta$$

Using the sine value in the cosine formula above, we obtain,

$$\mathbf{g} \cdot \mathbf{e}_1 = \pm [\mathbf{g} \cdot \mathbf{g}]^{\frac{1}{2}} \left[1 - \frac{\mathbf{h} \cdot \mathbf{h}}{\mathbf{g} \cdot \mathbf{g}}\right]^{\frac{1}{2}} = \pm [\mathbf{g} \cdot \mathbf{g} - \mathbf{h} \cdot \mathbf{h}]^{\frac{1}{2}}$$

Substituting the above into equation 27, we obtain the desired solution:

$$\mathbf{e}_1 = \frac{1}{2}\left(\mathbf{g} \times \mathbf{h} \pm \mathbf{g}\sqrt{\mathbf{g} \cdot \mathbf{g} - \mathbf{h} \cdot \mathbf{h}}\right) \tag{28}$$

The two values can be put in equation 24 to check the correctness of the value. The correct parameter will result in a intersection point that lies on the earth's surface and hence must have a distance of about 6371 km from the origin. We can eventually get the receiver coordinate using correct value of $\mathbf{e}_1$ as follows:

$$\rho = \mathbf{p}_1 + \tilde{r}_1 \mathbf{e}_1 \tag{29}$$

The Kleusberg's method is geometrically oriented and uses a minimum number of satellites. On the other hand, it cannot utilize more number of satellites even when they are available. This method is also dependent on the proper geometrical orientation of the satellites. Moreover, it often gives different results for different set of satellites and depending on the order of the satellites in solving the equations.

3.5 Paired measurement localization

In trilateration, the positioning works by simultaneous solution of three spherical LOP equations. Similar to the 2-D steps, we can equate two spherical LOP equations to find equation for a 2-D plane representing the planar locus of position. Analogous to 2-D case, three planar equations can be solved to find the ultimate receiver position.

As shown in section 2, the effect of noise will have detrimental impact on the aforementioned simple solution. On the other hand, instead of equating the two imprecise range equations we can maintain an equi-distant locus of position from two satellites as formulated in equation 21 for a hyperboloid LOP. This will be more accurate than a traditional 2-D planar LOP based positioning.

Solving the nonlinear hyperbolic/hyperboloid equations is difficult. Moreover, existing hyperbolic positioning methods proceed by linearizing the system of equations using either Taylor-series approximation (Foy, 1976; Torrieri, 1984) or by linearizing with another additional variable (Chan & Ho, 1994; Friedlander, 1987; Smith & Abel, 1987). However, while linearizing works well for existing approaches it is not readily adaptable for the proposed paired approach as linearizing is indeed pairing with an arbitrarily chosen hyperbolic LOP. The assumption of equal noise cannot be held for any arbitrary selection of pairs and hence alternate ways to solve such LOPs for paired measurement is now formulated.

3.6 PML with single reference satellite

(Chan & Ho, 1994) provided closed form least squares solution for non-linear hyperbolic LOPs by linearizing with reference to a single satellite. Analogous to their approach a closed form solution is found for PML using pairs having a common reference satellite in them. The solution is simpler than (Chan & Ho, 1994)'s approach as the effect of noise is considered early in the paired measurements formulations.

Let $r_{\widehat{ij}}$ represent the difference in the observed ranges for satellite pairs (i, j). In case of equal noise presence it follows:

$$\tilde{r}_{\widehat{ij}} = r_{\widehat{ij}} = r_i - r_j$$

$$\text{After squaring and rearranging,} \tag{30}$$

$$r_i^2 = r_{\widehat{ij}}^2 + 2 r_{\widehat{ij}} r_j + r_j^2$$

Hence, the actual spherical LOP can be transformed as follows:

$$(x - x_i)^2 + (y - y_i)^2 + (z - z_i)^2 = (r_{\widehat{ij}})^2 + 2 r_{\widehat{ij}} r_j + (r_j)^2 \tag{31}$$

Using (31) for pairs $(\mathbf{p}_i, \mathbf{p}_j) = (\mathbf{p}_k, \mathbf{p}_1)$ and $(\mathbf{p}_l, \mathbf{p}_1)$ and subtracting the second from the first,

$$-(x_k - x_l)x - (y_k - y_l)y - (z_k - z_l)y - (r_{\widehat{k1}} - r_{\widehat{l1}})r_1 = \frac{1}{2}\left((r_{\widehat{k1}})^2 - (r_{\widehat{l1}})^2 - ||\mathbf{p}_k||^2 + ||\mathbf{p}_l||^2\right) \quad (32)$$

where $||\mathbf{p}_k||^2 = (x_k^2 + y_k^2)$. The above formulation represents a set of linear equations with unknowns x, y, z and r_1 for all combination of two pair of satellites having satellite 1 in common. Let $x_{\widehat{ij}}, y_{\widehat{ij}}, z_{\widehat{ij}}$ represent the difference $x_i - x_j, y_i - y_j, z_i - z_j$ respectively, C_i represent the ith combination and m represent the total number of combinations with $\mathrm{C}_i = \{(\mathbf{p}_{k_i}, \mathbf{p}_1), (\mathbf{p}_{l_i}, \mathbf{p}_1)\}$. The system of linear equations for these m combinations can be concisely written as follows:

$$\mathbf{AX} = \mathbf{B} \quad (33)$$

where,

$$\mathbf{A} = -\begin{bmatrix} x_{\widehat{k_1l_1}} & y_{\widehat{k_1l_1}} & z_{\widehat{k_1l_1}} & -\left(r_{\widehat{k_11}} - r_{\widehat{l_11}}\right) \\ x_{\widehat{k_2l_2}} & y_{\widehat{k_2l_2}} & z_{\widehat{k_2l_2}} & -\left(r_{\widehat{k_21}} - r_{\widehat{l_21}}\right) \\ \cdot & \cdot & \cdot & \\ x_{\widehat{k_ml_m}} & y_{\widehat{k_ml_m}} & z_{\widehat{k_ml_m}} & -\left(r_{\widehat{k_m1}} - r_{\widehat{l_m1}}\right) \end{bmatrix},$$

$$\mathbf{X} = \begin{bmatrix} x \\ y \\ z \\ r_1 \end{bmatrix}, \mathbf{B} = \frac{1}{2}\begin{bmatrix} (r_{\widehat{k_11}})^2 - (r_{\widehat{l_11}})^2 - ||\mathbf{p}_{k_1}||^2 + ||\mathbf{p}_{l_1}||^2 \\ (r_{\widehat{k_21}})^2 - (r_{\widehat{l_21}})^2 - ||\mathbf{p}_{k_2}||^2 + ||\mathbf{p}_{l_2}||^2 \\ \cdot \\ (r_{\widehat{k_m1}})^2 - (r_{\widehat{l_m1}})^2 - ||\mathbf{p}_{k_m}||^2 + ||\mathbf{p}_{l_m}||^2 \end{bmatrix}$$

For $m \geq 3$, the system of equations can be solved. However, r_1 is related to x, y, z by (6). For pairing and equivalence of $\tilde{r}_i - \tilde{r}_1 = r_i - r_1$, observed ranges are always used in the equations and thus the system of equations are essentially independent of relationship between (x, y, z) and r_1. This is also verified by the iterative refinement of r_1 where $\tilde{r}_1$ is modified by obtained r_1 in successive runs. The results show no difference in position estimates (x, y, z) for successive iterations.

The equal noise assumption cannot be applied to any arbitrary selection of pairs while it is quite reasonable for satellites observing near equal ranges to have equal noise components. The selection of pairs with near equal ranges from a single reference satellite, may not be feasible for low visibility where only a very few satellites are available for positioning. This is the motivation for the next solution approach.

3.7 PML with refinement of the locus of positions

The linearization using one single reference satellite raises a performance issue and while it is superior to trilateration in most of the cases, occassionally it performs worse. In search for a positioning approach that can give consistently better estimates than basic trilateration, a locus refinement approach is now presented.

A refined and better approximation to planar form LOP is found from two imprecise planar form LOPs assuming equal noise presence due to receiver bias and ionospheric error in each pair and for specific instance of measurement as follows.

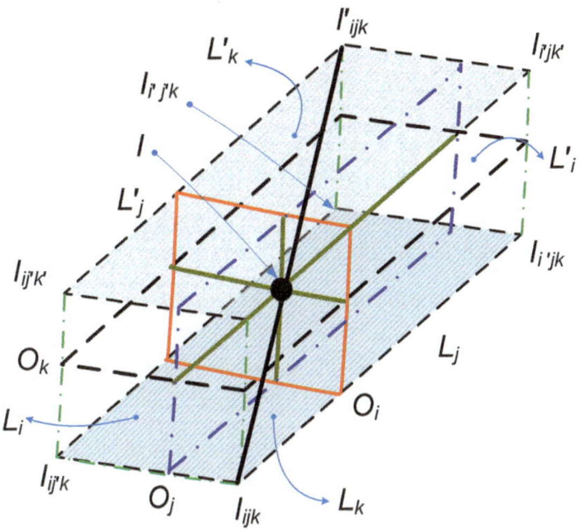

Fig. 5. Refining the locus of the receiver position under noisy measurement conditions.

Fig. 5 shows the ideal scenario where the position of the receiver to be determined, ρ, and the three respective planar form LOPs O_i, O_j and O_k are obtained from any three arbitrary satellite pairs $\mathbb{P}_i, \mathbb{P}_j$ and $\mathbb{P}_k$.

The equation for L_i, L_j, L_k can be found using (8). For specific measurement instance ξ is constant due to identical receiver clock bias and exposure to similar atmospheric noise. Hence, L_i, L_j, L_k vary from the ideal noise free LOPs O_i, O_j, O_k by the extra constant terms of $2\xi\left(r_1^i - r_2^i\right), 2\xi\left(r_1^j - r_2^j\right)$ and $2\xi\left(r_1^k - r_2^k\right)$ respectively. Crucially their slopes remain unchanged (Left hand side of (8)), and these are shown by the solid planes L_i, L_j, L_k parallel to O_i, O_j and O_k in Fig. 5. For non co-planar satellite pairs, L_i, L_j and L_k will have a physical intersection point $\mathbf{I}_{ijk} = \left(x_{ijk}, y_{ijk}, z_{ijk}\right)$.

Another plane $L_i^{'}$ parallel to L_i can be found as follows by modifying the term $2\xi\left(r_1^i - r_2^i\right)$ with $-q\left(r_1^i - r_2^i\right)$, where q is an arbitrary positive constant.

$$
\begin{aligned}
&\left(x_2^i - x_1^i\right)x + \left(y_2^i - y_1^i\right)y + \left(z_2^i - z_1^i\right)z = \\
&\tfrac{1}{2}\left(\|\mathbf{p}_2^i\|^2 - \|\mathbf{p}_1^i\|^2 + \left(r_1^i\right)^2 - \left(r_2^i\right)^2 - q\left(r_1^i - r_2^i\right)\right)
\end{aligned}
\tag{34}
$$

The original LOP O_i will then pass between the planes $L_i^{'}$ and L_i as the constants have opposite signs. A similar argument applies to $L_j^{'}, L_k^{'}$ so that the parallelopiped bounded by the planes O_i, L_i, O_j, L_j, O_k, L_k will have an aspect ratio $AR = \left(r_1^i - r_2^i : r_1^j - r_2^j : r_1^k - r_2^k\right)$ as L_i, L_j, L_k are $2\xi\left(r_1^i - r_2^i\right), 2\xi\left(r_1^j - r_2^j\right)$ and $2\xi\left(r_1^k - r_2^k\right)$ distances away from O_i, O_j and O_k respectively as they differ only by the constant terms in (8). The AR of the parallelopiped bounded by the planes $O_i, L_i^{'}, O_j, L_j^{'}, O_k, L_k^{'}$ will have exactly the same aspect ratio so indicating $\mathbf{I}_{ijk}$, $\mathbf{I}_{ijk}^{'}$ and $\mathbf{I}$ to be the diagonal points of the parallelopiped where $\mathbf{I}'_{ijk}$ denotes the intersection point of planes L'_i, L'_j and L'_k.

Hence, the equation of the actual LOP $\mathbf{I}_{ijk}\mathbf{I}'_{ijk}$ passing through $\mathbf{I}$ is found from the three intersection points $\mathbf{I}_{ijk}, \mathbf{I}_{ij'k}$ and $\mathbf{I}_{i'j'k'}$ which are available from equations (8) and (34) and analogous equations for LOPs L_j, L_k, L'_j and L'_k.

As the LOPs obtained in this way are expressed by linear equations with unknowns x, y and z, they can be solved using simple algebraic or least squares methods.

The locus refinement formulation assumes noise to be present in the formulae. However, if the noise is absent the diagonal points $\mathbf{I}_{ijk}$ and $\mathbf{I}'_{ijk}$ would be very close and during the calculation process whenever pairs having distance $< 2m$ are observed the estimated location is found as the mean of these two points.

The planar form LOP obtained from each satellite pair must be linearly independent so they do not represent either the same or a parallel planar LOP. Such satellite pairs are referred to as mutually independent, so a key objective is to identify such satellite pairs where each satellite has nearly similar distance from the receiver. PML may be intuitively viewed as positioning exploiting bearing measurements, as LOPs effectively denote a directional line. It is known that angular measurements are consistently more accurate compared to TOF range measurements and in (Chintalapudi et al., 2004) a combination of range and angular measurement has been shown to achieve better positioning results, providing a valuable insight as to why the LOP refinement furnishes better location estimation.

3.8 Selection of satellite pairs for PML

It is apparent from observation 1 that the existence of a pair of satellites having equal distance from the receiver position can have equal atmospheric noise exposure, with this prerequisite being relaxed and generalized by LOP refinement approach. Observation 1 highlights the significance of pairing the satellites for better noise cancellation and a better selection process can result in considerable improvement. With practical range estimations there is no explicit way to determine the best possible pairs following the observation. However, the range estimation ratios can be used as a rough measure for adhering to observation 1 which is the basis for the following empirically defined ranking criteria. The ranking criteria also considers the closeness of the satellites. If the two satellites are too close to each other they might have the best range estimation ratio while effectively they are like two satellites placed at the same place and hence providing no additional redundancy to help positioning. Utilizing, the above mentioned two principles the following empirical ranking criteria is introduced.

$$\Re = \frac{\tilde{r}_1}{\tilde{r}_2}\left(\frac{1}{\|\mathbf{p}_1\mathbf{p}_2\|}\right) \tag{35}$$

where $\tilde{r}_1$ and $\tilde{r}_2$ are the observed range estimates for satellite pair $(\mathbf{p}_1, \mathbf{p}_2)$ such that $\tilde{r}_1 \geq \tilde{r}_2$ and $\|\mathbf{p}_1\mathbf{p}_2\|$ is the Euclidean distance between the two satellites. The pairs having lower ranks ($\Re$) are preferred over ones with higher ranks. The complete satellite selection algorithm is given as follows.

Algorithm 1 searches all available satellites for a particular receiver so its computational complexity is $O(available\ satellites^2)$ if an exhaustive search is applied. This selection process

Algorithm 1 Satellite Selection for PML Refinement Approach

```
for all pair of neighboring satellites do
  Calculate rank (ℜ) for the pair (p_i, p_j);
  if (p_i, p_j) is co-planar with any previous selected pair and (ℜ) of present pair is lower
  than selected pair then
    Replace the previous co-planar pair with current pair
  else
    if Number of selected pair < Required number of pairs then
      Add the current pair to the selected pairs
    else
      Replace the worst ranking selected pair with the current pair
    end if
  end if
end for
```

can be run on-demand only when satellite positions are either changed or after considerable movement of the receiver. Given the small number of visible satellites in range, this will incur negligible cost.

Finally, as the new PML method itself is an analytical approach, the order of computational complexity is $O(1)$ once satellite selection has been completed.

Summarizing, PML approaches are improvement over basic trilateration in that it considers noisy measurement conditions in its formulation. Thus, this new strategy performs significantly better for real time GPS and tracking performance.

4. Conclusion

This chapter presented a detailed discussion on the analytical approaches for GPS positioning. Trilateration is the basis for most analytical positioning approaches and hence this chapter begins with fundamental discussion on trilateration. However, it performs poorly under noisy conditions which is analyzed in detail from theoretical and simulated scenarios. We also showed how difference of two range measurements can result in better positioning formulations. Subsequently, we present existing analytical approaches of Bancroft's method and Kleusberg's method that uses least squares and vector algebra respectively for solution of GPS equations. Later we present two newer approaches that are based on using better Locus Of Position (LOP) for the receiver than customary spherical locus in presence of noise. The first of these, called Paired Measurement Localization (PML) with single reference satellite uses hyperboloid planar locus of positions. The solution of these non-linear hyperboloids are found by linearizing with reference to a single satellite. The other PML approach obtains a better LOP from ordinary planar LOPs using a LOP refinement technique. Both of the PML based approaches have the advantage that they can utilize all the available satellites using least squares solution. If only four/three LOPs are used for PML single reference or PML LOP refinement respectively, the receiver position can be calculated by simple algebra. This has the advantage of avoiding matrix inversion for least squares solution and particularly suitable when the receiver has constraint computational support such as mobile embedded GPS receivers. Alternatively, when sufficient computational resources are available and better precision is needed full fledged least squares solution and further filtering techniques could be applied.

5. Acknowledgment

Part of this research is supported by University of Malaya high-impact research grant number UM.C/HIR/MOHE/FCSIT/04.

6. References

Bajaj, R., Ranaweera, S. & Agrawal, D. (2002). GPS: Location-tracking technology, *Computer* 35(4): 92–94.

Bancroft, S. (1985). An algebraic solution of the GPS equations, *IEEE Transactions on Aerospace and Electronic Systems* 21(6): 56–59.

Caffery, J. J. (2000). A new approach to the geometry of TOA location, *52nd Vehicular Technology Conference*.

Chaffee, J. & Abel, J. (1994). On the exact solutions of pseudorange equations, *IEEE Transactions on Aerospace and Electronic Systems* 30: 1021–1030.

Chan, Y. & Ho, K. (1994). A simple and effificient estimator for hyperbolic location, *IEEE Transactions on Signal Processing* 42: 1905–1915.

Chintalapudi, K. K., Dhariwal, A., Govindan, R. & Sukhatme, G. (2004). Ad-hoc localization using ranging and sectoring, *INFOCOM*.

Foy, W. H. (1976). Position-location solutions by Taylor-series estimation, *IEEE Trans. Aerosp. Electron. Syst.* 12: 187–194.

Friedlander, B. (1987). A passive localization algorithm and its accuracy analysis, *IEEE J. Ocean. Eng.* 12: 234–245.

Kleusberg, A. (1994). Analytical GPS navigation solution, pp. 1905–1915.

Rahman, M. Z. & Kleeman, L. (2009). Paired measurement localization: A robust approach for wireless localization, *IEEE Transactions on Mobile Computing* 8(8).

Smith, J. O. & Abel, J. S. (1987). Closed-form least-squares source location estimation from range-difference measurements, *IEEE Trans. Acoust., Speech, Signal Process.* 35: 1661–1669.

Strang, G. & Borre, K. (1997). *Linear Algebra, Geodesy, and GPS*, Wellesley-Cambridge.

Torrieri, D. J. (1984). Statistical theory of passive location systems, *IEEE Trans. Aerosp. Electron. Syst.* 20: 183–197.

12

Hybrid Positioning and Sensor Integration

Masahiko Nagai
Asian Institute of Technology
Thailand

1. Introduction

Utilization of a mobile platform is important for effectively acquiring spatial data over a wide area (Zhao & Shibasaki, 2000). Although mobile mapping technology was developed in the late 1980s, the more recent availability of Global Positioning Systems (GPSs) and inertial measurement units (IMUs), the latter being a combination of accelerometers and gyroscopes, has made mobile mapping systems possible, particularly for aerial surveys and ground vehicle surveys (Manandhar & Shibasaki, 2002). Remote sensors—such as image sensors or laser scanners—are instruments that gather information about an object or area from a distance. Using these sensors for surveying and collecting information from mobile platforms has become a valuable means of disaster mapping, environmental monitoring, and urban mapping, amongst others.

Trajectory tracking of a mobile platform is considered part of directing the movement of a platform from one place on Earth to another. Although GPS gives excellent trajectory tracking performance, it is not adequate to use for mobile mapping in terms of its lack of attitude information and low data acquisition frequency. On the other hand, an IMU is a closed system that is used to detect attitude and position with high frequency.

An IMU exhibits position errors, called drift errors that tend to increase with time in an unrestrained manner. This degradation is due to errors in the initialization of an IMU and inertial sensor imperfections such as accelerometers bias and gyroscope drift. By mitigating this growth and bounding, the errors update the inertial system periodically by fixing external reference sources. The combination of GPS and IMU has become increasingly common as the characteristics of these two mobile positioning technologies complement each other. Firstly, an IMU provides continuous positioning drifts, whereas GPS measurements do not drift, but are not continuously available. Also, GPS, as external data, is used not only for position updates but also for error correction of inertial components such as attitude, heading, velocity, gyro bias, and accelerometer bias. However, the integration of IMU and GPS is restricted due to the cost of high quality inertial components.

To obtain both the wide area coverage of remote sensors and the high levels of detail and accuracy of ground surveying at low costs, a mobile mapping system has been developed in this research. All the measurement tools are mounted on a mobile platform to acquire detailed information. This mobile platform integrates and combines equipment such as digital cameras, a small and cheap laser scanner, an inexpensive IMU, GPS, and VMS (Velocity Measurement System). These sensors are integrated by a high-precision positioning system designed for moving environments and they carry out a key role in hybrid positioning.

In this paper direct geo-referencing is achieved automatically from a mobile platform with hybrid positioning by multi-sensor integration. Here, direct geo-referencing means geo-referencing that does not require that the ground control points accurately measure ground coordinate values. Data are acquired and digital surfaces are modeled using equipment which is mounted on a mobile platform. This allows objects to be automatically rendered in rich shapes and detailed textures.

2. System design for hybrid positioning and sensor integration

The key attributes of the design of the system are low cost, ease of use, and mobility (Parra & Angel, 2005). Firstly, it utilizes a small laser scanner, commercially available digital cameras, and a relatively inexpensive IMU such as FOG (Fiber Optic Gyro), not a high-performance and expensive IMU like Ring Laser Gyro. The IMU and other measurement tools used are much cheaper than those in existing aerial measurement systems, such as Applanix's POS and Leica's ADS40 (Cramera, 2006). Moreover, these low-cost instruments are easily available on the market. Recent technological advances have also led to low-cost sensors such as micro electro mechanical system (MEMS) gyros. For example, it is considered that MEMS gyros will supplant FOG in the near future and that the price will be approximately one-tenth of that of FOG. For this reason, FOG was selected for this paper in an attempt to improve a low-cost system for the future. Secondly, "mobility" here means the item is lightweight and simple to modify. Such sensors allow the system to be borne by a variety of platforms: UAV (Unmanned Aerial Vehicle), ground vehicles, humans, and others. These sensors are generally low-performance, but they are light and low-cost while still meeting the specifications. These handy sensors are improved by integrating their data.

2.1 Sensors

In this paper a laser scanner, digital cameras, an IMU, a GPS, and a VMS are used to find the precise trajectory of sensors and to construct a digital surface model as a mobile mapping system. To automatically construct such a model, it is necessary to develop a high-frequency positioning system to determine the movement of the sensors in details. The integration of GPS and IMU data is effective for high-accuracy positioning of a mobile platform. A 3D shape is acquired by the laser scanner as point cloud data and texture information is acquired by the digital cameras all from the same platform simultaneously. The sensors used in this paper are listed in Table 1.

2.2 Sensors' calibration

Calibration of sensors is necessary for two reasons. One is to estimate the interior orientation parameter, such as lens distortion and focal length that are mechanical oriented parameters. The other reason calibration is necessary is to estimate exterior orientation parameters, such as a transformation matrix that has a relative position and attitude among sensors. All the sensors are tightly mounted on a platform and they have constant calibration parameters during the measurement. The purpose of calibration is chiefly to integrate all the sensors and positioning devices to a common single coordinating system, so that captured data can be integrated and expressed in terms of a common world coordinate system.

Sensors	Model	Specifications
Digital Camera	Canon EOS 10D	3,072×2,048 pixels Focal length: 24.0mm Price: $1,500US Weight: 500g
IR Camera	Tetracam ADC3	2,048×1,536 pixels Green, red, and NIR sensitivity with bands approximately equal to TM2, TM3, and TM4. Focal length: 10.0mm Price: $6,000US Weight: 500g
Laser Scanner	SICK LMS-291	Angular resolution: 0.25° Max. distance: 80m Accuracy (20m) : 10mm Price: $4,000US Weight: 4,000g
IMU	Tamagawa Seiki Co., Ltd. TA7544	Fiber optic gyro Accuracy Angle: ±0.1° Angle velocity: ±0.05°/s Acceleration: ±0.002G Price: $20,000US Weight: 1,000g
GPS	Ashtech G12	Accuracy differential: 30cm Velocity accuracy: 0.1(95%) Price: $4,000US Weight: 150g
VMS	Ono Sokki Co., Ltd. LC-3110	Range: -120~+250 km/h Resolution: 10 mm/P Price: $13,000US Weight: 1.7kg

Table 1. List of sensors on mobile platform

2.2.1 Calibration of digital camera

Initially, calibration of digital images is necessary due to the estimation of interior orientation parameters. Interior orientation is conducted to decide interior orientation parameters, principal point (x_0, y_0), focus length (f), and distortion coefficient (K_1). Control points for camera calibration are taken as stereo images several times. Camera calibration is performed by the bundle adjustment using target control points. In order to estimate appropriate lens distortion for a digital camera, lens distortion mode is shown in Eq. (1) and Eq. (2). These equations consider only radial symmetric distortion (Kunii & Chikatsu, 2001). Image coordinate of (x, y) is corrected and transferred to new image coordinates of (x_u, y_u). Interior orientation parameters that are computed in this calibration for Canon EOS 10D are shown in Table 2.

$$x_u = x' + x' (K_1 r^2) \tag{1}$$

$$y_u = y' + x' (K_1 r^2) \tag{2}$$

where $x' = x - x_0$; $y' = -(y - y_0)$; $r^2 = x'^2 + y'^2$; (x, y): image coordinate

x_0	1,532.9966 pixels	f	24.6906 mm
y_0	1,037.3240 pixels	K_1	1.5574e-008

Table 2. Interior orientation parameters of Canon EOS 10D

2.2.2 Calibration of laser scanner

Calibration of a laser scanner is not easy because the laser beam is invisible where the wavelength is approximately 905 nm±10 nm. Thus, in this research, a solar cell is utilized for laser beam detection. External parameters of the laser scanner are estimated by computing the scale factor, rotation matrix and shift vector by converting the laser scanner coordinate to the fiducial coordinate, which can be a common coordinate of the system.

3D Helmert's transformation equation, Equation (3), is used to estimate the laser scanner's external parameter. The laser scanner coordinates (Xl, Yl, Zl) are converted to the fiducial coordinates (Xt, Yt, Zt). Scale factor (s), rotation matrix (R), and translation matrix (Tx, Ty, Tz) are estimated by the least square method (Shapiro, 1978). According to this calibration methodology, external parameters of a laser scanner can be decided accurately, and it helps to combine a laser scanner with other sensors such as digital camera or IMU.

$$\begin{pmatrix} Xt \\ Yt \\ Zt \end{pmatrix} = sR \begin{pmatrix} Xl \\ Yl \\ Zl \end{pmatrix} + \begin{pmatrix} Tx \\ Ty \\ Tz \end{pmatrix} \tag{3}$$

2.2.3 Boresight offset measurement

Boresight offset must be estimated between the GPS and the IMU. In the hybrid positioning circulation, differences in position and velocity between the IMU and the GPS, and other sensors are used to estimate the severity of errors. If the vehicle only goes straight, this error amount is not affected because the relative movement is constant. However, if the vehicle turns, the error amount is not constant. The position and velocity of the near axis of gyration are small, although those of its far axis of gyration are large. In this paper, the boresight offset from the GPS to the IMU in the vehicle is obtained through direct measurement by using a total station.

The transformation matrix, which includes a rotation matrix and a translation matrix from the vehicle coordinate system to a world coordinate system, is calculated from positioning data where the origin of the common coordinate system considered as the center of IMU. A rotation matrix and a translation matrix are dependent on the instant posture and position of the vehicle when the vehicle is moving. On the other hand, the transformation matrix from the local coordinate system to the vehicle coordinate system is calculated based on the external calibration parameter measured physically as a boresight offset. This is a physical measurement, so it includes some measurement errors. However, this is initial information and it can be removed by further filtering.

3. Multi sensor integration

Navigation is the continuous positioning which is the process of monitoring and controlling the movement of a vehicle from one place to another. Inertial navigation is the self-determination of the instantaneous position and other parameters of motion of a vehicle by measuring specific force, angular velocity, and time in a previously selected coordinate system. The basic concept is to determine the vehicle velocity and position by real time integration of governing differential equations.

Figure. 1 shows an overview of data processing of sensor integration from navigation as a hybrid positioning to mapping by direct geo-referencing with a laser scanner. In this paper, the following data are acquired and those data are integrated: base station GPS data, remote station GPS data, IMU data, digital images, and laser range data. Although the data are acquired in different frequencies, they are synchronized with each other by GPS time.

First, differential or kinematic GPS post processing is conducted. Second, the processed GPS data and the IMU data are integrated by a Kalman filter to estimate the sensor trajectory. The bundle block adjustment (BBA) of the digital images is then made to acquire geo-referenced images and exterior orientations with the support of GPS and IMU data, which are the sensor position and attitude, as an external aid. Also, VMS and other sensors can be considered as external aids if GPS accuracy is not enough to support IMU in urban areas. Finally, GPS data, IMU data, and the external aides are combined to regenerate high-precision and time-series sensor position and attitude. Finally, these hybrid positioning data are used for the geo-referencing of the laser range data and for the construction of a digital surface model as 3D point cloud data.

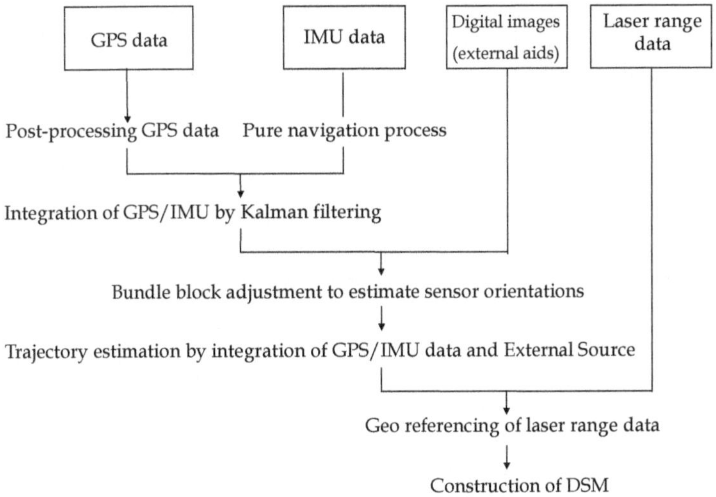

Fig. 1. Overview of data processing for multi sensor integration

4. Hybrid positioning

In general, IMUs exhibit position errors that tend to increase with time in an unbounded manner. This degradation occurs due to errors of initialization of IMUs and inertial sensor imperfections such as accelerometer bias and gyroscope drift. Hybrid positioning can mitigate the error by being updated periodically with external fixes, such as GPS, VMS, images, radio aids, or Doppler radar. Hybrid positioning is for finding the location of a mobile platform using or combining several different positioning technologies. The effect of fixing positions is that it allows for the reset or the correction of the position errors of the inertial system to the same level of accuracy inherent in the position fixing technology. The inertial system error grows at a rate equal to the velocity error. Therefore, external data is used not only for the position update but also the error correction of inertial components such as attitude, heading, velocity, gyro bias, and accelerometer bias. Furthermore, the error of the external data such as misalignment error, boresight error, and scale factor error is corrected in the same manner. Typical hybrid strapdown navigation is shown in Figure. 2.

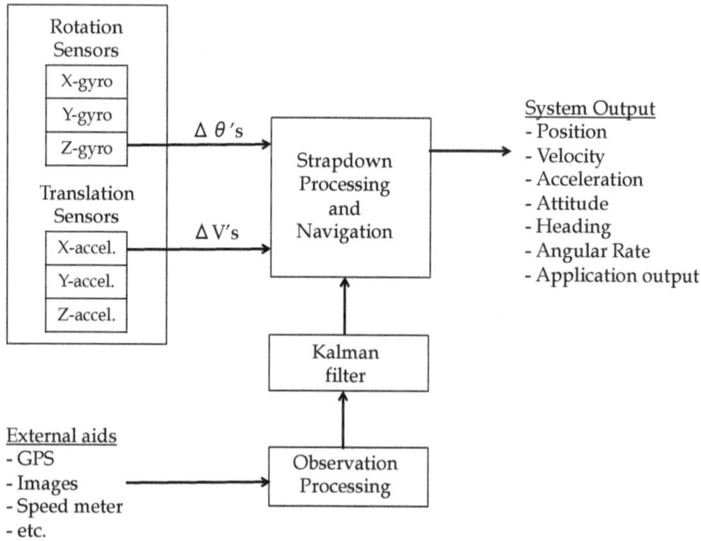

Fig. 2. Typical hybrid strap down navigation configuration

4.1 GPS and IMU data are integrated by Kalman filter

The Kalman filter can be used to optimally estimate the system states. One of the distinct advantages of the Kalman filter is that time varying coefficients are permitted in the model. With this filter, the final estimation is based on a combination of prediction and actual measurement. Figure 3 shows the pure navigation algorithm for deciding IMU attitude and IMU velocity step by step (Kumagai et al., 2002). Inertial navigation starts to define the initial attitude and heading based on the alignment of the system. It is processed and then it changes to the navigation mode. Over the years, the quality of IMUs has risen, but they are still affected by systemic errors. In this research, a GPS measurement is applied as an actual measurement to aid the IMU by correcting this huge drift error. With Kalman filtering, the sensor position and attitude are determined at 200 Hz.

Figure 4 shows the Kalman filter circulation diagram for the integration of the GPS and IMU data (Kumagai et al., 2000). Individual measurement equations and transition equations are selected, and covariance must be initialized in order to continue Kalman filtering circulation in response to the GPS data validation. The accuracy of the integration depends on the accuracy of the referenced GPS. In this case, it is approximately 30 cm.

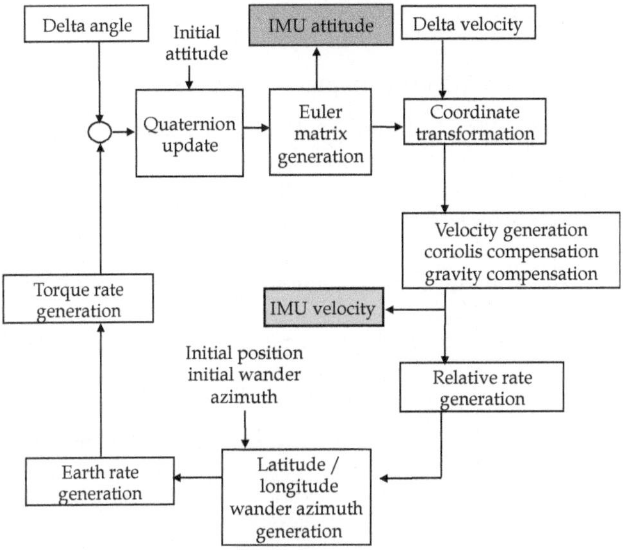

Fig. 3. Pure navigation block diagram expressed roughly step by step

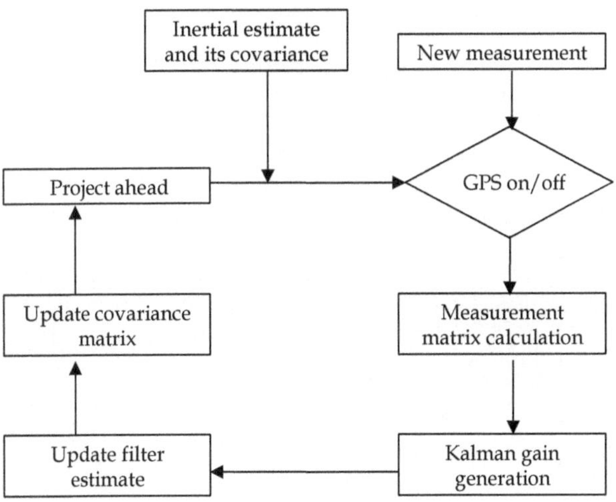

Fig. 4. Kalman filter circulation

This research adopts the Kalman filter and the following steps are included, as shown in Figure 4.

1. Initialization of covariance value and first estimation of each variable
2. Inspection of GPS validity and selection of measurements equation

3. Calculation of measurements
4. Calculation of the Kalman gain
5. Calculation of estimations
6. Calculation of next covariance
7. Updating of the covariance.

4.2 Bundle block adjustment (BBA) of digital images

The image exterior orientation is determined by the BBA for the mosaicked digital images where the BBA is a nonlinear least squares optimization method using the tie points of the inside block (Takagi & Shimoda, 2004). Bundle block configuration increases both the reliability and accuracy of object reconstruction. An object point is determined by the intersection of more than two images, which provides local redundancy for gross error detection and which consequently forms a better intersection geometry (Chen et al., 2003). Therefore, in this paper, the digital images are set to overlap by more than 50% in the forward direction, and by more than 30% on each side. The GPS and IMU data obtained in a previous step allow the automatic setting of tie points in overlapped images and reduce the time spent searching for tie points by limiting the search area based on the epipolar line. The epipolar line is the straight line of intersection between the epipolar plane and the image plane, and it is estimated by the sensor position and attitude, which is derived from GPS/IMU integration. It connects the image point in one image through the image point in the next image. Figure 5 shows an image orientation series with tie points that overlap each other. The image resolution is extremely high (approximately 1.5 cm), so it is easy to detect small gaps or cracks.

The accuracy of the image orientation (ba) is estimated by comparison with 20 control points (cp) as shown in Table 3. The average error of the plane is approximately 3 to 6 cm. The average error of the height is approximately 10 cm. That is, although the BBA is done automatically, the result is very accurate compared with the differential GPS or the GPS/IMU integration data, in which the average error is based on GPS accuracy. Moreover, the processing time is very short. Thus, the BBA's results aid Kalman filtering by initializing the position and attitude in the next step to acquire a greater accurate trajectory.

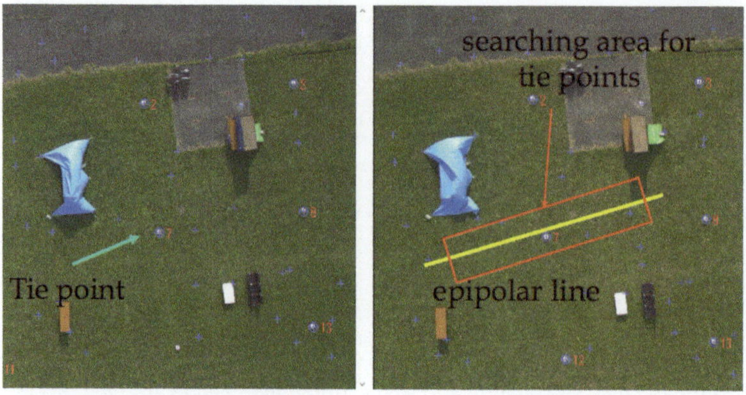

Fig. 5. Image orientation with tie points

unit: m

Num	X (cp:m)	Y (cp:m)	Z (cp:m)	X (ba:m)	Y (ba:m)	Z (ba:m)	error : X	error : Y	error : Z
1	0	0	-12.584	0.094	-0.059	-12.311	0.094	0.059	0.273
2	11.3105	0	-12.3825	11.293	-0.062	-12.48	0.0175	0.062	0.0975
3	20.8395	0.168	-12.4065	20.79	0.111	-12.515	0.0495	0.057	0.1085
4	32.588	0.2885	-12.441	32.527	0.229	-12.564	0.061	0.0595	0.123
5	46.196	0.5035	-12.5105	46.103	0.447	-12.518	0.093	0.0565	0.0075
6	0.074	-8.1735	-12.515	0.173	-8.145	-12.336	0.099	0.0285	0.179
7	11.3245	-7.905	-12.428	11.346	-7.891	-12.458	0.0215	0.014	0.03
8	20.5425	-7.703	-12.4345	20.525	-7.72	-12.499	0.0175	0.017	0.0645
9	30.677	-7.315	-12.406	30.622	-7.341	-12.575	0.055	0.026	0.169
10	46.7025	-7.81	-12.566	46.608	-7.849	-12.459	0.0945	0.039	0.107
11	0.4485	-14.9755	-12.473	0.551	-14.917	-12.376	0.1025	0.0585	0.097
12	11.6895	-15.058	-12.4075	11.734	-15.019	-12.483	0.0445	0.039	0.0755
13	20.3605	-14.902	-12.419	20.361	-14.891	-12.518	0.0005	0.011	0.099
14	30.447	-15.3555	-12.47	30.424	-15.347	-12.503	0.023	0.0085	0.033
15	46.3735	-15.455	-12.5715	46.289	-15.456	-12.401	0.0845	0.001	0.1705
16	0.3535	-24.139	-12.443	0.453	-24.072	-12.522	0.0995	0.067	0.079
17	11.911	-23.7855	-12.455	11.987	-23.721	-12.466	0.076	0.0645	0.011
18	20.594	-23.461	-12.453	20.623	-23.421	-12.507	0.029	0.04	0.054
19	30.176	-23.1665	-12.4505	30.165	-23.13	-12.491	0.011	0.0365	0.0405
20	46.258	-22.5005	-12.5545	46.2	-22.493	-12.39	0.058	0.0075	0.1645
						ave	0.05655	0.0376	0.09915

Table 3. Accuracy of the image orientation

4.3 Hybrid positioning by multi sensor integration

The position and attitude of the sensors are dictated by the integration of the GPS and IMU data, as well as by the image orientations that are acquired from digital cameras or digital video cameras. One of the main objectives of this paper is to integrate inexpensive sensors into a high-precision positioning system. Integration of the GPS (which operates at 1 Hz) with the IMU (200 Hz) has to be made with Kalman filtering for the geo-referencing of laser range data with a frequency of 18 Hz. The positioning accuracy of the GPS/IMU integration data is based on GPS accuracy. On the other hand, both position and attitude can be estimated with very high accuracy using the BBA as image orientations. However, the images are not taken frequently; in this case every 10 seconds.

Therefore, the combination of the BBA and Kalman filtering is conducted to increase accuracy, as shown in Figure 6. The BBA results are assumed to be true position values. They provide initial attitude and heading without any IMU alignment. The IMU is initialized by Kalman filtering using the BBA result every 10 seconds to avoid a culmination of errors. That is, after every computation of the BBA, the IMU data and its errors are corrected. Figure 6 shows the strapdown navigation algorithm for the GPS/IMU integration and the BBA result. The combination of GPS, IMU, and images can be a hybrid positioning.

As a result of the multi sensor integration, the trajectory of the hybrid positioning can assure sufficient geo-referencing accuracy for the images. The trajectory of the digital camera can be representative of the trajectory of the platform because the GPS and IMU data are initialized by camera orientation. Their coordinate is fitted to the digital camera coordinate. Figure 7 shows the hybrid position as trajectories of GPS/IMU/images and GPS/IMU. The coordinate system is JGD2000 (Japan Geodetic Datum 2000). The black solid line is the combination of GPS/IMU/images, and the red solid line is the ordinary combination of

GPS/IMU. On the one hand, with an ordinary GPS/IMU, the trajectory becomes notched because the position is revised forcibly by GPS due to drift error. The platform changes its attitude rapidly, especially in the corner, so the notched trajectory is very obvious. The drift error remains in the calculation until the alignment of IMU is complete. On the other hand, with GPS/IMU/images, the drift error of IMU is aligned by initialization from the bundle block adjustment. Moreover, the trajectory is very smooth in the corner.

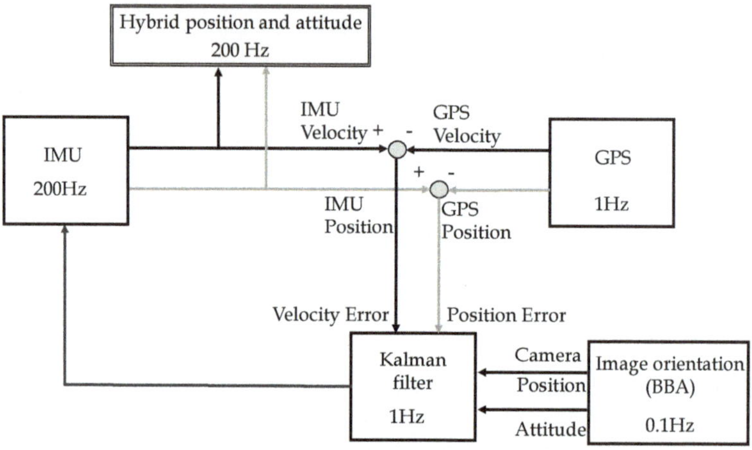

Fig. 6. Strapdown navigation diagram with images

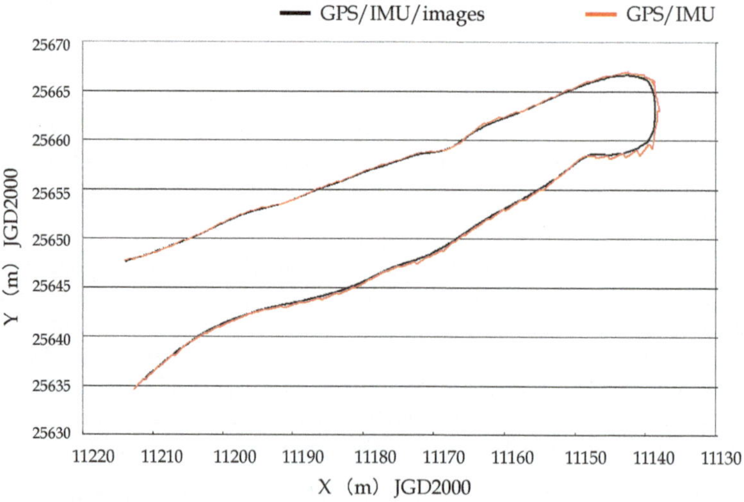

Fig. 7. Hybrid position

4.4 Evaluation of hybrid positioning

Trajectory tracking by ordinary GPS/IMU integration is compared to the combining of GPS/IMU and continual digital images, in order to validate this combination. Figure 8 shows the yaw angles of these two methods; the black solid line is the combination of GPS/IMU/images, and the red solid line is the ordinary integration of GPS/IMU. In the case of the GPS/IMU/images combination, an accurate azimuth angle, as a yaw angle, is recorded from the BBA. Thus, the yaw angle is then accurate from the beginning of the measurement. On the other hand, in the ordinary combination of GPS/IMU, the Kalman filter gradually estimates the state of a system from measurements which contain random

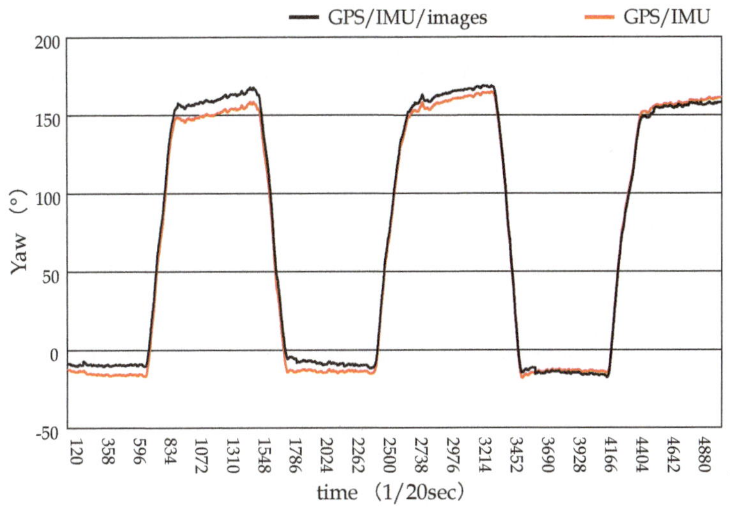

Fig. 8. Comparison of Yaw angle

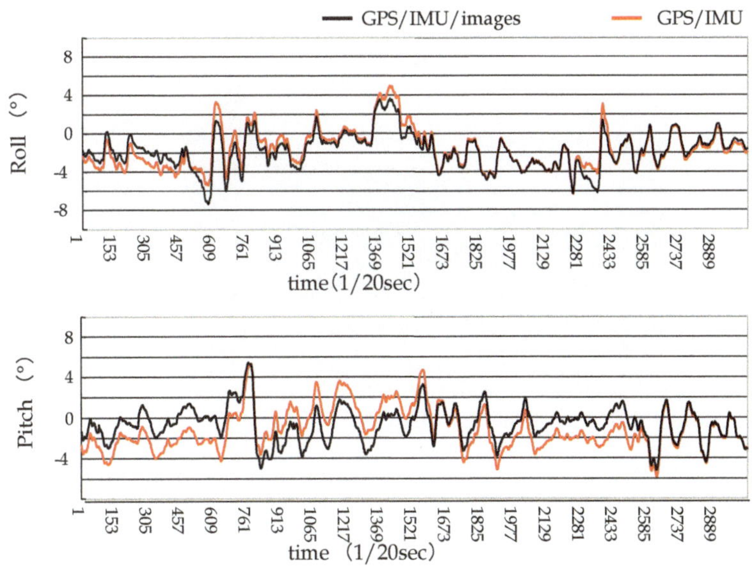

Fig. 9. Comparison of Roll and Pitch angle

errors. That is, error estimation is not enough at the beginning, and the yaw angle gradually improves. For that reason, in the case of ordinary GPS/IMU, it is necessary to have the system alignment before the measurement in order to estimate an accurate azimuth angle. Therefore, in this proposed method, the system alignment is not required. Figure 9 shows the roll and the pitch angle of the two methods. The phenomenon of the error for the roll and pitch angle is the same as the yaw angle.

5. Experiment

In order to appraise the characteristics and performance of the proposed algorithm, two experiments are conducted by using a UAV (Unmanned Aerial Vehicle) and ground vehicle (a car) as a platform. In the case of the UAV, images are used for external aid, whereas VMS is used as an external aid in the case of the ground vehicle.

5.1 UAV (Unmanned Aerial Vehicle) based mapping system

A UAV based mapping system is developed to obtain both the wide-area coverage of remote sensors and the high levels of detail and accuracy of ground surveying, at low cost. All the measurement tools are mounted under the UAV, which resembles a helicopter, to acquire detailed information from low altitudes, unlike high altitude systems in satellites or airplanes. The survey is conducted from the sky, but the resolution and accuracy are equal to those of ground surveying. Moreover, the UAV can acquire data easily as well as safely.

In this paper, all of the measurement tools are mounted under the UAV, which is a helicopter-like model RPH2 made by Fuji Heavy Industries, Ltd., and shown in Figure 10. All the sensors are mounted tightly to the bottom of the fuselage. The RPH2 is 4.1m long, 1.3m wide and 1.8m high. Table 4 shows its main specifications.

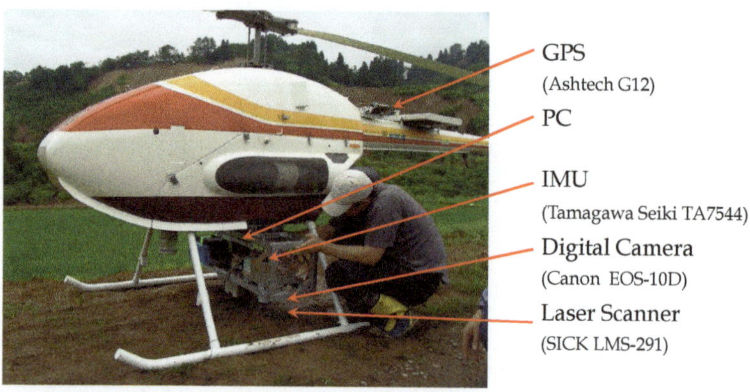

Fig. 10. UAV, model RPH2 made by Fuji Heavy Industries, Ltd.

As shown in Table 4, the RPH2 is a large UAV; however, it is considered a platform for the experimental development of a multi-sensor integration algorithm. The RPH2 has a large payload capacity; thus, it can carry large numbers of sensors, control PCs, and a large battery. After the algorithm is developed by a large platform, a small UAV system is implemented using selected inexpensive sensors for certain observation targets.

There are several advantages to utilizing a UAV. One of the most important advantages is that it is unmanned and therefore can fly over dangerous zones. This advantage suits the purpose of direct geo-referencing in this study. Direct geo-referencing does not require that ground control points have accurately measured ground coordinate values. In dangerous zones, it is impossible to set control points, unlike the case in normal aerial surveys. The addition of this direct geo-referencing method from a UAV could be an ideal tool for monitoring dangerous situations. Therefore, this UAV-based mapping system is perfectly suited for disaster areas such as landslides and floods, and for other applications such as river monitoring.

Weight	330kg
Payload	100kg
Motor	83.5 hp
Main rotor	2 rotors, diameter 4.8m
Tail rotor	2 rotors, diameter 0.8m
Operational	3km or more
Flight time	1 hour
Ceiling	2,000m

Table 4. Specification of RPH2

5.1.1 UAV based system

All the sensors are tightly mounted under the UAV to ensure that they have a constant geometric relationship during the measurement. The digital cameras and the laser scanner are calibrated to estimate the relative position and attitude. Moreover, all sensors are controlled by a laptop PC and are synchronized by GPS time, one pulse per second. Finally the sensors are set, as shown in Figure 10.

5.1.2 Digital 3D modeling

During measurement, the platform, including all of the sensors, is continuously changing its position and attitude with respect to time. For direct geo-referencing of laser range data, the hybrid positioning data are used. There are two coordinate systems; the laser scanner and the hybrid positioning, WGS84 (World Geodetic System 1984) based on GPS and the BBA data. It is necessary to transform the laser scanner coordinate into the WGS84 coordinate by geo-referencing. Geo-referencing of the laser range data is determined by the 3D Helmert's transformation, which is computed by the rotation matrix and translation matrix with the hybrid positioning data and calibration parameters as offset values, as shown in Equation (3), for calibration by the laser scanner. The offset values from the laser scanner to the digital cameras in the body frame are already obtained by the sensor calibration. Geo-referencing of laser range data and images is done directly.

Figure 11 shows the 3D point cloud data that are directly geo-referenced by the hybrid IMU data. In this research, the WGS84 is used as the base coordinate system for the 3D point cloud data. The UAV-based system in this research utilized a landslide survey by reconstructing a digital surface model. A digital camera and a laser scanner were mounted on a UAV to acquire detailed information from low altitude. The surveying is carried out from the sky, but the resolution and accuracy are the same level as a ground survey. Because of the utilization of a UAV, the data of the landslide site can be easily acquired collectively with safety and mobility. This new survey can be an intermediate method between aerial surveys and ground surveys.

Fig. 11. 3D point cloud model

5.2 Ground vehicle based mapping system

Understanding road environments has become increasing more important in recent years due to a wide range of applications, such as intelligent vehicles, driving assistance and sign inventory systems or route guidance systems for navigation assistance. For drivers, traffic signs/signals provide crucial information for safety and smooth driving; thus, they play an important role in all kinds of driver support systems. Much work on traffic signal/sign detection and recognition has been done in recent years and sensor systems now consist of three different types of sensors, including laser scanners for measuring object geometry, digital cameras for capturing scene texture, and the moving platform equipped with a GPS/IMU/VMS based hybrid positioning system.

5.2.1 Ground vehicle based system

Figure 12 shows the ground vehicle based system where all sensors are mounted on the roof of the vehicle. Two of the laser range scanners are mounted on the back and scan the horizontal plane. VMS is also mounted and is used to assist the navigation unit to locate vehicle positions when the GPS signal is unavailable. The other two laser scanners are placed on the front and rear of the vehicle's roof. The front one scans with an elevation of about 30 degrees to capture the front scene, especially the important urban spatial objects that assist navigation. For data measurement, all the sensors are under control of the vehicle-borne computers and synchronized by a GPS clock.

The navigation units used in the sensor system are composed of a DGPS, an IMU (FOG) and a VMS (Velocity Measurement System). The DGPS is responsible for measuring the vehicle's position using a satellite signal. The IMU, consisting of accelerometers and gyroscopes, measures the acceleration and direction changes of the vehicle, while VMS is in charge of measuring the vehicle's velocity with high accuracy. The combination of GPS/IMU/VMS is complementary as the velocity from VMS, and the acceleration and direction changes from IMU can be used to locate a vehicle's position when the GPS signal is unavailable. Moreover, the GPS can be used to rectify the output of IMU. VMS data is more accurate when compared with DGPS; thus, the estimation of VMS errors becomes possible when DGPS is valid. Therefore, it is possible to acquire a more precise positioning in the pure PDOP condition like an urban operation by means of blending VMS data. The strapdown navigation diagram for a ground vehicle mapping system is shown in Figure 13. The Kalman filter is processed every 10 seconds.

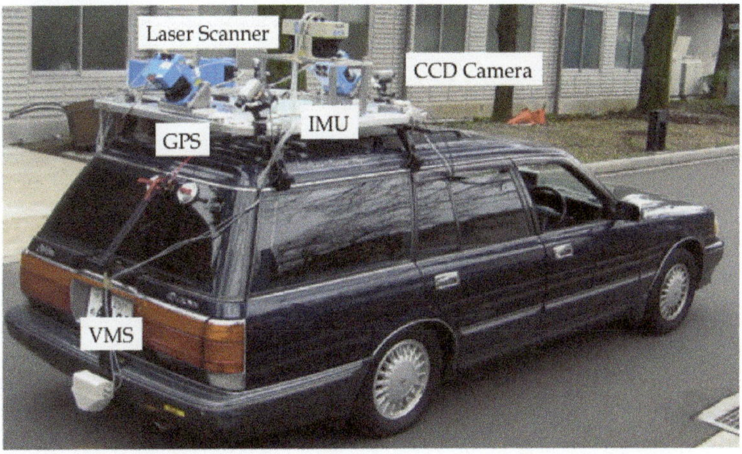

Fig. 12. Ground vehicle based system

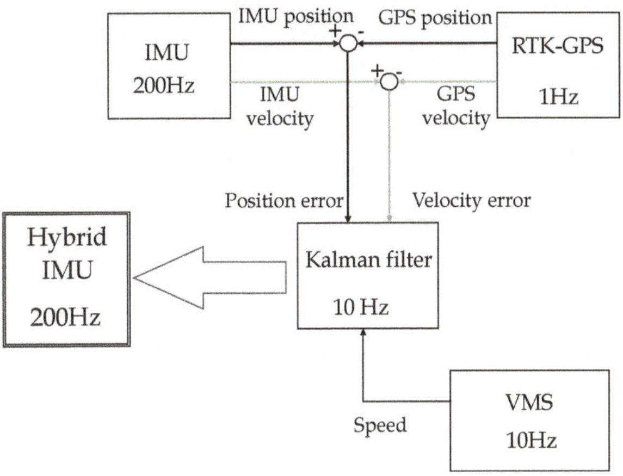

Fig. 13. Strapdown navigation diagram with VMS

5.2.2 Object extraction

Figure 14 shows the point cloud acquired by the laser scanner, which is geo referenced with hybrid positioning data and contains not only traffic signs/signals but also the surroundings (vegetation and buildings) beyond the road. Object segmentation and feature extraction of important objects, such as traffic signs/signals, are conducted. This is most common especially after most redundant range points have been detached from the point cloud after boundary extraction. Range points represent the geographic information of all the objects in the form of 3D discrete coordinates, which have no description of attribute and there is no topological relation among the data points. However, spatial features exist which can be used for object segmentation. It is clear that the traffic sign/signal has a strong linear feature after being projected onto the horizontal plane, while points belonging to other spatial objects (such as trees) are scattered on the horizontal plane without a dominant direction. This spatial feature can be utilized for segmentation.

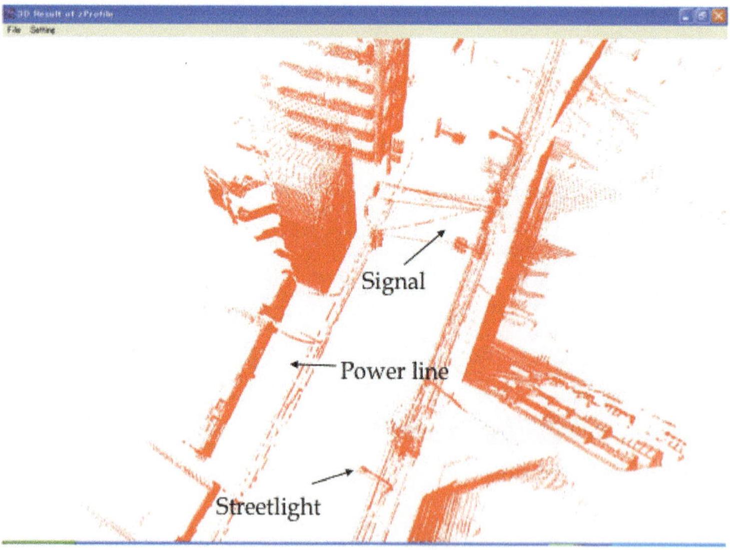

Fig. 14. 3D point cloud data from vehicle based system

6. Validation

In this mobile mapping system, multiple sensors are integrated, thus making it difficult to point out the origins of errors for positioning. Therefore, accuracy of positioning is assessed by an ordinary survey method, the result of which is compared with a digital surface model and control points from the oriented images; the accuracy is 3 to 10 cm, as shown in Table. 3. Those control points are considered as true values and selected feature points such as object corners, which can recognize both images and a digital surface model (DSM). As a result, the average error of the digital surface model is approximately 10 to 30 cm, as shown in Table. 5.

For this validation, DSM has been reconstructed and geo-referenced by using a hybrid position. The laser range data are acquired 50 m away from the object and the scan angler resolution is 0.25°; that is, the density of 3D laser points is approximately 20 cm per point. After comparing the mapping accuracy with the laser point density, it was found that the accuracy is good enough for mapping DSM. Therefore, the accuracy of hybrid positioning including their attitude is considered approximately 10 to 30 cm.

Unit: m

No.	Ground Control point (X)	(Y)	(Z)	DSM (X)	DSM (Y)	DSM (Z)	Error (X)	Error (Y)	Error (Z)
1	-11184.877	-25630.253	42.755	-11184.696	-25630.836	42.915	0.181	0.583	0.160
2	-11185.471	-25622.727	42.952	-11185.557	-25622.789	42.971	0.086	0.062	0.019
3	-11167.603	-25670.474	42.391	-11168.282	-25670.312	42.406	0.679	0.162	0.015
4	-11177.107	-25634.721	42.704	-11177.262	-25634.918	42.523	0.155	0.197	0.181
5	-11152.866	-25641.753	42.029	-11152.172	-25641.036	42.071	0.694	0.717	0.042
6	-11176.511	-25625.571	42.824	-11176.467	-25625.426	42.767	0.044	0.145	0.057
7	-11153.911	-25643.823	42.534	-11154.375	-25643.041	42.075	0.464	0.782	0.459
8	-11150.564	-25631.724	42.340	-11150.887	-25631.869	42.296	0.323	0.145	0.044
9	-11176.771	-25635.344	43.992	-11176.394	-25635.308	44.082	0.377	0.036	0.090
10	-11186.666	-25631.657	44.289	-11186.417	-25631.888	44.202	0.249	0.231	0.087
						Ave. Error	0.325	0.306	0.115

Table 5. Positioning accuracy assessment from DSM

7. Conclusion

In this paper, robust trajectory tracking by hybrid positioning was developed and a digital surface model was reconstructed with multi-sensor integration using entirely inexpensive sensors, such as a small laser scanner, digital cameras, an inexpensive IMU, a GPS, and a VMS. A new method of direct geo-referencing was proposed for laser range data and images by combining a Kalman filter and the BBA or VMS. Because the result of BBA avoids the accumulation of drift errors in the Kalman filtering, the geo-referenced laser range data and the images were automatically overlapped properly in the common world coordinate system. Hybrid positioning data is acquired by using or combining several different positioning technologies. Since this paper focused on how to integrate the sensors into a mobile platform, all the sensors and instruments were assembled and mounted under a mobile platform such as a UAV or ground vehicle in this experiment. Finally, the precise trajectory, including attitude of the sensors, was computed as the hybrid positioning for direct geo-referencing of a laser scanner. The hybrid positioning data is used to reconstruct digital surface models.

8. References

Zhao, H. & Shibasaki, R. (2000). *Reconstruction of Textured Urban 3D Model by Ground-Based Laser Range and CCD Images*, IEICE Trans. Inf.&Syst., vol.E83-D, No.7

Manandhar, D. & Shibasaki, R. (2002). *Auto-Extraction of Urban Features from Vehicle-Borne Laser Data*, ISPRS, GeoSpatial Theory, Processing and Application, Ottawa

Parra, S. & Angel, J. (2005). *Low cost navigation system for UAV's*, Aerospace Science and Technology, Issue 6, Volume 9, pp.504-516

Cramera, M. (2006). *The ADS40 Vaihingen/Enz geometric performance test*, ISPRS Journal of Photogrammetry and Remote Sensing, Volume 60, Issue 6, pp. 363-374

Kunii, Y. & Chikatsu, H. (2001). *Application of 3-Million Pixel Armature Camera for the 3D Modeling of Historical Structures*, Asian Journal of GEOINFORMATICS Vol. 2, No. 1, pp. 39-48

Shapiro, R. (1978). *Direct linear transformation method for three-dimensional cinematography*, Res. Quart. 49, pp. 197-205

Kumagai, H., Kubo, Y., Kihara, M. & Sugimoto, S. (2002). *DGPS/INS/VMS Integration for High Accuracy Land-Vehicle Positioning*, Journal of the Japan Society of Photogrammetry and Remote Sensing, vol.41, no.4 pp. 77-84

Kuamgai, H., Kindo, T., Kubo, Y. & Sugimoto, S. (2000). *DGPS/INS/VMS Integration for High Accuracy Land-Vehicle Positioning*, Proceedings of the Institute of Navigation, GPS-2000, Salt Lake

Takagi, M. & Shimoda, H. (2004). *Handbook of image analysis*, University of Tokyo Press

Chen, T., Shibasaki, R. & Murai, S. (2003). *Development and Calibration of the Airborne Three-Line Scanner (TLS) Imaging System*, Journal of the American Society for Photogrammetry and Remote Sensing PE&RS, vol.69, No.1,pp. 71-78

13

Emerging New Trends in Hybrid Vehicle Localization Systems

Nabil Drawil and Otman Basir
Department of Electrical and Computer Engineering, University of Waterloo
Canada

1. Introduction

Over the last decade, vehicle localization has been attracting attention in a wide range of applications. A number of localization techniques have been developed to serve a variety of applications Al-Bayari & Sadoun (2005); Aono et al. (1998); Bouju et al. (2002); Cramer (1997); Dao et al. (2002); Drawil & Basir (2010); Jabbour, Cherfaoui & Bonnifait (2006); Lai & Tsai (2003); Nishimura et al. (1996); Sliety (2007); Stockus et al. (2000). In recent years, the focus has been on localization accuracy improvement – an issue considered crucial, specially in mission critical applications. For instance, for emergency response systems, such as the eCall system, to deliver on their task they need reliable and accurate localization capabilities. These capabilities are becoming as important in other applications, including, accident avoidance and management, navigation systems, location sensitive billing systems, location based services.

The focus of much recent research in localization has been on improving accuracy through the use of multiple localization modalities. This chapter provides a review on multi-modality based localization techniques and establishes a categorization of such techniques based on the type of measurement and the strategy employed to fuse measurements from multiple localization sources.

Although these techniques have demonstrated significant performance improvement, there remain situations that give rise to degraded localization accuracy. Moreover, current localization systems lack in their ability to reliably quantify the accuracy of localization estimates, neither the means by which sources of localization information are properly discounted based on reliability/accuracy merits.

In this chapter, a novel framework is proposed to tackle the aforementioned issues. The proposed framework fuses different localization techniques in order to improve their location estimates, and provides a location reliability assessment that captures the integrity of the estimates. Knowledge about estimate integrity allows the system to plan the use of its localization resources so as to match the target accuracy of the application. The proposed framework provides the tools that would allow for modeling the impact of the operation conditions on estimate integrity, as such it enables more robust system performance.

2. Motion and GPS measurement data fusion

Differential GPS (DGPS) and Assisted GPS (A-GPS) are two advanced types of GPS technologies that provide a high level of accuracy and fast retrieving rate. Nevertheless, using a GPS receiver as the sole vehicle localization measurement source may turn to be unreliable, especially in urban canyons and other areas where the satellite signal can be distorted or lost. A number of solutions have been reported in the literature that proposed augmenting GPS measurements with information about the vehicle's motion in order to improve localization accuracy. In what follows we provide a summary of a number of such solutions.

2.1 Dead Reckoning (DR) and GPS integration

A DR is a localization method that estimates the next location of a mobile object over a series of short time intervals, given the object's direction, speed, and previous location. DR is simple and known for producing incremental error and hence needs to be reset periodically. It is therefore suitable for use over short periods of time.

One approach to resetting the accumulative localization error is to combine DR with GPS whereby GPS measurements are used to reduce the DR accumulative error; when the GPS measurement is unavailable, the DR estimates the location using sensors such as wheel odometers, a flux-gate compass, a gyroscope, and an accelerometer Kao (1991).

2.2 Inertial Navigation System (INS) and GPS fusion

Basically, INS operates as a DR system. INS employs a computing unit and motion sensors to estimate its location without relying on any external reference once it is initialized using for example a GPS measurement. To avoid the accumulated error caused by the measurements of internal sensors in INS, the INS location estimate is fused with measurement data from other sources. As discussed in Skog & Handel (2009) fusing INS and GPS can take the form of a loosely or tightly coupled system architecture.

An example of a system that fuses INS and GPS is the real-time kinematic global positioning system (RTK GPS) Bouvet & Garcia (2000) which uses an Extended Kalman Filter (EKF) to fuse data. In this system, GPS latency is defined as the time required for the satellite signals to travel to Earth and the time required for the computation of the location; GPS latency varies with the number of observed satellites. Therefore, the GPS latency is encapsulated in the EKF state so that the fusion of the INS and GPS data is synchronized with the readings of the sensors.

It is possible to fuse standard GPS and INS by means of a KF as well Honghui & Moore (2002). In this case the computational complexity of the EKF can be reduced by preprocessing the INS measurements and inputting them into the KF as a linear component. However, preprocessing the INS measurement adds to the computational cost of the solution.

2.3 Other motion sensors and DGPS fusion

Integrating the INS of a dynamic model with a DGPS is also investigated inRezaei & Sengupta (2005). To deal with the nonlinearity of the dynamic model, an EKF is used. Due to the accelerometer noise other motion sensors, such as six wheel-speed encoders, a steering angle encoder, and an optical yaw rate gyro, are used instead. Localization accuracy of 0.9 m on 100 m driving track was reported for situations where the system relies on the dynamic model more than it does on the GPS measurements. The multipath effect is not addressed as the experiment was conducted in an open space environment.

In Aono et al. (1998) a method of positioning a vehicle on undulating ground by fusing DGPS data and motion sensor data is proposed. A fibre optic gyro, a roll pitch sensor, and wheel encoders are used as motion sensors. The positioning accuracy is improved by compensating for the error for each sensor. The error is determined by means of a KF, which is also utilized as a fusion unit.

In Sharaf et al. (2005) an Artificial Neural Network (ANN) is chosen as a tool for detecting errors and noises in INS measurements using a DGPS as a guide to the true location of the vehicle during a training phase. The work reported in Sharaf et al. (2005) is similar to that reported in Bouvet & Garcia (2000) in that preprocessing operations are performed on the measurements before they are fused. An assumption that is made in this method is that the DGPS data is always either available or unavailable due to an outage in satellite signal. However, in urban areas, satellite signals are often available but quite often are contaminated by multipath noises, which effects the quality of the ANN learning.

3. Fusion of landmark, INS, and GPS measurements

Detecting and recognizing landmarks provide spatial information related to the local environment. It is therefore possible to integrate spatial information with localization measurements from DR and GPS in order to improve localization accuracy Fuerstenberg & Weiss (2005); Jabbour, Bonnifait & Cherfaoui (2006); Jabbour, Cherfaoui & Bonnifait (2006); Rae & Basir (2007); Weiss et al. (2005). Two approaches for detecting and augmenting landmarks to vehicle localization systems are presented next along with another localization technique that attempts to detect visible satellites for use in the positioning process.

3.1 Laser scanners, digital maps, and GPS/DR

Due to the accumulated error caused by the long satellite outages in GPS/DR localization systems, digital maps are utilized to perform localization during such outages Weiss et al. (2005). A laser scanner mounted on a vehicle scans major objects in the vehicle environment. The system matches these landmarks with other landmarks in the digital map that represent the region of interest. If there is a match, the vehicle location is estimated by correlating the identified landmarks.

However, segmentation is not a trivial job specially in situations where landmarks are merged with background objects. Moreover, the system must be trained by having it traverse the regions of interest Fuerstenberg & Weiss (2005) to extract landmarks (features, such as traffic signs and the posts of traffic lights) that can later be used as a reference points.

In Jabbour, Bonnifait & Cherfaoui (2006), a vehicle equipped with an autonomous navigation system and a laser scanner is reported. The laser scanner is used to detect the edges of sidewalks and estimate the distance between the edge of the sidewalk and the vehicle. Distance measurements are utilized to improve the accuracy of a localization system that comprises GPS, DR, and Geographic Information System (GIS). The GIS data contains digitized information such as abstract road maps, road edges, and other landmarks. Landmark information is created through a learning stage. During the testing stage, the EKF fusion technique produces an innovation value from which the system determines whether to accept the fusion location estimate. If the GPS data is corrupted by multipath signals or is unavailable, only the DR location estimate utilized. The vehicle location estimate is used to select the region of interest from the GIS database that contains the landmark information. To improve the vehicle location estimate, a matching scheme is performed to compare the GIS-extracted landmarks (i.e., sidewalk edges) with those extracted by the laser scanner, and the estimated distances between the sidewalk edge and the vehicle are then used in fixing the vehicle location. Although the memory constraints are overcome by using the GIS, the accuracy of the estimate of the distances is not consistent due to occluding objects between the laser scanner and the edge of the sidewalk. The training phase required for any traversed region is also not insignificant.

3.2 Vision, digital maps, and GPS/DR

Visual data is also utilized in localization techniques since digital images can provide a wide range of information about the surrounding environment. Due to the time required for image processing Jabbour, Cherfaoui & Bonnifait (2006), only key images are maintained and linked to the GIS database Jabbour, Bonnifait & Cherfaoui (2006). Again, both GPS/DR are used and the proximity of the vehicle location estimate to the roads in the GIS database is examined. The road segment closest to the location estimate is then selected, and key images of that road are extracted in order to compare their features with the features of the images taken during the navigation stage. The weakness of this strategy appears when the curvature of the vehicle's path is significant, especially when the vehicle turns in orthogonal intersections.

Visual features can, however, be blended with other location measurements, such as GPS and DR data in the EKF formulation Rae & Basir (2007). The main advantage of this strategy is that the uncertainty of all the information sources is kept local to the EKF, namely, in the error covariance matrix, which guarantees a minimum mean square error estimates. In Rae & Basir (2007), the EKF structure is derived and validated where the curvature of the roads is employed as a visual feature. It is shown that when the roads are curvy, the vehicle location estimate is dramatically improved. On the other hand, if the road traversed is not curved, then the accuracy of the location estimate remains the same as that produced by the GPS/DR fusion localization technique.

3.3 Satellite visibility and DGPS

In urban areas, GPS multipath signals cause unpredictable localization errors due to the NLOS satellite signals. Another approach is the localization system which is driven by tracking visible GPS satellites using an infrared camera. An omni-directional infrared camera mounted on the top of a vehicle is used to recognize obstacles and their height and to detect visible satellites by observing their positions with a satellite orbit simulator Meguro et al. (2009). This method allows the system to exclude any radio waves emitted by invisible satellites to improve the localization accuracy.

The vehicle localization system used in this approach has high degree of accuracy since it employs a DGPS receiver. However, in high rise building areas, the availability of location estimates is low due to the lack of enough visible satellites, and even with enough visible satellites, the geometric configuration of the constellation may result in a high Dilution of Precision (DOP).

4. Cooperative localization

Cooperative Localization is a recent location estimation approach that has been implemented in vehicular positioning and wireless communication systems. This localization scheme is suitable for scenarios which involve the coexistence of several entities that independently provide location information. The goal is to localize a mobile node or to enhance its location estimate given that it shares relative spatial information with nearby nodes (e.g., other vehicles or mobile network towers).

4.1 Radio signal measurement data fusion

Radio localization methods have been studied extensively for cellular networks in a wide range of applications (e.g., for CDMA networks see Al-Jazzar & Caffery (2004); Caffery & Stuber (1994; 1998); Caffery (2000); Le et al. (2003); McGuire et al. (2003); Porretta et al. (2008); Sayed et al. (2005); Venkatraman et al. (2002); Wang et al. (2003); Wylie & Holtzman (1996) and for GSM networks see Chen et al. (2006)). An example of these systems is a localization system that estimates the locations of emergency calls initiated by cellular phones. The system operates on the principle that measurements from different Base Stations (BS's) are combined in order to compute the location of a Mobile Station (MS). The BS's typically have different levels of uncertainty in their measurements, which are minimized as a result of the fusion process. The relative spatial information in this system is based on the measurements from radio signals, such as Time of Arrival (TOA), Time Difference of Arrival (TDOA), Angel of Arrival (AOA), Received Signal Strength (RSS). In some of these GPS-less approaches, a mix of two or more different types of radio signal measurements is utilized in order to relax constraints such as the synchronization of the BS's.

In the following subsections detailed models for some of these techniques are given. (x_m, y_m) signifies the MS location. The locations of n base stations: $(\mathrm{BS}_1, \mathrm{BS}_2, \mathrm{BS}_3, \ldots, \mathrm{BS}_n)$ are denoted by $\{(x_1, y_1), (x_2, y_2), (x_3, y_3), \ldots, (x_n, y_n)\}$, respectively. For simplicity and without loss of generality, locations are represented by two coordinates, x and y, in the Cartesian coordinate system.

4.1.1 TOA data fusion

Time of arrival measurements are based on the time of flight of a signal as it travels between a source and a destination. Since the signal travels at the speed of light (c), it is possible to compute the distance between the two points as follows:

$$d_i = (t_i - t_m)c \tag{1}$$

where t_m signifies the signal sending time from the MS, t_i signifies the signal arrival time at the BS_i, and i signifies the BS's index (i.e., $i = \{1, 2, \ldots, n\}$).

According to Caffery & Stuber (1998), the TOA technique can be employed using three BS's, the minimum number of reference points in two dimensions (Figure 1), in order to estimate the MS location by computing the distances between each BS and the MS (i.e., d_1, d_2, d_3), as per Equation 1, and then formulating the following optimization problem:

$$\hat{x}_m, \hat{y}_m = arg \min_{x_m, y_m} \sum_{i=1}^{3} \left(d_i - \sqrt{(x_i - x_m)^2 + (y_i - y_m)^2} \right)^2 \tag{2}$$

Nevertheless, due to possible NLOS propagation conditions, the actual Euclidean distances between the MS and the BS_i is less than or equal to $(t_i - t_m)c$. This inequality creates more than one solution for the optimization problem in 2, all of which reside in a bounded area, as shown in Figure 1. A constrained version of the optimization problem in 2 is proposed in Caffery (1999); Porretta et al. (2008) in order to increase the localization accuracy; however, the

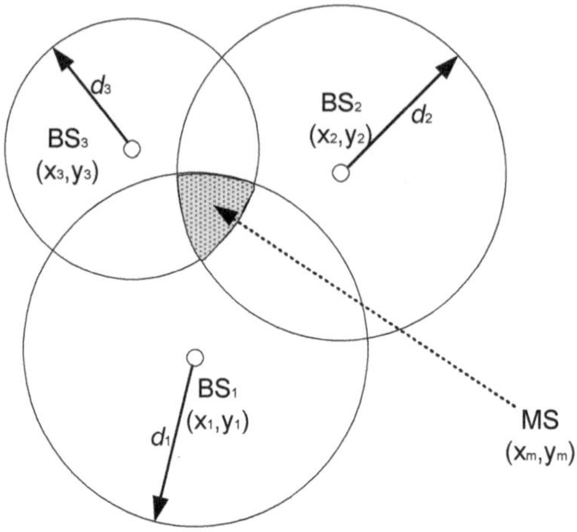

Fig. 1. The TOA localization method.

geometric arrangement of the BS's may produce a poor location estimates due to the shape of the bounded area that contains the MS. This shortcoming might be avoided using more BS's. The next method described below utilizes more than three BS's in estimating the MS location so that ambiguity in the distance computation is reduced.

In Caffery (2000); Sayed et al. (2005) the Cartesian coordinate system is represented as follows. The location of one of the base stations is assumed to be the origin (e.g., BS_1 be the origin: $(x_1, y_1) = (0,0)$) and the locations of the other objects in the network are computed with respect to the origin. Hence, the distances $(d_1, d_2, d_3 \ldots, d_n)$ can be used to estimate the location of the MS by solving the following set of equations:

$$\begin{aligned} d_1^2 &= x_m^2 + y_m^2 \\ d_2^2 &= (x_2 - x_m)^2 + (y_2 - y_m)^2 \\ d_3^2 &= (x_3 - x_m)^2 + (y_3 - y_m)^2 \\ &\vdots \\ d_n^2 &= (x_n - x_m)^2 + (y_n - y_m)^2 \end{aligned} \tag{3}$$

After rearranging terms, the above equations can be written as follows:

$$\begin{bmatrix} x_2 & y_2 \\ x_3 & y_3 \\ \vdots & \vdots \\ x_n & y_n \end{bmatrix} \begin{bmatrix} x_m \\ y_m \end{bmatrix} = \frac{1}{2} \begin{bmatrix} k_2^2 - d_2^2 + d_1^2 \\ k_3^2 - d_3^2 + d_1^2 \\ \vdots \\ k_n^2 - d_n^2 + d_1^2 \end{bmatrix} \tag{4}$$

where $k_i^2 = x_i^2 + y_i^2$. Equation 4 can be expressed in a matrix form

$$\mathbf{Hx} = \mathbf{b} \tag{5}$$

where $\mathbf{H} = \begin{bmatrix} x_2 & y_2 \\ x_3 & y_3 \\ \vdots & \vdots \\ x_n & y_n \end{bmatrix}$, $\mathbf{x} = \begin{bmatrix} x_m \\ y_m \end{bmatrix}$, and $\mathbf{b} = \frac{1}{2}\begin{bmatrix} k_2^2 - d_2^2 + d_1^2 \\ k_3^2 - d_3^2 + d_1^2 \\ \vdots \\ k_n^2 - d_n^2 + d_1^2 \end{bmatrix}$.

Equation 5 represents an overdetermined system (i.e., $n > 2$). Practically, such a system has no exact solution. Therefore a linear least squares method is used to estimate the location of the MS as follows:

$$\hat{\mathbf{x}} = \left(\mathbf{H}^T\mathbf{H}\right)^{-1}\mathbf{H}^T\mathbf{b} \tag{6}$$

where $(.)^T$ signifies matrix transpose and $(.)^{-1}$ signifies matrix inverse.

Alternative techniques, such as the maximum likelihood are reported in McGuire et al. (2003); Wang et al. (2003).

4.1.2 TDOA data fusion

TDOA is preferable to the TOA due to the fact that TDOA does not require synchronization between the MS and BS's, Figure 2. Instead, it takes advantage of the synchronization of the CDMA cellular network BS's to compute the difference between the time of arrivals of the MS

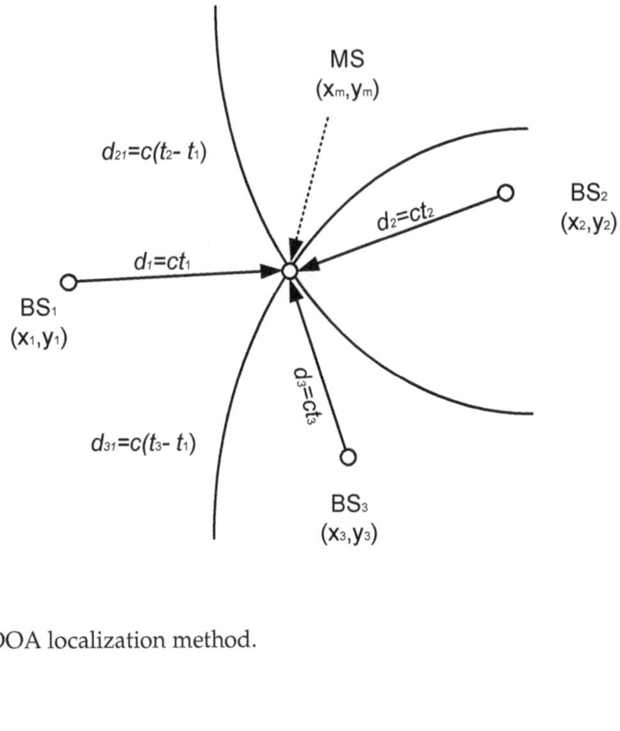

Fig. 2. The TDOA localization method.

signal at the BS_i and BS_1, where $i \in \{2,3,\ldots,n\}$. The difference in the distance is therefore defined as follows:

$$\begin{aligned} d_{i1} &\equiv d_i - d_1 \\ &= (t_i - t_m)c - (t_1 - t_m)c \\ &= (t_i - t_1)c \end{aligned} \tag{7}$$

It can be seen that the difference is not affected by errors in the MS clock time t_m. Substituted Equation 7 in Equation 3, and then expanding and rearranging the terms produce the following set of equations:

$$\begin{bmatrix} x_2 & y_2 \\ x_3 & y_3 \\ \vdots & \vdots \\ x_n & y_n \end{bmatrix} \begin{bmatrix} x_m \\ y_m \end{bmatrix} = d_1 \begin{bmatrix} -d_{21} \\ -d_{31} \\ \vdots \\ -d_{n1} \end{bmatrix} + \frac{1}{2} \begin{bmatrix} k_2^2 - d_{21}^2 \\ k_3^2 - d_{31}^2 \\ \vdots \\ k_n^2 - d_{n1}^2 \end{bmatrix} \tag{8}$$

which can be expressed in a matrix form as follows:

$$\mathbf{H}\mathbf{x} = d_1\mathbf{c} + \mathbf{r} \tag{9}$$

where $\mathbf{H} = \begin{bmatrix} x_2 & y_2 \\ x_3 & y_3 \\ \vdots & \vdots \\ x_n & y_n \end{bmatrix}$, $\mathbf{c} = \begin{bmatrix} -d_{21} \\ -d_{31} \\ \vdots \\ -d_{n1} \end{bmatrix}$, and $\mathbf{r} = \frac{1}{2} \begin{bmatrix} k_2^2 - d_{21}^2 \\ k_3^2 - d_{31}^2 \\ \vdots \\ k_n^2 - d_{n1}^2 \end{bmatrix}$.

Similarly, Equation 9 can be solved using the following linear least squares formulation:

$$\hat{\mathbf{x}} = \left(\mathbf{H}^T\mathbf{H}\right)^{-1} \mathbf{H}^T \left(d_1\mathbf{c} + \mathbf{r}\right) \tag{10}$$

The solution of Equation 10 is determined in two steps. First, the estimate of the MS is determined in terms of d_1, which is substituted in the quadratic expression $d_1^2 = x_m^2 + y_m^2$ to compute d_1. Second, the value of d_1 is substituted back in Equation 10 to solve for $\hat{\mathbf{x}}$ Sayed et al. (2005).

4.1.3 AOA data fusion

AOA techniques estimate the location of an MS by measuring the angle of signal arrival from the MS at several BS's by means of an antenna array. The MS location is then estimated through the intersection of the straight paths leaving from at least two BS's, as depicted in Figure 3. However, combining only two AOA measurements may introduce a large amount of uncertainty with respect to the MS location estimate, especially when the MS is close to the line connecting the two BS's. Moreover, this localization method requires the MS to be in LOS with the participating BS's, since reflected or diffracted signals result in misleading information. For this reason, it is preferable for the AOA to be combined with another localization method, such as TOA or TDOA.

4.1.4 RSS data fusion

RSS based localization is a method that employs mathematical models that describe the path loss as a function of distance. Since these models translate the received signal power into a

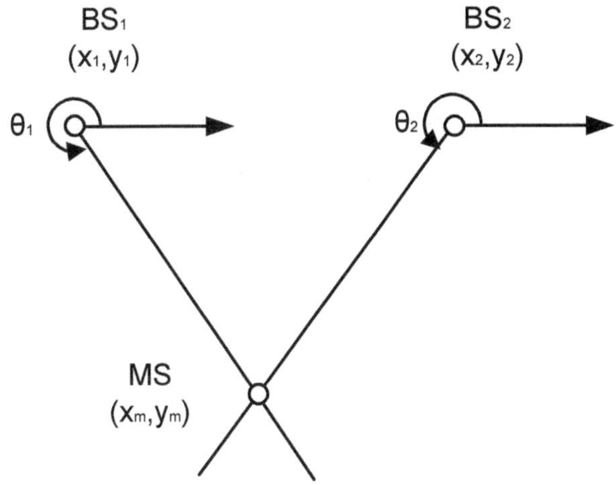

Fig. 3. The AOA localization method.

distance between an MS and a BS, the MS must lie on a circle centered at the BS. Employing three or more BS's provides an estimate for the MS location.

RSS is well known for being drastically affected by multipath fading and shadowing (multipath signals). The error caused by multipath signals can be reduced by using prior knowledge available on the contours of the signal strength centered at the BS's Smith (1991). However, such knowledge assumes a specific surrounding environment that can change due to change in whether, moving objects, such as trucks, as well as new buildings and other barriers.

4.1.5 Fingerprinting

This localization method is a pattern recognition, or pattern matching, technique. The underlying concept of fingerprinting is that the radio signal propagation characteristics of an MS are unique in terms of TOA, AOA, and RSS when captured at different BS's Chen et al. (2006); Porretta et al. (2008). These characteristics can therefore be used as a signature to indicate the location of an MS. The fingerprinting method has two phases: a training phase and localization phase. In the training phase, a database is created to index the different patterns in the characteristics of the radio signal propagation. In the localization phase, the signature of the MS is matched with the patterns in the database. The challenging aspect of this method is assuring that the system can distinguish between similar patterns that represent different locations.

Of course, the more exhaustive is the training phase (i.e., recording a signature for every small area in the environment), the more accurate is the MS location estimate. The main drawback of this method is the requirement to continually update the database as the configuration of the BS's changes when BS's are removed or new ones are added. Nevertheless, this method is becoming more attractive for indoor applications because the database creation can be more comprehensive and manageable.

4.2 VANET localization using relative distances measurements

This approach takes advantage of the emerging VANET environments. The distances between VANET nodes are estimated and exchanged among vehicles along with preliminary estimates of the vehicles' locations. Vehicles can then use this information to construct local relative position maps that contain the vehicles and their neighbours. This strategy has initially emerged in Wireless Sensor Networks (WSN), but recently, a number of solutions have been proposed for use in VANET Benslimane (2005); Drawil & Basir (2008); Parker & Valaee (2007)

4.2.1 Vehicle localization in VANET

A VANET based localization method was introduced in Benslimane (2005) for localizing vehicles with no GPS receivers, or those whose location can not be determined because satellite signals have been lost, for instance, in a tunnel. With this method, vehicles that are not equipped with GPS determine their own locations by relying on information they receive from vehicles that are equipped with GPS. Vehicles within transmission range can measure the distances between each other using one of the radio-location methods presented in Caffery & Stuber (1998). By finding its closest three neighbours the unequipped vehicle can compute its position using trilateration.

4.2.2 Cooperative vehicle position estimation

The work reported in Parker & Valaee (2006) presents a method of distributed vehicle localization in VANET. The method utilizes RSS measurements to estimate the distances between one vehicle and others in its coverage area. It is assumed that vehicles initially estimate their own locations using a GPS receiver and then exchange their location information so that they can perform an optimization technique in order to improve their location estimates.

This technique demonstrates robustness of location estimates. However, it lacks the ability to detect and avoid the effect of multipath signals in the GPS measurements, which drastically degrades the localization accuracy in multipath environments (e.g., urban canyons).

In Drawil & Basir (2010) an algorithm called InterVehicle-Communication-Assisted Localization (IVCAL) is proposed to mitigate the multipath effect in the location estimates of vehicles in VANET. A KF and an inter-vehicle-communication system collaborate in order to increase the robustness and accuracy of the localization of every vehicle in the network. The two main components that allow the inter-vehicle-communication system and the KF to interact are the Multipath Detection Unit (MDU), which detects the existence of a multipath effect in the output of the KF, and the Localization Enhancement Unit (LEU), which obtains the neighbours' information from the inter-vehicle-communication system and feeds an optimized location estimate back to the KF (Figure 4). As in Jabbour, Cherfaoui & Bonnifait (2006) and Jabbour, Bonnifait & Cherfaoui (2006), KF innovation is used as an indication of the contamination of the GPS measurement, and it has therefore been used as a learning pattern for the MDU in IVCAL. An uncertainty measure is also utilized in order to specify a subset of the most accurate network neighbours that can be used as anchors to enable vehicles to improve their location estimates.

Lack of adequate location anchors and/or prolonged multipath conditions remain unsolved problems that continue to degrade localization accuracy.

5. Multi-level fusion approaches

As it has been reported above, a verity of multi-modality localization methods have evolved in recent years. Typical modalities include satellite signals, VANET communication, vision features, laser rays, etc. This variety of information has motivated the concept of multi-level fusion.

For instance, in Boukerche et al. (2008), a data-fusion model is proposed in the form of a three-level fusion localization system. In the first level, a variety of location information is gathered as row data and processed separately using local filters that are suitable for each type of location information. As with the system in Skog & Handel (2009), the second level combines the output of the first level and produces a better location estimate. In Boukerche et al. (2008), the results are then fused in the third level based on contextual information (e.g., digital maps and traffic information). In this scheme, the final location estimate is fed back to the second level in order to improve future estimations.

Multi Level fusions aims to tackle data fusion as a hierarchical process so as to allow for combining measurements at various levels of abstraction in a simple manner. Nevertheless, if the estimates in the lowest-level filters are evaluated for reliability, the fusion of these estimates in higher-level filters will then be more robust.

6. Integrity of localization systems

Due to the inherent errors in the positioning information, a level of uncertainty in location estimates is inevitable. Therefore, it is essential to measure the reliability of the positioning information in order to identify any hidden anomalies. To achieve this task, a level of trust, integrity, in every estimate must be determined.

In the last two decades, a significant effort has been made in aviation to develop integrity monitoring systems Hewitson (2003); Walter & Enge (1995). Integrity is defined as a measure

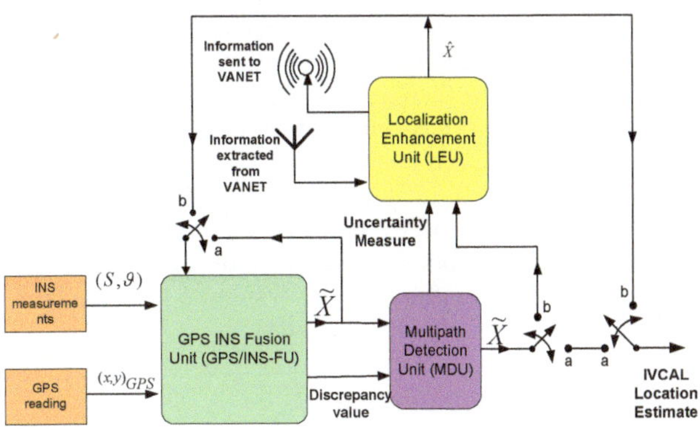

Fig. 4. Block diagram of the inter-vehicle-communication-assisted localization technique.

of the trust which can be placed in the correctness of the information supplied by the total system; integrity includes the ability of a system to provide timely and valid measurements to users ESA (n.d.). Three key components have been proposed for integrity monitoring: 1) fault detection, 2) fault isolation, and 3) removal of faulty measurement sources from the estimates Hewitson et al. (2004). The European Geostationary Navigation Overlay Service (EGNOS) and the Wide Area Augmentation System (WAAS), Hewitson (2003), are developed to form a redundant source of information for the Global Navigation Satellite Systems (GNSS) in order to perform integrity monitoring by providing correction information.

During the last decade, monitoring the integrity of land-vehicles' localization has attracted attention due to the increasing demand for highly reliable accurate location data. Since roving in dense urban environments may limit access to the signals from augmentation systems such as EGNOS or WAAS, other means of measuring integrity have been proposed Schlingelhof et al. (2008).

For instance, Toledo-Moreo et al. (2006) presents a localization solution based on the fusion of GNSS and INS sensors. In this fusion process an interactive multimodel method is used. Different covariance matrices are used as a response to change in the noise behaviour. The proposed integrity measure is based on the covariance matrix of the EKF estimation error.

Relying on the error covariance matrix can be misleading especially when experiencing unmodeled environment noise. In other words, it is not possible in many cases to detect, isolate, and remove the estimation faults, let alone the unavoidable false alarms.

Also, in Jabbour et al. (2008) a binary integrity decision-maker is proposed for a map-matching localization technique in which multihypothesis road-tracking method combines proprioceptive sensors (odometers and gyrometers) with GPS and map information. In this work, the integrity represents high or low confidence of the location estimate. The candidate tracks or roads are associated with a probability that is computed using the multihypothesis road-tracking method. If one credible road exists and the normalized innovation is below a prespecified threshold, the technique declares high confidence location estimate. However, the lack of granularity in the integrity measure limits the range of the integrity-level based application that can use this method.

Integrity monitoring of map-matching localization has also been proposed and tested in Quddus (2006). However, in this work three indicators has been monitored to achieve this task: distance residuals, heading residuals, and an indicator related to uncertainty of the map matched position. Due to the linguistic nature of these indicators, they have been combined using a fuzzy inference model to produce a value between 0 to 100 to indicate the integrity of the system. The integrity threshold has been determined experimentally to be 70, where the type of the environment experienced during the experiment was not specified. The value of the threshold thus can be considered specific to the environment of the experiment. Therefore, the approach might not guarantee a robust integrity monitoring. In other words, it is possible to come across an environment that influences the system to produce both an integrity value above the threshold and a location estimate mismatch.

7. Performance criteria and benchmarking

From the discussion above it is clear that vehicle localization is an increasingly growing area of research. Nevertheless, there is a number of outstanding issues that still need to be addressed.

In order to put these outstanding issues in practical context the following performance criteria are proposed.

7.1. ***Accuracy:*** *Accuracy of a vehicle location estimate is defined as the degree of closeness of a vehicle's location estimate to its actual (true) location.*

7.2. ***Availability:*** *Availability of a vehicle location estimate is defined as the ratio of the number of estimates produced to the number of estimates expected per one unit of time.*

7.3. ***Response Time:*** *Response time is the time required by a localization technique to produce a location estimate.*

7.4. ***Integrity:*** *Integrity is defined as the level of confidence that can be placed in the correctness of the location estimate* Bakhache & Nikiforov (2000); ESA (n.d.); Quddus (2006).

Based on the above performance criteria, a benchmark can be established in order to compare the performance of different localization techniques based on reported best achievable accuracy localization performance. Localization performance is compared with respect to reliability as well. Table 1 provides a summary of the comparison in terms of modality used, best case accuracy, environmental constraints, synchronization requirements, and dependency on infrastructure. Table 2 reports emerging applications and their requirements

Modality(ies)	Best Case Accuracy (m)	Availability	Synch.	Infr.str.
GPS	10-20 Hoshen (1996); Leva (1996)	Out Door-Open Sky	Yes	No
DeadReckoning (DR)	Worsen with time Kao (1991)	Anywhere	No	No
DGPS with Visible Satellites	0.01-7.6 Meguro et al. (2009)	suburban-Open Sky	Yes	Yes
DGPS+DR+Map Matching	0.5-5 Lahrech et al. (2005)	Out Door-Open Sky	N/A	Yes
GPS+Vision+Map Matching	0.5-1 Chausse et al. (2005); Jabbour, Bonnifait & Cherfaoui (2006)	Out Door-Open Sky	No	Yes
Cellular Localization	90-250 Chen et al. (2006); 25-69 Porretta et al. (2008)	Under Network Coverage	Yes	Yes
Location Services	Submeter Zhang et al. (2008)	In Door	N/A	Yes
Relative Ad hoc Localization	2-7 (Simulation Drawil & Basir (2008); Parker & Valaee (2006))	Suburban	Yes	No

Table 1. Specifications of Localization Techniques.

with respect to localization accuracy. It is evident from Tables 1 and 2 that current localization techniques do not live up to the required integrity and availability performance. In other words, the delivered performance of the localization techniques listed in Table 1 is not always above the target performance specified by the applications, and that is due to the unavailability of their measurements or the decrease in their accuracy in some environments, such as urban canyons, foggy weather, and dark areas.

Hence, performance needed by applications can constitute a challenging issue in the fusion process of a multi-sensory system. Therefore task driven integrity issues relevant to vehicle localization are highlighted next.

8. Task driven localization integrity

From the above discussion it is obvious that for localization systems to meet the expectations of emerging applications it is imperative that they employ diverse location measurement sources and effective strategies to fuse these sources so as to achieve the Quality of Service expected of them. Of course this Quality of Service is multi-dimensional as it pertains to expected accuracy, availability, response time and integrity. The Quality of Service as a function of these performance criteria is application and task dependent. The more stringent is the required Quality of Service with respect to a given performance criterion, the more resources are needed and the higher is the computational cost. This presents a challenge for

Application	Required Accuracy		
	Low(10-20 m)	Medium (1-5 m)	High (less than 1 m)
Message Routing (VANET)	X		
Data Dissemination	X		
Map Localization	X		
Coop. Cruise Control		X	
Coop. Intersection Safety		X	
Blind Crossing		X	
Platooning		X	
Collision Warning Sys.			X
Vision Enhancement			X
Automatic Parking			X
Road Pricing			X

Table 2. Applications Requirement for Location Estimates Boukerche et al. (2008).

the system as calls for effective use of resources to achieve the target Quality of Service. For example, there are applications where accuracy can be traded for faster response time. On the other hand, there are applications where response time is not as important as accuracy (offline vehicle track mapping). There are also applications where both requirements, accuracy and response time, can not be compromised for any other gain.

Indeed, task or goal driven localization is about effective allocating system resources and planning of localization tasks such that the system mission is achieved with maximum integrity possible. This strategy to performance is a key issue to the new trends of hybrid localization systems. In order for this strategy to work it is imperative that the impact of the environment is not ignored. Without modeling the impact of the environment on the system, the system can not be guaranteed to achieve its target performance, and even worst as it may falsely determine its task is accomplished. Thus, modeling the impact of the environmental conditions on the system is a central issue to the following proposed framework.

9. Task-driven localization through integrity assessment and control

It is well understood that the reported techniques can estimate the location of vehicles relatively accurately in some situations if they are given adequate time to perform the task. However, they may not perform as well in other situations. The deficiencies of these localization techniques are uncorrelated as they are expected to be of diverse phenomena, and/or utilize different algorithmic paradigms. This motivates the development of systems that can take advantage of this diversity to achieve a reliable and accurate performance.

In this section, a high level concept of a novel framework for fusing different localization techniques is proposed, Figure 5. What distinguishes this framework from existing ones is its ability to take in account the impact of the measurement conditions on the individual techniques. Thus, it is able to optimize the fusion process so as to maximize the accuracy and integrity of the localization estimates. The framework consists of three logical layers: (1) Primary Localization layer which provides preliminary location estimates using the available localization techniques; (2) Integrity Monitoring layer which computes the reliability of the vehicle's location estimates produced by the Primary Localization layer- a process that captures the impact of measurement conditions; and (3) Estimate Fusion and Management layer which interacts with the application task to ensure that the task's expected localization

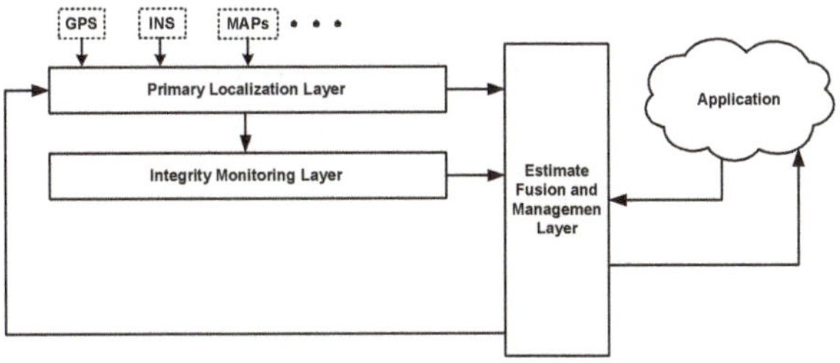

Fig. 5. The structure of the proposed framework.

accuracy and integrity are achieved by executing a proper fusion scheme. In what follows a further description of the framework layers functionality.

9.1 Primary localization layer

The primary localization layer comprises of the system's localization techniques which are partitioned in the form of a set of Primary Localization Units (PLUs), as can be seen in Figure 6. Any localization technique, such as those mentioned above, can be used in any given PLU. These primary localization units receive localization requests from the Estimate Fusion and Management layer. Each PLU is constructed from techniques that are based on different phenomena/algorithms to ensure minimum correlation. A primary localization unit can share its information sources with other units; it can constitute a single modality or multiple-modalities. An example of a single modality PLU is one that estimates the vehicle

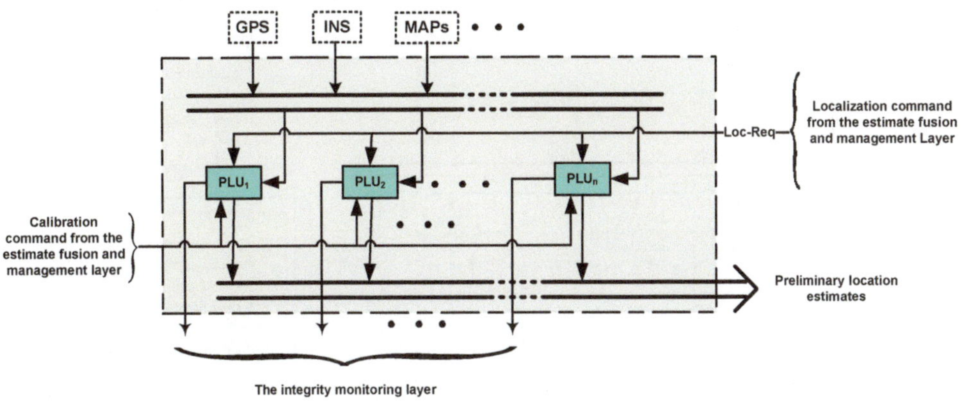

Fig. 6. Primary localization layer.

location from a GPS information source. IVCAL is an example of a PLU that utilizes three modalities: GPS, INS, and Inter-Vehicle-Communication.

9.2 Integrity Monitoring layer

Central to the proposed framework is the integrity monitoring layer. Here, an Integrity Monitoring unit (IMU) is used to monitor the performance of a primary localization unit (Figure 7). The monitoring process takes in consideration the impact of the measurement conditions on the PLU. For example, to indicate the reliability of an estimate DOP measure

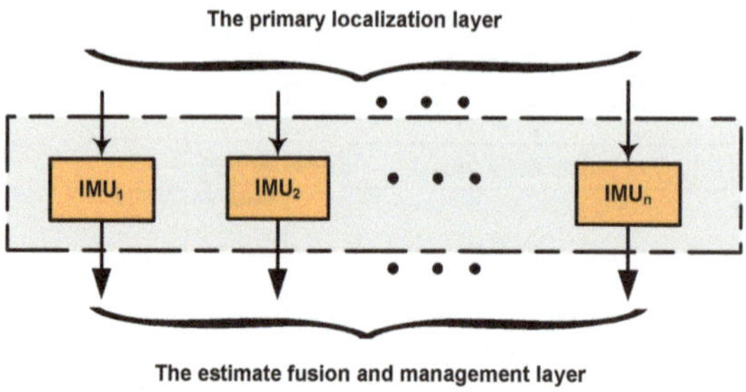

Fig. 7. Integrity monitoring layer.

and/or the signal to noise ratio can be utilized when a GPS receiver is used, light intensity can be utilized when vision features are used, and KF innovation can be utilized when IVCAL is used. Various tools can be employed in this layer based on the type of the localization technique. Fuzzy inference systems and probabilistic models for reliability are two examples of these tools.

9.3 The Estimate Fusion and Management Layer

The Estimate Fusion and Management layer (EFM) is responsible for determining an effective integration (Meta-Fusion) strategy for fusing the estimates produced by the different primary localization units so as to achieve the required localization accuracy and integrity (Figure 8). The estimate fusion and management processes the location estimates produced by the

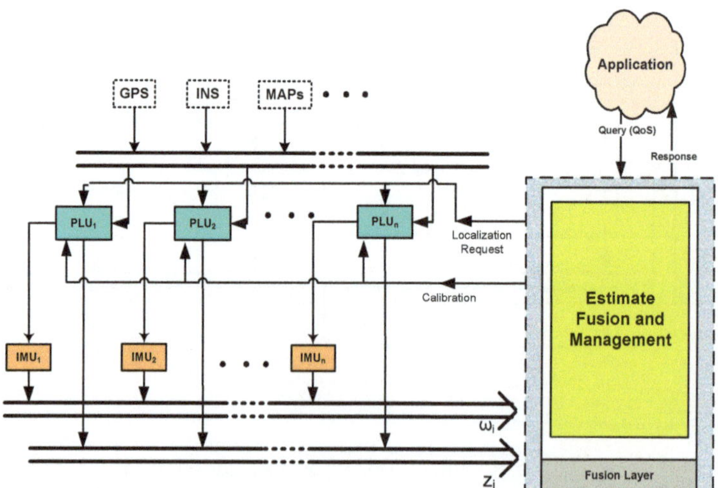

Fig. 8. Estimate fusion and management layer.

different primary localization units in conjunction with their integrity assessments. Since the vehicle is expected to be performing localization while moving, it is imperative for the fusion process to perform spatial and temporal alignment of the estimates produced by the different PLUs. Therefore, this layer employs a synchronization handler to manage timing issues among the different PLUs. Given the task's target accuracy and integrity, as well as that of the various PLUs, a management and fusion scheme is computed such that the scheme produces a location estimate that meets the task requirements.

To overcome the problems of this layer, first of all, PLU estimates should be time-stamped as close as possible to a common time base. Of course the allowable synchronization error would depend on factors such as the speed of the vehicle relative to the PLU response time. It is also affected by the system's desired spatial precision and detection frequency. The tighter time synchronization is achieved with respect to the common time base, the greater precision is possible in the tracking of the vehicle.

Second of all, since the fusion process is task driven, an optimal fusion strategy is the one that achieves the target accuracy and integrity within the constraints of the task deadline. This gives rise to the challenge of optimal estimate fusion and reliability aggregation. Both fuzzy reasoning and evidential reasoning are a tentative tools to be investigated as the bases for constructing the meta-fusion model. Fuzzy reasoning can be used for representing uncertainty in the estimates as well as for representing linguistic task requirements. Since some PLUs may employ probabilistic (Bayesian) estimators, it will be interesting to study how probabilistic estimates and fuzzy estimates are represented in a unified uncertainty framework.

Bayesian theory based fusion techniques have been evolving in fields such as process control, target tracking and object recognition. Nonetheless, effective fusion performance can only be achieved if adequate and appropriate priori and conditional probabilities are available. Although, at least in some situations, assumptions can be made with respect to priori and posteriori probabilities, these assumptions can turn to be unreasonable in many other situations, especially if we are to allow for non-probabilistic estimators in the PLU layer. One possible solution is using the Dempster-Shafer (DS) evidence theory as an extension to the Bayes theory. DS belief and plausibility functions can be used to quantify evidence and unify uncertainty of the PLU estimates. DS evidence theory can also model how the uncertainty of a given location estimate diminishes as pieces of evidence accumulate during the localization process. One important aspect of this theory is that reasoning or decision making can be carried out with incomplete or conflicting pieces of evidence – a reality that is quit common in localization problems.

10. Conclusions

In this chapter, a variety of reported localization techniques are presented and classified based on the type of the measurement of the location information used.

Although, techniques that incorporate fusion of motion sensory data with GPS localization have demonstrated improvement in performance, there are still situations that can have a negative effect on their localization accuracy. Incremental localization errors in motion-sensor data and the multipath effect in urban canyon environments contribute significantly to such location estimate errors, which necessitates augmenting the initial location data with other sources of location information in order to overcome these shortcomings.

Digital maps and visual features enhance GPS-DR localization by recognizing landmarks in the surrounding environment and matching them with others in a reference GIS map. A key problem associated with this scheme is that the landmark segmentation process is complex and ill conditioned process.

Multi-level fusion schemes are promising as they employ multiple location measurement phenomena. However, these schemes have given birth to new challenges in the localization problem in terms of resource synchronization, resource management, and task driven performance.

A novel framework for vehicle localization is presented. The aim is to develop a vehicle localization system that can optimize and plan the use of its resources so as to achieve the performance requirements of the localization task or application. The main components of the proposed framework are key research issues.

11. References

Al-Bayari, O. & Sadoun, B. (2005). New Centralized Automatic Vehicle Location Communications Software System Under GIS Environment, *INTERNATIONAL JOURNAL OF COMMUNICATION SYSTEMS* 18(9): 833.

Al-Jazzar, S. & Caffery, J., J. (2004). NLOS Mitigation Method for Urban Environments, *IEEE 60th Vehicular Technology Conference* 7: 5112–5115.

Aono, T., Fujii, K., Hatsumoto, S. & Kamiya, T. (1998). Positioning of Vehicle on Undulating Ground using GPS and Dead Reckoning, *IEEE International Conference on Robotics and Automation* 4: 3443–3448.

Bakhache, B. & Nikiforov, I. (2000). Reliable detection of faults in measurement systems, *International Journal of Adaptive Control and Signal Processing* 14(7): 683–700.

Benslimane, A. (2005). Localization in Vehicular Ad hoc Networks, *Systems Communications. Proceedings* pp. 19–25.

Bouju, A., Stockus, A., Bertrand, R. & Boursier, P. (2002). Location-Based Spatial Data Management in Navigation Systems, *IEEE Intelligent Vehicle Symposium* 1: 172–177.

Boukerche, A., Oliveira, H., Nakamura, E. & Loureiro, A. (2008). Vehicular Ad hoc Networks: A New Challenge for Localization-Based Systems, *Computer Communications* 31(12): 2838–2849.

Bouvet, D. & Garcia, G. (2000). Improving the Accuracy of Dynamic Localization Systems using RTK GPS by Identifying the GPS Latency, *IEEE International Conference on Robotics and Automation* 3: 2525–2530 vol.3.

Caffery, J. (1999). *Wireless Location in Cdma Cellular Radio Systems*, Kluwer Academic Pub.

Caffery, J. & Stuber, G. (1994). Vehicle Location and Tracking for IVHS in CDMA Microcells, *5th IEEE International Symposium on Personal, Indoor and Mobile Radio Communications* 4: 1227-1231.

Caffery, J. & Stuber, G. (1998). Overview of Radiolocation in CDMA Cellular Systems, *IEEE Communications Magazine* 36(4): 38 – 45.

Caffery, J.J., J. (2000). A New Approach to the Geometry of TOA Location, *IEEE 52nd Vehicular Technology Conference* 4: 1943–1949.

Chausse, F., Laneurit, J. & Chapuis, R. (2005). Vehicle Localization on a Digital Map using Particles Fltering, pp. 243–248.

Chen, M., Sohn, T., Chmelev, D., Haehnel, D., Hightower, J., Hughes, J., LaMarca, A., Potter, F., Smith, I. & Varshavsky, A. (2006). Practical Metropolitan-scale Positioning for GSM Phones, *Lecture Notes in Computer Science* 4206: 225.

Cramer, M. (1997). GPS/INS Integration.

Dao, D., Rizos, C. & Wang, J. (2002). Location-Based Services: Technical and Business Issues, *GPS Solutions* 6(3): 169–178.

Drawil, N. & Basir, O. (2008). Vehicular Collaborative Technique for Location Estimate Correction, *IEEE 68th Vehicular Technology Conference* pp. 1–5.

Drawil, N. & Basir, O. (2010). Intervehicle-communication-assisted localization, *Intelligent Transportation Systems, IEEE Transactions on* 11(3): 678 –691.

ESA (n.d.). Making EGNOS Work for You,CD-ROM.

Fuerstenberg, K. & Weiss, T. (2005). Feature-Level Map Building and Object Recognition for Intersection Safety Applications, *Proceedings of IEEE Intelligent Vehicles Symposium* pp. 490–495.

Hewitson, S. (2003). GNSS receiver autonomous integrity monitoring: A separability analysis, *Proc. ION GPS* pp. 1502–1509.

Hewitson, S., Kyu Lee, H. & Wang, J. (2004). Localizability Analysis for GPS/Galileo Receiver Autonomous Integrity Monitoring, *The Journal of Navigation* 57(02): 245–259.

Honghui, Q. & Moore, J. B. (2002). Direct Kalman Filtering Approach for GPS/INS Integration, *IEEE Transactions on Aerospace and Electronic Systems* 38(2): 687 – 693.

Hoshen, J. (1996). The GPS Equations and the Problem of Apollonius, *Aerospace and Electronic Systems, IEEE Transactions* 32(3): 1116 – 1124.

Jabbour, M., Bonnifait, P. & Cherfaoui, V. (2006). Enhanced Local Maps in a GIS for a Precise Localisation in Urban Areas, *IEEE Intelligent Transportation Systems Conference* pp. 468–473.

Jabbour, M., Bonnifait, P. & Cherfaoui, V. (2008). Map-Matching Integrity using Multi-Sensor Fusion and Multi-Hypothesis Road Tracking, *Journal of Intelligent Transportation Systems Technology Planning and Operations* 12(4): 189–201.

Jabbour, M., Cherfaoui, V. & Bonnifait, P. (2006). Management of Landmarks in a GIS for an Enhanced Localisation in Urban Areas, *Intelligent Vehicles Symposium, 2006 IEEE* pp. 50–57.

Kao, W. (1991). Integration of GPS and Dead-Reckoning Navigation Systems, *Vehicle Navigation and Information Systems Conference, 1991* 2: 635–643.

Lahrech, A., Boucher, C. & Noyer, J.-C. (2005). Accurate Vehicle Positioning in Urban Areas, *31st Annual Conference of IEEE Industrial Electronics Society* p. 5 pp.

Lai, C.-C. & Tsai, W.-H. (2003). Location Estimation and Trajectory Prediction of Moving Lateral Vehicle using Two Wheel Shapes Information in 2-D Lateral Vehicle Images by 3-D Computer Vision Techniques, *IEEE International Conference on Robotics and Automation* 1: 881–886.

Le, B. L., Ahmed, K. & Tsuji, H. (2003). Mobile Location Estimator With NLOS Mitigation using Kalman Filtering, *Wireless Communications and Networking, IEEE* 3: 1969–1973.

Leva, J. L. (1996). An Alternative Closed-Form Solution to the GPS Pseudo-Range Equations, *Aerospace and Electronic Systems, IEEE Transactions* 32(4): 1430 – 1439.

McGuire, M., Plataniotis, K. & Venetsanopoulos, A. (2003). Location of Mobile Terminals using time Measurements and Survey Points, *IEEE Transactions on Vehicular Technology Conference* 52(4): 999–1011.

Meguro, J.-i., Murata, T., Takiguchi, J.-i., Amano, Y. & Hashizume, T. (2009). GPS Multipath Mitigation for Urban Area Using Omnidirectional Infrared Camera, *IEEE Transactions on Intelligent Transportation Systems* 10(1): 22–30.

Nishimura, Y., Tanahashi, I., Taniguchi, S., Matsumoto, N. & Nakamura, K. (1996). A New Concept for Vehicle Localization of Road Debiting System, *Proceedings of the IEEE Intelligent Vehicles Symposium* pp. 93–98.

Parker, R. & Valaee, S. (2006). Vehicle Localization in Vehicular Networks, *IEEE 64th Vehicular Technology Conference* pp. 1–5.

Parker, R. & Valaee, S. (2007). Cooperative Vehicle Position Estimation, *IEEE International Conference on Communications* pp. 5837–5842.

Porretta, M., Nepa, P., Manara, G. & Giannetti, F. (2008). Location, Location, Location, *Vehicular Technology Magazine, IEEE* 3(2): 20–29.

Quddus, M. (2006). *High integrity map matching algorithms for advanced transport telematics applications*, PhD thesis, Citeseer.

Rae, A. & Basir, O. (2007). A Framework for Visual Position Estimation for Motor Vehicles, *4th Workshop on Positioning, Navigation and Communication* pp. 223–228.

Rezaei, S. & Sengupta, R. (2005). Kalman Filter Based Integration of DGPS and Vehicle Sensors for Localization, *Mechatronics and Automation, IEEE International Conference* 1: 455–460.

Sayed, A., Tarighat, A. & Khajehnouri, N. (2005). Network-Based Wireless Location: Challenges Faced in Developing Techniques for Accurate Wireless Location Information, *Signal Processing Magazine, IEEE* 22(4): 24–40.

Schlingelhof, M., Betaille, D., Bonnifait, P. & Demaseure, K. (2008). Advanced Positioning Technologies for Co-operative Systems, *Intelligent Transport Systems, IET* 2(2): 81–91.

Sharaf, R., Noureldin, A., Osman, A. & El-Sheimy, N. (2005). Online INS/GPS Integration with A Radial Basis Function Neural Network, *IEEE Aerospace and Electronic Systems Magazine* 20(3): 8–14.

Skog, I. & Handel, P. (2009). In-Car Positioning and Navigation TechnologiesâĂŤA Survey, *IEEE Transactions on Intelligent Transportation Systems* 10(1): 4–21.

Sliety, M. (2007). Impact of Vehicle Platform on Global Positioning System Performance in Intelligent Transportation, *Intelligent Transport Systems, IET* 1(4): 241–248.

Smith, W.W., J. (1991). Passive Location of Mobile Cellular Telephone Terminals, *IEEE International Carnahan Conference on Security Technology* pp. 221–225.

Stockus, A., Bouju, A., Bertrand, F. & Boursier, P. (2000). Web-Based Vehicle Localization, *Proceedings of the IEEE Intelligent Vehicles Symposium* pp. 436–441.

Toledo-Moreo, R., Zamora-Izquierdo, M. & Gomez-Skarmeta, A. (2006). A Novel Design of a High Integrity Low Cost Navigation Unit for Road Vehicle Applications, pp. 577–582.

Venkatraman, S., Caffery, J., J. & You, H.-R. (2002). Location using LOS Range Estimation in NLOS Environments, *Vehicular Technology Conference, IEEE 55th* 2: 856–860.

Walter, T. & Enge, P. (1995). Weighted RAIM for precision approach, *PROCEEDINGS OF ION GPS*, Vol. 8, Citeseer, pp. 1995–2004.

Wang, X., Wang, Z. & O'Dea, B. (2003). A TOA-Based Location Algorithm Reducing the Errors due to Non-Line-of-Sight (NLOS) Propagation, *IEEE Transactions on Vehicular Technology Conference* 52(1): 112–116.

Weiss, T., Kaempchen, N. & Dietmayer, K. (2005). Precise ego-Localization in Urban Areas using Laserscanner and High Accuracy Feature Maps, *Proceedings of IEEE Intelligent Vehicles Symposium* pp. 284–289.

Wylie, M. & Holtzman, J. (1996). The Non-Line of Sight Problem in Mobile Location Estimation, *5th IEEE International Conference on Universal Personal Communications* 2: 827–831.

Zhang, G., Krishnan, S., Chin, F. & Ko, C. C. (2008). UWB Multicell Indoor Localization Experiment System with Adaptive TDOA Combination, pp. 1–5.

14

Indoor Positioning with GNSS-Like Local Signal Transmitters

Nel Samama
Institut Telecom / Telecom SudParis
France

1. Introduction

After more than ten years of research into indoor positioning and localisation techniques, whose aim has been to provide real continuity of service, as with GNSS outdoors, one has to conclude that no solution has yet been found.

1.1 A very brief history

The real story started a little bit more than ten years ago, in the context of the Galileo project, with the very interesting idea of the so-called *"local elements"*. The question was to do better than the future competitor GPS in designing a real positioning service for the twenty first century: technology transparency to the end user, simple and intuitive operation, performance and of course continuity of the positioning in all possible environments that the modern citizen will face with his/her mobile phone.

One technology followed another: Ultra Wide band (UWB) was the first candidate, at the end of the 20th century. But, facing the problem of considering the proposed approaches as a real *"indoor GPS"*, Assisted-GPS (A-GPS, shortly followed by the Assisted-GNSS) was the next one, typically between 2003 and 2007. It was at that time that the positioning community seemed to realise that the problem was really hard and that a huge research effort would be necessary. For instance, this was the time that ubiquitous positioning was no longer described as imminent and works being carried out in many directions now had a chance to be heard. A few industrial partners, often small organisations, proposed various technical solutions, from the well-known WiFi to the use of TV (television) signals for example.

On the other hand, the market of *"Location Based Services"* developed very slowly, probably due to the complexity and diversity of the environments to be addressed: "one" is still waiting for THE FREE technological solution (as in the case of GPS : this system is one example of the numerous modern "costly free" services) (Kupper 2005). Current techniques proposed in order to provide this continuity of service are mainly oriented, for commercially available solutions, towards WiFi. Some R&D partners also propose inertial sensors or vision based approaches.

1.2 Applications and services

The potential applications and services likely to use such ubiquitous positioning systems are numerous. Of course, the first kind is clearly related to guidance and navigation, as currently for outdoors and GNSS related services, which is the natural extension of the most popular applications. But now that the citizen is considered, through his/her mobile phone, the new services are not only individual (same as the car navigation system, designed for a single user), but also for the community with, for example, the "group" approaches developed by so-called social networks. There is probably a historical parallel that can be drawn between the introduction of the portable clock, about two hundred and fifty years ago, and the development of the navigation capabilities: from individual to collective and from collective to individual. Maybe the advent of these ubiquitous positioning devices will lead to social transformations similar to those induced by the portable clock ... but this is another story. Note also that for these collective approaches, telecommunications systems are required (and in that way, this is now probably the "right time"): this is evidence that the two domains, telecommunication and positioning, are so closely linked. Another very important point to consider, when addressing the mobile phone of a user, is that there are then no constraints on the displacements of the citizen (as was the case for a car for instance) and that current positioning devices, namely mainly GNSS ones, are placed in far more difficult environments and uses (this latter point is the most important for the discussion): thus, new techniques, new devices and new services must be imagined and designed.

It is also possible to cite the classical asset management and various surveillance applications, but which must now work in many different environmental conditions. Once again, the individual and collective approaches are one of the important new features. Multimodal transportation, a desire not yet realised, of a world that would like to be able to reduce its energy consumption, clearly needs the ability to position in real-time all the actors and the various components: pedestrians are indoors more than seventy percent of a typical day and are in constant mobility (and in addition have a potential problem of energy), when vehicles will have to be precisely monitored in order to manage not only their locations, but also their energy, their availability, their reservation, to check the payments, etc. Self-service car locations or co-driving applications fit naturally in this same category.

In a totally different domain, certification and security applications can be envisaged on a geographical basis but ubiquity must be reached (current performance of GNSS are not enough). Following the privacy issues, the conditional liberty of prisoners could be largely extended: currently, due to the limitations of positioning systems (coverage indoors), the prisoners are not allowed to take the underground for example (at least in France). The large scale deployment of ubiquitous systems could allow substantial improvements of the capabilities.

The next generation of applications could be in the domain of social networks. The developments of these networks have been huge and the permanent exchanges between people and connected groups are enhanced when geographical data are associated. Note that our imagination could easily apply this approach to objects, of course.

1.3 The main radio based approaches

In terms of technologies for indoor positioning[1], numerous candidates are almost available, some of them being proposed as commercial products and solutions. A fundamental point to understand is that one is always looking for a positioning system that is globally the continuation of GPS in all environments, i.e. a few meters of accuracy, free for the users and with no specific infrastructure to be deployed by any commercial operator. Hence the various directions of works carried out in recent years: indoor GNSS through Assisted-GNSS, although this is not a solution to the problem (see the first lines of this sections), WiFi because one considers that the required infrastructure will be deployed anyway for telecommunication purposes[2] and inertial approaches that really don't need any specific infrastructure. The accuracy being sought eliminates candidates such as the GSM (Global System for Mobile) or UMTS (Universal Mobile Telecommunications System), whatever the technique envisaged.

Among a few others, it is possible to list the following global categories:

- Wireless Local Area Networks (WLANs, such as WiFi) or Wireless Personal Area Networks (WPANs, such as UWB or Bluetooth) based: the main idea is to use these telecommunication networks for positioning purposes. The main problems for translating the GNSS time of flight measurements lie in the non-synchronised nature of these networks and the complexity of the indoor propagation environment. Thus, the usually implemented technique is based on so-called fingerprinting, described in the next section. An exception to this rule is the Ultra Wide Band that fundamentally works in the time domain, thus could potentially allow us to carry out time measurements. Technological developments are still on-going and initial promises have not yet been met.
- Wireless Mobile Networks (such as GSM or UMTS). The use of mobile networks leads to the same basic difficulty as WLAN or WPAN. Although non-synchronisation is a problem, propagation characteristics are probably the largest difficulty. Performances are not at a sufficient level in order to allow a real continuity with outdoor GNSS. Nevertheless, some services are available which implement the so-called Cell-Id (Identification of the telecom Cell the mobile is associated with). This technique allows a mobile terminal to know the area it is in by analysing the base station it is associated with. The accuracy is rather poor, ranging from a few hundreds of meters in densely populated areas to several kilometres.
- Inertial systems have typically three problems: time related shift of the accuracy, distance related shift of the accuracy and the cost of the terminal. Recent smart phones have embedded inertial sensors but positioning remains a challenge. Nowadays, techniques are mainly oriented in two directions: integration of the measurements provided by the sensors (accelerometers, gyrometers and magnetometers) or modelling

[1] Note that indoor positioning is seen as the ultimate difficulty in order to cope with ubiquity since this seems to include all the most difficult phenomena. This is of course not the only environment where GNSS are not very efficient: so-called urban canyons are also important to be dealt with. Nevertheless, the topic of this chapter is clearly limited to indoor techniques.

[2] This assertion is not 100% right with current proposed solutions since it is almost always necessary to distribute additional access points to existing networks in order to create the required redundancy.

the walking of an individual based on the detection of some very specific instances, such as the precise time the foot touches the ground. Then, the method consists in counting the number of footsteps. These approaches are not yet mature for mass-market applications but research is still being carried out.

- GNSS based systems. In addition to Assisted-GNSS, which is once again not a solution for ubiquitous positioning, the following sections will deal specifically with this problem. Various approaches have been proposed with rather good accuracy results: the remaining problem is clearly the need for an additional infrastructure that needs to be deployed locally. Operators are not ready for this and although very good results are reported, very few systems are really available.

The last category is related to sensor networks. Many systems have been proposed in the last fifteen years, but the lack of standardisation and the high number of sensors that need to be deployed are currently a real drawback.

1.4 The perceived and real needs

If we take a little break to try to analyze the needs (i.e. requirements) for the continuity of service definition, it will quickly become apparent that it greatly depends on the targeted applications and services. But if you ask anybody, the answer will very often be given in terms of positioning accuracy, availability and latency: it should be accurate to better than one meter, available everywhere and instantaneously in real-time. Curiously, the fact that it should be available in three dimensions will almost never be mentioned. Although it really depends on the application (the requirement is not the same for the guidance of a robot in a nuclear reactor and for finding the nearest restaurant), one should be able to distinguish between the positioning "engine" and the resulting services. For instance, GPS does not provide a one meter positioning everywhere, even outdoors, but car navigation systems are very accurate for the delivered service, thanks to map matching and Kalman filtering. The same should apply to ubiquitous positioning. Nevertheless, a good rule of thumb could be to consider that the major difference, in terms of environments, between outdoors and indoors is that indoors is typically a 3D environment, thus requires full 3D positioning capabilities. In that sense, the accuracy should probably be enough to allow the floor level to be determined, i.e. an accuracy of typically half the height of a given floor. In most buildings this means roughly one meter.

Following this general presentation of the indoor field, this chapter is going to focus on radio positioning solutions, and more specifically on GNSS-based radio approaches. The second paragraph is dedicated to an introduction to radio positioning. It is followed by three paragraphs dedicated to GNSS-based architectures: pseudolites, repeaters and repealites. The chapter ends with a synthesis and some hints for the possible future, as seen by the author.

2. The concepts of indoor positioning using radio transmitters

Not all the techniques proposed have, of course, been based on radio techniques, but they are the most important ones for two main reasons: their level of development and maturity on the one hand and their ability to "cross" or to "get around" obstacles such as walls, furniture or people on the other hand. Optical based techniques, like laser based distance measurements or vision based (camera) scene analysis systems present some real advantages in terms of measurement accuracy (a few millimetres for the former) or orientation determination (very useful for any guidance system, available for the latter). Unfortunately, the foreseen use of positioning devices being mainly dedicated to pedestrians in urban environments, optical obstacles are numerous. These latter techniques are then considered as potential hybridisation[3] candidates. Many types of sensors have also been studied for positioning, such as infrared or ultrasound. Once again, although accuracy can reach centimetre values, the environmental constraints are not compatible with the ubiquitous systems being sought. Another category is, of course, inertial systems which could be a valuable alternative to radio systems: time and distance associated position drifts are not yet sufficiently mastered and the given positioning is relative[4], which means the need for "something else" in order to provide the user with an absolute location. The object of this section is to focus on radio based approaches.

2.1 Measurement techniques

There are mainly four techniques that are used for radio positioning. In fact they come from the history of mathematics and have been improved over the centuries, thanks to the development of instrumentation (Samama 2008). In chronological order there are *angle measurements, fingerprinting, time of flight measurements* and *cell-id.*

Angle measurement is the basis of triangulation used by geodesists for measuring the earth. For positioning purposes, the technique is a little bit different and is illustrated in figure 1. The main idea is to measure the absolute direction of a signal received from a transmitter (at the mobile terminal). The reference usually used is the magnetic north which can be obtained from a compass. Thus, with a single measurement, the terminal knows that it is somewhere on the line L1 (see figure 1). Of course, this is not accurate enough, so it is necessary to carry out a second measurement from another transmitter, say T2. This second measurement allows the terminal to know that it is somewhere on line L2. The combination of both measurements gives the location of the terminal, at the intersection of lines L1 and L2. This kind of approach, combining multiple measurements in order to find the location geometrically, is often applied. Two measurements give a location in two dimensions.

This technique can be applied in 3D but requires a 3D angle measurement, hence two angles (azimuth and elevation): this is possible with 2D receiving antennas. Two 3D angle measurements, hence four angles, lead to a location in 3D. Note that when, in 2D, three measurements are available, there is the need for an additional method in order to determine the location considered, as can be seen in figure 1 (right). In the present case, it is often chosen to consider the centre of the inner circle of the triangle that is formed by the intersections of the three lines.

This technique can be quite efficient since angle of arrival measurements are usually based on phase differences which can be measured with rather high precision. Unfortunately, in

[3] Hybridization is the approach that consists in coupling two or more techniques in order to provide the device with improved performance, either in terms of accuracy or in coverage or availability.

[4] Relative positioning refers here to a position that is given with reference to the previous one. Thus, there is the need to know the first position in order to be able to give an absolute positioning (given in a known reference frame).

indoor environments, the difficulty comes from the fact that the propagation is characterised by a large number of reflected path (from walls and all reflective objects), called multipath. Those multipath are even sometimes more powerful than the direct signal, which can in turn also be absent. Thus, even if the angle measurement is accurate, the environmental conditions are likely to mislead the positioning algorithms.

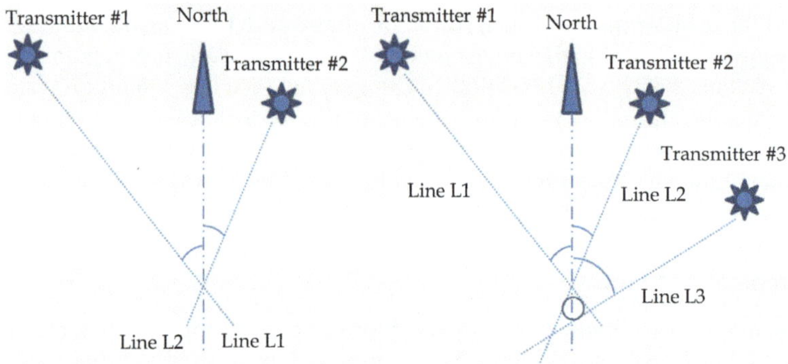

Fig. 1. Angle measurements

The second technique is called *fingerprinting*. The first idea of this method was reported around the sixteenth century when a solution to the longitude problem was being sought. Some scientists had the idea to make a complete geographical cartography of the magnetic field of the earth: if there is a unique link between the location on earth and the value of the magnetic field, then one can consider that the magnetic field value is a perfect indicator for finding a location. Unfortunately, the magnetic field is not a good candidate for such a purpose. This idea came back to engineers with the development of wireless networks: the complexity of the indoor environment for propagation led to the revival of the fingerprinting approach: the received power of the radio signal is now the physical value that is measured. The indoor environment is then cut into squares and the fingerprints (the received power) measured at each intersection of the grid (see figure 2): the "map" associated with transmitter #1 (a data base indeed) is created. The problem is now that many different fingerprints are identical for different locations. The method of multiple measurements is once again implemented: in this case, a second (and more, if required) transmitter is added and a second map is filled in. The location is no longer characterised by a single value but now by a couple of values. In the case of n transmitters, then all calibrated locations are characterised by a vector of length n.

The way in which positioning is then achieved in real-time is quite simple: the mobile terminal carries out received power measurements from all the "radio visible" transmitters in its environment and fills in its own vector. The location is obtained by finding the nearest neighbour in the complete set of maps (data bases) available. The need for this "calibration" phase is clearly a drawback of the method because it is time consuming and, moreover, because it is not a stable operating mode, since the power received is bound to be modified by any movement of any obstacle (including people for instance). Thus, techniques have been proposed in order to manage in real-time (or for longer periods of time) the variation of the maps in comparison with the reference maps. Note also that more measurements should lead to a more accurate positioning.

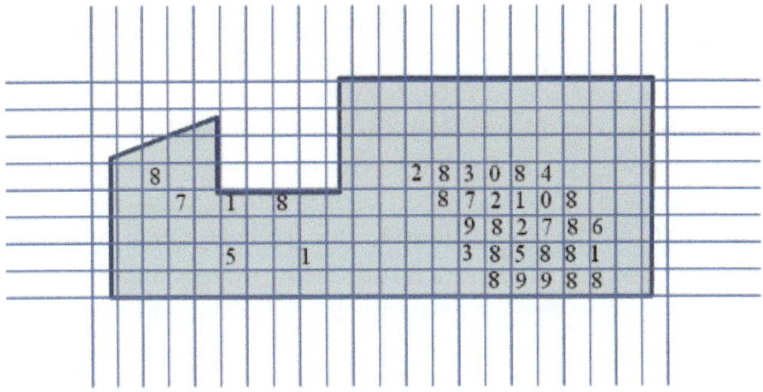

Fig. 2. The fingerprinting approach

Time of flight measurements are quite simple in principle but require acceptable propagation models (Kaplan 2006), (Parkinson 1996). The basic idea, shown in figure 3, is based on the measurement of the time required by a signal to propagate from a transmitter to a receiver. Once obtained, this time is usually converted into a distance. In the case of radio signals, it simply consists in multiplying the time by the speed of light, typically $3x10^8$m/s. Of course, this model is too simple in real cases, so the modelling of the propagation is an essential step. Once one has the distance between the transmitter and the receiver, it means that the receiver is somewhere on the surface of a sphere whose centre is the transmitter and the radius the above mentioned distance. It appears clearly that this is not enough for positioning. Thus, we use additional measurements in order to reduce, geometrically, the uncertainty. A second measurement from a second transmitter (see figure 3) allows us to reduce the set of possible locations to a circle, while a third one reduces the set to two points and finally a fourth measurement leads to a unique location[5]. In case of more than four measurements, techniques such as least square are usually applied in order to find the optimal location in a set of superabundant equations.

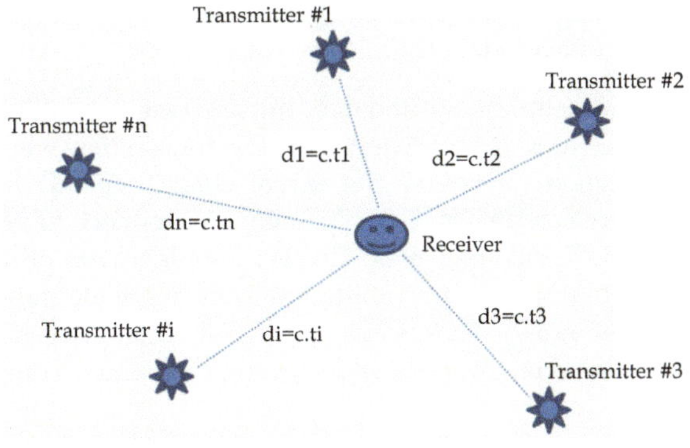

Fig. 3. Time of flight positioning

[5] Note that here we are dealing with the real world and we know that the location exists. Thus, even if four spheres do not have an intersection in mathematics, we are sure that in the present case they do have one, the location of the receiver. The positioning algorithms must implement mechanisms that are able to obtain such a location even when considering unavoidable measurement errors.

There is a really difficult problem in this time of flight measurements: the synchronisation between transmitters and receivers. There are indeed two different synchronisation problems: the first concerns the synchronisation between transmitters (since multiple measurements are carried out from different transmitters) and the second concerns the receiver with the various transmitters. The two problems are not equivalent since if it is possible (not necessarily simple) to imagine "wiring" the various transmitters, it is often not possible to have a link from the transmitters to the receiver, other than the radio link Radio synchronisation is possible but requires a bandwidth proportional to the accuracy needed. In practice, synchronisation to the nanosecond[6] is not achieved through radio links. In the case of GNSS, this synchronisation is achieved by adding an additional measurement, from an additional satellite, in order to solve this new unknown variable. In previous systems, such as Decca[7], the synchronisation between transmitters and the receiver was not carried out: instead, differences of time measurements from two transmitters were carried out. In such a case, the synchronisation unknown disappears (because of the difference) and the positions of the receiver, characterised by a given difference of flight times, are located on a hyperboloid whose foci are the transmitters. Once again, multiple difference measurements are needed for positioning.

Note that the complexity of synchronisation of radio systems comes from the speed of light. Ultrasound based approaches do not have the same problem since the speed of the signal is reduced by a factor of nearly one million. In such a case, synchronisation to the millisecond is comparable to the nanosecond requirement of the radio system.

For the inter-transmitter synchronisation, two generic approaches have been implemented. The first one uses cables in order to create a real physical link between transmitters: then, a simple calibration phase, once only, is carried out in order to know the exact synchronisation. The second one, implemented in GPS for instance, is to use very slow drift clocks[8] and to carry out a multitude of measurements from known locations in order to inverse the positioning problem and to determine the non-synchronisation variables (one for each transmitter). Of course, this approach is expensive and cannot be followed when designing low cost indoor positioning solutions.

The *Cell-id* approach is the simplest one and does not need any modelling (see figure 4). As a matter of fact, a coverage area is associated with the transmitter, whose shape is usually considered to be a hexagon (of course the actual shape depends highly on the radio environment). When the receiver is "simply" able to connect to the transmitter, one considers that it is within the coverage area. This is a simple way to provide a location. This is not very accurate for high power transmitters that have a wide radio range, but can be very good for very low range devices. Of course, in this latter case, the number of transmitters should be high if one wants a wide coverage. As usual, compromises have to be made.

---

[6] One nanosecond at the speed of light is equivalent to 30cm. When a typical positioning accuracy of one meter is wanted, such a synchronization precision is needed.
[7] Decca was a terrestrial positioning system. Propagation models were developed and it appeared that a better performance was obtained over sea rather than over land.
[8] Please note that using atomic clocks is not enough for synchronization purposes. These clocks are used for the low rate of their drift, hence the larger time interval required between synchronization updates.

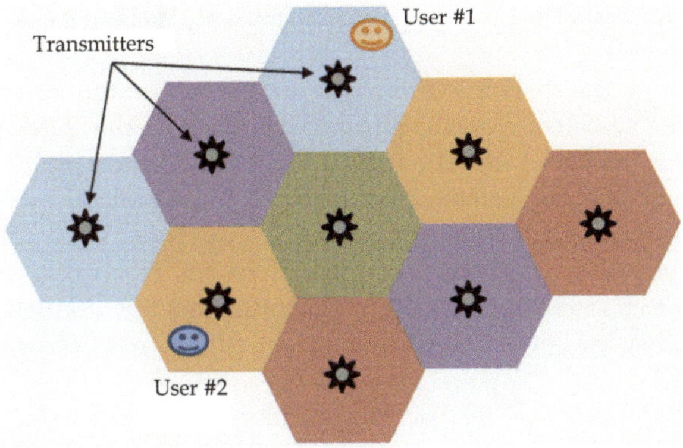

Fig. 4. The cell-id approach

2.2 Main differences with outdoor techniques

Let us come back to the specific case of indoors: some major differences have to be kept in mind in comparison with outdoors. Let us also discuss the case of GNSS since this chapter is dedicated to indoor GNSS-based solutions. First of all, the various techniques are based on time of fight measurements, the same as outdoors, but consider the following parameters for discussion.

- *Propagation environments*: indoors is a very difficult environment and acceptable models are not available. This means that signal processing must solve problems that are either not present, or less difficult to solve, outdoors, the most challenging being multipath. Another problem is related to the possibility of Non Line of Sight (NLOS) path from transmitters to receiver, which happens more often than outdoors. The same kind of techniques could be envisaged but outdoors they are usually based on a certain redundancy of available signals, which is not the usual case indoors.
- *Dilution Of Precision (DOP)*: the geometrical distribution[9] of the transmitters is a very important point to consider when dealing with positioning systems that use distances in order to carry out the calculation. Outdoors, for a location on earth, with GNSS for instance, there is disequilibrium between the horizontal DOP (HDOP), calculated in the horizontal plane, and the vertical DOP (VDOP), calculated in the vertical plane. This discrepancy is due to the fact that when the distribution can be really uniform horizontally (all the satellites being uniformly distributed around the receiver), leading to a good HDOP, this distribution cannot be so good vertically since only satellites above the radio horizon (which is quite similar to the geometrical horizon in the present case) are visible. Thus, the HDOP is usually better than the VDOP. Indoors, things are quite different since one can decide the location of the transmitters: it is very important to locate at least one transmitter below the receiver in order to reduce the VDOP (Vervisch-Picois and Samama 2006). Evaluations have shown a dramatic

[9] This DOP allows the receiver to give a real-time estimation of the accuracy provided to the user (the User Estimated Range Error in GPS for example): it is of uppermost importance for any application or service.

improvement in the VDOP values, leading to a much better estimation of the user location accuracy.

- *Distances*: indoors, the distances are much smaller than outdoors and new problems arise such as the so-called near-far effect. Depending on the codes that are used (case of GPS), there is a limit of detection of two signals with too high a power difference. The lower one will be undetectable because it is impossible to extract from the noise. Once again, this situation is almost impossible outdoors since the transmitters (the satellites in the case of GPS) are very far from the receiver and the difference in distances from two satellites can reach a maximum of a few decibels only. Indoors, this difference can reach a few tens of decibels: specific signal processing techniques are then required.
- *Initial point in the calculations*. Classical algorithms of calculation of location are based on iterative techniques that require an initial estimation of the user position. In the case of GNSS, only three measurements are necessary for geometrical purposes[10] since the intersection of the surfaces of three spheres gives two points, one of which is above the plane that includes the three centres (the satellites indeed) of the three spheres, the other one being below. When the receiver is on the surface of the earth, only the location that is below the plane is possible: thus, only three satellites are required from a geometrical point of view. Consequently, the initial location estimation is usually taken somewhere on the earth's surface, and this is sufficient. Indoors, the situation is a little bit different since the two resulting locations (above and below the plane) are rather close to each other and the choice of the initial estimation is fundamental in the convergence of the algorithms. Thus, either one chooses to use five transmitters (instead of four satellites) or to keep four transmitters and choose an initial location of the user that is inside the building (which is not such an easy task).
- *Immobility of the transmitters*: in the sky, GNSS satellites are non-stationary. This feature causes some troubles in the way one needs to calculate their locations each time one wants to carry out positioning, but offers some interesting features that are no longer available indoors where transmitters are stationary. The Doppler shifts are only due to the displacements of the mobile terminal, but in case of multipath it is not possible to wait in the same place for a while in order to average the results considering that only multipath will be varying, since if nothing moves around the propagation conditions have no reasons to change. Thus, static positioning is much more difficult indoors.

2.3 Main existing approaches

Many positioning systems have been proposed with radio transmitters. All the above mentioned techniques have been implemented and this paragraph proposes a sort of classification depending on the technique. Table 1 gives provides a non-exhaustive summary of them. A few references are provided concerning UWB (Fontana 2004), Bluetooth (Takada et al. 2003), WiFi (Wang et al. 2004) or TV (Martone and Metzler 2005) signals.

---

[10] A fourth measurement is required for "synchronization" purposes as long as the receiver is on the earth's surface.

	Technique			
System	Angle measurements	Fingerprinting	Time of flight measurements	Cell-Id
GNSS			✓	
WiFi		✓	✓i	
Bluetooth		✓		
UWB			✓	
GSM/UMTS	✓ii		✓iii	✓iv
RFID				✓v
TV			✓vi	

Table 1. Summary of a few radio based positioning systems

i. Since wireless local area networks are not synchronised, distance (and not time) measurements are used. The distance is estimated through a power level measurement and a model of propagation (typically modified Friis formulae where the power of the distance is between 2.5 and 4 depending on the environment). This is not really accurate and too dependent on the fluctuations of the environment.
ii. Angle measurements are already carried out at base stations in order to allow the use of the same frequency channel for transmissions in different directions. Thus, those measurements are available, but the limitations discussed in previous sections still apply.
iii. These networks are not synchronised and mainly differences of time of flight have been proposed (but direct times of flight have also been proposed). Unfortunately, the propagation models are not well suitable and the best reported performance is around one hundred metres outdoors, and can rise to a few hundreds of metres indoors.
iv. Cell-Id is used by networks in order to route communications: once again, this is already implemented in mobile networks because it is needed. The accuracy is typically a few hundreds of metres but it is completely free and available. Many telecom operators propose services based on GSM/UMTS cell-id positioning.
v. Many definitions of RFID (Radio Frequency IDentification) are proposed: let us consider this is a short range technology that allows two radio transmitters to exchange data, and identification, for instance. A simple way to carry out positioning (but not the only one) is to consider the cell-id model. The coverage area (or range) of a given transmitter is approximately known: when a second transmitter can connect to it, then it is located in the coverage area. In case of a very short range (say one metre or less), the accuracy of the positioning is thus better than one metre. The consequence is that the positioning is no longer a continuous process in space and time (as for GNSS for example), but becomes typically discrete.
vi. Television signal are available almost everywhere in modern countries: why not use them in order to position a receiver? This idea was developed a few years ago and an accuracy of around ten metres has been reported through time of flight measurement, even indoors.

3. The first GNSS signal approach using pseudolites

Although this chapter is dedicated to infrastructure based GNSS systems, other solutions have been investigated by the GNSS community. For instance, High Sensitivity GNSS, HS-

GNSS, had the objective to provide continuity of service with no additional infrastructure. The simple underlying idea is that the signals are still present indoors, but even lower in the noise than outdoors. Thus, if one is able to design a very highly sensitive receiver, it should be possible to locate indoors. A similar, but not identical, idea led to the design of the so-called Assisted-GNSS (Duffett-Smith and Rowe 2006). The initial goal was also to provide indoor positioning by "aiding" the receiver to find the signals in difficult environments. In such situations, one major problem with stand alone receivers is the impossibility of decoding the navigation message (too long to envisage having good radio conditions for such a long duration). Thus, a solution could be to send the navigation message through telecommunication networks that are widely available indoors. Thus, knowing the message, the receiver is able to use the high-sensitivity in order to acquire the GNSS signals and then is able to calculate a position since all the parameters needed (from the navigation message) are available. High sensitivity and assisted approaches are thus quite complementary.

Unfortunately, with a higher sensitivity, the receiver is now jammed with reflected signals in such a large amount that positioning, although possible, is really bad because there is too much interference. Thus, even if real improvements have been proposed in environments where the signals were just at the detection limit, these approaches are clearly not the ultimate solutions for indoor positioning and continuity of service. One has to move to infrastructure-based techniques.

3.1 Technical historical introduction

In the early 1980s the first ideas of GPS-like signals transmitters arose from the considerations of the obvious limitations of the original system. How to use a GPS receiver when fewer than three or four satellites are available? What kind of approaches could be imagined to position the Mars rover? How to improve the VDOP of the constellation in case a good vertical accuracy is needed? Etc.

One answer could be to increase the number of satellites by a factor of two or three but the associated cost for the relatively reduced increment in performance was judged to be non-viable. One has to find another way. The idea of implementing GPS-like signal generators that could be locally deployed came out: the pseudolites were born.

3.2 The concept of pseudo-satellites

A pseudolite (which comes from the contraction of pseudo and satellites) is a generator that transmits GPS signals but which is not a satellite. Such a generator can easily be deployed on earth in places where the number of visible satellites is too low to allow standard positioning (Klein and Parkinson 1986). The first applications were thus naturally oriented towards open cast mines for optimisation purposes. Indeed, as the mine is dug, the view of the sky is reduced and the optimal number of satellites reduces. Adding a pseudolite allows a continuity of the positioning service to the mine to be provided.

A similar idea was developed in the context of so-called Local Area Augmentation Systems (LAAS) where the problem was to provide a good vertical accuracy to landing planes, for example. We know that this vertical accuracy is linked to the VDOP and that locating a satellite below the plane would greatly improve the VDOP. Since it is not possible, the use of a pseudolite seems once again a good idea (Bartone and Van Graas 2000).

Similar to the open cast mine, the case of modern so-called "urban canyons" are complex environments for GNSS signals (see figure 5). A receiver located between large buildings has some difficulties acquiring a sufficient number of satellites. When having additional signals from judiciously located pseudolites, a normal situation can be obtained, leading to the positioning of the receiver in these kinds of environments.

In the previous three examples, the pseudolite is used in order to "augment" the GPS system, its coverage or its accuracy. But one can push forward the concept towards a completely new system: this was imagined for positioning the Mars rover. A complete set of several pseudolites was deployed on the surface of the planet and the signals used for positioning, the same way it is achieved with GPS signals from space. Based on this idea, it was thought that an indoor positioning system could be designed.

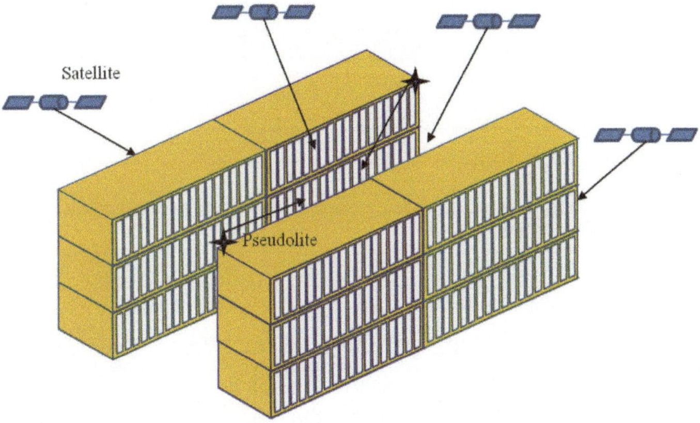

Fig. 5. The urban canyon configuration

3.3 The system for indoor positioning

The basic idea is indeed very simple and is based on the construction of a local terrestrial constellation of GNSS-like signal generators (Kee et al. 2003). They are located at the corners of the building in order to simulate satellites. Figure 6 is a typical distribution although not optimal since the DOP is not very good (please refer to the discussion in previous sections). This is nevertheless a good basis for understanding the concept.

Some major differences apply with comparison to satellites, the most important ones being the immobility of the pseudolites and the shorter distances between the pseudolites and the receiver (leading to unambiguous code for instance, as will be discussed in the next section).

As discussed in previous sections, one has to take care of the initial location considered in the computations of the receiver location since the two possible solutions[11] are not so far

[11] Remember that four transmitters are used for geometrical (three) and synchronization (one) purposes. The intersection of the surfaces of three spheres gives two points located symmetrically apart from a plane that includes the three transmitters (this comes from the form of the equations that are non-linear). Thus, in case of local transmitters, the final location obtained depends on the initial guess: if it is above the final location it will be the point above the plane, if it is below, the final location will be the point below the plane.

away from each other if one uses the optimal number of transmitters (i.e. four in a 3D positioning system).

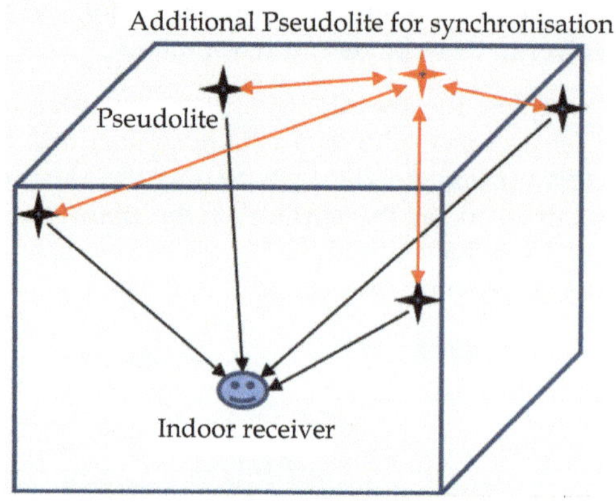

Fig. 6. Pseudolite indoor positioning system

3.4 Advantages and main drawbacks

Such an indoor positioning system is not widely deployed because of numerous major drawbacks, despite some fundamental advantages. Let us list the most important features and comment on whether they are an advantage or a drawback (Kanli 2004).

Continuity with outdoor GNSS: this is obviously a major advantage of the proposed system. Moreover, the continuity is obtained by using the same hardware as for outdoors (since GNSS are clearly a very good candidate when the satellites are visible and is almost free[12]). Note that using GNSS-like signals means that current receivers are already capable, with a software update, of processing them. This fact constitutes a second major advantage. The first drawback is the need for a local infrastructure.

Synchronisation between pseudolites is required. Satellites include atomic clocks or masers in order to reduce significantly the time drift but require a terrestrial infrastructure for synchronisation purposes. In the case of pseudolites, two approaches have been proposed: synchronous and asynchronous systems. In the latter case, the pseudolites are not synchronised and the measurement technique must carry out a sort of synchronisation: the method used is the double differencing that allows us to get rid of the synchronisation of the transmitters. The major drawback is then the need for a reference receiver that should be in radio visibility of the transmitters. Apart from the deployment complexity that this adds, a data link has to exist between the two receivers. This first approach is not intended to be selected for indoor positioning purposes. The other approach uses synchronous pseudolites. Several methods have been proposed: the simplest one in theory, but not in practice, is to link the various transmitters by wire. In such a case a sort of calibration phase is required in order to know precisely the delay between pseudolites. An implementation of this approach used a master receiver located in a known location with respect to all the pseudolites.

[12] A GNSS receiver integrated into a modern device is estimated to cost a few dollars.

Calculations are then carried out at the master receiver and synchronisation values are sent back, through wires, to the pseudolites. Another way consists in transmitting these synchronisation data through a wireless link, leading this time to latency problems and potential interference (but this is an interesting approach). In addition to this concept, one imagined working the other way round by placing the receiver (which listens to the signals) in the same place as the pseudolites. By considering that one (or several) pseudolites are "pilot(s)", the receiver can synchronise its own pseudolite if it knows the distance(s) that separate(s) it from the pilot(s) pseudolite(s). The difference between received times for two different pseudolites (indeed the associated receivers) allows the synchronisation of the pseudolites. This once again requires data links. Of course, these solutions are clearly adding cost and complexity to the system.

Another simple approach consisted in locating pseudolites in places where the GNSS signals are available, namely outdoors, and to use the constellation time to synchronise the transmitters.

Code and carrier phase measurements are possible. In the first case, code phase measurements are carried out: the positioning accuracy of the pseudolites needs to be in the range of a few decimetres. The resulting positioning is intended to reach a few meters, as outdoors. Note that multipath are bound to largely degrade this very optimistic goal (discussion follows). The other approach described is based on carrier phase measurements (Kee et al. 2001, Rizos et al. 2003). We know that this kind of measurement is much more accurate but suffers from the ambiguity resolution problem. Nevertheless performances reported are in the range of a few centimetres[13]: the requirement in terms of pseudolite location accuracy is also increased to typically one centimetre (this task is not so easy to carry out).

Ambiguity is no longer such a difficult problem. In the case of code phase measurements, ambiguity is totally suppressed since indoor distances are much smaller than three hundred kilometres. In the case of carrier phase, ambiguity is still present but is not so high: typically fifty metres for indoor distances, the carrier phase ambiguity for frequency L1 is around 260. Current works are evaluating the possibility to use classical code phase ambiguity resolution methods for the carrier phase resolution indoors.

A *potential accuracy of a few centimetres* is achievable with the carrier phase approach, even if these measurements are probably not the most important ones for the foreseen applications looking forward to the continuity of the positioning for mobile phones. Nevertheless this is significant of the capabilities of the principles.

Near-Far effect is a new propagation concern (Madhani et al. 2003). Since the deployment complexity of the pseudolites must be reduced, their number should be reduced to a minimum. As a corollary, the distance between pseudolites should be increased to a maximum. Unfortunately, the Pseudo Random Noise (PRN) codes used in the case of GPS, for instance, have auto correlation functions that present some secondary peaks. These secondary peaks can have amplitudes of about -24 decibels (dB) in comparison to the main peak. This is very good for outdoors where the difference of distances from various satellites

---

[13] Techniques similar to high accuracy methods for outdoors are used together with the associated problems such as the determination of the initial location.

can reach a maximum of less than 2dB, but is a real problem indoors. In terms of phenomenon, the problem is related to the fact that if secondary peaks of a transmitter are greater than the main peak of another, then this second one will appear as noise and will not be detectable. This 24dB margin in power is reached as soon as the ratio between the closer and the farther transmitters reaches four: this is not an unusual situation indoors. Thus, a few solutions have been proposed, among which: 1/ pulsed transmissions consisting in allocating between 10 and 20 percent of the time to a particular pseudolite (this has shown to provide an additional margin of about 10dB corresponding to nearly an additional factor of two in distance), 2/ frequency shifts in order to almost eliminate the near-far effect, but at the cost of a substantial increase in the terminal complexity or 3/ in sophisticated mitigation algorithms that successively suppress the more powerful signals to finally extract the lowest one.

Interferences with outdoor signals. Another advantage of pseudolites is the ability to decide the power level to be transmitted, depending on the required coverage and performance, and of course on the environments. This advantage becomes a major drawback when thinking in terms of cohabitation with the outdoor world (Glennon et al. 2007, Yang and Morton 2009). If one takes the case of GPS (but this is true whatever the system considered), using GPS-like signals for indoor transmission is susceptible to create interference with the signals that could be received by an outdoor receiver receiving signals from the satellites. As a matter of fact, the same phenomenon as described indoors for the near-far may occur. Thanks to GPS project management, some specific PRN codes have been reserved for pseudolite operation[14] at the early stages and this interference problem is slightly relaxed, but is still a real concern for GPS authorities. A specific section, at the end of this chapter, is dedicated to the regulations restricting the power levels allowed to be transmitted for indoor operations.

Finally, *multipath* are a major issue. Mitigation techniques must be found in order to imagine a proper operation of the code phase pseudolite system. This topic is such a challenge that the next section is dedicated to it.

3.5 The specific problem of multipath in indoor environments

As already discussed in previous sections, indoor environments are characterized by the presence of many reflectors in the path from the transmitter and the receiver. All these reflected signals are going to combine at the receiver end and produce the really received signal on the receiver antenna. This signal is the one that the receiver is going to deal with since this is the real physical received signal. As this is not only the direct signal from the transmitter, and depending on the signal processing techniques used, the distance finally measured can be erroneous (remember that as a matter of fact, this is a time that is measured and not a distance).

From a physical point of view, the situation can be seen as follows: the physical quantity that is transmitted is indeed an electric field, given in V/m. It is furthermore characterized by a frequency, an amplitude, a phase and a delay, in comparison, say, with the first

[14] PRN 1 to 32 are reserved for so-called space vehicles, the satellites, and PRN 33, 34, 35, 36 and 37 are reserved for terrestrial transmitters.

arriving signal. Let us consider that the frequencies of all the reflected signals are identical[15]. Then, the physical phenomenon that occurs is simply an addition, in amplitude, phase and delay, between all the reflected signals (and the first one, which should be the direct path[16]). The problem is now to be able to get rid of all contributions except the first one (which should be the direct path under our assumption). Such a time discrimination is somehow equivalent to the synchronisation problem and requires theoretically a radio bandwidth proportional to the time discrimination interval wanted: in our case, where nanoseconds are sought, this bandwidth is too large and other approaches must be found.

Let us now come back to the specific problem of multipath in GNSS, and to GPS for illustration. The way time separation is obtained, from the transmitter to the receiver, is based on the famous auto correlation function (ACF) of the codes. A typical such function is given in figure 7.

In case of multipath, we are interested by the main lobe of the ACF. Let us consider only one reflected path (in addition to the direct path) for simplicity of explanations, knowing that this is clearly not a real situation. If the reflected path is delayed by more than one and an half chip, the ACF of the incident signal (which is composed of the direct and reflected paths) with the receiver generated replica has the shape given in figure 8. Remember that a GPS chip length is given by 1/1023 milliseconds, hence 977.5 nanoseconds, which in turn corresponds to 293 meters. Thus, figure 8 is characteristic of a reflected path delayed by more than 440 meters. The receiver will be able without any problem to find the direct path considering (this is the assumption that is classically made) that the first peak of the ACF is the value being sought.

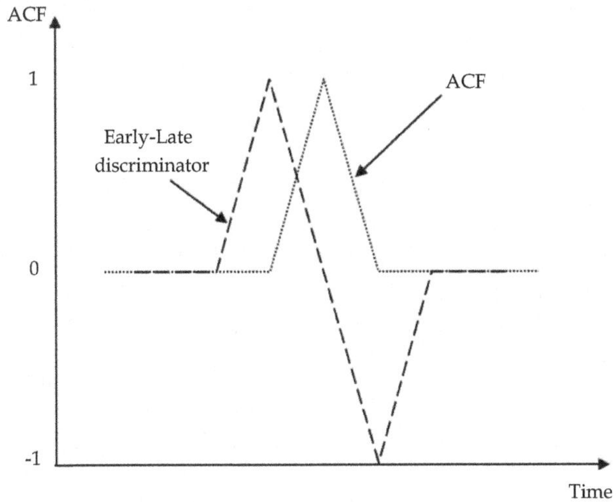

Fig. 7. Typical autocorrelation function of a GPS code

[15] This could not be true in case of reflection on moving objects, such as cars for example. But in our indoor case, we are going to consider this hypothesis as correct.
[16] The direct path could not be present and then the first received signal would also be a reflected path. This situation is one that we are not going to deal with in this chapter.

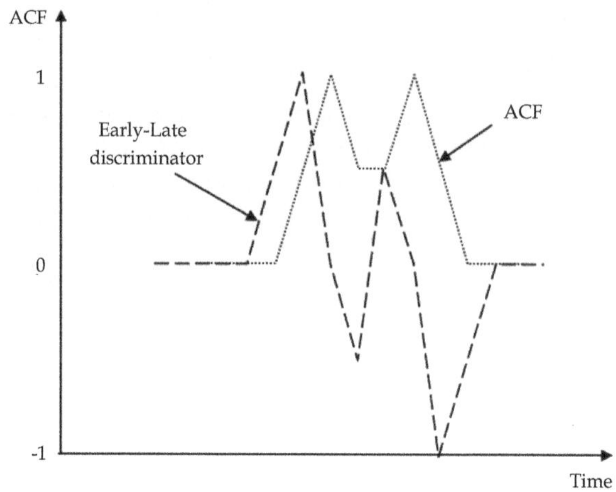

Fig. 8. Typical autocorrelation function for a large delayed multipath

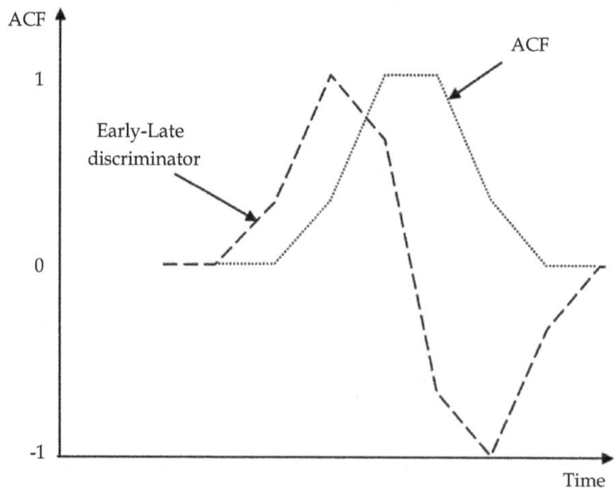

Fig. 9. Typical autocorrelation function for a small delayed multipath

Indoors, reflected path have delays that are indeed much smaller. In such a case, the ACF is completely disturbed (see figure 9) and can take many different shapes. The problem is now that the receiver will be fooled when detecting the maximum of the ACF which is no longer at the time of arrival of the direct path. Note also that this maximum now depends on the relative phases, delays and amplitudes of the direct and reflected paths.

The classical way this multipath effect is characterised is given in figure 10. This curve allows the comparison of various multipath mitigation techniques, as illustrated in figure 10 for the Standard Digital Locked Loop (SDLL) and the so-called Narrow Correlator (NC). Note the reading of the figure: considering a direct path and a single reflected path of amplitude half that of the direct path (only suitable for comparison purposes and certainly not for evaluation purposes, since this situation is clearly not representative of reality), the envelope of the resulting error in pseudo-range measurement is drawn. The upper curve corresponds to reflected path in phase with the direct signal, when the lower curve is related to the case where the reflected path is out of phase with the direct path. Note that SDLL is absolutely not suitable for indoor environments since errors as high as 60 meters are possible[17]. The same almost applies to the Narrow Correlator[18] since errors of 10 metres are still possible: if the goal of accuracy is in the range of a few metres, this approach is also not viable.

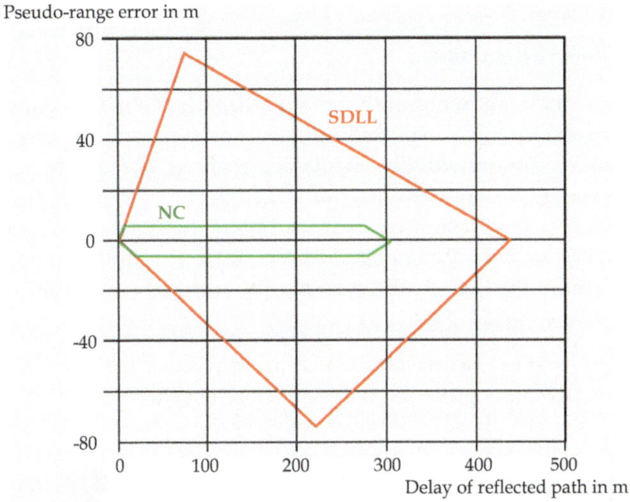

Fig. 10. Multipath effect of the pseudo-range measurement

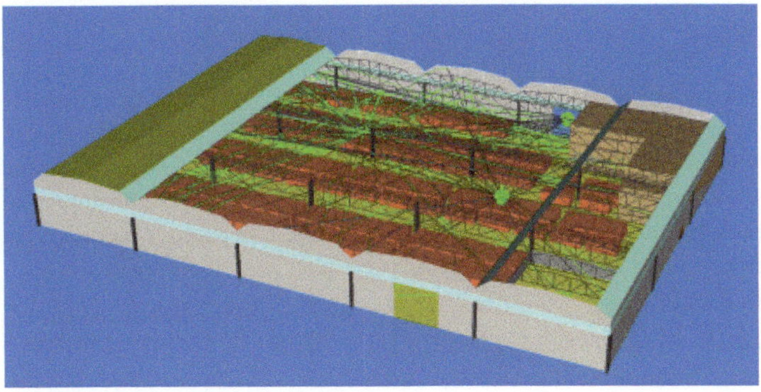

Fig. 11. Typical multipath environment indoors

[17] Here one has the explanation why no commercial solutions are available in the field of code phase pseudolites!

[18] Another constraint of the Narrow Correlator is the need for a receiving bandwidth of at least 8 MHz, which is not the current standard. Nevertheless, the standard is bound to evolve with the advent of Galileo in the frequency band L1/E1 since the current 2 MHz are too narrow for an acceptable detection of their signals.

Let us now come back to real situations where several (many indeed) multipath are present. In order to give an idea of such configurations, we consider the environment described in figure 11 which is a large car park. The structure is made of metallic beams and concrete walls. Cars are also modelled as red parallelepipeds. In figure 11 are also shown the path from transmitters to a receiver that is located in the centre of the building. The black paths are direct ones, while the green ones are reflected path: the conclusion is quite clear! In such cases, one can easily imagine that the ACF is even more disturbed than the ones proposed in figures 8 and 9.

3.6 The performances attainable

The preceding pages have shown that the only multipath problem is enough to disqualify the pseudolite approach which uses code phase measurements. A few other multipath mitigation techniques are potentially available, such as the Strobe Correlator or the Double Delta correlator, but both require rather a high signal to noise ratio (SNR) in order to properly function[19]. This is not so easy to obtain indoors since reflected paths are bound to reduce significantly the signal to noise ratio (by destructively combining the electric fields). Thus, this kind of systems is not yet available with acceptable performance.

The other possibility is to use carrier phase measurements that we know are less sensitive to multipath (because the ambiguity is reduced to nineteen centimetres instead of three hundred metres for code phase). Unfortunately, carrier phase based systems are more complex to use in practice because they require both an initial location which is accurate to a few decimetres and the carrier phase to be followed continuously, which is much more difficult than to follow code phase. Such systems exist but are not widely deployed for these additional reasons (in conjunction with the need for infrastructure).

3.7 Short synthesis

Pseudolite systems require an infrastructure deployment and synchronisation, and have to cope with near-far and multipath but provide full continuity with the technical approach for outdoors, GNSS, with only minor modifications to the receiver. This is a good candidate if no solution without infrastructure can be found but the community of service and application providers is not yet ready to accept such a solution, except in situations where an installation cost is counterbalanced by already well identified revenues.

4. The first step in overcoming some pseudolite linked problems: The repeaters

Following pseudolites, one tries to propose ameliorations to the main drawbacks (Im et al. 2006, Jee et al. 2004). Since it is based on transmitters, the infrastructure is still present, but some approaches reduce its complexity by the introduction of the concept of a "common signal" to all the transmitters (Caratori et al. 2002).

[19] The Narrow Correlator is the only one that does not degrade the SNR while improving the multipath behaviour.

4.1 Introduction to the basic idea

The first simplification concerns the synchronisation. In a similar way that outdoor pseudolites can synchronise themselves using GNSS signals, the idea here is to put an outdoor antenna on the roof of the building in order to obtain the constellation signals. Note, that in this case the antenna is probably (certainly indeed) receiving several satellite signals. Here is taken into account the second new idea that consists in forwarding this signal to the transmitters: the innovation lies in the fact that the same signal (which is probably made up of many satellite signals, as mentioned) will then be transmitted from the various transmitters, now called repeaters[20]. In this way, an obvious problem appears: if the repeaters are transmitting simultaneously, the same signals will be transmitted from different locations and, once received by the terminal, will certainly be considered as reflected paths. Since the principle is to carry out time measurements, and thus distance measurements, it is clearly not acceptable. Thus, the transmission is now achieved in a sequential manner with always only one repeater transmitting at a given time. This presents another interesting advantage: the near-far effect is now removed[21].

4.2 The systems proposed

Two measurement systems are then possible.

- The first one carries out the computation of the location, at the receiver's end, for each transmitter successively. At each corresponding time the fourth coordinate (the so-called clock bias[22]) of the navigation solution vector is recorded. As soon as four successive computations have been obtained, it is possible to compute the indoor distances through the calculations of the differences between the fourth coordinates considered at different time. These differences give a new system of three independent equations that can be solved classically. The resolution gives the indoor location of the receiver. A short demonstration of this principle is given below.
- The second one carries out some differences of pseudo-range measurements at the precise instant of the transitions from one repeater to the next (Fluerasu et al. 2009, Fluearsu and Samama 2009). At these instants, the difference of the pseudo-ranges that are measured just before and just after the transition shows the value of the difference of distances between the two repeaters and the receiver. In order to obtain the indoor distances, a second difference is needed, as briefly explained below. Note that this approach also removes all the effects whose second derivative is zero, including

[20] Please note that if the term is appropriate since transmitters are just "repeating" the outdoor received signal, it should not be confused with the classical repeater that is used for demonstration purposes or just for having outdoor signals available indoors. Here repeaters represent a new approach for indoor positioning and should be seen more as a means of improving some aspects of pseudolite rather than just forwarding signals. It is so true that all the sections could have been written with a signal generator instead of the outdoor antenna.

[21] Only the dynamic range is now a limitation when the receiver is processing signals from two successive repeaters.

[22] This is clearly not the clock bias, but indeed the sum of all contributions that are common to all the satellites that are considered for the resolution: thus, this included the free space indoor distance that we want to obtain.

atmosphere propagation or the major part of clock drifts. This new differential mode is also susceptible to increasing the positioning accuracy.

Let us deal briefly with the mathematics of the first approach based on clock bias analysis. The method is based on the use of the clock bias coordinates. As described above, once one has carried out four receiver location computation (one for each repeater), a new vector is available, where the ct_i are the calculated fourth coordinates, the $ct_r(i)$ are the real clock bias of the receiver at each transmission times and the d_i the distances separating the repeaters from the receiver.

$$\begin{bmatrix} ct_1 \\ ct_2 \\ ct_3 \\ ct_4 \end{bmatrix} = \begin{bmatrix} ct_r(t_1) + d_1 \\ ct_r(t_2) + d_2 \\ ct_r(t_3) + d_3 \\ ct_r(t_4) + d_4 \end{bmatrix} \quad (1)$$

The unknown variables are now the d_i, but the problem appears to be the real clock bias of the receiver which is naturally not a constant. Thus, in (1), one has not only the four d_i unknowns, but also the four clock biases. The technique consists indeed in estimating the clock bias difference between instant t_2 and instant t_1 by the way of the clock drift computation carried out through Doppler measurements by the receiver. Thus, the idea is to consider that:

$$ct_r(t_j) = ct_r(t_i) + \sum_{k=i+1}^{j} cdt_k \quad (2)$$

Where cdt_k is the clock bias rate (called the clock drift) at time t_k. The various $ct_r(t_i)$ of (1) are now reduced to a single unknown, $ct_r(t_1)$. In addition, one knows that the four distances d_i are characterised by only three spatial coordinates, x, y and z of the receiver once the coordinates of the repeaters are known: this is a system requirement to provide the receiver with these coordinates. The indoor location computation is then carried out typically through hyperboloid intersection, as soon as the receiver is able to determine which repeater is transmitting at any given time. This is achieved through synchronisation which is made possible since the signals transmitted by all the repeaters are identical (thus, there is just the need for an initial calibration of the wire delays between the signal generator, or the outdoor antenna, and the repeaters).

On the other hand, the need to estimate the clock drift is somehow a constraint since the final performances will greatly depend on the quality of the receiver clock. Thus, another approach was proposed, based simply on classical measurements carried out by all current receivers: the raw pseudo-ranges. When one draws the difference of pseudo-ranges from one instant to the next in a repeater like system, the curve of figure 12 is obtained (note that in this example only three repeaters are deployed, leading to a 2D positioning).

Clear skips, called "transitions" in the figure, can be seen: they correspond to the difference of distances, d_j-d_i, that characterises the increase or decrease of the distance from repeaters to the receiver when the transmitted signal switches from repeater i to repeater j. It is positive when the distance increases, and negative otherwise. Note also that two additional phenomena are present: 1/ a slow constant increase in the equilibrium value (which represents the remaining contributions whose first derivatives are not zero, for example the clock acceleration) and 2/ a characteristic shape of the curve just after the skips, which is due to the receiver's loop that tends to come back to the equilibrium after this destabilisation. Note that the sum of the transitions, for a complete cycle should be zero. Of course, due to measurement errors, this is usually not the case, and the choice of the best transitions to be considered for positioning has to be carried out.

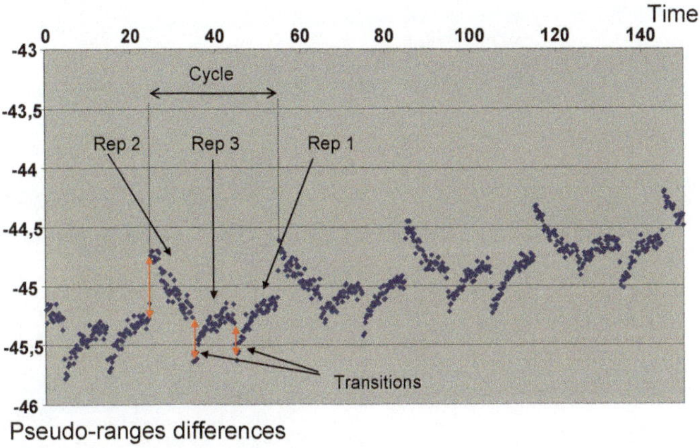

Fig. 12. Typical response of a receiver

The curve of figure 12 is a single difference of raw measurements. In order to extract the differences of distances mentioned above, there is the need to carry out, at the precise instant of transition, a second difference between two successive single differences. Thus, a process of double differencing is the basis of this proposed approach to repeater positioning. Following these measurement steps, the computations are similar to those described for the clock bias based approach.

4.3 The performance achieved

The most often implemented approach is the second one because it is simply based on classical measurements of GNSS receivers and that no additional computation errors affect the positioning. Tests have been carried out in various environments: each time, the system was deployed and positioning carried out with different receivers. Note that the receivers used are so-called software defined radio (SDR) receivers since the method is affected by multipath, in a similar way that pseudolite based systems are. Thus, a specific mitigation technique was implemented (described in a following section) which required the tracking loops to be slightly modified. Since proprietary receivers do not allow such modifications, an SDR receiver was required. It should be pointed out that transmitters are located in such a way that walls are included in the propagation path from the transmitters to the receiver. These environments, together with their "Ergospace[23]" representations, are as given in figures 13 to 16 below.

[23] Ergospace is the electromagnetic propagation software used for the deployment phase. The main goal is to evaluate the multipath related effects.

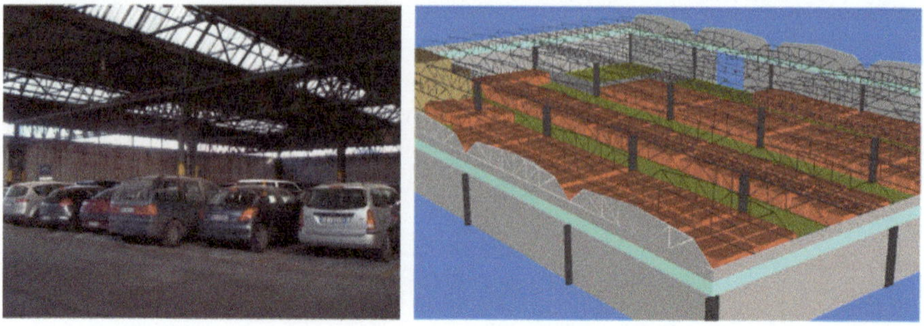

Fig. 13. A car park

Fig. 14. An entrance hall

Fig. 15. Classrooms

Fig. 16. An amphitheatre

The system used for these experiments consists of a few (typically four) transmitters which are located indoors and which transmit a signal provided by a GNSS-like signal generator (we used both an AeroFlex GPS-101 and a Spirent GSS6560). Note that only one such signal is required since the approach proposed is based on the transmission of the same signal through the various transmitters deployed. Note also that in order to satisfy the ongoing various regulations (both in the US and in Europe, briefly described in a following section) the power transmitted is limited (from -80dBm to -65dBm). The principle of the approach is given in the figure 17. The transmitting antennas had a radiating pattern with a maximal gain of around 3dBi.

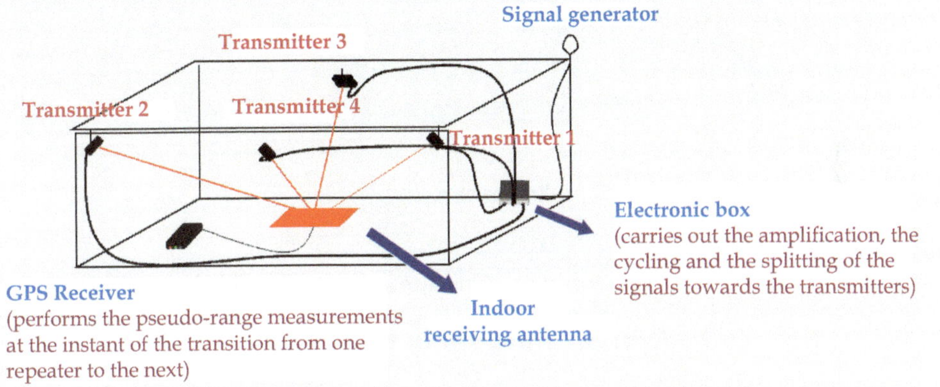

Fig. 17. The system as it was deployed

A summary of the results obtained, all environments included, is given in figures 18 and 19. The first figure shows the results obtained in classrooms, an amphitheatre and an entrance hall. About 20 different locations have been tested in these environments. The various curves represent different ways to filter the resulting fixes obtained. The "unfiltered" curve takes into account all the fixes, with no filtering at all. The other three curves, named "-xm", give the resulting fixes obtained once we remove the ones that are outside the largest rectangle defined by the locations of the transmitters by more than x metres. Note that this is achieved for two main reasons: outside this rectangle, the DOP values increase very rapidly and the positioning algorithms sometimes do not converge.

Figure 19 is a summary of the results obtained in all experiments and with various receivers. In red in the figure are the results obtained in the car park, and the two blue curves are the results obtained in the other environments described. The two curves have been obtained with -80dBm and -65dBm respectively.

The main conclusion is that the current performances are roughly in the range of 3 to 4 metres for 80% of the fixes. It is of uppermost importance to understand that this can be considered as really raw fixes since calculations are carried out totally independently from one fix to the next. It is highly probable that basic smoothing or filtering (applied on pseudoranges or locations) would lead to a significant improvement. In addition, a complete continuity with outdoor GNSS is achieved since velocity computations are also possible (Samama and Vervisch-Picois 2005).

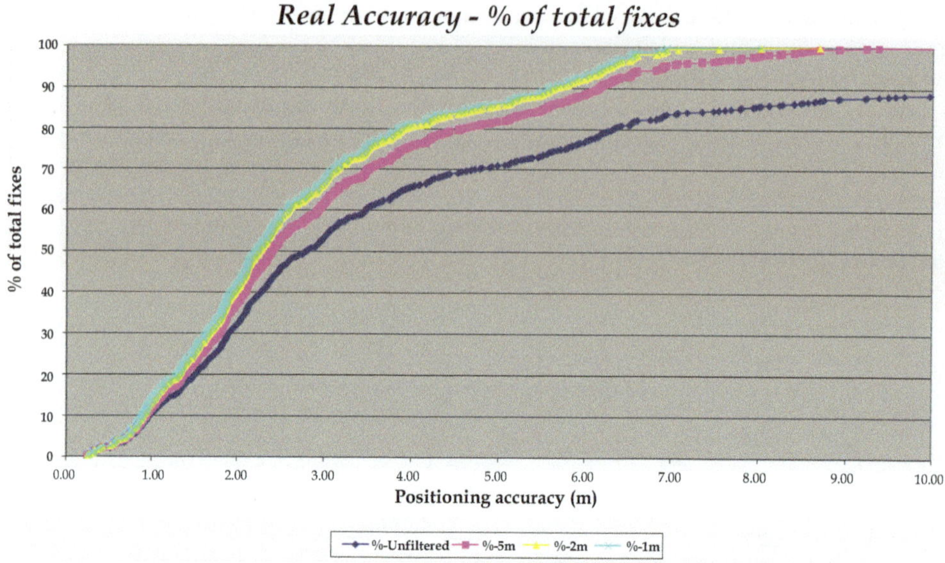

Fig. 18. Results obtained in classrooms, amphitheatre and hall

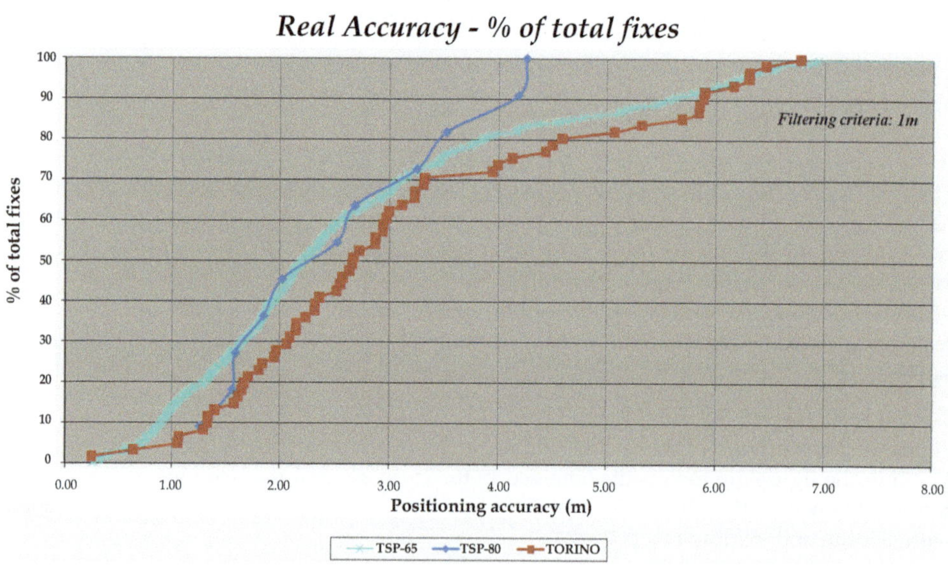

Fig. 19. Summary of all the results

4.4 The main limitations

Synchronisation, absence of near-far effect and implementation of a differential approach are the main competitive advantages of repeaters over pseudolites. Unfortunately, they go with a few disadvantages, described below.

- Carrier phase measurements are no longer possible (or in reality certainly very complex to carry out) since the skips that are the basis of the method, mean that the phases are lost at each transition, leading to the need for a new search for the integer ambiguity number at each transition. Thus a few meters of accuracy is the goal of this system: enough for the continuity of service, but improvement directions will not be easy to find.
- The sequential scheme is a problem when one wants to address dynamic positioning since the time the cycle takes should be taken into account in the displacement. This is quite complex to implement and only slow movements can be dealt with. This is acceptable for pedestrians in a commercial mall, but not for a car in a tunnel. This sequential technique is very interesting for time based double differencing, but not for dynamics where additional errors are present.

In addition, the multipath problem (Kaplan 2006) is not solved by the repeater concept and since code phase measurements are typically carried out, it has to be solved: this is the topic dealt with in the next section.

4.5 The multipath mitigation technique developed

This paragraph addresses a "short multipath insensitive code loop" (SMICL) mitigation technique, developed in the context of the repeater based positioning system: the goal is to mitigate multipath (Jardak and Samama 2010). For this, a new discriminator function has been proposed which is insensitive to multipath signals having relative delays of less than 146.5 m, equivalent to half a chip length. The standard discriminator used by the Standard DLL (SDLL, the DLL having a correlator spacing of 1 chip), has a non zero steady state error in the presence of multipath signals. This is due to the non-symmetrical behaviour of the composite ACF. As a result, when the early autocorrelation value equals the late autocorrelation value, the prompt replica is not synchronized with the direct signal, but rather with the composite signal. Consequently, another discriminator function was found: the proposed code discriminator compares the early correlation value to an adjusted version of the prompt one. The result is that the new discriminator expression yields zero when the prompt reaches the delay of the direct signal component even in presence of multipath rays of relative delays less than half a chip.

The proposed new expression of the discriminator is given by:

$$D = \left(IE^2 + QE^2\right) - \left(IP'^2 + QP'^2\right) \tag{3}$$

Where

$$\begin{cases} IP' = IP - \dfrac{\Delta}{2}\dfrac{IE + IL}{2 - \Delta} \\ QP' = QP - \dfrac{\Delta}{2}\dfrac{QE + QL}{2 - \Delta} \end{cases} \tag{4}$$

Δ is the correlator spacing and IE, QE, IL, QL, IP and QP are respectively the in phase and in-quadrature phase of the Early, Late and Prompt classic correlators. Note that modified prompt correlators are introduced, IP' and QP', as described. Expression (3) is based on the fact that the left part of the ACF is the one that is the least modified by multipath, but that in addition the prompt replica is modified by the presence of multipath. Thus, the new discriminator uses the Early correlator that is less modified and a modified form of the prompt correlator. Expressions (4) represent the way the prompt correlator is modified and are in fact obtained from the analysis of the general form of the multipath contribution to the discriminator. Indeed, for multipath of less than half a chip, one can show that

$$\begin{cases} IE + IL = (2-\Delta) \sum\limits_{0 \le k \le N} A_k \cos\left(\theta_k - \hat{\theta}\right) \\ QE + QL = (2-\Delta) \sum\limits_{0 \le k \le N} A_k \sin\left(\theta_k - \hat{\theta}\right) \end{cases} \tag{5}$$

The various sums in (5) being the multipath contributions, considering there are N reflected paths of amplitudes A_k and delay $(\theta_k-\theta)$. The limitation of the efficiency of the method to reflected paths of less than half a chip is due to the validity domain of these approximations.

Let us now give the main results obtained for multipath mitigation. The proposed code loop is compared to the standard code loop and the Narrow Correlator (NC). The signal received is assumed to be the sum of a direct signal and a single reflected signal whose amplitude is half that of the direct signal. The following curves show the envelopes of the pseudo-range errors in a similar way as in figure 10.

With an unlimited front-end bandwidth receiver, the results are given in figure 20. The half chip limit is quite clear for the SMICL. Nevertheless, performances are better than SDLL and NC for short multipath[24].

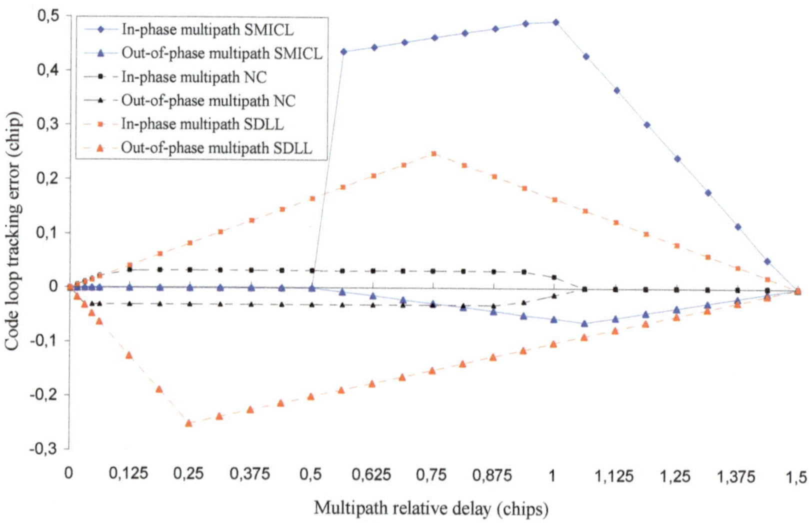

Fig. 20. Comparison of discriminators for an unlimited bandwidth

[24] Many simulations, carried out with Ergospace, have shown that this assumption concerning the delays of the reflected paths indoors is acceptable almost all the time.

With a 2 MHz front-end bandwidth receiver, the current standard for GPS receivers, things are a little bit different[25]. The performances are in this case reduced (the efficiency is not as good for mitigation), and a typical result is an equivalence between the SMICL (at 2 MHz) and the NC (at 8 MHz). Thus, the SMICL allows one to obtain performances of the NC with the current available bandwidth. This is a nice result but it is not sufficient since we showed that 10 to 12 metres of accuracy is not enough indoors. Thus, a 2 MHz bandwidth is not sufficient.

With an 8 MHz front-end bandwidth receiver, which is an intermediate plausible value for future GNSS receivers (including Galileo), the ACF is very close to that obtained with the theoretical unlimited bandwidth. The performance of the SMICL is then acceptable, as shown in figure 21 which compares NC and SMICL. Note that the vertical axis is now given in "chip" (0.01 is equivalent to approximately 3 metres). Based on this figure, multipath errors are reduced with the SMICL to three meters in the worst case (very short out-of-phase multipath) and to 0.7 m when the relative delay is between 0.1 and 0.5 chip.

Please keep in mind the fact that these results are obtained with only one reflected path. Some other simulations were carried out in the case of a typical environment involving several multipath rays and showed that the code measurement error due to multipath is also significantly reduced when the SMICL is considered.

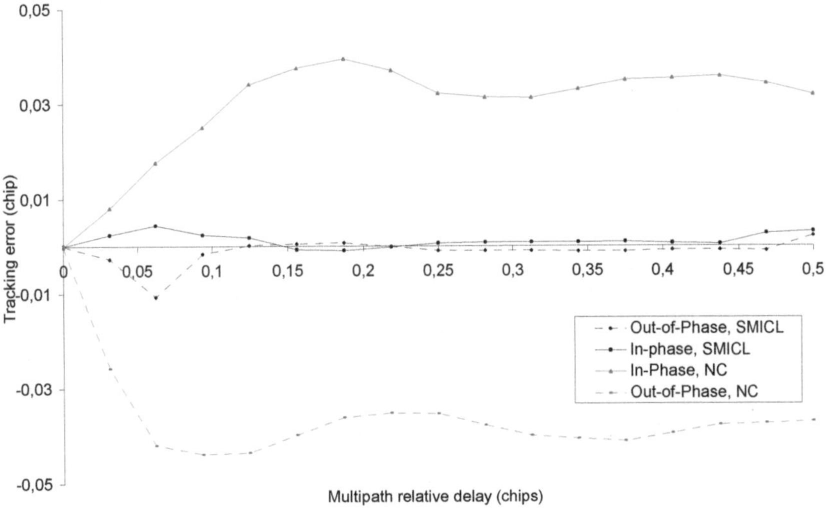

Fig. 21. Comparison of SMICL and NC for an 8 MHz bandwidth

4.6 Discussion

If one combines all the advantages of both pseudolites and repeaters, only the need for a local infrastructure and the multipath effects are not dealt with. The repeater based

[25] Once again, there is a direct link between multipath mitigation efficiency and bandwidth.

infrastructure is still required, although using only one signal distributed to all the transmitters clearly constitutes a huge improvement (also in terms of synchronisation). On another hand, multipath mitigation with the SMICL has shown impressive results that have been validated experimentally, as can be seen through the experimental results. But even with the SMICL, the repeater approach has two major limitations: the difficulty to carry out carrier phase measurement, hence limiting the accuracy attainable (although this is sufficient for the continuity with GNSS outdoors), and poorer performance in dynamic modes. The goal of the next step presented is to propose a synthesised approach that could be the way to overcome these last limitations.

5. The repealite concept: Mixing the advantages of both pseudolites and repeaters

The cycling approach, implemented until now, has a great disadvantage: carrier phase measurements are almost impossible. In order to improve the indoor accuracy, a new approach is proposed based on the so-called "repealites[26]" approach which tries to cumulate the advantages of both repeaters and pseudolites (i.e. carrier phase measurements and same signal transmitted through all the transmitters). First theoretical works have shown a potential of less than one metre accuracy by implementing classical code measurement smoothing techniques using carrier phase measurements. The remaining problem is that repealites are now transmitting simultaneously, which leads to the near-far effect. Thus works have also been carried out concerning this effect.

5.1 Introduction to the idea

It is rather simple in principle: synchronisation is advantageously carried out when the same single signal is transmitted by all the repealites and simultaneous transmissions allow us to implement carrier phase measurements (Vervisch-Picois et al. 2010). Multipath is always a problem but the SMICL, developed in the context of the repeater system, appears to be quite an efficient answer. The pseudolite double differencing approach is probably a little bit too complex for mass market devices (this could be discussed) thus the goal is simply to smooth the code phase measurements with carrier phase measurements, following the classical way of many current GNSS receivers.

The only remaining difficulty is now the near-far effect: a solution to this problem is proposed. Note that when both multipath effects and near-far effects have found a solution, one could consider that the pseudolite system is well suited, since two major problems are solved. As a matter of fact, this is quite true except for synchronisation purposes. Thus, the repealite approach seems to be rather an acceptable compromise.

5.2 The proposed system architecture

The proposed method comes from the transmitting approach of the repeated system, but instead of the sequential mode, the transmission on each antenna is delayed in such a way that the transmitted signals on each repealite do not interfere once they arrive at the receiver

[26] Repealite is a contraction of Repeater and Pseudolite.

antenna. A new problem arises: since a high level of interference can occur because of simultaneous broadcasting of different signals. This can induce severe interference that can disrupt the signal. If one observes the ACF at the receiver end (see figure 22), there is no longer one maximal peak for each code length, but N peaks if N is the number of transmitting repealites (assuming that all the transmissions are included in one code length, but note that this is necessary for the system).

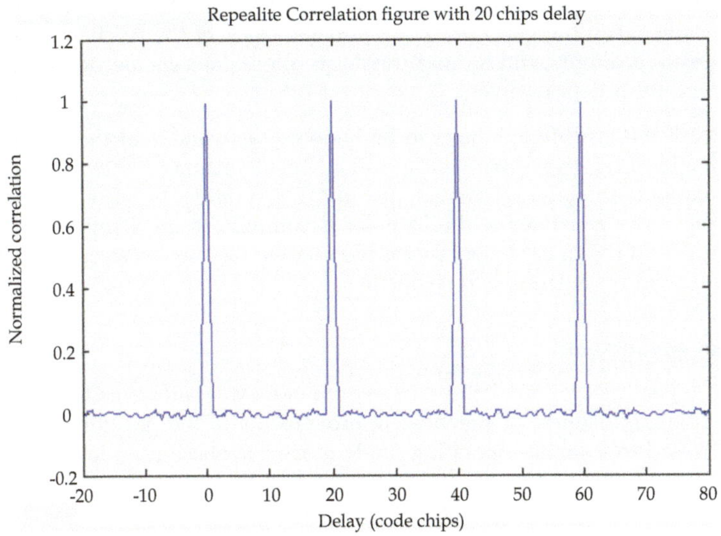

Fig. 22. The resulting auto-correlation function at the receiver

The system shown in figure 23 uses a signal generator that ensures the synchronisation. A single signal is sufficient, as in the case of repeaters.

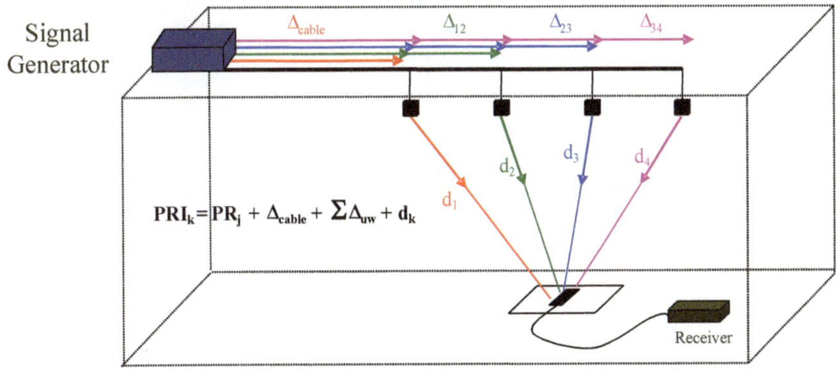

Fig. 23. The repealite system

With 4 delayed channels the terminal is able to carry out 4 indoor pseudo-range measurements. These measurements lead to the equations of the system (the notations of figure 23 are used):

$$\begin{cases} PR_1 = d_1 + \Delta_{cable} \\ PR_2 = d_2 + \Delta_{cable} + \Delta_{12} \\ PR_3 = d_3 + \Delta_{cable} + \Delta_{12} + \Delta_{23} \\ PR_4 = d_4 + \Delta_{cable} + \Delta_{12} + \Delta_{23} + \Delta_{34} \end{cases} \tag{6}$$

Where the PR_k are the indoor pseudo-ranges measured by the receiver, Δ_{cable} is the common part of the delay in the cable between the generator and the first repealite (including error and clock bias between the generator clock and the clock of the receiver), the Δ_{uw} are the delays between repealite R_u and R_w and the d_k are the indoor geometric distances between repealite R_k and the indoor receiver.

The locations of the transmitters have to be known[27], as usual, and the indoor position is computed in a local referential with a classical GNSS algorithms. Note that the velocity can also be calculated in the local referential, just like GNSS outdoors, since the contribution of the clock drift of the generator to the Doppler is common to the 4 repealites and that the only contribution of the signal to Doppler is the relative velocity between the antenna of the indoor receiver and the antenna of repealite R_i.

5.3 The main advantages

The fact that repealites are transmitting in a continuous way allows us to follow the carrier phase of the signal, a source of potential improvements in the positioning accuracy. This feature could lead to a similar operating mode to carrier phase pseudolites, but this is not the main objective here. Another interesting improvement compared to repeaters is the ability to carry out dynamic positioning with no restriction since instantaneous measurements and calculations are carried out. It is also noticeable that dynamic positioning is bound to be of better quality since the receiver movement will have a direct impact on the average multipath distribution, leading to a more efficient averaging of their effects.

The continuity with outdoor GNSS is even simplified in comparison with a repeater where a switch between the outdoor mode and the indoor mode and its cycling scheme was required. With repealites, this switch only concerns the PRN number used which should be characteristic of indoors: the same apply to pseudolites.

The last main advantage is associated to synchronisation. The fact of using a single signal is an advantage in comparison to pseudolites, but does not allow the synchronisation problem to be completely removed since transmitters still have to be synchronised. This is currently achieved through wire connections, either by coaxial cables or by the way of optical fibres[28]. The synchronisation of the system is obtained once several measurements are carried out at known locations.

5.4 The remaining limitations and the ways they are dealt with

The two most important remaining limitations are respectively the multipath and the near-far effect. Multipath effects are dealt with through the use of the SMICL. Note that good

[27] Some works are under consideration in order to propose methods for auto-positioning the transmitters.

[28] Optical fibres are also considered for the physical realization of the time delays between repealites.

pseudo-range measurements are a must if one wants the smoothing of the code by the carrier to be efficient: thanks to the SMICL, this is possible.

We have seen that the ACF of the various codes used in GNSS present secondary peaks that are the origin of the near-far problem. In the case of the repealite based system, this problem is enhanced since the same signal is repeated N times (in the case of N repealites transmitting simultaneously). Thus, the interferences are of uppermost importance, in particular when defining the delays between repealites (since superposing the repealite signal to a secondary peak of the preceding repealite would be a particularly bad idea). Thus, a proper choice of the delays has to be carried out in coordination with the code used and the size of the indoor environment (because the signals should not interfere at the receiver).

A few approaches have been proposed in order to reduce the near-far effect in the case of repealite systems, depending on the codes used. For the GPS codes, it appears that the appropriate delays are obtained when the ACF is close to zero (see figure 24): such "locations" are numerous but depend on the chosen code (the locations are not identical for all codes). In order to reduce the near-far, a double transmission technique is proposed: it consists indeed in modifying the shape of the transmitted signal in order to allow the receiver to carry out differences that could allow it to remove the most powerful signal which is the cause of the near-far. The signal sent is composed of the initial code to which is added, in opposite phase, the same signal delayed by half a chip. Improvements of up to 30dB in comparison with solutions where no near-far mitigation techniques are implemented have been reported. Note that this means still 20dB of improvement in the power that can be managed in comparison with a pulsed pseudolite system.

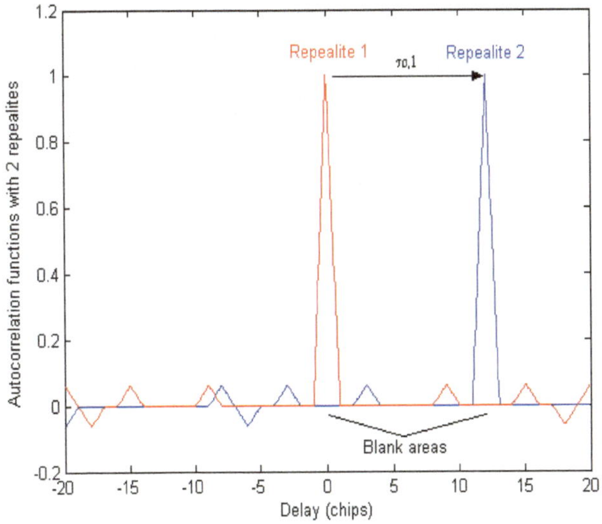

Fig. 24. Optimal determination of the delays between repealites

The drawback of this approach is that it requires a specific signal to be sent and in turn a modification of the software of the receivers which have to be aware of this specific mode. Nevertheless, the efficiency theoretically demonstrated may be worth implementation.

Another interesting proposition concerns the potential use of maximal sequences that have the advantage of providing us with a unique value of auto-correlation outside the main peak (Vervisch-Picois and Samama 2009). Thus, it is possible to carry out differences without the need for a half chip delay for the additional signal. The implementation is then quite easy and can be applied to an almost unlimited number of repealites.

In these cases, interference with outdoors is a very interesting and challenging topic since regulations are appearing in order to "preserve" the GNSS bands. The fact of using similar codes to those used outdoors is a real concern which could find an elegant solution through the use of originally designed sequences. Of course, a frequency shifted approach would definitively solve the interference problem with outdoors, but would require new frequency resources in the case of the modern Code Division Multiple Access (CDMA) GNSS systems.

5.5 A few preliminary estimated performances

The smoothing of the code with the carrier phase is a very classical operation in GNSS. It consists of using the low noise carrier phase measurements in order to smooth the pseudo-range measurements. It is very efficient in order to reduce thermal noise but not really for multipath. Thus, the coupling of the SMICL with this smoothing technique is a very nice combination. The Kalman filter implemented is then nearly optimal. Note that indoors, the main error source comes from multipath, since no atmospheric contributions or clock bias errors of transmitters (in this repealite based configuration) are present. In the present case, the filter uses the carrier phase measurement in order to carry out its estimation of the future state.

Simulations have been carried out considering a circular displacement of a pedestrian in a place where a severe multipath (only one) is present, sometimes of even greater amplitude than the direct path from a transmitter. This is achieved through a perfect reflector located in the close vicinity of the trajectory. As can be seen in figure 25, the repealites are located in

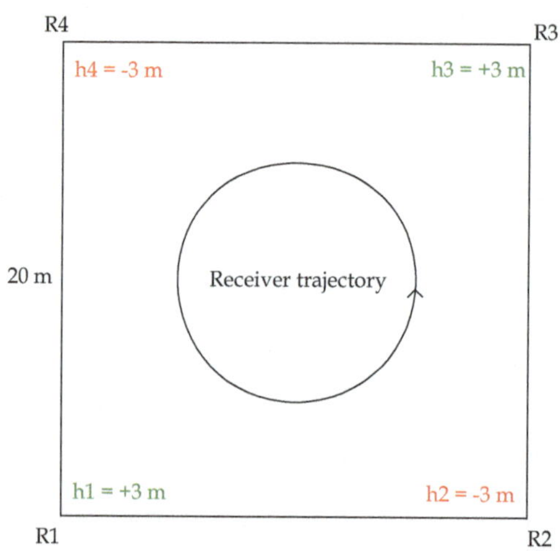

Fig. 25. Considered trajectory and repealite distribution

the corners of a square that includes the complete trajectory. Their exact locations are given in the figure (note that the altitude of the repealites are also given and allow for quite a nice indoor VDOP). The receiver is considered to be at an altitude of zero meters.

The speed of the receiver is set at 1m/s and the multipath delay appeared to vary between 0.07 and 0.17 chips, equivalent to between 20 and 50 metres roughly. Note that since the SMICL is more sensitive to noise than the SDLL (or the NC), the simulations were carried out using 50dB-Hz for the C/N0 value. This is rather a high value for outdoors, but not impossible indoors since one decides the indoor power transmitted (except that regulations are limiting the maximum allowed).

The results are given in figure 26 for 2D and 3D positioning. These simulations show a few decimeters accuracy range for the whole trajectory, and results a little bit better for 2D than 3D. Some skips can be seen in figure 26 which are the ambiguity skips: thus, these skips are typically a multiple of nineteen centimeters. This allows us to evaluate the efficiency of the estimation of this ambiguity. Note that it is calculated every second and is based on the SMICL assisted measurement of the code phase. This once again confirms the very good performance of the SMICL approach.

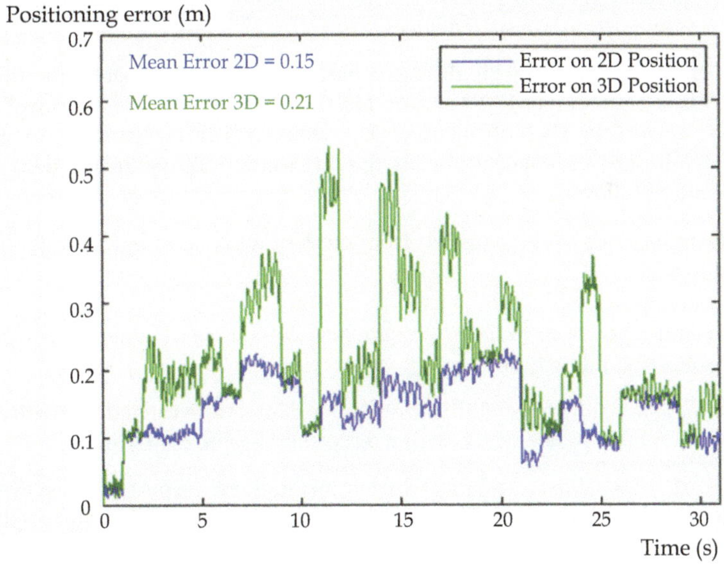

Fig. 26. Positioning accuracy obtained with a repealite system

6. Regulatory issues for L1/E1

The problem of using the same frequency band as the outdoor GNSS is that interference may occur. Of course, when a single system is deployed, these interferences should be very limited and only disturb locally the outdoor receivers. Nevertheless, if no regulations exist, there is a potential danger for GNSS. Thus, some countries have worked towards the development of constraints on the power allowed to be transmitted.

6.1 General introduction

The problem is due to the inter correlation functions (ICF) of the various code sequences that are used. As a matter of fact, these ICF have small peaks, comparable to the secondary peaks of the ACF. If the number of the ground based transmitters is too high or if the total power is too high, then the addition of these secondary peaks is likely to generate interferences to an unacceptable level for outdoor receivers.

Two different cases have been considered by the regulatory authorities: the repeaters and the pseudolites. The repeater case corresponds to a transmitter which uses the outdoor available signals and, after amplification, retransmits them indoors. The ICF between indoor signals and outdoor ones can been considered as being indeed ACF, thus leading to potentially higher interferences. Thus, the maximal acceptable power associated with repeaters is lower than for pseudolites[29].

6.2 The case of the repeaters

In the United States it is not legal to sell GPS repeaters and only the Federal government or agencies operating under its direction, parties that would have received either a Special Temporary Authority (STA) or an Experimental License, or parties operating in an anechoic chamber are authorised to use such devices.

In Europe, things are a little bit different and regulations are based on the Electronic Communications Committee (ECC) report 145 (ECC report 145), dated May 2010. Studies were carried out on the base on interference evaluations in the various GNSS associated frequency bands. Let us concentrate on the L1 band (1559 to 1610 MHz). The global conclusions are as follows:

- The maximum gain of the repeater, from outdoor antenna to indoor antenna should be limited to 45dB.
- The radiated power[30] should not exceed -77dBm.
- The maximum power re-radiated that are not GNSS signals should be less that -20dBm.
- The repeater should include filtering.

Some experimental results presented in previous sections were carried out with -80dBm and have shown acceptable performance within a typical range of 20 metres.

In addition to the above technical recommendations, report 145 states that any authorisations should include guidance instructions in order to help the applicant in the deployment phase of the repeaters. Also, particular attention is recommended for installations close to airports or to military sites.

Finally, the report proposes that any uses of repeaters should be subject to individual authorisation and that no mobile use should be permitted.

---

[29] Please note that the various indoor positioning systems proposed in this chapter have to be considered as « pseudolite based » for regulation purposes, although the so-called « repeater based » approach could also be implemented using repeaters (in the sense of the regulations), and then fall into the corresponding regulation, of course.

[30] The so-called eirp (Equivalent isotropically radiated power).

6.3 The case of indoor pseudolites

The GPS predicted the need for terrestrial generators when reserving the specific codes, PRN 33 through 37, for ground transmitters. Galileo also included the possibility of using such transmitters. Note that since codes are different from satellite's ones, the limitations are a little bit relaxed in comparison to repeaters.

The following lines are based on report 168 of the ECC (ECC report 168), dated May 2011, and relate to indoor pseudolites. Similar to the case of repeaters, computations were carried out on the base on interference evaluations in the various GNSS associated frequency bands. For the L1 band, the main conclusions are as follows:

- The radiated power should not exceed -50dBm.
- The antenna of the pseudolite should point at the ground and be directed towards the inside of the building.
- The radiated power for an elevation angle superior to 0 degree should be reduced by more than 6dB.
- The radiated power should be reduced to -59dBm in airport areas and specific mitigation techniques implemented when aircraft are in their parking stands.

Note that the power level is rather high in comparison to repeaters and largely sufficient in order to have all the techniques described in the chapter implemented in real conditions with good performance. As a matter of fact, the estimated range with -60dBm is around one hundred metres in real environments, i.e. including walls and multiple floor levels (ceilings). The remaining 10dB margin could be used in order to provide the receiver with a high SNR, required for the SMICL for instance. On the other hand, the interesting feature that consists in positioning a pseudolite on the ground pointing at the top of the building (in order to substantially increase the VDOP) will have to be implemented with a 6dB reduced maximal power.

In addition, report 168 states the same as for repeaters concerning individual authorisations, insertion of guidance instructions in order to help the applicant in the deployment and interdiction of mobile pseudolites. It is also proposed that some authorities (military, government and meteorological services) be allowed to apply for specific site limitations.

Moreover, the report mentioned that longer codes could improve both the compatibility with non-participative receivers and the performance of participative ones. Note that research works are on-going in this direction.

7. Synthesis and future trends

The GNSS-like signal indoor positioning systems, either based on pseudolites, repeaters or repealites are a real alternative in order to provide users with a continuous service, at the cost of deploying a local infrastructure. This is now possible in particular thanks to multipath and near-far effect mitigation techniques. Performance attainable is in the metre range through rather good quality measurements and elementary computation algorithms. In comparison, such solutions as WiFi based ones, are based on low quality measurements (power level typically) and complex computation algorithms.

A classical way to cope with the continuity of service is to consider GNSS for outdoors and another solution for indoors, say WiFi, UWB or inertial systems. These types of approaches are called hybridisation. Another approach, being currently investigated, is to find a combination of techniques that would complement each other depending on the type of environments, but not based on a dichotomy between indoors and outdoors. Indeed, a specificity of positioning is that environments, indoors as well as outdoors, are much more complex than imagined.

An example of the approach could be a coupling between repealites and an inertial system, deployed in a very large building, such as warehouses or office blocks. In such a way, the three techniques are used in turn where appropriate, and this does not mean just indoors or outdoors. Outdoors where the sky is free, GNSS is used, but as soon as obstacles are present, in urban canyons for example, a coupling with inertial is carried out. In places where too few satellites are available, one or two additional repealites could be used. Indoors, the same applies: a repealite system is deployed in rather a large area where one meter accuracy is enough for direction determination and the propagation environment is not so important that good SNR are easy to obtain. When a user is leaving these “great halls” and entering offices or corridors, the inertial system is once again activated. Such a system is efficient in all possible environments.

8. References

Bartone C, Van Graas F., (2000), Ranging airport pseudolite for local area augmentation. *IEEE Trans Aerosp Electron Syst* 36(1), pp 278–286.

Caratori J., François M., Samama N., (2002), "Universal Positioning Theory Based on Global Positioning System - Upgrade", *InLoc2002*, Bonn, Germany.

Duffett-Smith P, Rowe R., (2006), Comparative A-GPS and 3G-MATRIX testing in a dense urban environment. *ION GNSS 2006*, Forth Worth (TX).

ECC report 145, (2010), Regulatory framework for GNSS repeaters, St. Petersburg.

ECC report 168, (2011), Regulatory framework for indoor GNSS pseudolites, Miesbach.

Fontana RJ., (2004), Recent system applications of short-pulse ultra-wideband (UWB) technology. *IEEE Trans Microwave Theory Tech*, pp 2087–2104.

Fluerasu A., Jardak N., Vervisch-Picois A., Samama N., (2009), "GNSS Repeater Based Approach for Indoor Positioning: Current Status", *ENC-GNSS2009*, Naples, Italy.

Fluerasu A., Samama N., (2009), "GNSS transmitter based indoor positioning systems - Deployment rules in real buildings", *13th IAIN World Congress*, Stockholm, Sweden.

Glennon E. P., Bryant R. C., Dempster A. G., Mumford P. J. , (2007), "Post Correlation CWI and Cross Correlation Mitigation Using Delayed PIC" *ION GNSS*, Forth Worth, USA.

Im S-H, Jee G-I, Cho YB., (2006), An indoor positioning system using time-delayed GPS repeater. *ION GNSS 2006*, Forth Worth, (TX).

Jardak N., Samama N., (2010), "Short Multipath Insensitive Code Loop Discriminator", *IEEE Trans. on Aerospace and Electronic Systems*, Vol. 46, PP.278-295.

Jee GI, Choi JH, Bu SC., (2004), Indoor positioning using TDOA measurements from switched GPS repeater. *ION GNSS 2004*, Long Beach (CA).

Kanli M.O., (2004), "Limitations of Pseudolite Systems using off-the-shelf GPS receivers", *The International Symposium on GNSS/GPS*, Sydney, Australia.

Kaplan ED, Hegarty C., (2006), Understanding GPS: principles and applications. 2nd ed. *Artech House*, Norwoood, MA, USA.

Kee C, Yun D, Jun H, Parkinson B, Pullen S, Lagenstein T., (2001), Centimeter-accuracy indoor navigation using GPS-like pseudolites. *GPS World*.

Kee C, Jun H, Yun D, (2003), "Indoor Navigation System using Asynchronous Pseudolites", *Journal of Navigation*, 56, pp 443-455.

Klein D. and Parkinson B. W., (1986), "The Use of Pseudolites for Improving GPS Performance." *Global Positioning System*, volume 3. Institute of Navigation, Washington, DC.

Kupper A., (2005), Location based services — fundamentals and operation. *John Wiley and Sons*.

Madhani P.H, Axelrad P., Krumvieda K., Thomas J., (2003), "Application of Successive Interference Cancellation to the GPS Pseudolite Near-Far Problem" *IEEE Transaction on Aerospace and Electronic System*, vol. 39, no 2, pp. 481–487.

Martone M, Metzler J., (2005), Prime time positioning: using broadcast TV signals to fill GPS Acquisition gaps. *GPS World 2005*, pp 52–59.

Parkinson BW, Spilker Jr. JJ., (1996) Global positioning system: theory and applications. American Institute of Aeronautics and Astronautics.

Rizos C., Barnes J., Wang J., Small D., Voigt G. and Gambale N., (2003), "LocataNet: Intelligent Time-Synchronised Pseudolite Transceivers for cm-Level Stand-Alone Positioning", *11th IAIN World Congress*, Berlin, Germany.

Samama N., Vervisch-Picois A., (2005), "3D Indoor Velocity Vector Determination Using GNSS Based Repeaters", *ION GNSS 2005*, Long Beach, USA.

Samama N, (2008), "Global Positioning - Technologies and Performance", *Wiley InterScience*, Hoboken, USA.

Takada Y, Kishimoto M, Kawamura N, Komoda N, Yamazaki T, Oiso H, Masanari T., (2003), An information service system using Bluetooth in an exhibition hall. *Annales des Telecommunications* 2003, ¾, pp 507–530.

Vervisch-Picois A, Samama N, (2006), "Analysis of 3D Repeater Based Indoor Positioning System - Specific Case of Indoor DOP ", *ENC-GNSS 2006*, Manchester, UK.

Vervisch-Picois A., Samama N., (2009), "Interference Mitigation In A Repeater And Pseudolite Indoor Positioning System", *IEEE Journal of Specific Topics on Signal Processing*, Vol. 3, N°5, PP.810-820.

Vervisch-Picois A., Selmi I., Gottesman Y., Samama N., (2010), "Current Status of the Repealite Based Approach - A Sub-Meter Indoor Positioning System", *IEEE-NAVITEC 2010*, Noordwijk, The Netherlands.

Wang Y, Jia X, Rizos C., (2004), Two new algorithms for indoor Wireless Positioning System (WPS). *ION GNSS 17th International Technical Meeting of the Satellite Division*, Long Beach (CA).

Yang C., Morton J., (2009), "Adaptive Replica Code Synthesis for Interference Suppression in GNSS Receivers", *ION ITM*, Anaheim, USA.

Part 3

GNSS Errors Mitigation and Modelling

15

Ionospheric Propagation Effects on GNSS Signals and New Correction Approaches

M. Mainul Hoque and Norbert Jakowski
German Aerospace Center (DLR), Institute of Communications and Navigation
Neustrelitz,
Germany

1. Introduction

The ionosphere is the ionized part of the earth's atmosphere lying between about 50 km and several earth radii (Davies, 1990) whereas the upper part above about 1000 km height up to the plasmapause is usually called the plasmasphere. Solar extreme ultraviolet (EUV) radiation at wave lengths < 130 nm significantly ionizes the earth's neutral gas. In addition to photoionisation by electromagnetic radiation also energetic particles from the solar wind and cosmic rays contribute to the ionization. The ionized plasma can affect radio wave propagation in various ways modifying characteristic wave parameters such as amplitude, phase or polarization (Budden, 1985; Davies, 1990). The interaction of the radio wave with the ionospheric plasma is one of the main reasons for the limited accuracy and vulnerability in satellite based positioning or time estimation.

A trans-ionospheric radio wave propagating through the plasma experiences a propagation delay / phase advance of the signal causing a travel distance or time larger / smaller than the real one. The reason of the propagation delay can be realized considering the nature of the refractive index which depends on the density of the ionospheric plasma. The refractive index ($n \neq 1$) of the ionosphere is not equal to that of free space ($n = 1$). This causes the propagation speed of radio signals to differ from that in free space. Additionally, spatial gradients in the refractive index cause a curvature of the propagation path. Both effects lead in sum to a delay / phase advance of satellite navigation signals in comparison to a free space propagation.

The variability of the ionospheric impact is much larger compared to that of the troposphere. The ionospheric range error varies from a few meters to many tens of meters at the zenith, whereas the tropospheric range error varies between two to three meters at the zenith (Klobuchar, 1996). The daily variation of the ionospheric range error can be up to one order of magnitude (Klobuchar, 1996).

After removal of the Selective Availability (SA, i.e., dithering of the satellite clock to deny full system accuracy) in 2000, ionosphere becomes the single largest error source for Global Navigation Satellite Systems (GNSS) users, especially for high-accuracy (centimeter - millimeter) applications like the Precise Point Positioning (PPP) and Real Time Kinematic (RTK) positioning. Fortunately, the ionosphere is a dispersive medium with respect to the radio wave; therefore, the magnitude of the ionospheric delay depends on the signal frequency. The advantage is that an elimination of the major part of the ionospheric refraction through a linear combination of dual-frequency observables is possible. However, inhomogeneous plasma distribution and anisotropy cause higher order nonlinear effects which are not removed in this linear approach. Mainly the second and third order ionospheric terms (in the expansion of the refractive index) and errors due to bending of the signal remain uncorrected. They can be several tens of centimeters of range error at low elevation angles and during high solar activity conditions.

Brunner & Gu (1991) were pioneers to compute higher order ionospheric effects and developing correction for them. Since then higher order ionospheric effects have been studied by different authors during last decades, e.g., Bassiri & Hajj (1993), Jakowski et al. (1994), Strangeways & Ioannides (2002), Kedar et al. (2003), Fritsche et al. (2005), Hawarey et al. (2005), Hoque & Jakowski (2006, 2007, 2008, 2010b), Hernández-Pajares et al. (2007), Kim & Tinin (2007, 2011), Datta-Barua et al. (2008), Morton et al. (2009), Moore & Morton (2011). The above literature review shows that higher order ionospheric terms are less than 1% of the first order term at GNSS frequencies. Hernández-Pajares et al. (2007) found sub-millimeter level shifting in receiver positions along southward direction for low latitude receivers and northward direction for high latitude receivers due to the second order term correction. Fritsche et al. (2005) found centimeter level correction in GPS satellite positions considering higher order ionospheric terms. Elizabeth et al. (2010) investigated the impacts of the bending terms described by Hoque & Jakowski (2008) on a Global Positioning System (GPS) network of ground receivers. They found the bending correction for the dual-frequency linear GPS L1-L2 combination to exceed the 3 mm level in the equatorial region. Kim & Tinin (2011) found that the systematic residual ionospheric errors can be significantly reduced (under certain ionospheric conditions) through triple frequency combinations. All these studies were conducted to compute higher order ionospheric effects on GNSS signals for ground-based reception. Recently Hoque & Jakowski (2010b, 2011) investigated the ionospheric impact on GPS occultation signals received onboard Low Earth Orbiting (LEO) CHAMP (CHAllenging Minisatellite Payload) satellite.

In this chapter, the first and higher order ionospheric propagation effects on GNSS signals are described and their estimates are given at different level of ionospheric ionization. Multi-frequency ionosphere-free and geometry-free solutions are studied and residual terms in the ionosphere-free solutions are computed. Different correction approaches are discussed for the second and third order terms, and ray path bending correction. Additionally, we have proposed new approaches for correcting straight line of sight (LoS) propagation assumption error, i.e., ray path bending error for ground based GNSS positioning. We have modelled the excess path length of the signal in addition to the LoS path length and the total electron content (TEC) difference between a curved and LoS paths as functions of signal frequency, ionospheric parameters such as TEC and TEC derivative with respect to the elevation angle. We have found that using the TEC derivative in addition to the TEC information we can improve the existing correction results.

2. Ionospheric propagation effects

Quantitatively, the propagation of a radio wave through the ionospheric plasma is described by the refractive index of the ionosphere (Appleton-Hartree formula). At high frequencies (> 100 MHz), the refractive index mainly depends on the electron density, the strength and direction of the geomagnetic field in relation to the ray path. Thus, the spatial distribution of the electron density along the ray path and corresponding geomagnetic field relationships determine the ionospheric impact on the electromagnetic wave.

2.1 Ionospheric refractive index

For high frequency (HF) radio waves with frequencies f > 100 MHz the phase refractive index n can be derived from the Appleton - Hartree formula as (Appleton, 1932; Bassiri & Hajj, 1993)

$$n = 1 - \frac{f_p^2}{2f^2} \pm \frac{f_p^2 f_g \cos\Theta}{2f^3} - \frac{f_p^2}{4f^4}\left[\frac{f_p^2}{2} + f_g^2\left(1+\cos^2\Theta\right)\right] \tag{1}$$

in which

$$f_p^2 = n_e e^2 / \left(4\pi^2 \varepsilon_0 m\right)$$

$$f_g = eB / (2\pi m)$$

where f_p is the plasma frequency, f_g is the gyro frequency, ε_0 is the free space permittivity, B is the geomagnetic induction, Θ is the angle between the wave propagation direction and the geomagnetic field vector $\boldsymbol{B}$, and e, n_e, m are the electron charge, density and mass, respectively. The wave with the upper (+) sign in Eq. (1) is called the ordinary wave and is left-hand circularly polarized, whereas the wave with the lower (-) sign is called the extraordinary wave and is right-hand circularly polarized (Hartmann & Leitinger, 1984). The GPS signals are transmitted in right-hand circular polarization (Parkinson & Gilbert, 1983).

Equation (1) indicates that the phase refractive index is less than the unity resulting in a phase velocity that is greater than the speed of light in vacuum (i.e., phase advance). Therefore, the integration of n along a signal path gives a measure of the range / travel time between a receiver and a satellite that is smaller than the geometric distance / travel time in the vacuum.

To compute group delay measurements, the group refractive index n_{gr} should be considered. The expression for n_{gr} can be determined by the relationship $n_{gr} = n+f(dn/df)$.

$$n_{gr} = 1 + \frac{f_p^2}{2f^2} \mp \frac{f_p^2 f_g \cos\Theta}{f^3} + \frac{3f_p^2}{4f^4}\left[\frac{f_p^2}{2} + f_g^2\left(1+\cos^2\Theta\right)\right] \tag{2}$$

Equation (2) indicates that the group refractive index is greater than the unity resulting in a group velocity that is less than the speed of light. Therefore, the integration of n_{gr} along a signal path gives a measure of the range / travel time that is greater than the geometric distance / travel time in the vacuum. Therefore, when GNSS signals propagate through the ionosphere, the carrier-phase experiences an advance and the code experiences a group delay. The carrier-phase pseudoranges are measured too short and the code pseudoranges are measured too long compared to the geometric range between a satellite and a receiver.

Considering ionospheric refraction the geometric distance (Euclidean line) or true range ρ between a transmitting satellite S and a receiver R can be written in units of length as

$$\rho = L + \int_S^R (1-n)ds - d_I^{len} \tag{3}$$

where the optical distance $L = \int_S^R n ds$ is the line integral of the refractive index between the satellite and the receiver along the ray path, $\int_S^R (1-n)ds$ is the ionospheric group delay and d_I^{len} is the excess path length of the signal in addition to the geometric path length caused by the ray path bending and defined by

$$d_I^{len} = \int_S^R ds - \rho \tag{4}$$

where $\int_S^R ds$ is the curved path length in the vacuum. The travel time of the signal can be computed dividing the expression of ρ (Eq. 3) simply by the speed of light.

2.2 Group delay and phase advance

Assuming a right hand circularly polarized signal, the ionospheric group delay d_{Igr} and carrier phase advance d_I can be written in units of length as (using Eqs. (1) and (2))

$$d_{Igr} = d_{Igr}^{(1)} + d_{Igr}^{(2)} + d_{Igr}^{(3)} = \int_S^R \left(n_{gr} - 1\right) ds = \frac{p}{f^2} + \frac{q}{f^3} + \frac{u}{f^4} \tag{5}$$

$$d_I = d_I^{(1)} + d_I^{(2)} + d_I^{(3)} = \int_S^R (1-n) ds = \frac{p}{f^2} + \frac{q}{2f^3} + \frac{u}{3f^4} \tag{6}$$

$$p = K \int n_e ds = K \cdot TEC = K\left(TEC_{LoS} + \Delta TEC_{bend}\right) \tag{7}$$

$$q = 2.2566 \times 10^{12} \int n_e B \cos\Theta \cdot ds \tag{8}$$

$$u = 2437 \int n_e^2 ds + 4.74 \times 10^{22} \int n_e B^2 (1 + \cos^2\Theta) ds \tag{9}$$

where $K = e^2/(8\pi^2\varepsilon_0 m) = 40.3\ m^3s^{-2}$, the integration of n_e along signal paths $\int n_e ds$ is called the total electron content TEC and measured in TEC units (1 TECU = 10^{16} electrons/m^2). The terms $d_{Igr}^{(1)}$ / $d_I^{(1)}$, $d_{Igr}^{(2)}$ / $d_I^{(2)}$ and $d_{Igr}^{(3)}$ / $d_I^{(3)}$ in Eq. (5) / (6) are the first, second and third order ionospheric group delays / phase advances, respectively. Due to the dispersive nature of the ionosphere, satellite signals transmitted on different frequencies travel along different ray paths through the ionosphere on their way to a receiver and thus the TEC along a f_1 path will be different from that along a f_2 path and also from that along the LoS path. Considering this, TEC in Eq. (7) is separated into TEC_{LoS} and ΔTEC_{bend} where TEC_{LoS} is the TEC along the LoS and ΔTEC_{bend} is the difference between TECs along a curved path and the LoS path. The term ΔTEC_{bend} represents TEC contribution due to ray path bending only, i.e., the second and third order terms are not considered in TEC estimation by Eq. (7).

2.3 Ionospheric effects on GNSS observables

The observables are travel time or ranges which are deduced from measured time or phase differences based on a comparison between received signals and receiver generated signals. Thus, the ranges are biased by satellite and receiver clock errors, instrumental biases and atmospheric effects, and therefore, called pseudoranges. The code pseudorange (Ψ) and carrier-phase pseudorange (Φ) at a selected frequency can be described by observation equations in units of length as

$$\Psi = \rho + c(dt - dT) + d_{Igr} + d_A + (d_{MP})_{\Psi} + dq + dQ + \varepsilon_{\Psi} \tag{10}$$

$$\Phi = \rho + c(dt - dT) - d_I + d_A + (d_{MP})_{\Phi} + dq + dQ + N\lambda + \varepsilon_{\Phi} \tag{11}$$

where ρ is the geometric distance between a satellite and a receiver, c is the velocity of light, dt and dT are the satellite and receiver clock errors, respectively, d_I and d_{Igr} are the ionospheric effects on carrier-phase and code pseudoranges, respectively, d_A is the atmospheric (tropospheric delay) effect, $(d_{MP})_{\Psi}$ and $(d_{MP})_{\Phi}$ are multipath effects on code and carrier-phase pseudoranges, respectively, dq and dQ are the instrumental biases of the satellite and the receiver, respectively, λ is the carrier wavelength, N is the integer carrier-phase ambiguity, and ε_{Ψ} and ε_{Φ} are the rest errors. The carrier-phase pseudorange is expressed in units of length (meters) instead of cycles. However, it can be expressed in cycles dividing simply by the signal's wave length (λ meter/cycle).

For simplicity we confine our interest to only ionospheric effects. Thus, the code and carrier-phase pseudoranges can be simplified as

$$\Psi = \rho + d_{Igr} + d_I^{len} = \rho + \frac{p}{f^2} + \frac{q}{f^3} + \frac{u}{f^4} + d_I^{len} \tag{12}$$

$$\Phi = \rho - d_I + d_I^{len} = \rho - \frac{p}{f^2} - \frac{q}{2f^3} - \frac{u}{3f^4} + d_I^{len} \tag{13}$$

where f is the signal frequency. In case of GPS L1, L2 and L5 signals f = 1575.42, 1227.6 and 1176.45 MHz, respectively. To take into account the ray path bending on observables, the term d_I^{len} is introduced in Eqs. (12) and (13).

2.4 Multi-frequency combinations

2.4.1 First order ionosphere-free combination

As already mentioned, ionosphere is a dispersive medium, i.e., the ionospheric propagation delay is frequency dependent. Therefore, one very popular way to get rid of ionospheric effects is to compute the so called first order ionosphere-free combination of carrier-phase or code pseudoranges measured on two frequencies. However, the second and third order ionospheric terms and errors due to bending of the signal remain uncorrected in this approach. Such a dual-frequency combination can be written in units of length as (combining code / carrier-phase pseudoranges Eq. (12) / Eq. (13) measured on f_1 and f_2 frequencies and substituting p by Eq. (7), for details see Hoque & Jakowski, 2008)

$$\frac{f_1^2}{f_1^2 - f_2^2}\Psi_1 - \frac{f_2^2}{f_1^2 - f_2^2}\Psi_2 = \rho \underbrace{- \Delta s_{TEC} - 2\Delta s_2 - 3\Delta s_3 - \Delta s_{len}}_{RRE_{gr}} \tag{14}$$

$$\frac{f_1^2}{f_1^2 - f_2^2}\Phi_1 - \frac{f_2^2}{f_1^2 - f_2^2}\Phi_2 = \rho \underbrace{+ \Delta s_{TEC} + \Delta s_2 + \Delta s_3 - \Delta s_{len}}_{RRE} \tag{15}$$

$$\Delta s_{TEC} = \frac{K(TEC_2 - TEC_1)}{\left(f_1^2 - f_2^2\right)} = \frac{K(\Delta TEC_{bend2} - \Delta TEC_{bend1})}{\left(f_1^2 - f_2^2\right)} \tag{16}$$

$$TEC_{1,2} = \int n_e ds = (TEC_{LoS} + \Delta TEC_{bend1,2}) \tag{17}$$

$$\Delta s_2 = \frac{q}{2 f_1 f_2 \left(f_1 + f_2\right)} \tag{18}$$

$$\Delta s_3 = \frac{u}{3 f_1^2 f_2^2} \tag{19}$$

$$\Delta s_{len} = \frac{d_2^{len} f_2^2 - d_1^{len} f_1^2}{\left(f_1^2 - f_2^2\right)} \tag{20}$$

where Ψ_1, Ψ_2 and Φ_1, Φ_1 are the measured code and carrier-phase pseudoranges on f_1 and f_2 frequencies, respectively, Δs_2 and Δs_3 are the dual-frequency second and third order residual terms, respectively. The TEC along a f_1 path will be different from that along a f_2 path due to ray path bending. Due to the same reason the excess path length will not be the same for both signals. Therefore, they will not be cancelled out in the ionosphere-free solution. Thus, the terms Δs_{TEC} and Δs_{len} in Eq. (14) and (15) refer to the dual-frequency residual errors due to TEC difference and excess path length, respectively. Their expressions are given by Eqs. (16) and (20). The quantities ΔTEC_{bend1} and ΔTEC_{bend2} are the differences between TECs along curved and LoS paths and the quantities d_1^{len} and d_2^{len} are the differences between curved and LoS path lengths for f_1 and f_2 signals, respectively. The RRE and RRE_{gr} are the total residual range errors in the carrier-phase and code combinations, respectively.

The disadvantages of such combinations (Eqs. 14, 15) are that i) the observation noise is increased by a factor depending on frequencies involved in the combination, ii) the ambiguity term of the carrier-phase combination is no more an integer value and iii) only the first order term is eliminated, i.e., higher order terms remain uncorrected. Moreover, this method cannot be applied to single-frequency receivers.

Assuming the same measurement noise on each signal, it can be shown that the carrier-phase or code noise will be amplified by a factor of 2.98 for the GPS L1-L2 combination, whereas for the L1-L5 and L2-L3 combinations amplification factors are 2.59 and 16.64, respectively (see Hoque & Jakowski, 2010a). The amplification factor is inversely proportional to the separation of combination frequencies. Since the frequency separation is relatively large for the L1-L5 combination, the amplification factor is the smallest.

Since the first order ionospheric effect on carrier-phase and code pseudoranges (see Eq. 12 and 13) is the same in magnitude but opposite in sign, computing the sum of carrier-phase and code pseudoranges would theoretically eliminate the first order ionospheric term in single frequency measurements. However, the resulting observable would inherit the high code noise and the carrier-phase ambiguity and is therefore, practically not suitable.

2.4.2 Second order ionosphere-free combination

Receiving signals on three coherent frequencies will allow triple-frequency combinations to eliminate the first and the second order ionospheric terms. The third order ionospheric term and errors due to ray path bending are not fully removed in this approach. Such a triple-frequency combination can be written as (combining code / carrier-phase pseudoranges Eq. (12) / Eq. (13) measured on f_1, f_2 and f_3 frequencies and substituting ρ by Eq. (7), for details see Hoque & jakowski, 2010a).

$$\frac{1}{C}\left[A\left(\Psi_1 f_1^2-\Psi_2 f_2^2\right)-B\left(\Psi_1 f_1^2-\Psi_3 f_3^2\right)\right]=\rho\ \underbrace{+\left(\Delta s_{TEC}\right)_{tr}+3\left(\Delta s_3\right)_{tr}+\left(\Delta s_{len}\right)_{tr}}_{\left(RRE_{gr}\right)_{tr}} \tag{21}$$

$$\frac{1}{C}\left[A\left(\Phi_1 f_1^2-\Phi_2 f_2^2\right)-B\left(\Phi_1 f_1^2-\Phi_3 f_3^2\right)\right]=\rho\ \underbrace{-\left(\Delta s_{TEC}\right)_{tr}-\left(\Delta s_3\right)_{tr}+\left(\Delta s_{len}\right)_{tr}}_{(RRE)_{tr}} \tag{22}$$

In which

$$\left(\Delta s_{TEC}\right)_{tr}=\frac{K}{C}\left[B\left(\Delta TEC_{bend3}-\Delta TEC_{bend1}\right)-A\left(\Delta TEC_{bend2}-\Delta TEC_{bend1}\right)\right] \tag{23}$$

$$\left(\Delta s_3\right)_{tr}=\frac{u}{3C}\frac{\left(f_2-f_3\right)}{f_2 f_3} \tag{24}$$

$$\left(\Delta s_{len}\right)_{tr}=\frac{1}{C}\left[B\left(f_3^2 d_3^{len}-f_1^2 d_1^{len}\right)-A\left(f_2^2 d_2^{len}-f_1^2 d_1^{len}\right)\right] \tag{25}$$

$$\left.\begin{aligned} A&=\frac{f_1 f_2}{f_1-f_2}\\ B&=\frac{f_1 f_3}{f_1-f_3}\\ C&=f_1\left(f_2-f_3\right)\left(f_1+f_2+f_3\right)\end{aligned}\right\} \tag{26}$$

where K = 40.3 m^3s^{-2}, Ψ and Φ are the code and carrier-phase pseudoranges, and their subscripts correspond to measured signals on f_1, f_2 and f_3 frequencies, $(\Delta s_3)_{tr}$ is the third order residual term and $(\Delta s_{TEC})_{tr}$ and $(\Delta s_{len})_{tr}$ are the residual terms due to TEC difference and excess path length, respectively. The quantities ΔTEC_{bend} and d_I^{len} are the TEC and path length differences between curved and LoS paths and and their subscripts correspond to received signals on frequencies f_1, f_2 and f_3. The $(RRE)_{tr}$ and $(RREgr)_{tr}$ are the total residual range errors in the triple-frequency carrier-phase and code pseudorange combinations, respectively.

However, as already mentioned, such a multiple frequency combination amplifies all uncorrelated errors or noises (multipath and noise). Assuming the same measurement noise on each signal, it can be shown that the noise will be amplified by a factor of 33.7 in the GPS L1-L2-L5 combination (see Hoque & Jakowski, 2010a).

The Galileo system will transmit signals on four frequencies E2-L1-E1, E5a, E5b and E6 (1575.42, 1176.45, 1207.14 and 1278.75 MHz, respectively). Simultaneous reception of four signals will allow quadruple-frequency combinations to eliminate the first, second and third order ionospheric terms. Such a combination would theoretically eliminate higher order ionospheric terms successfully from the range equation. However, the noise will be amplified by a factor of about 626.13 in the E2L1E1-E5a-E5b-E6 combination (assuming the same measurement noise on each signal) which is about two orders larger than a dual-frequency factor. Therefore, a quadruple-frequency combination is barely pragmatic. However, if the frequency separation is large (e.g., combinations between 4-8 GHz C band and 1-2 GHz L band frequencies), the amplification factor will be small. In such cases, measurements on four frequencies may be useful.

2.4.3 Geometry-free combination

When microwave signals are transmitted on two frequencies, all the nondispersive effects, e.g., tropospheric delay, satellite and receiver clock offsets, antenna phase centre offsets and variations etc., manipulate the signals on both frequencies in the same way - apart from the ionosphere. Therefore, by differencing code / carrier-phase pseudoranges measured on two frequencies, all non-dispersive terms including ρ will be cancelled out giving the estimate of TEC along ray paths as (combining code / carrier-phase pseudoranges Eq. (12) / Eq. (13) measured on f_1 and f_2 frequencies and substituting ρ by Eq. (7) and neglecting the second and higher order terms)

$$TEC = \frac{f_1^2 f_2^2}{K\left(f_1^2 - f_2^2\right)}\left[\left(\Psi_2 - \Psi_1\right) + noise_{\Psi_2 - \Psi_1}\right] \tag{27}$$

$$TEC = \frac{f_1^2 f_2^2}{K\left(f_1^2 - f_2^2\right)}\left[\left(\Phi_1 - \Phi_2\right) + B_{ambiguity} + noise_{\Phi_1 - \Phi_2}\right] \tag{28}$$

in which $B_{ambiguity} = \lambda_2 N_2 - \lambda_1 N_1$ is the carrier-phase ambiguity constant where λ_1, λ_2 are wave lengths and N_1, N_2 are integer ambiguities measured on f_1 and f_2 frequencies, $noise_{\Psi_2 - \Psi_1}$ and $noise_{\Phi_1 - \Phi_2}$ are noises (e.g., thermal noise etc.) in code and carrier-phase combinations, respectively. For simplicity different terms such as inter-frequency satellite and receiver biases and multipath effects are not considered.

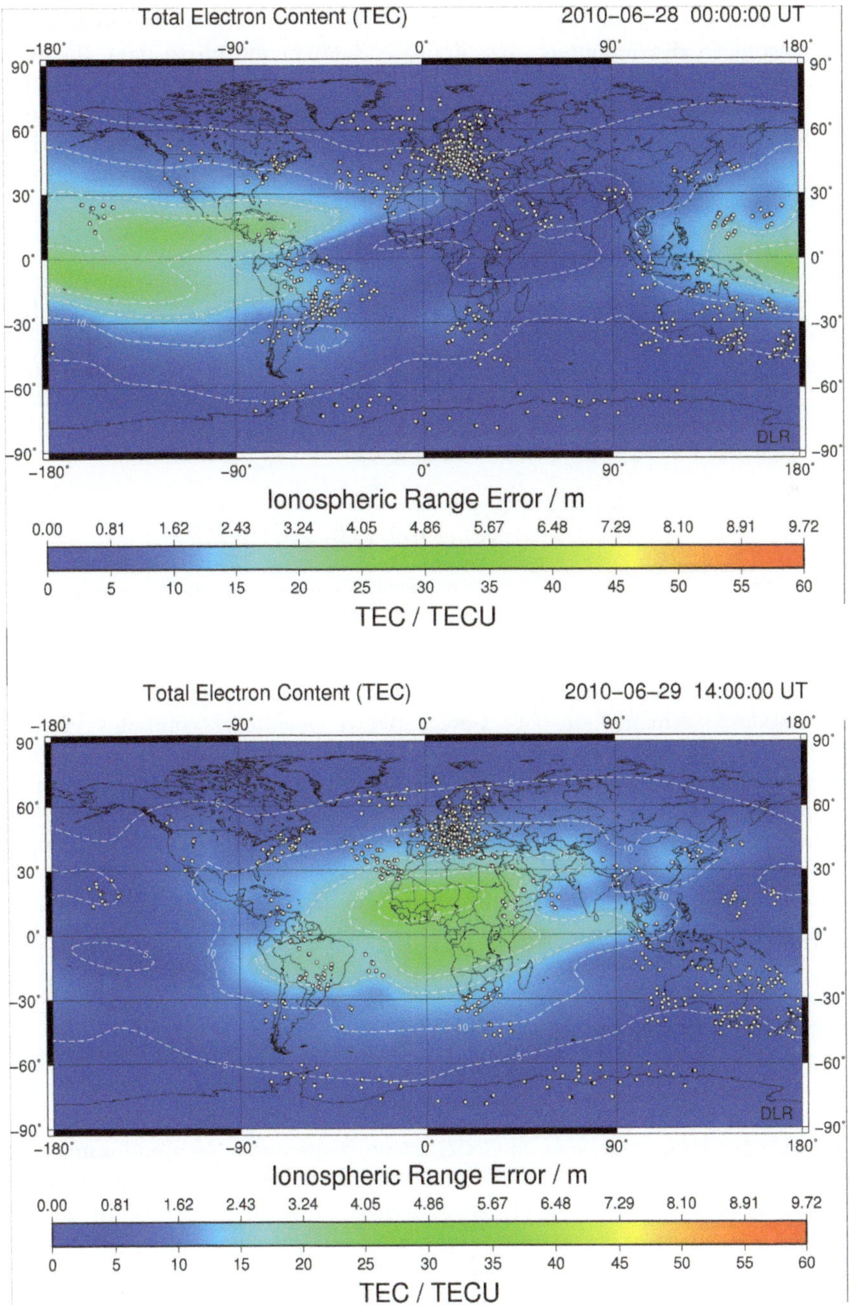

Fig. 1. Examples of global TEC maps and corresponding ionospheric range errors at GPS L1 during night time (0 UT) and day time (14 UT) (http://swaciweb.dlr.de). The dots represent IPP locations

In case of cycle slips (a jump in carrier-phase ambiguity constant due to loss of signal tracking results in discontinuous arcs of phase data) in the phase data, the wide-lane combination method of Blewitt (1987) can be applied for the correction. While the TEC estimated by the carrier-phase difference Eq. (28) is precise and smooth but biased by an unknown phase ambiguity constant, the TEC estimated by the code pseudorange difference Eq. (27) is noisy and less precise but not ambiguous. In order to obtain an absolute and precise estimate of TEC, the accurate phase measurements needs to be levelled to the calibrated absolute code measurements by a least square method.

To derive an elevation independent vertical TEC from a slant TEC measurement, the ionosphere is assumed to be composed of a single thin layer at a representative height of about 350, 400 or 450 km from the earth's surface. The intersection point between a slant ray path and the thin layer is called an ionospheric piercing point (IPP). A mapping function is used to convert the slant STEC to vertical VTEC at the IPP or vice versa (details of the derivation is given in Hoque & Jakowski, 2008).

$$STEC / VTEC \approx \frac{(h_m + R_E)}{\sqrt{(h_m + R_E)^2 - (R_h + R_E)^2 \cos^2 \beta}} \tag{29}$$

where h_m is the height of maximum electron density (varies between 250 -450 km), R_E is the earth's mean radius (~ 6371 km), R_h is the receiver height above the earth's surface and β is the elevation angle.

Based on similar techniques using observation from more than hundred worldwide GNSS ground stations, German Aerospace Center (DLR) Neustrelitz computes vertical TEC estimates at numerous IPPs worldwide. Thus, TEC maps are produced by assigning IPP measurements to homogeneous latitude and longitude grid points as shown in Fig. 1. European and global TEC maps and 1-hour-ahead forecasts are distributed via the operational space-weather and ionosphere data service SWACI (Space Weather Application Center Ionosphere, http://swaciweb.dlr.de, see also Jakowski et al., 2011) to the international community with an update rate of 5 minutes. The advantage of such services is that single frequency GNSS users can correct the ionospheric propagation effect in near real time.

3. Estimation of ionospheric effects

3.1 First- and higher-order ionospheric terms

Equations (12) and (13) indicate that the signal delay caused by the first order term is equal in magnitude but opposite in sign on GNSS carrier-phase and code pseudoranges, i.e., the carrier-phase pseudorange is advanced while code pseudorange is retarded. The first order term is directly proportional to the TEC encountered by the satellite signal during its travel through the ionosphere and inversely proportional to the square of the signal frequency. The first order term includes about 99% of the total ionospheric effect. Therefore, if the frequency and link related slant TEC are known, the first order propagation effect can easily be computed and corrected. If the TEC map is available, the slant TEC can be computed simply multiplying the vertical TEC at the IPP by the mapping function (e.g., Eq. 29).

The vertical TEC may vary between 1 TECU and 300 TECU depending on a number of factors such as local time, geographic location, season, solar activity level etc. The frequency dependence of the first order ionospheric term has been plotted for elevations 5° and 30° in Fig. 2 at different levels of ionospheric ionization characterized by vertical TECs such as i) 250 and 150 TECU correspond to TEC during extreme space weather conditions, ii) 50 TECU corresponds to mid latitude day time and iii) 5 TECU corresponds to mid latitude night time TEC.

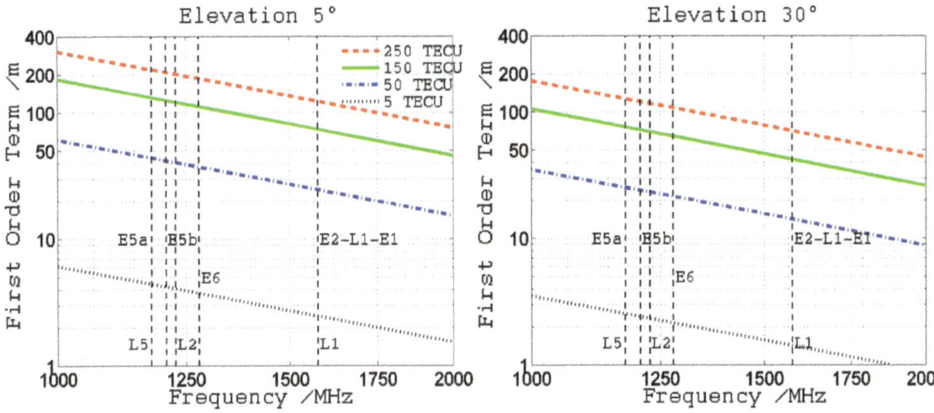

Fig. 2. Frequency dependence of the first order term at different levels of ionospheric ionization and elevation angles.

As Fig. 2 demonstrates, the first order ionospheric term can be more than 100 m at GNSS L-band frequencies (1 - 2 GHz) during times of high TEC at low elevation angles.

3.1.1 Second order term

Higher order ionospheric terms include the second and third order ionospheric terms, and the excess path length. Equations (5, 6, 8) indicate that the second order term depends on the electron density n_e and as well as on the geomagnetic induction B. The electron gyro frequency $f_g = eB/(2\pi m)$ is usually less than 1.4 MHz. The value of B can be derived as ~5x10^{-5} Tesla for f_g = 1.4 MHz and considered constant throughout the propagation. Thus, for the worst case condition with f_g = 1.4 MHz and $\Theta = 0$, the second order term can be simplified as (using Eqs. 5 and 8)

$$d_{Igr}^{(2)} = \frac{11.28 \times 10^{7}}{f^{3}} TEC \tag{30}$$

where $d_{Igr}^{(2)}$ is measured in meters, TEC in electrons/m^2 and f in Hz. Using the above approximation, the frequency dependence of the second order term at different levels of ionospheric ionization and elevation angles has been plotted in Fig. 3.

Figure 3 shows that during the worst case conditions the second order ionospheric term on code observables can be as big as about 500 millimeters at GNSS L-band frequencies. Due to the dependency on B field, the second order term depends on the receiver's geographic / geomagnetic position and direction of the signal reception. Such dependencies are given in Fig. 4. The simulation has been made using a two dimensional ray tracing program (Hoque & Jakowski, 2008) which includes International Geomagnetic Reference Field (IGRF) (Mandea & Macmillan, 2000) model for magnetic field computation along ray paths. A single layered Chapman profile (Eq. 40, see Rishbeth & Garriott, 1969) with a maximum ionization of 4.96×10^{12} m^{-3} at 350 km altitude and atmospheric scale height of 70 km has been used, and the corresponding vertical TEC is found 143 TECU.

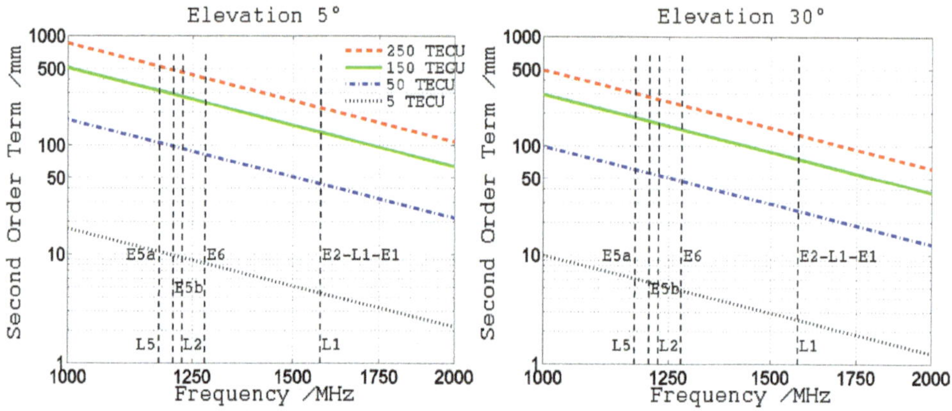

Fig. 3. Frequency dependence of the second order term $d_{Igr}^{(2)}$ at different levels of ionospheric ionization and elevation angles

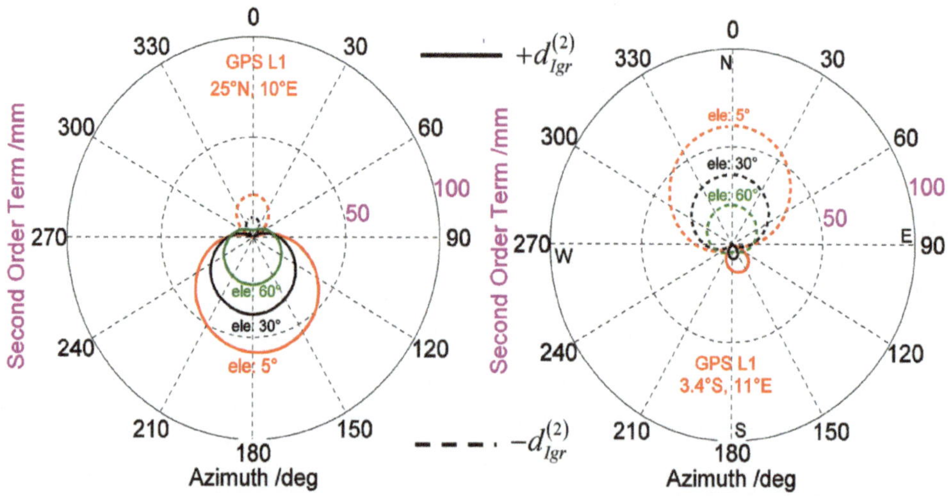

Fig. 4. Azimuth dependency of the second order term at GPS L1 frequency for elevations 5°, 30° and 60° and VTEC = 143 TECU. The receiver position is considered at geographic 25° N, 10° E and at its geomagnetic conjugate position 3.4° S, 11° E. The symbols N, E, S and W correspond to the geographic north, east, south and west directions, respectively

For a GNSS user in the northern hemisphere the magnitude of the second order term is the largest when the signal is received from a satellite in southward direction. However, for a user in the southern hemisphere the scenario is reversed, i.e., the largest effect is observed when the signal is received from a satellite in northward direction. Figure 4 shows that the magnitude of the second order term and its sign differ depending on the user location on the earth and direction of the signal reception. Therefore, such non systematic effects cannot be cancelled out by averaging GNSS measurements over long period at a certain user location.

3.1.2 Third order term

The third order term depends on the integral $\int n_e^2 ds$ (see Eq. 9) which can be simplified as $0.6577 N_m TEC$ (obtained by analytical integration of the Chapman layer, Hoque & Jakowski, 2008, see also Brunner & Gu, 1991; Hartmann & Leitinger, 1984; Leitinger & Putz, 1988) where N_m is the maximum ionospheric ionization. Therefore, assuming the worst case condition with f_g = 1.4 MHz and $\Theta = 0$, the third order term can be simplified as (using Eqs. 5 and 9)

$$d_{Igr}^{(3)} = \left(1602.81 N_m + 2.37 \times 10^{14}\right) \frac{TEC}{f^4} \tag{31}$$

where $d_{Igr}^{(3)}$ is measured in meters, TEC is the slant TEC and measured in electrons/m^2, f in Hz and N_m is measured in m^{-3}. If the vertical TEC is known, N_m can be computed assuming a Chapman profile for the ionosphere by the following expression (Hoque & Jakowski, 2007).

$$VTEC = 4.13 H N_m \tag{32}$$

where VTEC is the TEC in vertical direction and H is the atmospheric scale height. The parameter H can be assumed as 70 km for a rough estimation of the third order ionospheric term. Using the above approximations (Eqs. 31 and 32) the frequency dependence of the third order term at different levels of ionospheric ionization and elevation angles has been plotted in Fig. 5.

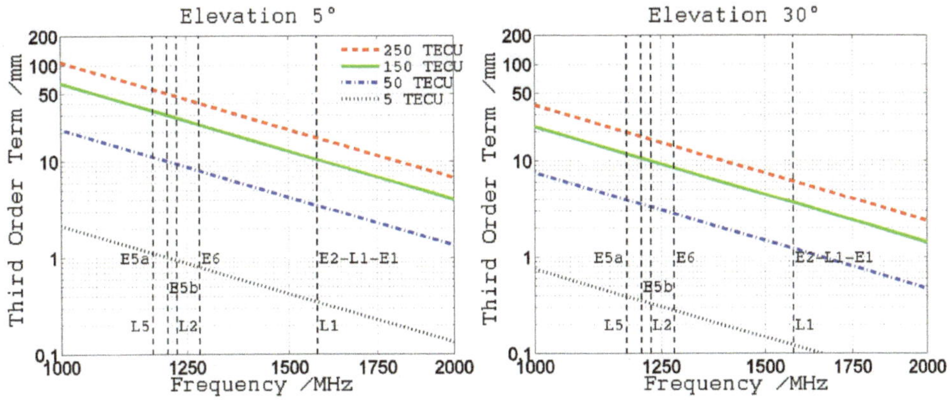

Fig. 5. Frequency dependence of the third order term $d_{Igr}^{(3)}$ at different levels of ionospheric ionization and elevation angles

Figure 5 shows that the third order term on code observables can be as big as 50 mm at low elevation angles during times of high TEC.

3.2 Estimate of LoS propagation assumption error

Due to the ray path bending satellite signals propagate through curvature paths instead of straight line of sight paths. However, a curvature path length and the corresponding LoS path length are not equal rather the curvature path is slightly longer than the LoS one. The difference between them is defined as the excess path length and it can be computed by the following formula given by Jakowski et al. (1994).

$$d_I^{len} = \frac{b_1}{f^4}\left(\frac{1}{\left(1 - b_2 \cos^2 \beta\right)^{1/2}} - 1\right) TEC^2 \tag{33}$$

where $b_1 = 2.495 \times 10^8$, $b_2 = 0.8592$ and β is the elevation angle. The excess path length d_I^{len} will be estimated in millimeters when β is measured in radians, f is in MHz and TEC is in TEC units. The frequency dependence of the excess path length has been plotted in Fig. 6.

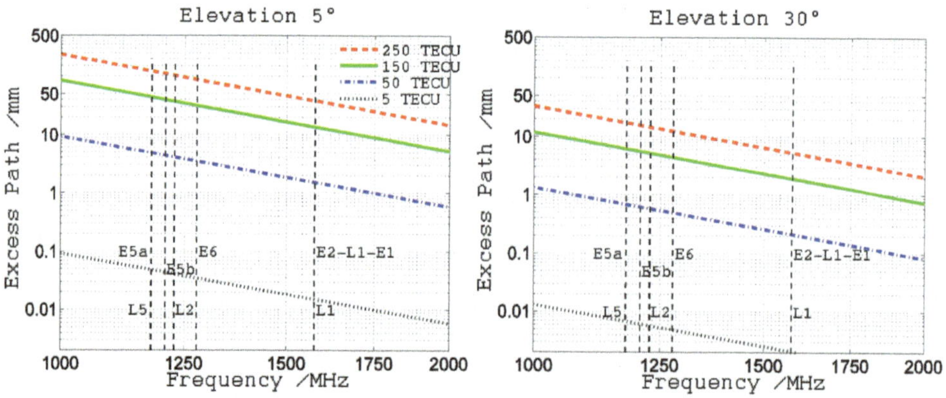

Fig. 6. Frequency dependence of the excess path length at different levels of ionospheric ionization and elevation angles

Figure 6 shows that at the L2 frequency, the excess path length can be as big as 100 mm at low elevation angles during times of high TEC such as VTEC = 250 TECU.

3.3 Estimates of residual terms in the ionosphere-free solution

Although residual terms in ionosphere-free solutions are less than 1% of the first order ionospheric effect, they cannot be ignored if centimeter / millimeter level accuracy is required in GNSS positioning and timing applications. A plot showing comparison of dual-frequency GPS L1-L2 residual terms is given in Fig. 7 for better understanding of their relative influences on precise range estimation. For this, the ray tracing tool has been used in which the ionosphere is modelled by a Chapman layer with a peak density of 7.75×10^{12} m^{-3} at 350 km altitude and scale height of 78 km, and corresponding VTEC = 250 TECU.

We see that at low elevation angle (< 15°), Δs_{TEC} is the largest and it decreases very rapidly with increasing the elevation angle. The second order term Δs_2 is determined for an azimuth angle 180° at a receiver position at geographic 50° N and 15° E. Although Δs_2 is less than the Δs_{TEC} at low elevation angles, it exceeds Δs_{TEC} at higher elevation angles (> 25°). The Δs_2 does not reduce significantly with increasing the elevation angle and therefore, it cannot be ignored even at zenith. The excess path length Δs_{len} decreases with increasing the elevation angle very rapidly and vanishes at zenith. The third order term Δs_3 is small (< 5 mm) but it can be bigger than Δs_{TEC} and Δs_{len} at very high (> 60°) elevation angles. We find that the magnitude of the RRE_{gr} is much higher than the RRE. This is mainly due to different signs of Δs_{TEC} and Δs_{len} in the RRE and RRE_{gr} expressions. In case of the code combination Eq. (14), Δs_{len} is additive to other terms whereas it is subtractive in the carrier-phase combination Eq. (15). Additionally, the Δs_2 and Δs_3 are two and three times higher in the code combination compared to the phase combination.

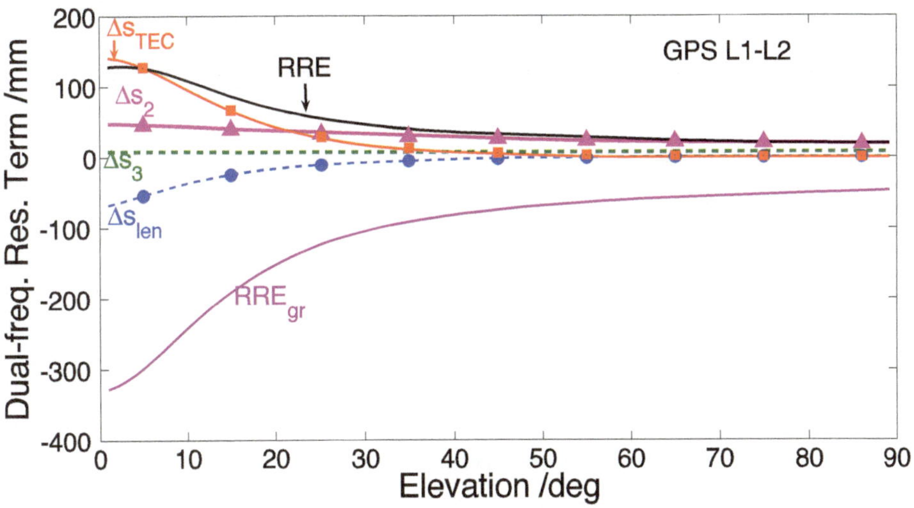

Fig. 7. Residual terms in the dual-frequency GPS L1-L2 combination for an ionospheric ionization of VTEC = 250 TEC units

For the same ionospheric ionization, the residual terms in the triple-frequency GPS L1-L2-L5 combination (Eqs. 21-22) are plotted in Fig. 8. It shows that the magnitude of $(RRE_{gr})_{tr}$ is much higher than the magnitude of $(RRE)_{tr}$. The reason is already discussed for the dual-frequency case.

We find that the GPS L1-L2 residual terms Δs_3, Δs_{TEC} and Δs_{len} are about 2.4 times higher than the GPS L1-L2-L5 residual terms $(\Delta s_3)_{tr}$, $(\Delta s_{TEC})_{tr}$ and Δs_{len}. The sum of all residual terms, i.e., RRE and RRE_{gr} are found to be more than three times higher for the L1-L2 combination than the L1-L2-L5 combination.

Comparing the dual- and triple-frequency carrier-phase combinations Eq. (15) and Eq. (22), we see that the signs of Δs_{TEC} and Δs_3 are positive in the dual-frequency combination whereas their signs are negative in the triple-frequency combination. However, the sign of

Δs_{len} is negative in the dual-frequency combination and positive in the triple-frequency combination. Again, the magnitude of Δs_{TEC} is higher than the magnitude of Δs_{len} for both combinations. As a result, the triple-frequency $(RRE)_{tr}$ is found to be negative, i.e., the corrected ρ is longer than the uncorrected one whereas the dual-frequency RRE is found to be positive, i.e., the corrected ρ is shorter than the uncorrected ρ. Similarly, it can be shown that $(RRE_{gr})_{tr}$ is positive in the triple-frequency combination and RRE_{gr} is negative in the dual-frequency combination. These relations are true for combination frequencies $f_1 > f_2 > f_3$.

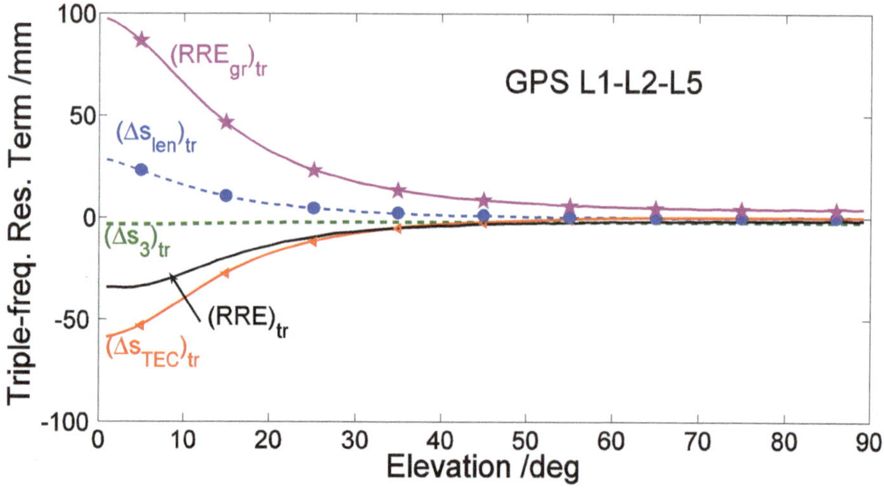

Fig. 8. Residual terms in the triple-frequency GPS L1-L2-L5 combination for an ionospheric ionization of VTEC = 250 TEC units

4. Correction of higher order ionospheric terms

4.1 Second order term correction

The estimation of the second order term requires computation of the geomagnetic induction and its direction with respect to the propagation direction along ray paths. Since this computation is very cumbersome, a common practice is to assume the ionosphere as a single thin layer at a certain altitude and compute $B\cos\Theta$ at the IPP and consider it constant throughout the propagation. Thus, the second order term coefficient q (Eq. 8) can be written as

$$q \approx 2.2566 \times 10^{12} B^* \cos\Theta^* TEC \tag{34}$$

where TEC is the total electron content along ray paths, B^* is the magnitude of B, and Θ^* is the angle between the magnetic field vector and the wave direction at the IPP (the symbol * denotes the values at the single layer).

Bassiri and Hajj (1993) were the first to propose such a single thin layered ionosphere for the second order ionospheric correction; they choose the 300 km as a representative global average peak height. Later Kedar et al. (2003) assumed the ionosphere as a single layer at a 400 km altitude for estimating the effect of the second order GPS ionospheric correction on receiver positions. Hernandez-Pajares et al. (2007) considered the ionosphere as a single layer at a 450 km altitude to estimate the impact of the second order ionospheric term on geodetic estimates.

The computation of $B\cos\Theta$ along ray paths requires the knowledge of the ionospheric profile shape which is not available to the GNSS users; they only have TEC information along ray paths. Therefore, assumptions of a thin ionospheric layer and $B\cos\Theta$ computation at the IPP are very suitable for practical use. However, such assumptions lead up to 2 mm errors in the second order ionospheric term computation (Hoque & Jakowski, 2008).

As an alternative approach, we (Hoque & Jakowski, 2007) assume an average value $\overline{B\cos\Theta}$ for the magnetic field component and consider it constant throughout the propagation. Based on simulation studies we derived a correction formula for the $\overline{B\cos\Theta}$ computation along any receiver-to-satellite link geometries inside European geographic latitude 30 - 65° N and longitude 15° W - 45° E.

$$\Delta s_2 = \frac{1.1283\times 10^{12}}{f_1 f_2(f_1+f_2)}\cdot \overline{B\cos\Theta}\cdot TEC \tag{35}$$

In which

$$\overline{B\cos\Theta} = -y_1\cos\alpha + \left|\sqrt{r_1^2 - y_1^2\sin^2\alpha}\right| - 2r_2\cos\alpha' \tag{36}$$

$$\begin{aligned} r_1 &= \varsigma(\beta,\phi,\lambda,a_i) \\ r_2 &= \varsigma(\beta,\phi,b_i) \\ y_1 &= \varsigma(\beta,\phi,c_i) \end{aligned} \tag{37}$$

The parameters r_1, r_2 and y_1 are the functions of the receiver-to-satellite elevation angle β, geographic latitude φ and longitude λ at the receiver position. The quantity α is the receiver-to-satellite azimuth angle and α' is the modified azimuth angle. The quantities a_i, b_i and c_i are the polynomial coefficients. Thirty polynomial coefficients have been derived for the European region (30° - 65° N, 15° W- 45° E) by least squares fitting of ray tracing results. Inside the ray tracing program, the IGRF model has been used to compute $B\cos\Theta$ along ray paths. For details and values of the polynomial coefficients we refer to Hoque & Jakowski (2007). Using such a correction formula and knowing the TEC value, the second order term can be corrected to the 2-3 millimeter accuracy level for a vertical TEC level of 100 TEC units. The formula can be adapted for other geographic regions too after deriving new set of polynomial coefficients.

4.2 Third order term correction

It has been found that the second term of Eq. (9) is less than the first term by about 1-2 orders of magnitude. As already discussed in the section 3.1.2, the integral $\int n_e^2 ds$ can be simplified as $0.6577N_mTEC$. Thus, the third order residual term can be approximated by the first term only as (using Eqs. 9 and 19)

$$\Delta s_3 = \frac{534.27}{f_1^2 f_2^2} N_m TEC \tag{38}$$

The third order term Δs_3 will be measured in meters when f is measured in Hz and the maximum ionization N_m and TEC in m^{-3} and electrons/m^2, respectively.

4.3 New approaches for correcting LoS propagation assumption errors

4.3.1 Excess path length correction

As we have seen in the section 3.2, the excess path length d_I^{len} can be computed by Eq. (33). There is another formula published by Hoque & Jakowski (2008) for the excess path length computation.

$$d_I^{len} = \frac{7.5 \times 10^{-5} \exp(-2.13\beta) TEC^2}{f^4 H(hm)^{1/8}} \tag{39}$$

where d_I^{len} is measured in meters, TEC is in TEC units, frequency f in GHz, atmospheric scale height H and maximum ionization height hm in kilometers and elevation β in radians. Comparing both formulas we see that Eq. (33) requires only TEC and elevation information as inputs whereas Eq. (39) additionally requires ionospheric parameters H and hm. However, these parameters are not easy to estimate in practical cases.

Both the correction formulas are derived based on simulation studies using Chapman profiles for the ionosphere. The Chapman profile (Rishbeth & Garriott, 1969) has been proved very useful for modeling ionospheric correction. It describes the electron density distribution n_e as a function of height h in the ionosphere as

$$n_e(h) = Nm \exp(0.5(1 - z - \exp(-z))) \tag{40}$$

where Nm is the maximum ionization and $z = (h - hm)/H$ in which hm is the height of maximum ionization and H is the atmospheric scale height.

We have found that the correction by Eq. (33) shows the best performance for the atmospheric scale height H = 70 km. However, when the scale height is too low (e.g., H = 60 km) or too large (e.g., H = 80 km), its performance degrades especially at low elevation angles (see Fig. 9). Our present investigation shows that its performance can be improved by taking into account the d_I^{len} dependency on the rate of change of TEC with respect to the elevation angle. In order to find their dependencies, the excess path length has been computed by the ray tracing program considering Chapman profiles with different H = 60 and 80 km. The signal frequency f = 1227.6 MHz, parameters hm = 350 km and Nm = 4.96×10^{12} m^{-3} are kept constant in each case. The total electron content in the vertical direction is found 123 and 164 TEC units, respectively. The obtained d^{len}, TEC, and the first and second order TEC derivatives with respect to the elevation angle $dTEC/d\beta$ and

$d^2TEC/d\beta^2$ have been plotted as functions of elevation angle in Fig. 9. The $dTEC/d\beta$ has been calculated dividing the TEC difference between two measurement epochs by the corresponding elevation angle difference. Then, $d^2TEC/d\beta^2$ has been calculated dividing the $dTEC/d\beta$ difference between two measurement epochs by the corresponding elevation angle difference.

Comparing plots in Fig. 9, we see that although the dependency of d^{len} on the $dTEC/d\beta$ is not straight forward, its dependency on the $d^2TEC/d\beta^2$ is obvious at low elevation angles (< 20°). Thus, the magnitude of the d^{len} depends on the magnitude of TEC as well as on the magnitude of $d^2TEC/d\beta^2$. Considering this, functional dependencies have been studied separately for different parameters to develop correction formulas. For this, ray tracing calculation has been carried out for different geometrical and ionospheric conditions varying elevation and Chapman layer parameters H, Nm and hm. Thus, the following formula has been obtained for the d_I^{len} correction.

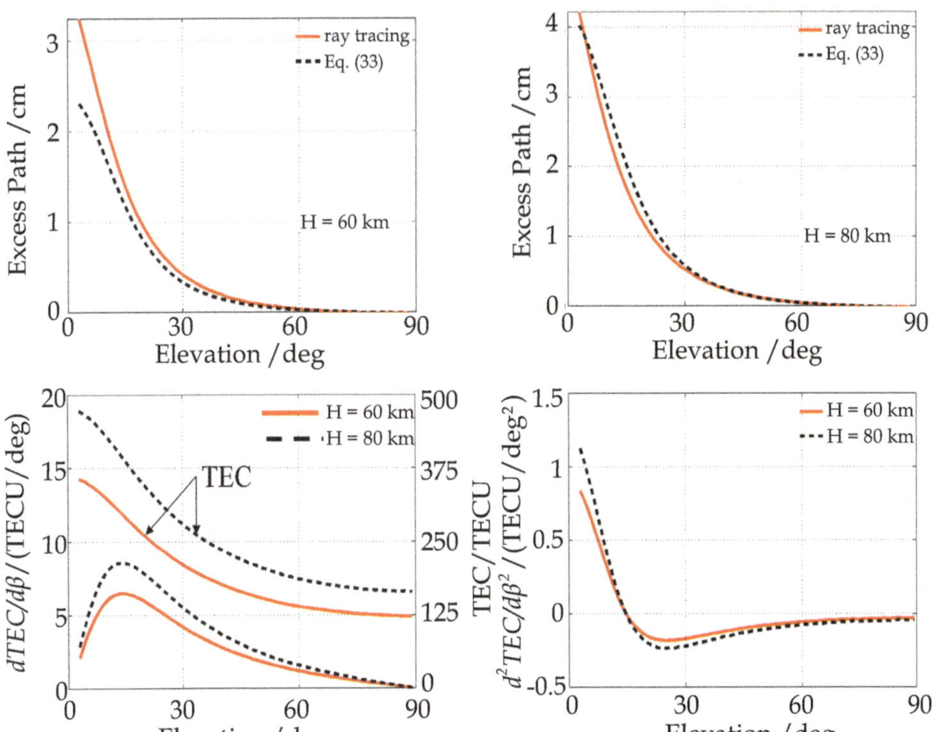

Fig. 9. Elevation angle dependence of d_I^{len}, TEC (see right scale), $dTEC/d\beta$ and $d^2TEC/d\beta^2$ for the Chapman layer parameter H = 60 and 80 km

$$d_I^{len} = \frac{a_1}{f^4}\left(\frac{1}{\left(1-a_2\cos^2\beta\right)^{1/2}}-1\right)TEC^2 + a_3\left(\frac{d^2TEC}{d\beta^2}\right)^2\cos\beta \qquad (41)$$

where $a_1 = 2.6123\times10^8$, $a_2 = 0.8260$, $a_3 = 6.64$. The d_I^{len} will be computed in millimeters when β is measured in radians, f is in MHz, TEC is in TEC units and $d^2TEC/d\beta^2$ in TECU/deg^2. The polynomial coefficients are derived based on a nonlinear fit with ray tracing results in least square senses.

The elevation angle dependence of d_I^{len} has been plotted in Fig. 10 using the proposed correction formula Eq. (41) as well as by Eqs. (33) and (39). In addition, ray tracing results are plotted for comparisons. Comparing d_I^{len} computed by the Eq. (33) and Eq. (41) with ray tracing results, we see that at higher H values (e.g., H = 80 km) the correction given by the Eq. (41) performs better. However, its performance degrades at lower H values (e.g., H = 60 km), especially around 7 - 21° elevation angle.

We find that the Eq. (39) gives the best performance. However, it requires ionospheric parameters H and hm as inputs which are not known to the GNSS users. Inaccurate assumption of ionospheric parameters may give erroneous estimation of d_I^{len}. We see that at H = 80 km the correction given by the new approach Eq. (41) is even comparable to the correction given by Eq. (39).

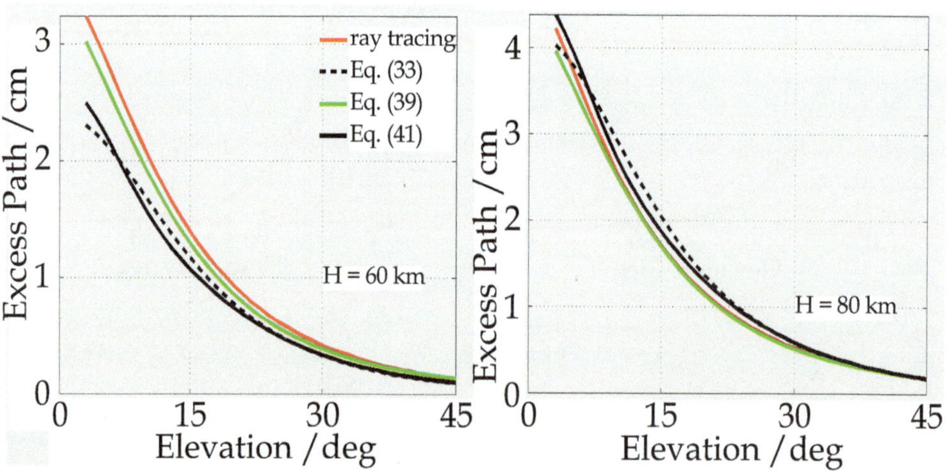

Fig. 10. Comparison of excess path length correction formulas with ray tracing results

While using the proposed correction Eq. (41), it should be remembered that due to its dependency on the $d^2TEC/d\beta^2$ term, it is very sensitive to TEC gradients or irregularities. In such cases the correction given by Eq. (33) is recommended for use.

4.3.2 ΔTEC_{bend} correction

Due to the ray path bending satellite signals propagate in curvature paths instead of straight LoS paths. However, TECs along a curvature path and the corresponding LoS path are not the same rather the TEC along the curvature path is slightly larger than the LoS one. The difference between them is defined as the ΔTEC_{bend} (see Eq. 7). The ΔTEC_{bend} can be computed by the following formula given by Hoque & Jakowski (2008).

$$\Delta TEC_{bend} = \frac{1.108\times10^{-3}\exp(-2.1844\beta)TEC^2}{f^2 H hm^{0.3}} \tag{42}$$

where ΔTEC_{bend} is measured in TECU, atmospheric scale height H is in km, the maximum ionization height hm is in km, signal frequency f is in GHz, TEC is in TECU and elevation angle β is in radians. Again, it requires the knowledge of the ionospheric parameters H and hm. If actual parameters are not known, the formula may not be useful in practical purposes. Therefore, in the present work, we have looked for a correction formula depending only on the TEC, elevation angle and second order derivative of TEC with respect to the elevation. We have found that the formula Eq. (41) can be used for such purposes multiplying simply by f^2 and determining new coefficients.

$$\Delta TEC_{bend} = \frac{c_1}{f^2}\left(\frac{1}{\left(1-c_2\cos^2\beta\right)^{1/2}}-1\right)TEC^2 + c_3\left(\frac{d^2TEC}{d\beta^2}\right)^2\cos\beta \tag{43}$$

where $c_1 = 1.2963$, $c_2 = 0.8260$, $c_3 = 0.0496$. The ΔTEC_{bend} will be computed in TEC units when β is measured in radians, f is in MHz and TEC is in TEC units and $d^2TEC/d\beta^2$ in TECU/deg^2. The polynomial coefficients are derived based on a nonlinear fit with ray tracing results in least square senses as before.

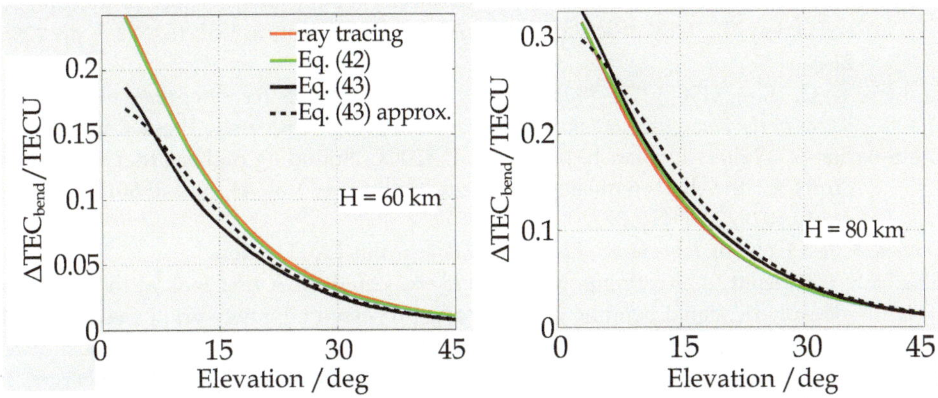

Fig. 11. Comparison of ΔTEC_{bend} correction formulas with ray tracing results

The elevation angle dependence of ΔTEC_{bend} has been plotted in Fig. 11 for the proposed correction formula Eq. (43) as well as for the Eq. (42). Also ray tracing results are plotted for comparisons. Comparing ΔTEC_{bend} computed by the Eq. (43) and Eq. (42) with ray tracing results, we see that at higher H values (e.g., H = 80 km) both correction results are comparable. However, the performance of the new approach significantly degrades at lower H values (e.g., H = 60 km).

As already mentioned, the derivative $d^2TEC/d\beta^2$ is very sensitive to TEC gradients. Considering this, another set of coefficients have been determined excluding the derivative term in Eq. (43). In this case, we have found that $c_1 = 1.4563$, $c_2 = 0.8260$ and c_3 is set to zero.

The elevation angle dependency for such an approximation of Eq. (43) is also plotted in Fig. 11.

As already mentioned, after removal of the Selective Availability, the ionosphere becomes the single largest error source for GNSS error budgets. Fortunately, a dual-frequency ionosphere-free combination can remove about 99% of the ionospheric effects; thanks to the dispersive nature of the ionosphere. Although higher order residual terms are less than 1% of the first order term, they can be many centimeters during times of high TEC and represent large errors in geodetic measurements especially in precise point positioning. This chapter gives estimation of higher order ionospheric terms at different level of ionospheric ionization and discusses different correction approaches for them.

5. References

Appleton, E. V. (1932). Wireless studies of the ionosphere, *Proceeding of Instn. Elect. Engrs.*, Vol. 7, No. 21, pp. (257-265), 10.1049/pws.1932.0027

Bassiri, S. & Hajj, G. A. (1993). Higher-order ionospheric effects on the global positioning system observables and means of modeling them. manuscripta geodaetica, Vol. 18, No. 6, pp. (280-289)

Blewitt, G. (1987). New approaches to GPS carrier-phase ambiguity resolution. Proceeding of XIX General Assembly of the IUGG, Vancouver, Canada, August 10-22

Brunner, F. K. & Gu, M. (1991). An improved model for the dual frequency ionospheric correction of GPS observations. manuscripta geodaetica, Vol. 16, No. 3, pp. (205-214)

Budden, K. G. (Ed.). (1985). *The Propagation of Radio Waves: the theory of radio waves of low power in the ionosphere and magnetosphere*, Cambridge University Press, Cambridge.

Datta-Barua, S., Walter, T., Blanch, J. & Enge, P. (2008). Bounding higher-order ionosphere errors for the dual-frequency GPS user. Radio Sci., Vol. 43, No. RS5010, pp. (15), doi:10.1029/2007RS003772

Davies, K. (Ed.). (1990). *Ionospheric Radio*, Peter Peregrinus Ltd, London.

Elizabeth, J. P., Matt, A. K., Philip, M. & David, A. L. (2010). A first look at the effects of ionospheric signal bending on a globally processed GPS network. J Geod, Vol. 84, pp. (491–499), DOI 10.1007/s00190-010-0386-2

Fritsche, M., Dietrich, R., Knöfel, C., Rülke, A., Vey, S., Rothacher, M. & Steigenberger, P. (2005). Impact of higher-order ionospheric terms on GPS estimates. Geophys Res Lett, Vol. 32, No. 23, L23311, DOI 10.1029/2005GL024342

Hartmann, G. K. & Leitinger, R. (1984). Range errors due to ionospheric and tropospheric effects for signal frequencies above 100 MHz. Bull. Geod, Vol 58, No. 2, pp. (109-136)

Hawarey, M., Hobiger, T. & Schuh, H. (2005). Effects of the 2nd order ionospheric terms on VLBI measurements. Geophys Res Lett, Vol. 32, No. 11, L11304, DOI 10.1029/2005GL022729

Hernandez-Pajares, M., Jaun, J. M., Sanz, J. & Orus, R. (2007). Second order ionospheric term in GPS: implementation and impact on geodetic estimates. Journal of Geophysical Research, Vol. 112, No. B08417, doi:10.1029/2006JB004707

Hoque, M. M. & Jakowski, N. (2006). Higher-order ionospheric effects in precise GNSS positioning. Journal of Geodesy, Vol. 81, No. 4, pp. (259-268), DOI 10.1007/s00190-006-0106-0

Hoque, M. M. & Jakowski, N. (2007). Mitigation of higher order ionospheric effects on GNSS users in Europe. GPS Solut., Vol. 12, No. 2, doi: 10.1007/s10291-007-0069-5

Hoque, M. M. & Jakowski, N. (2008). Estimate of higher order ionospheric errors in GNSS positioning, Radio Sci., Vol. 43, No. RS5008, doi: 10.1029/2007RS003817

Hoque, M. M. & Jakowski, N. (2010a). Higher Order Ionospheric Errors in Modernized GPS and Future Galileo Systems. In: Global Positioning Systems, Asphaug V. & Sorensen E. (Eds.), pp. (1-28), Nova Science Publishers, Inc., ISBN 978-1-60741-012-6

Hoque, M. M. & Jakowski, N. (2010b). Higher order ionospheric propagation effects on GPS radio occultation signals. J. Adv. Space Res., Vol. 46, No. 2, pp. (162-173), doi:10.1016/j.asr.2010.02.013

Hoque, M. M. & Jakowski N. (2011). Ionospheric bending correction for GNSS radio occultation signals. Radio Sci, Vol. 46, No. RS0D06, pp. (9), doi:10.1029/2010RS004583

Jakowski, N., Porsch, F. & Mayer, G. (1994). Ionosphere-Induced-Ray-Path Bending Effects in Precise Satellite Positioning Systems. Zeitschrift für Satellitengestützte Positionierung, Navigation und Kommunikation, Vol. SPN 1/94, pp. (6-13)

Jakowski, N., Mayer, C., Hoque, M. M. & Wilken, V. (2011). TEC Models and Their Use in Ionosphere Monitoring. Radio Sci., doi:10.1029/2010RS004620

Kedar, S., Hajj, G., Wilson, B. & Heflin, M. (2003). The effect of the second order GPS ionospheric correction on receiver positions. Geophys Res Lett, Vol. 30, No. 16, pp. (1829), DOI 10.1029/2003 GL017639

Kim, B. C. & Tinin M. V. (2007). Contribution of ionospheric irregularities to the error of dual-frequency GNSS positioning. J Geod, Vol. 81, pp. (189–199), DOI 10.1007/s00190-006-0099-8

Kim, B.C. & Tinin, M.V. (2011). Potentialities of multifrequency ionospheric correction in Global Navigation Satellite Systems. J Geodesy, Vol. 85, No. 3, pp. (159-169), doi: 10.1007/s00190-010-0425.

Klobuchar, J. A. (1996). Ionospheric Effects on GPS, In: *Global Positioning System: Theory and Applications*, Vol I, Parkinson, B. W. & Spilker, J. J. (Eds.), pp. (485-515), American Institute of Aeronautics & Astronautics, ISBN 156347106X

Leitinger R., Putz, E. (1988). Ionospheric refraction errors and observables, In: *Atmospheric effects on the geodetic space measurements*, Brunner, F. K. (Ed.), pp. (81-102), Monograph 12, School of Surveying, UNSW, Sydney

Mandea M. & Macmillan, S. (2000). International Geomagnetic Reference Field—the eighth generation. *Earth Planets Space*, Vol. 52, No. 12, pp. (1119–1124)

Moore, R. C. & Morton, Y.T. (2011). Magneto-ionic polarization and GPS signal propagation through the ionosphere, Radio Sci, Vol. 46, No. RS1008, doi: 10.1029/2010RS004380

Morton, Y. T., Zhou, Q. & van Graas, F. (2009). Assessment of second-order ionosphere error in GPS range observables using Arecibo incoherent scatter radar measurements, Radio Sci., Vol. 44, No. RS1002, doi:10.1029/2008RS003888

Parkinson, B. W. & Gilbert, S. W. (1983). NAVSTAR: Global Positioning System - Ten Years Later. Proc IEEE, Vol. 71, No. 10, pp (1177-1186), 10.1109/PROC.1983.12745

Rishbeth, H. & Garriott O. K. (Eds.). (1969). *Introduction to ionospheric physics*, Academic, New York

Strangeways, H. J. & Ioannides, R. T. (2002). Rigorous calculation of ionospheric effects on GPS Earth-Satellite paths using a precise path determination methods. Acta Geod Geoph Hung, Vol. 37, No. 2-3, pp. (281-292)

16

GNSS Atmospheric and Ionospheric Sounding – Methods and Results

Shuanggen Jin
Shanghai Astronomical Observatory, Chinese Academy of Sciences
China

1. Introduction

The GPS atmospheric and ionospheric delays have been considered as an error source for a long time. In 1992 when the GPS became fully operational, Ware (1992) suggested limb sounding the Earth atmosphere using GPS atmospheric delay signals. In April 1995, the small research satellite of Microlab-1 was successfully put into a Low Earth Orbit (LEO) to validate the GPS radio occultation method (Feng and Herman, 1999). Since then, the GPS/Meteorology Mission (GPS/MET) has been widely used to produce accurate, all weather pressure, temperature, density profiles in the troposphere and the ionospheric total electron content (TEC) as well as electron density profiles (Rocken, 1997; Hajj and Romans, 1998; Syndergaard, 2000), to improve weather analysis and forecasting, monitor climate change, and monitor ionospheric events. While traditional observing instruments, e.g. water vapour radiometer (WVR), incoherent scatter radars (ISR), ionosonde, topside sounders onboard satellites, in situ rocket and satellite observations, are expensive and also partly restricted to either the bottomside ionosphere or the lower part of the topside ionosphere (usually lower than 800 km), such as ground based radar ionospheric measurements. While GPS satellites in high altitude orbits (~20,200 km) are capable of providing details on the structure of the entire atmosphere, even the plasma-sphere. Moreover, GPS is a low-cost, all-weather, near real time, and high-temporal resolution (1~30s) technique. Therefore, GPS is a powerful tool to sound the atmosphere and ionosphere as well as their application in meteorology, climate and space weather.

2. Tropospheric sounding

The tropospheric delay of GPS signal through the neutral atmosphere was one of major error sources in navigation and positioning, which contributes a bias in height of several centimetres (Tregoning et al. 1998). Nowadays, GPS has been used to determine the zenith tropospheric delay (ZTD) (Jin and Park, 2005) through mapping functions (Niell, 1996). The ZTD is the integrated refractivity along a vertical path through the neutral atmosphere:

$$ZTD = c\tau = 10^{-6}\int_0^{\infty} N(s)ds \quad (1)$$

where c is the speed of light in a vacuum, τ is the delay measured in units of time and N is the neutral atmospheric refractivity. The N is empirically related to standard meteorological variables as (Davis et al. 1985)

$$N = k_1 \rho + k_2 \frac{P_w}{Z_w T} + k_3 \frac{P_w}{Z_w T^2} \tag{2}$$

where $k_i (i = 1,2,3)$ is constant, ρ is the total mass density of the atmosphere, P_w is the partial pressure of water vapor, Z_w is a compressibility factor near unity accounting for the small departures of moist air from an ideal gas, and T is the temperature in degrees Kelvin. The integral of the first term of equation (2) is the hydrostatic component (N_h) and the integral of the remaining two terms is the wet component (N_w). Thus, ZTD is the sum of the hydrostatic or dry delay (ZHD) and non-hydrostatic or wet delay (ZWD), respectively. The dry component ZHD is related to the atmospheric pressure at the surface, while the wet component ZWD can be transformed into the precipitable water vapor (PWV) and plays an important role in energy transfer and in the formation of clouds via latent heat, thereby directly or indirectly influencing numerical weather prediction (NWP) model variables (Bevis et al, 1994; Tregoning et al. 1998; Manuel et al., 2001). Therefore, the Zenith Tropopospheric Delay (ZTD) is an important parameter of the atmosphere, which reflects the weather and climate processes, variations, and atmospheric vertical motions, etc.

In the last decade, ground-based GPS receivers have been developed as all-weather, high spatial-temporal resolution and low-cost remote sensing systems of the atmosphere (Bevis et al., 1994; Manuel et al., 2001), as compared to conventional techniques such as satellite radiometer sounding, ground-based microwave radiometer, and radiosondes (Westwater, 1993). With independent data from other instruments, in particular water vapor radiometers, it has been demonstrated that the total zenith tropospheric delay or integrated water vapor can be retrieved using ground based GPS observations at the same level of accuracy as radiosondes and microwave radiometers (Elgered et al. 1997; Tregoning et al. 1998). Currently, the International GPS Service (IGS) has operated a global dual frequency GPS receiver observation network with more than 350 permanent GPS sites since 1994 (Beutler et al., 1999). It provides a high and wide range of scales in space and time to study seasonal and secular variations of ZTD as well as its possible climate processes.

2.1 GPS data and analysis

The IGS (International GPS Service) was formally established in 1993 by the International Association of Geodesy (IAG), and began routine operations on January 1, 1994 (Beutler et al. 1999). The IGS has developed a worldwide network of permanent tracking stations with more than 350 GPS sites, and each equipped with a GPS receiver, providing raw GPS orbit and tracking data as a data format called Receiver Independent Exchange (RINEX). All available near-real-time global IGS observation data archived in the Global IGS Data Center, which contributes to geodesy and atmosphere research activities in a global scale. Here the globally distributed 150 IGS sites with better continuous observations are selected with spanning at least four years of measurements (Figure 1), and most sites observations are from 1994 to 2006.

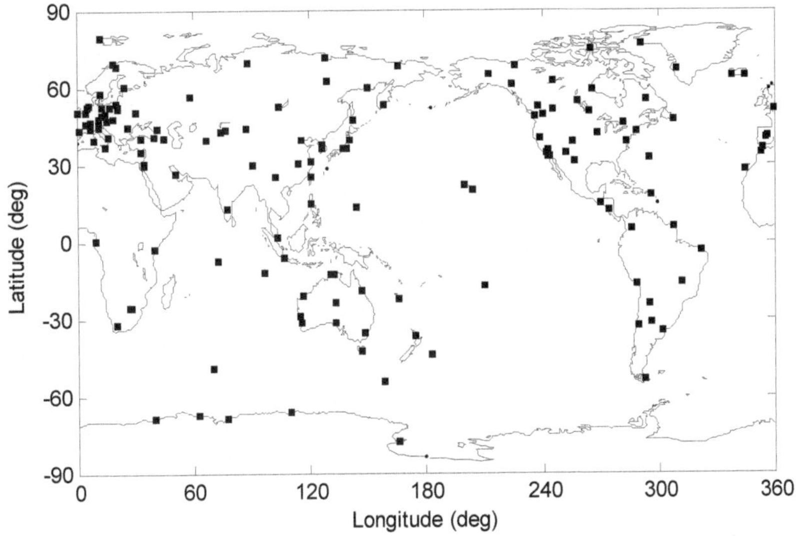

Fig. 1. The distribution of global IGS GPS sites

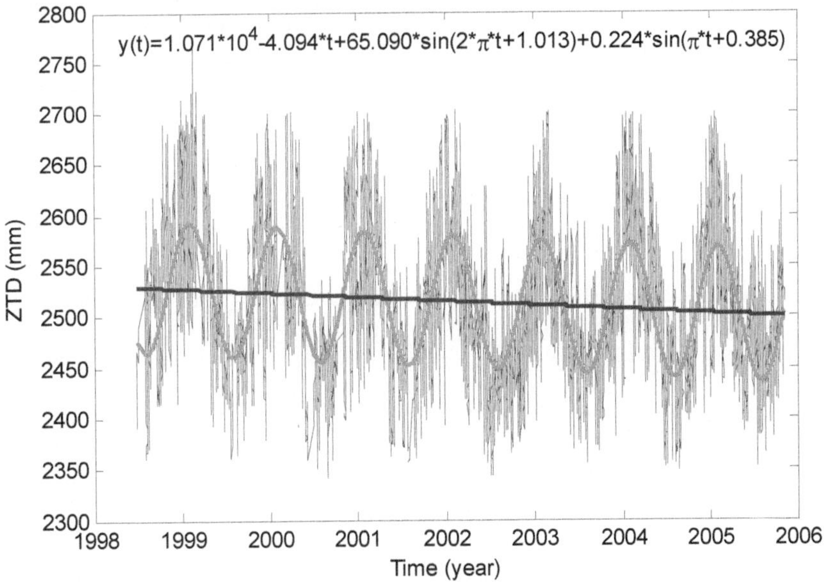

Fig. 2. ZTD time series at TOW2 station, Australia. The solid line is the fitting results, consisting of a linear decrease and seasonal components

2.2 Zenith tropospheric delay retrieval

The processing software must resolve or model the orbital parameters of the satellites, solve for the transmitter and receiver positions, account for ionospheric delays, solve for phase cycle ambiguities and the clock drifts in addition to solving for the tropospheric delay parameters of interest. We use the GAMIT software (King and Bock 1999), which solves for the ZTD and other parameters using a constrained batch least squares inversion procedure. In addition, this study uses the newly recommended strategies (Byun et al. 2005) to calculate ZTD time series with temporal resolution of 2 hours from 1994 to 2006. The GAMIT software parameterizes ZTD as a stochastic variation from the Saastamoinen model (Saastamoinen 1972), with piecewise linear interpolation in between solution epochs. GAMIT is very flexible in that it allows a priori constraints of varying degrees of uncertainty. The variation from the hydrostatic delay is constrained to be a Gauss-Markov process with a specified power density of $2\text{cm}/\sqrt{hour}$, referred to below as the "zenith tropospheric parameter constraint". We designed a 12-hour sliding window strategy in order to process the shortest data segment possible without degrading the accuracy of ZTD estimates. The ZTD estimates are extracted from the middle 4 hours of the window and then move the window forward by 4 hours. Finally, the ZTD time series from 1994 to 2006 are obtained at globally distributed 150 IGS sites with temporal resolution of 2 hours. For example, Figure 2 shows the times series of zenith total delay (ZTD) (upper) at TOW2 station, Australia.

2.3 Global mean zenith tropospheric delay

The ZTD consists of the hydrostatic delay (ZHD) and wet delay (ZWD). The ZHD can be well calculated from surface meteorological data, ranging 1.5-2.6 meters, which accounts for 90% of ZTD. It derives from the relationship with hydrostatic equilibrium approximation for the atmosphere. Under hydrostatic equilibrium, the change in pressure with height is related to total density at the height h above the mean sea level by

$$dp = -\rho(h)g(h)dh \tag{3}$$

where $\rho(h)$ and g(h) are the density and gravity at the height h. It can be further deduced as

$$\text{ZHD=kp0} \tag{4}$$

where k is constant (2.28 mm/hPa) and p_0 is the pressure at height h_0 (Davis et al. 1985). It shows that the ZHD is proportional to the atmospheric pressure at the site. The ZWD is highly variable due possibly to varying climate, relating to the temperature and water vapour. The mean ZTD values at all GPS sites are shown in Figure 3 as a color map. It has noted that lower ZTD values are found at the areas of Tibet (Asia), Andes Mountain (South America), Northeast Pacific and higher latitudes (Antarctica and Arctic), and the higher ZTD values are concentrated at the areas of middle-low latitudes. In addition, the ZTD values decrease with increasing altitude, which is due to the atmospheric pressure variations with the height increase. Atmospheric pressure is the pressure above any area in the Earth's atmosphere caused by the weight of air. Air masses are affected by the general atmospheric pressure within the mass, creating areas of high pressure (anti-cyclones) and low pressure (depressions). Low pressure areas have less atmospheric mass above their locations, whereas high pressure areas have more atmospheric mass above their locations. As elevation increases, there are exponentially, fewer and fewer air molecules. Therefore, atmospheric pressure decreases with increasing altitude at a decreasing rate. The following relationship is a first-order approximation to the height (http://www.chemistrydaily.com/chemistry/Atmospheric_pressure):

$$\log_{10} P \approx 5\text{-}\frac{h}{15.5} \tag{5}$$

where P is the pressure in Pascals and h is the height in millimeters. Based on the Eq. (4), ZHD can be expressed as $2.28 * 10^{(5-h/15.5)}$. Therefore, ZTD at all GPS sites can be approximately expressed as:

$$ZTD = 2.28 * 10^{(5-h/15.5)} \tag{6}$$

where the units of ZTD and h are in millimeters, respectively. Comparing GPS-derived ZTD with the empirical formula estimations, it has shown a good consistency.

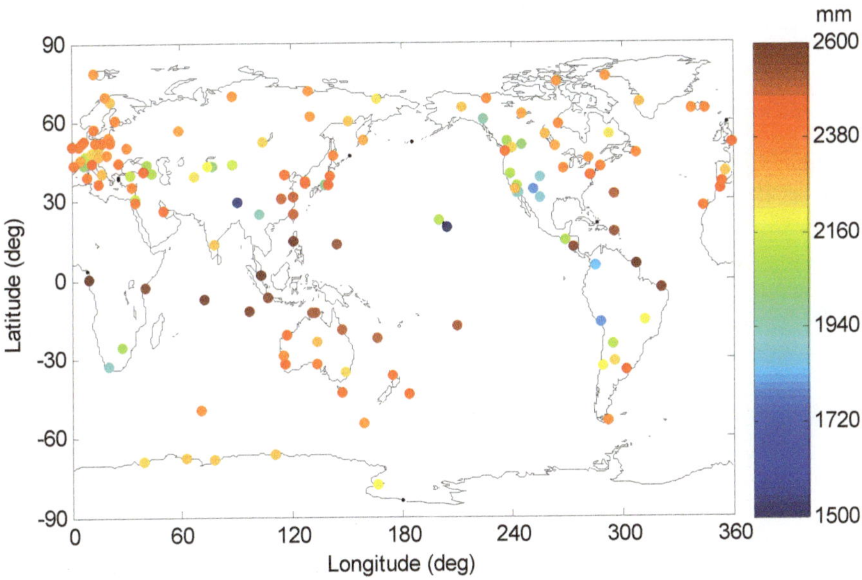

Fig. 3. Distribution of mean ZTD at global IGS sites

2.4 Trend analysis

The GPS ZTD time series have been analyzed for 4-12 years at globally distributed 150 GPS sites. Using the least square the fitting parameters of all GPS site are obtained, including trend and seasonal variation terms (Jin et al., 2007). The mean secular ZTD variation trend is about 1.5±0.001 mm/yr. Figure 4 shows the distribution of the secular ZTD variation trends at all GPS sites as the yearly increase or decrease. It can be seen that the trends are positive in most parts of the Northern Hemisphere and negative in most parts of the Southern Hemisphere (excluding positive in Antarctic), corresponding to a systematic increase or decrease of ZTD. It is interesting to note that the downtrend in Australia is larger than other regions. This downtrend of ZTD is probably due to the highly deserted in Australia. In addition, the ZTD variation trend decreases with increasing altitude, and furthermore, the ZTD trends are almost symmetrical with altitude. This indicates that the secular ZTD variations are larger at the lower altitude and at the higher altitude the secular ZTD variations hardly increase or decrease. In addition, the sum of downward and upward trends at globally distributed GPS sites is almost zero, which possibly indicates that the secular variation is in balance at a global scale, but subjecting to unevenly distributed GPS stations, etc. It need further be confirmed with much denser GPS network in the future. These secular ZTD variation characteristics reflect the total variations of surface atmospheric pressure, temperature and relative humidity, atmospheric vertical motions, etc.

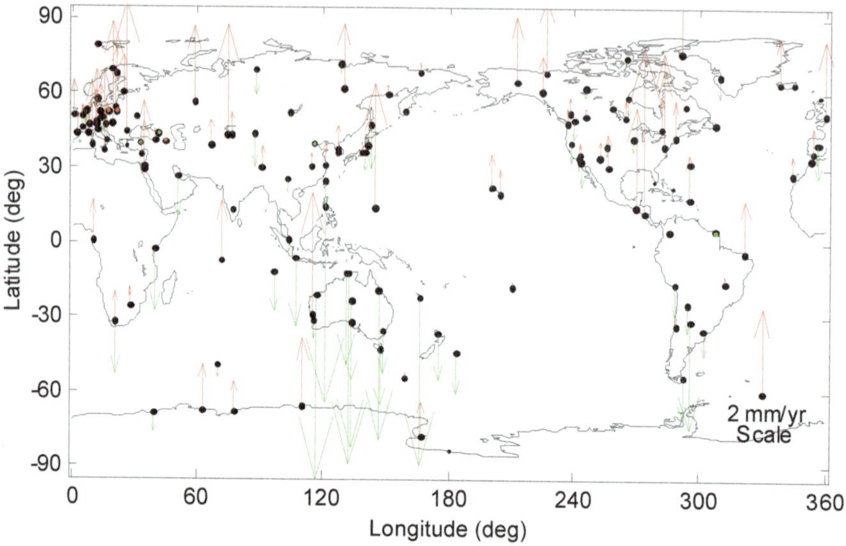

Fig. 4. Secular trend of ZTD variations at global IGS sites. The red upwards arrows represent the increase of secular ZTD variations and the green downwards arrows stand for the decrease of secular ZTD variations

2.5 Seasonal cycles

Meanwhile, the seasonal components are also obtained using least square at annual and semi-annual scales, which can be used to study the seasonal cycle, including amplitude and phase shift. The fitted phase shift is used to determine in which month the seasonal maximum takes place. The annual variation of ZTD ranges from 25 to 75 mm depending on the site, and the average amplitude is about 50 mm at most sites (Fig. 5). The annual variation amplitudes of ZTD at the IGS sites near Oceanic coasts are generally larger than in the continental inland. In addition, larger amplitudes of annual ZTD variation are mostly found at middle-low latitudes (near 20S° and 40N°), and the amplitudes of annual ZTD variation are especially smaller at higher latitudes (e.g. Antarctic and Arctic) and the equator areas (see Fig. 6). Sites on the eastern Atlantic and northeast Pacific coasts have lower annual variations, probably because of the moderating effect of the ocean on climate. Sites on the lee side of the Alps have higher annual variation, possibly due to the combined effects of a rain shadow in the winter and high moisture from the Mediterranean in the summer (Haase et al., 2003; Deblonde et al., 2005). Figure 7 shows the annual phase distribution with the latitude, where phase values are counted as clockwise from the north. It can be seen that the phase of annual ZTD variation is almost found at about 60° (about February, summer) in the Southern Hemisphere and at about 240° (about August, summer) in the Northern Hemisphere, which is just a half-year difference.

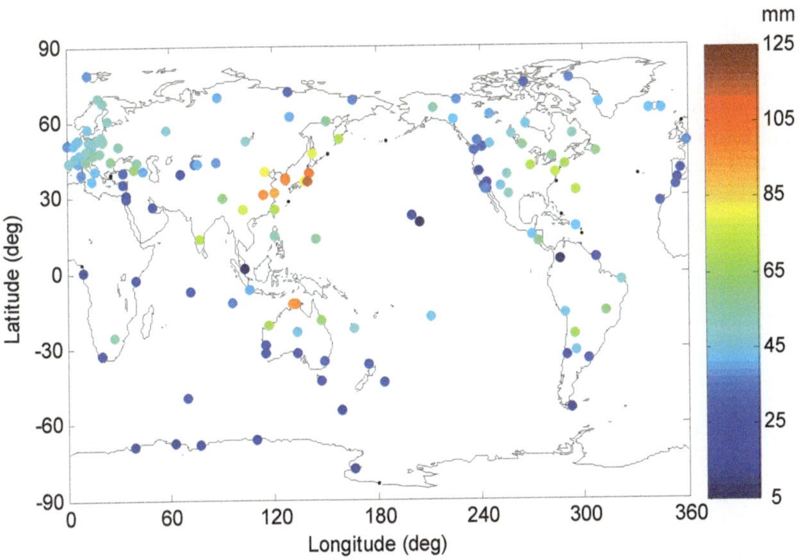

Fig. 5. Annual variation amplitude of ZTD at globally distributed 150 GPS sites

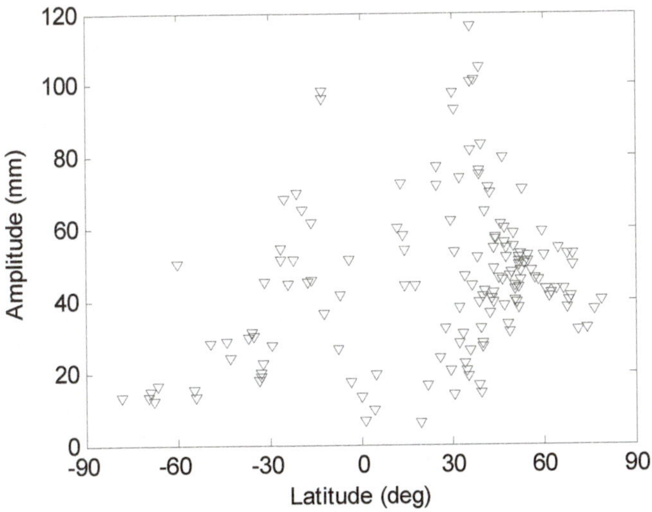

Fig. 6. Distribution of annual ZTD variation amplitude with the latitude

The mean amplitude of semiannual ZTD variations is much smaller than annual variations with about 10 mm. The amplitudes of the semiannual oscillations are much smaller on the Southern Hemisphere than on the Northern Hemisphere. The distribution of the semiannual variation phase with the latitude has no clear symmetry. For example, at the latitudes of 40°N-50°N in the Northern Hemisphere, the semiannual phase is about 30° (about January), while at the latitudes of 40°S-50°S in the Southern Hemisphere, the semiannual phase is about 330° (November).

Although the ZWD is small accounting for about 10 percent of the total zenith tropospheric delay (ZTD), the seasonal cycle in ZTD is due primarily to the wet component (ZWD). Furthermore, the seasonal variation phases of ZTD are almost consistent with surface temperature variations with the correlation coefficient of about 0.8. This reflects that annual and semiannual variations of ZTD are due mainly to the ZWD variations, about 80% in the surface temperature and 20% mainly in water vapour variations.

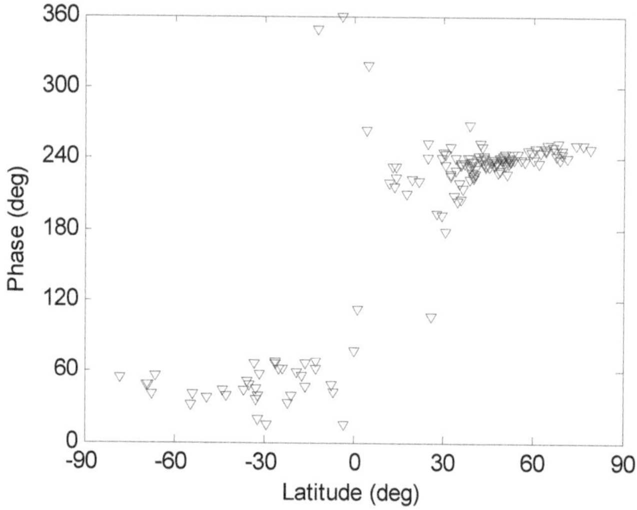

Fig. 7. Distribution of annual variation phase with the latitude. The phases are counted as clockwise from the north

2.6 Diurnal and semidiurnal ZTD cycles

The 4-7 years of GPS-derived ZTD time series has been used to analyze the diurnal and semidiurnal ZTD cycles and their features. Fig. 8a shows a colour coded map of diurnal ZTD amplitudes derived from the GPS data. The diurnal cycle (S1) has amplitudes between 0.2 and 10.9 mm with an uncertainty of about 0.5 mm. The diurnal ZTD amplitudes reduce with increasing latitude with the largest amplitudes appearing in the low-latitude equatorial areas, in particular in tropical Asia and the Gulf of Mexico. At these low latitudes, amplitudes of up to 10.9 mm are observed, while the high latitude areas reveal generally lower diurnal ZTD amplitudes. The peak values of the diurnal cycles occur spreading over the whole day (Fig. 9a). For the European stations there appears to be a preference for the second half of the day. At the semidiurnal cycle (S2), amplitudes between 0.1 and 4.3 mm with an uncertainty of about 0.2 mm are observed. Similar to the diurnal results mentioned above, the largest semidiurnal amplitudes are also found in low-latitude equatorial areas. The first peak of the semidiurnal cycle occurs typically around local noon. These diurnal and semidiurnal cycles of ZTD may be due to certain short time scale physical processes such as diurnal convection, atmospheric tides, general circulation and the coupling between the lower and the middle and upper atmosphere.

From Eq. (4), the zenith hydrostatic delay (ZHD) can be written as ZHD=2.28 p0. The scale factor k varies less than 1% even under severe weather conditions. As the hydrostatic

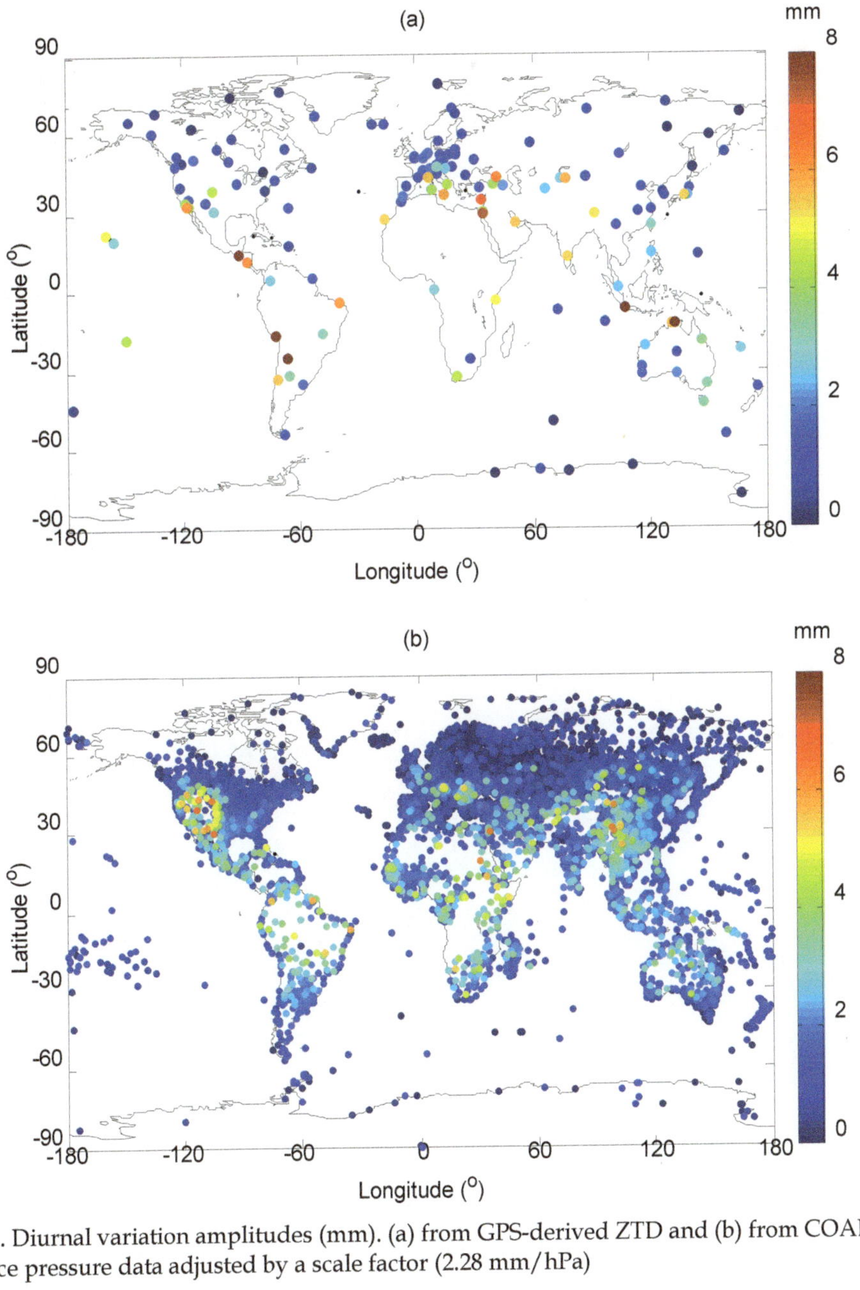

Fig. 8. Diurnal variation amplitudes (mm). (a) from GPS-derived ZTD and (b) from COADS surface pressure data adjusted by a scale factor (2.28 mm/hPa)

component ZHD accounts for approximately 90% of ZTD, ZTD is strongly correlated with surface pressure p0 at the site. It can be seen that if subdiurnal surface pressure varies by 1 hPa, the scale factor predicts a subdiurnal ZTD variation with amplitude of 2.28 mm. The

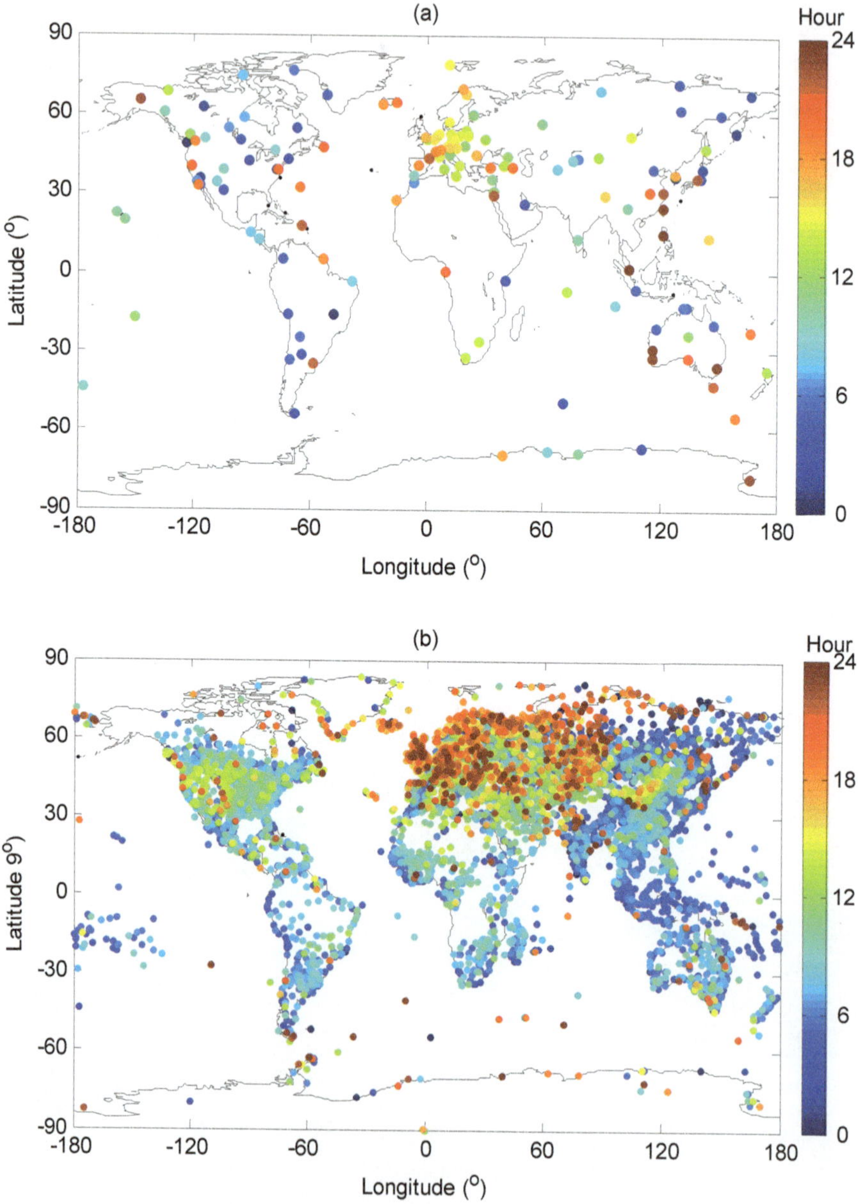

Fig. 9. Time of diurnal peak values at local time (LT: hour) where at each GPS sites longitude the Sun is at its highest elevation at 12:00 LT. (a) from global IGS GPS observations and (b) from COADS surface pressure data

GPS-derived S1 and S2 signals in ZTD are compared with three-hour surface synoptic pressure observations from 1997 to 2007 that are archived at the National Center for

Atmospheric Research (NCAR) (DS464.0; (http://dss.ucar.edu/datasets/ds464.0). Notably, these results are adjusted by a scale factor 2.28 mm/hPa, which are accounted for in the comparison. These pressure data origin from more than 8000 land and ocean weather stations including the Global Telecomuncation System (GTS) and marine reports from the Comprehensive Ocean-Atmosphere Data Set (COADS) (Dai and Wang 1999). The plots of diurnal and semidiurnal ZTD cycles show general similarities, indicating that the diurnal and semidiurnal atmospheric tides are probably the main driver of the diurnal and semidiurnal ZTD variations derived from GPS (Jin et al., 2009).

3. Ionospheric sounding

The GPS consists of a constellation of 24 operating satellites in six circular orbits 20,200 km above the Earth at an inclination angle of 55° with a 12-h period. The satellite transmits two frequencies of signals (f1 = 1575.42 MHz and f2 = 1227.60 MHz). The equations of carrier phase (L) and code observations (pseudorange P) of double frequency GPS can be expressed as follows:

$$L_{k,j}^{i} = \lambda_k \phi_{k,j}^{i} = \rho - d_{ion,k,j}^{i} + d_{trop,j}^{i} + c(\tau^{i} - \tau_j) - \lambda_k (b_{k,j}^{i} + N_{k,j}^{i}) \tag{7}$$

$$P_{k,j}^{i} = \rho + d_{ion,k,j}^{i} + d_{trop,j}^{i} + c(\tau^{i} - \tau_j) + d_{q,k}^{i} + d_{q,k,j} + \varepsilon_j^{i} \tag{8}$$

where superscript i and subscript j represent the satellite and ground-based GPS receiver, respectively, ρ is the distance between satellite i and GPS receiver j, d_{ion} and d_{trop} are the ionospheric and tropospheric delays, respectively, c is the speed of light in vacuum space, τ is the satellite or receiver clock offset, b is the phase delay of satellite and receiver instrument bias, d_q is the code delay of satellite and receiver instrumental bias, λ is the carrier wavelength, ϕ is the total carrier phase between the satellite and receiver, N is the ambiguity of carrier phase, and ε is the other residuals. From Eq.(7) and (8), the ionospheric delay can be determined, which is useful for ionospheric delay correction and space weather.

3.1 2-D ionospheric imaging

The ionospheric delay can be determined from the double frequency GPS phase and code (pseudorange) observations as

$$L_4 = \phi_{1j}^{i} - \phi_{2j}^{i} = -40.3\left(\frac{1}{f_1^2} - \frac{1}{f_2^2}\right) F(z) VTEC(\beta, s) + B_4 \tag{9}$$

$$P_4 = P_{1j}^{i} - P_{2j}^{i} = 40.3\left(\frac{1}{f_1^2} - \frac{1}{f_2^2}\right) F(z) VTEC(\beta, s) + b_4 \tag{10}$$

where $F(z)$ is the mapping function, B_4 is $\left(B_4 = -\lambda_1(b_{1j}^{i} + N_{1j}^{i}) + \lambda_2(b_{2j}^{i} + N_{2j}^{i})\right)$, and b_4 is $(dq_{1j} - dq_{2j}) + (dq_1^{i} - dq_2^{i})$. The Differential Code Biases (b_4) can be obtained through GPS carrier phase observations, and B_4 can be obtained through the formula, $\sum_{i=1}^{N}(p_4 + L_4 - b_4)/N$, where N is the epoch of GPS observation (Jin et al., 2008). For the TEC representation, a single layer model (SLM) ionosphere approximation was used. SLM assumes that all the free electrons are contained in a shell of infinitesimal thickness at altitude H (generally 350 km above the Earth). A mapping function is used to convert the slant TEC into the vertical TEC (VTEC) as shown:

$$F(z) = \sqrt{(1 - R\cos(90 - z)/(R + H))} \tag{11}$$

where R is Earth radius, H is SLM height, and z is satellite zenith angle. When using the above mapping function, F(z), one can obtain VTEC values at the ionosphere pierce points (IPPs). The GPS-derived TEC can correct ionospheric delay for microwave techniques and monitor space weather events.

3.2 3-D ionospheric tomography reconstruction

The STEC is defined as the line integral of the electron density as expressed by:

$$STEC = \int_{R_{receiver}}^{R_{satellite}} N_e(\lambda,\varphi,h)ds \tag{12}$$

where $N_e(\lambda,\varphi,h)$ is the ionospheric electron density, λ, φ and h are the longitude, latitude and height, respectively. To obtain N_e, the ionosphere is divided into grid pixels with a small cell where the electron density is assumed to be constant, so that the STEC in Eq.(4) along the ray path i can be approximately written as a finite sum over the pixels j as follows:

$$STEC_i = \sum_{j=1}^{M} a_{ij} n_j \tag{13}$$

where a_{ij} is a matrix whose elements denote the length of the path-pixel intersections in the pixel j along the ray path i, and n_j is the electron density for the pixel j. Each set of STEC measurements along the ray paths from all observable satellites at consecutive epochs are combined with the ray path geometry into a linear expression:

$$Y = Ax + \varepsilon \tag{14}$$

where Y is a column of m measurements of STEC, x is a column of n electron density unknowns for cells in the targeted ionosphere region, and A is an m × n normal matrix with elements a_{ij}. The unknown electron densities x can be estimated by the ionospheric tomographic reconstruction technique. Many tomography algorithms are used in different ways, e.g. algebraic reconstruction technique (Gordon et al., 1970). One of the most common approaches is the algebraic reconstruction technique (ART), which was first introduced in Computerized Ionospheric Tomography (CIT) by Austen et al. (1986). This is an iterative procedure for solving a linear equation. A modified version of ART is the so-called multiplicative ART (MART), where the correction in each iteration is obtained by making a multiplicative modification to x (Raymund et al., 1990; Tsai et al., 2002). The ART generally produces estimates of the unknown parameters by minimization of the L2 norm, while the MART follows maximum entropy criteria and thus underlies different statistics. In addition, the MART performs a multiplicative modification in each iteration, and thus the inversion results are always positive. Therefore, MART has the advantage over ART in determining the electron densities that avoid unreasonable negative values and is the one used in this study. Basically, the MART algorithm is iterated cyclically:

$$x_j^{k+1} = x_j^k \cdot \left(\frac{y_i}{\langle a_i, x^k \rangle} \right)^{\lambda_k a_{ij}} \tag{15}$$

where yi is the ith observed STEC in a column of m measurements, xj is the jth resulted cell electron density in a column of n unknowns, aij is the length of link i that lies in cell j, λ_k is the relaxation parameter at the kth iteration with $0 < \lambda_k < 1$, and the inner product of the vectors x and ai is the simulated STEC for the ith path. The electron density matrix x is therefore corrected iteratively by the ratio of the measured STEC and the simulated STEC with a relaxation parameter of λ_k until the residual does not change (see Figure 10). This relaxation parameter value is chosen from experience in which the best λ_k value is identified where the solution converges quickly with a reasonable number of iterations and the residuals are a minimum. Here λ_k =0. 01 has been chosen for all iterations. In addition, it is noted that any iterative algorithm requires an initial condition before the iteration begins. Due to the poor STEC geometry, the initialization could be extremely important for the tomographic reconstruction. In practice, the closer the initial condition is to the true electron density distribution, the more accurate the reconstruction will be. Here the latest IRI-2007 model (http://nssdcftp.gsfc.nasa.gov/models/ionospheric/iri/iri2007) is used as an initial guess for the reconstruction iteration.

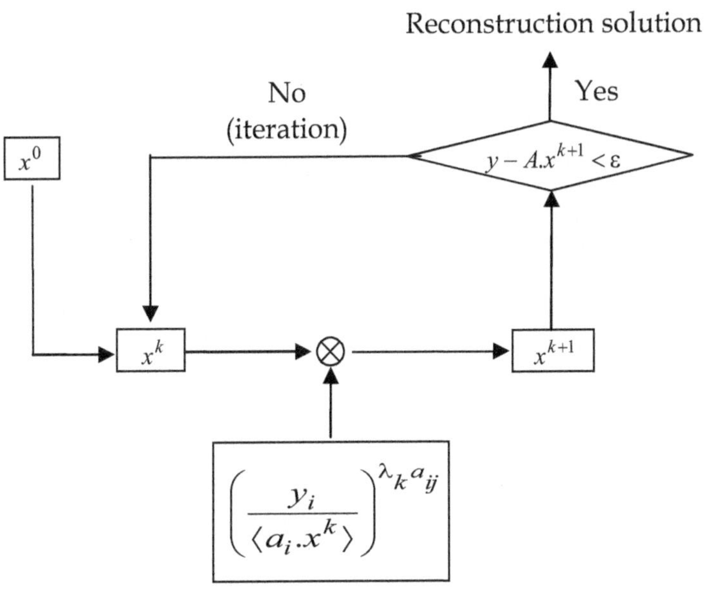

Fig. 10. Flow chart of MART

The ionospheric reconstruction algorithm MART can integrate the STEC from all available GPS receivers to all GPS satellites visible from each site of the KGN network above a user-specified elevation cut-off angle (usually 15°). The unknown electron density profile is expressed in 4-D (longitude-latitude-height and time) voxel basis functions over the following grid: longitude 124E°-130E° in 1° increments, latitude 33N°-39N° in 0.5° increments, altitude 100 km - 1000 km in 25 km increments and time: 1 h increments of linear change in the electron density per voxel. As there are a sparse number of ions in the ionosphere above 1000 km, the effect on the inversion for ionospheric electron density profiles is very small and the ionosphere is only considered up to an altitude of 1000 km. Furthermore, it is faster to invert the unknown ionospheric density parameters because of the reduced number of unknown variables. In addition, the fewer leaving rays from the ionospheric space of above defined latitude and longitude range are not used, but are useful to further obtain ionospheric profiles outside the latitudinal/longitudinal boundaries space. Using the STEC of all ray paths passing the ionospheric grid cells from the Korean GPS network, the 4-D ionospheric electron density profiles can be derived through the tomography reconstruction algorithm. To verify the reliability of GPS ionospheric tomography reconstruction results, the available ionosonde station (Anyang) in South Korea provides an independent comparison with the GPS tomographically reconstructed electron density profiles. The electron density profiles at 25 km height steps from GPS reconstruction and ionosonde data match well in October and November with a root-mean-square (RMS) of $0.3 \times 10^{11} el / m^3$. For example, Figure 11 shows a comparison of the GPS reconstructed electron density profile at 9:00 UT on 1 October 2003 with the available ionosonde data at Anyang station (37.39°N, 126.95°E) and the profiles from the IRI-2007 model. It can be seen that the GPS tomographically reconstructed density profile is in a good agreement with ionosonde data and the IRI-2007 model, but is closer to the ionosonde, which confirms the validity of our GPS ionospheric reconstruction approach (Jin et al., 2006 and 2008).

3.3 F-2 layer ionospheric response to storm

It is well known that geomagnetic storms may profoundly affect the global ionosphere and upper atmosphere, inducing great variations in such parameters as the Total Electron Content (TEC), the F2-layer peak density (NmF2) and its height (hmF2). These influences vary with location, season, local time and solar activity. The responses of the ionosphere to geomagnetic storms have been studied for several decades using moderately priced to expensive instrumentation, such as ionosondes and Incoherent Scatter Radar (ISR) (Lei et al. 2004). However, it is well known that ionospheric storms have a global impact on ionization, and under very disturbed conditions the ionospheric response to severe storms often presents significant changes in the distribution of ionization with latitude and altitude. Furthermore, ionosondes cannot measure the topside ionosphere and sometimes suffer from absorption during storms, whereas Incoherent Scatter Radars have geographical limitations. Nowadays the GPS satellites, being in high altitude orbits (~20,200 km), are very useful for studying the structure of the entire ionosphere, even the plasmasphere. Moreover, GPS is a low-cost, all-weather, near real time, and high-resolution atmospheric sounding technique. Therefore, GPS has been widely used to monitor the ionosphere (e.g. Jin et al. 2004-2007; Yin et al., 2004).

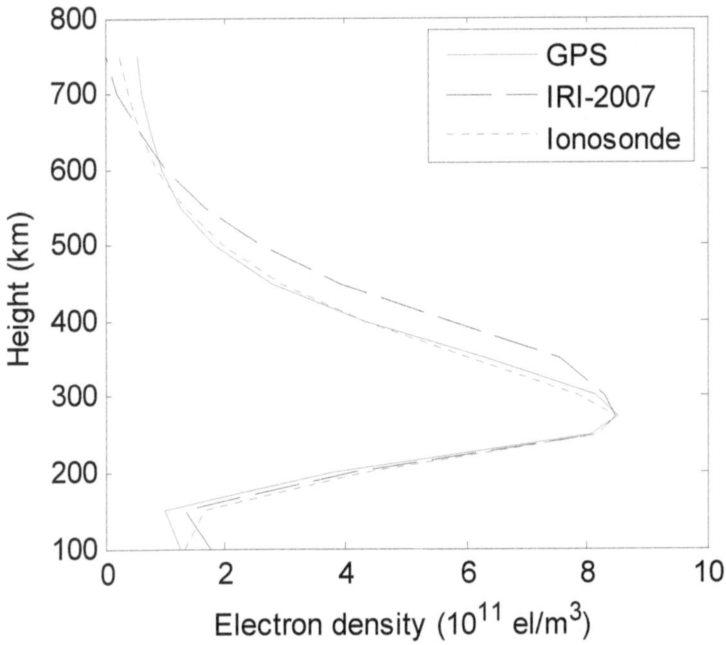

Fig. 11. Comparison of the electron density profiles derived from the ground-based GPS tomography reconstruction (solid line), ionosonde observation at Anyang stations (37.39°N, 126.95°E) (dot line) and IRI-2007 estimation (dashed line) at 9:00 UT on 1 October 2003

As the F2-layer peak electron density value (denoted as NmF2, proportional to the square of the F2 layer critical frequency foF2) is of great influence on the shape of the ionospheric electron density profile Ne(h), and also probably related to various physical processes of ionospheric activities as well as the F2-layer peak height (hmF2), the NmF2 and hmF2 are essential parameters for monitoring ionospheric activities and understanding the nature of the ionosphere. Although many studies of geomagnetic storms have been carried out in the US and Europe (e.g. Foster et al. 2004; Yin et al. 2004; Goncharenko et al. 2007), investigations on ionospheric behaviour to storms in Asia are relatively few due to a lack of dense sets of GPS observations, etc. In this paper, the responses of the GPS-derived NmF2 and hmF2 to the super geomagnetic storm (20 November 2003) over South Korea using data from the dense Korean GPS Network are described. First, ground-based dual-frequency GPS observations are used to produce electron density profiles using the ionospheric tomography technique, which are verified by independent ionosonde data. Then the responses of the key ionospheric F2-layer parameters NmF2 and hmF2 to the 20 November 2003 geomagnetic storm are investigated over South Korea to gain insights into the effects of different physical conditions and processes.

In the following, the 3-D ionospheric disturbances during the large November 20th 2003 geomagnetic storm are investigated and analyzed using GPS data in South Korea. The geomagnetic storm Dst, Kp and AE indices on 20-21 November 2003 obtained from the World Data Center in Kyoto (http://swdcdb.kugi.kyoto-u.ac.jp/) showed a strong geomagnetic storm on November 20th, 2003 (Figure 2).

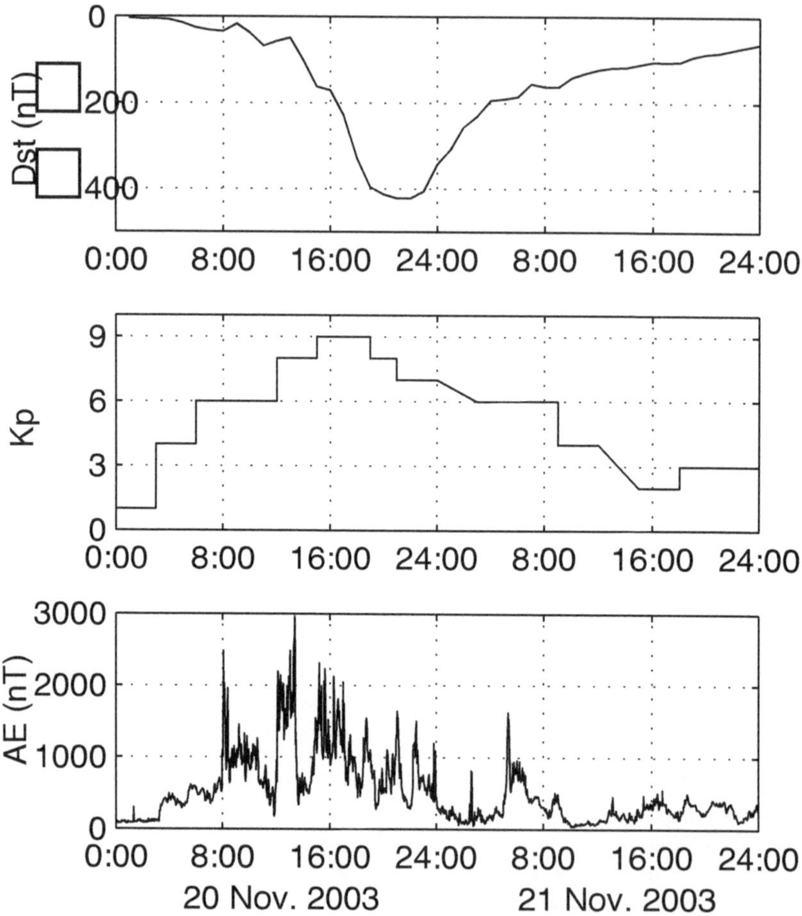

Fig. 12. The geomagnetic storm index (Dst) (upper), Kp (middle) and AE indices (bottom) on 20-21 November 2003

As the F2-layer peak electron density (NmF2) and its height (hmF2) are main parameters of the ionospheric electron density profile Ne (h), the behaviour of the ionospheric F2-layer to the storm was investigated over South Korea in terms of the NmF2 and hmF2. Here the peak density (NmF2) and its corresponding height (hmF2) are obtained from the ground-based GPS observations using the MART reconstructed tomography technique. The monthly median value of GPS reconstructed electron density profiles during the quiet days is regarded as the reference and the deviation of ionospheric NmF2 and hmF2 can reflect the ionospheric behaviors during the geomagnetic storm. It has shown that the GPS-derived NmF2 has a disturbance at 9:00 UT and then increases from 10:00 UT until 19:00 UT. The corresponding hmF2 also suddenly rises from 8:00 UT when the storm just started, and reaches the maximum height at about 16:00 UT with a maximum Kp value of 9, and then gradually descends until 21:00 UT (Fig.13), which are also supported by anther independent ionosonde measurement at Anyang station.

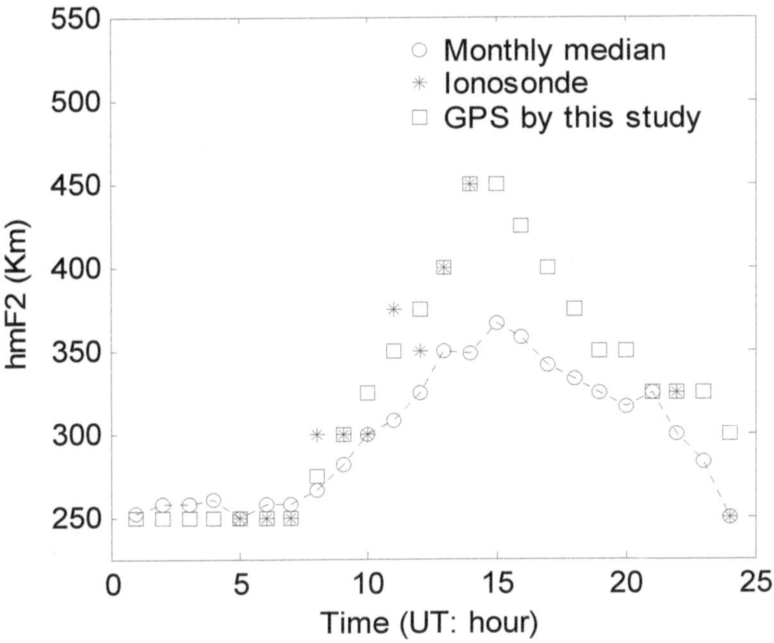

Fig. 13. The hmF2 variations from monthly median, ionosonde and GPS observations on 20 November 2003

Normally the increase/loss rate of F region electron density depends mainly on the molecular nitrogen concentration [N2] and atomic oxygen concentration [O] [8]. However, the O/N2 ratio obtained by the GUVI instrument on board the TIMED satellite doesn't show significant changes in South Korea where the increased NmF2 was observed, indicating that the increased NmF2 in South Korea is not caused by changes in neutral composition, and other possible non-chemical effects, such as dynamical changes of vertical ion motions induced by winds and E×B drifts, tides and waves in the mesosphere/lower thermosphere (MLT) region, which can be dynamically coupled upward to generate ionospheric perturbations and oscillations.

4. Conclusion

The GPS signals are delayed due to the effects of dry gas and water vapor when propagating through the neutral atmosphere. The hydrostatic delay is proportional to the surface pressure and the wet delay is a key parameter in atmospheric radiation, hydrological cycle, energy transfer and the formation of clouds via latent heat. Thereby, the total tropospheric delay (ZTD) is an important parameter of the atmosphere, which directly or indirectly reflects the weather and climate processes, variations, and atmospheric vertical motions, etc. Comparing the time series of the zenith tropospheric delay (ZTD), zenith hydrostatic delay (ZHD), zenith wet delay (ZWD), surface temperature, pressure and relative humidity, it has been noted that the ZHD is highly proportional to the atmospheric pressure at the site and relatively stable and the ZWD is positively correlated with the temperature and correlated with the relative humidity. The mean correlation coefficient between ZWD and surface temperature is about 0.80 and the correlation coefficient between ZTD and ZWD is about 0.95, reflecting a good agreement between ZTD and ZWD variations. Therefore, the seasonal cycles of the ZTD are due primarily to the wet component (ZWD), especially in the surface temperature, even though the wet delay is only 10% of the total delay (ZTD).

The lower mean ZTD values are located at the areas of higher altitude (e.g. Tibet, Asia and Andes Mountain, South America) and higher latitude areas (Antarctica and Arctic), and the higher mean ZTD values are concentrated at the areas of middle-low latitudes. The mean ZTD decreases with increasing altitude at an exponentially decreasing rate due to the atmospheric pressure decreasing with the height increasing. The mean secular ZTD variation trend is about 1.5±0.001 mm/yr at all GPS sites. The secular variations are systematically increasing in most parts of the Northern Hemisphere and decreasing in most parts of the Southern Hemisphere (excluding increasing in Antarctic). The ZTD trends are almost symmetrically decreasing with increasing altitude, while the sum of trends at globally distributed GPS sites is almost zero, possibly reflecting that the secular ZTD variation is in balance at a global scale. The annual variation of ZTD ranges from 25 to 75 mm depending on the site, and the mean amplitude is about 50 mm at most sites. The annual variation amplitudes of ZTD at the IGS sites near Oceanic coasts are generally larger than in the continental inland. Larger amplitudes of annual ZTD variation are mostly found at middle-low latitudes (near 20S° and 40N°), and the smaller amplitudes of annual ZTD variation are located at higher latitudes (e.g. Antarctic and Arctic) and the equator areas. The phase of annual ZTD variation is almost about 60° (about February, summer) in the Southern Hemisphere and at about 240° (about August, summer) in the Northern Hemisphere. The mean amplitude of semiannual ZTD variations is about 10 mm, much smaller than annual variations. The significant semiannual variations with a consistent phase of about 30° (about January) are at above 30°N in the Northern Hemisphere and the amplitudes of semiannual variations in other parts are not significant. In addition, the higher frequency variability (RMS of the ZTD residuals) ranges from 22 to 40 mm of delay, once again primarily due to the wet component. The variability depends on altitude of the station. Inland stations tend to have lower variability and stations at ocean and coasts have higher variability. This is because these stations in particular are located in a region well known for large abrupt changes in the weather, such as Indian, West Pacific and West Atlantic oceans.

In addition, the responses of the key ionospheric F2-layer parameters (NmF2 and hmF2) to the 20 November 2003 super storm have been studied using the GPS ionospheric tomography technique over South Korea. A strong increase of NmF2 during this storm has been found, accompanied by a significant hmF2 uplift, which is also confirmed by independent ionosonde observations. The uplift of the F2 layer is mainly associated with a strong eastward electric field. The increase of electron density in the F2-layer peak depends mainly on the molecular nitrogen concentration [N2] with some contribution from molecular oxygen concentration [O2], while the production rate depends on the atomic oxygen concentration [O]. However, the O/N2 ratio from the GUVI instrument on board the TIMED satellite shows no significant change during this geomagnetic storm. It suggests that the increase in NmF2 is not caused by changes in neutral composition, but is related to other possible non-chemical effects, such as dynamical changes of vertical ion motions induced by winds and E×B drifts, tides and waves in the mesosphere/lower thermosphere (MLT) region, which can be dynamically coupled upward to generate ionospheric perturbations and oscillations.

Ground-based and space borne GPS observations have been widely used in atmospheric sounding, including sensing tropopsheric precipitable water vapor (PWV), ionospheric total electron content (TEC) and atmospheric profile information (e.g., pressure, temperature, humidity, tropopause and ionospheric electron density). These observations have facilitated greater advancements in meteorology, climatology, numerical weather model, atmospheric science and space weather (e.g., Jin et al., 2007; Jin et al., 2009; Schmidt, et al. 2010). For example, the dual frequency ground GPS array could detect ionospheric response and its processes during large geomagnetic storms (Jin et al., 2008). Meanwhile, ground GPS also observed the plasma bubbles and retrieved reliable propagation characteristics of the depletions without assumptions about the mapping of the depletion along magnetic field lines to large latitudinal distances, comparable with airglow data (Haase et al., 2011).

Additionally, space-borne GPS receivers have proven very successful in making high vertical resolution and global atmospheric measurements using the radio occultation (RO) technique The existing GPS radio occultation (RO) missions have been widely used to estimate the detailed vertical profile information, including pressure, temperature, gravity waves and sporadic E-layers as well as their variation characteristics, particularly six satellites of the Taiwan/US FORMOSAT-3/COSMIC (FORMOsa SATellite mission-3/Constellation Observing System for Meteorology, Ionosphere and Climate) mission with more than 2000 radio occultation profiles per day. Schmidt et al. (2010) was the first to observe upper tropospheric warming and lower stratospheric cooling using GPS RO data (2001-2009). Although a number of progresses in atmospheric and ionospheric sensing have been made using GPS RO missions in the past few years, they still do not satisfy actual requirements for short-time scales and higher temporal-spatial resolution monitoring together with ground GNS observations. For instance, the tropospheric or ionospheric profile information cannot be directly estimated from GPS tomography due to the lack of enough line-of-sight GPS signals passing each grid cell (Jin et al., 2006 and 2008; Nesterov and Kunitsyn, 2011). Moreover, most current RO satellite missions are approaching their end of operations. With the increase of future GNSS satellite constellations and more GNSS RO missions, the goal of improved temporal-spatial resolution will enable more detailed profile information and evolution processes of the atmosphere and ionosphere.

5. Acknowledgement

This work was supported by the Shanghai Pujiang Talent Program, National Natural Science Foundation of China (Grant No.11043008) and Key Direction Project of Chinese Academy of Sciences (Grant No. KJCX2- EW-T03).

6. References

Austen, J. R.; S. J. Franke, and C. H. Liu, Application of computerized tomography techniques to ionospheric research, in Radio Beacon contributions to the study of ionization and dynamics of the ionosphere and to corrections to geodesy and technical workshop, A. Tauriainen, Eds. pp. 25-35, Ouluensis Universitas, Oulu, Finland, 1986.

Beutler G., M. Rothacher, S. Schaer, T.A. Springer, J. Kouba, R.E. Neilan (1999), The International GPS Service (IGS): An Interdisciplinary Service in Support of Earth Sciences, Adv. Space Res. 23(4) 631-635, 1999.

Bevis M., S. Businger, S. Chiswell, et al. (1994), GPS Meteorology: Mapping Zenith Wet Delays Onto Precipitable Water, J. App. Meteor., 33, 379-386.

Byun, S.H., Y. Bar-Sever, and G.Gendt (2005), The new tropospheric product of the International GNSS service, paper presentaed at 2005 ION GNSS conference, Inst. of Navig., Long Beach, Calif.

Dai A, Wang J (1999) Diurnal and Semidiurnal Tides in Global Surface Pressure Fields. J Atmos Sci 56, 3874-3891.

Dai A, Wang J, Ware RH, Van Hove T (2002) Diurnal variation in water vapor over North America and its implications for sampling errors in radiosonde humidity. J Geophys Res 107(D10): 4090, doi:10.1029/ 2001JD000642.

Davis, J.L., T.A. Herring, I. Shapiro, A. Rogers, and G.Elgered (1985), Geodesy by radio interferometryL Effects of atmospheric modeling errors on esimatetes of baseline length, Radio Sci., 20(6), 1593-1607.

Deblonde G., S. Macpherson, Y. Mireault , et al.(2005), Evaluation of GPS precipitable water over Canada and the IGS network, Journal of Applied Meteorology, 44 (1): 153-166.

Elgered, G., et al. (1997), Measuring Regional Atmospheric Water Vapor Using the Swedish Permanent GPS Network, Geophysical Research Letters, 24, 2663-2666.

Feng, D., Herman B.M. Remotely sensing the earth's atmosphere using the Global Positioning System (GPS)- The GPS/MET data analysis. Journal of Atmospheric and Oceanic Technology, 16, 989-1002, 1999.

Foster J, Coster A, Erickson P, Rich F, Sandel B (2004) Stormtime observations of the flux of plasmaspheric ions to the dayside cusp/magnetopause, Geophys Res Lett 31: doi:10.1029/2004GL020082.

Goncharenko LP, Foster J, Coster A, Huang C, Aponte N, Paxton L(2007) Observations of a positive storm phase on September 10, 2005, J Atmos Solar Terr Phys 69: 1253-1272.

Gordon R, Bender R, Therman G (1970) Algebraic Reconstruction Techniques (ART) for three Dimensional Electron Micoscopy and X-ray Photography. J Theor.Biol 29: 471-481

Haase, J., M. Ge, H.Vedel and E. Calais (2003),Accuracy and Variability of GPS Tropospheric Delay Measurements of Water Vapor in the Western Mediterranean, Journal of Applied Meteorology,42(11): 1547-1568.

Hagemann S., L. Bengtsson and G. Gendt (2003), On the determination of atmospheric water vapor from GPS measurements, J. Geophys. Res., 108 (D21): 4678, doi:10.1029/2002JD003235.

Hajj, G.A., Romans, L.L. Ionospheric Electron Density Profiles Obtained with the Global Positioning System: Results from the GPS/MET Experiment. Radio Science, 33(1): 175-190, 1998.

Haase, J.S., T. Dautermann,, M.J. Taylor, N. Chapagain, E. Calais,, D. Pautet, Propagation of plasma bubbles observed in Brazil from GPS and airglow data. Adv. Space Res. doi: 10.1016/j.asr.2010.09.025, 2011.

Jin SG, Wang JL, Zhang HP, Zhu WY (2004), Real-time monitoring and prediction of the total ionospheric electron content (TEC) by means of GPS. Chinese Astronomy and Astrophysics 28(3): 331-337.

Jin S.G., and P.H. Park (2005), A new precision improvement of zenith tropospheric delay estimates by GPS, Current Science, 89 (6): 997-1000.

Jin SG, Park J, Wang J, Choi B, (2006), Electron density profiles derived from ground-based GPS observations. Journal of navigation 59(3): 395-401.

Jin SG, Park J, Cho J, Park P (2007), Seasonal variability of GPS-derived Zenith Tropospheric Delay (1994-2006) and climate implications. J Geophys Res 112: D09110, doi: 10.1029/2006JD007772.

Jin, S.G., O. Luo, and P. Park (2008), GPS observations of the ionospheric F2-layer behavior during the 20th November 2003 geomagnetic storm over South Korea, J. Geod., 82(12), 883-892, doi: 10.1007/s00190-008-0217-x.

Jin, S.G., O. Luo, and S. Gleason (2009), Characterization of diurnal cycles in ZTD from a decade of global GPS observations, J. Geod., 83(6), 537-545, doi: 10.1007/s00190-008-0264-3.

King R.W. and Y. Bock (1999), Documentation for the GAMIT GPS Analysis Software, Mass. Inst. of Technol., Cambridge Mass.

Lei J, Liu L, Wan W, Zhang S (2004) Modeling the behavior of ionosphere above Millstone Hill during the September 21-27, 1998 storm. Journal of Atmospheric and Solar-Terrestrial Physics 66: 1093-1102.

Manuel H., J. Juan, J. Sanz, et al. (2001), A New Strategy for Real-Time Integrated Water Vapor Determination in WADGOPS Networks, Geophy. Res. Lett., 28(17), 3267-3270.

Nesterov, I.A., V.E., Kunitsyn, GNSS radio tomography of the ionosphere: The problem with essentially incomplete data, Advances in Space Research, doi: 10.1016/j.asr.2010.11.034, 2011.

Niell A.E. (1996), Global Mapping Functions for the Atmospheric Delay at Radio Wavelengths, Journal of Geophysical Research, 101(B2), 3227-3246.

Raymund TD, Austen JR, Franke SJ (1990) Application of computerized tomography to the investigation of ionospheric structures. Radio Science 25: 771-789.

Rocken, C. Analysis and validation of GPS/MET data in the neutral atmosphere. Journal of Geophysical. Research, 102: 29849-29866, 1997.

Saastamoinent J. (1973), Contribution to the theory of atmospheric refraction, Bulletin Geodesique, 107, 13-34.

Schmidt, T., Wickert, J., and Haser, A. (2010), Variability of the upper troposphere and lower stratosphere observed with GPS radio occultation bending angles and temperatures, Adv. Space Res., 46(2),150-161,doi:10.1016/j.asr.2010.01.021

Syndergaard, S. On the Ionosphere Calibration in GPS Radio Occultation Measurements. Radio Science, 35(3): 865-883, 2000.

Tsai LC, Liu CH, Tsai WH, Liu CT (2002) Tomographic imaging of the ionosphere using the GPS/MET and NNSS data. J. Atmos. Sol. Terr. Phys. 64, 2003-2011.

Tregoning P., R. Boers and D. O'Brien (1998) Accuracy of Absolute Precipitable Water Vapor Estimates from GPS Observations, Journal of Geophysical Research, 103(28), 701-710.

Wagner, C.. Kloko, J. The value of ocean reflections of GPS signals to enhance satellite altimetry: data distribution and error analysis. Journal of Geodesy, 74, 128-138, 2003.

Westwater, E. R. (1993), Ground-based microwave remote sensing of meteorological variables. Atmospheric Remote Sensing by Microwave Radiometry, M. A. Janssen, Ed., J. Wiley and Sons, Inc., 145-213.

Yin P, Mitchell CN, Spencer PS, Foster JC (2004) Ionospheric electron concentration imaging using GPS over the USA during the storm of July 2000. Geophys Res Lett 31: L12806, doi: 10.1029/2004GL019899.

17

Multipath Mitigation Techniques for Satellite-Based Positioning Applications

Mohammad Zahidul H. Bhuiyan and Elena Simona Lohan
Department of Communications Engineering, Tampere University of Technology
Finland

1. Introduction

The ever-growing public interest on location and positioning services has originated a demand for a high performance Global Navigation Satellite System (GNSS), such as the Global Positioning System (GPS) or the future European satellite navigation system, Galileo. The performance of GNSS is subject to several errors, such as ionosphere delay, troposphere delay, receiver noise and multipath. Among all these errors, multipath is the main limiting factor in precision-oriented GNSS applications. The reception of multipath creates a bias into the time delay estimate of the Delay Lock Loop (DLL) of a conventional navigation receiver, which eventually leads to an error in the receiver's position estimate. In order to mitigate the multipath influence on navigation receivers, the multipath problem has been approached from several directions. Among them, the use of special multipath limiting antennas (i.e., choke ring or multi-beam antennas), the post-processing techniques to reduce carrier multipath, the carrier smoothing to reduce code multipath, and the code tracking techniques based on receiver internal correlation function are the most prominent approaches.

In this chapter, the discussion is mainly focused on the correlation-based multipath mitigation techniques at the receiver side; since the correlation-based multipath mitigation approach is by far the most convenient and popular way to deal with multipath error for a stand-alone GNSS receiver. The classical correlation-based code tracking structure used in GNSS is based on a feedback delay estimator and is implemented via a feedback loop. The most known feedback delay estimator is the Early-Minus-Late (EML) DLL technique, where two correlators spaced at one chip from each other are used in the receiver in order to form a discriminator function, whose zero crossings determine the path delays of the received signal Baltersee et al. (2001), Bischoff et al. (2002), Chen & Davisson (1994), Fine & Wilson (1999), Fock et al. (2001), Laxton (1996). The classical EML fails to cope with multipath propagation Dierendonck et al. (1992), Simon et al. (1994). Therefore, several enhanced EML-based techniques have been introduced in the literature during the last two decades in order to mitigate the impact of multipath, especially in closely spaced path scenarios. One class of these enhanced EML techniques is based on the idea of narrowing the spacing between the early and late correlators, i.e., narrow EML (nEML) or narrow correlator Dierendonck et al. (1992), Irsigler & Eissfeller (2003), McGraw & Braasch (1999). The choice of correlator spacing depends on the receiver's available front-end bandwidth along with the associated sampling frequency Betz & Kolodziejski (2000). Correlator spacings in the range of 0.05 to 0.2 chips are commercially available for nEML based GPS receivers Braasch (2001).

Another family of discriminator-based DLL variants proposed for GNSS is the so-called Double-Delta (ΔΔ) technique, which uses more than 3 correlators in the tracking loop (typically, 5 correlators: two early, one in-prompt and two late) Irsigler & Eissfeller (2003). The ΔΔ technique offers better multipath rejection in medium-to-long delay multipath Hurskainen et al. (2008), McGraw & Braasch (1999) in good Carrier-to-Noise density ratio (C/N_0). Couple of well-known particular cases of ΔΔ technique are the High Resolution Correlator (HRC) McGraw & Braasch (1999), the Strobe Correlator (SC) Garin & Rousseau (1997), Irsigler & Eissfeller (2003), the Pulse Aperture Correlator (PAC) Jones et al. (2004) and the modified correlator reference waveform Irsigler & Eissfeller (2003), Weill (2003). One other similar tracking structure is the Multiple Gate Delay (MGD) correlator Bello & Fante (2005), Bhuiyan (2006), Fante (2003), Fante (2004), where the number of early and late gates and the weighting factors used to combine them in the discriminator are the parameters of the model, and can be optimized according to the multipath profile as illustrated in Hurskainen et al. (2008). While coping better with the ambiguities of Binary Offset Carrier (BOC) correlation function, the MGD provides slightly better performance than the nEML at the expense of higher complexity and is sensitive to the parameters chosen in the discriminator function (i.e., weights, number of correlators and correlator spacing) Bhuiyan (2006), Hurskainen et al. (2008).

Another tracking structure closely related to ΔΔ technique is the Early1 / Early2 (E1/E2) tracker, initially proposed in Dierendonck & Braasch (1997), and later described in Irsigler & Eissfeller (2003). In E1/E2 tracker, the main purpose is to find a tracking point on the correlation function that is not distorted by multipath. As reported in Irsigler & Eissfeller (2003), E1/E2 tracker shows some performance improvement over ΔΔ technique only for very short delay multipath for GPS L1 Coarse / Acquisition (C/A) signal.

Another feedback tracking structure is the Early-Late-Slope (ELS) Irsigler & Eissfeller (2003), which is also known as Multipath Elimination Technique (MET) Townsend & Fenton (1994). The simulation results performed in Irsigler & Eissfeller (2003) showed that ELS is outperformed by HRC with respect to Multipath Error Envelopes (MEEs), for both Binary Phase Shift Keying (BPSK) and Sine BOC(1,1) (SinBOC(1,1)) modulated signals.

A new multipath estimation technique, named as A-Posteriori Multipath Estimation (APME), is proposed in Sleewaegen & Boon (2001), which relies on a-posteriori estimation of the multipath error tracking. Multipath error is estimated independently in a multipath estimator module on the basis of the correlation values from the prompt and very late correlators. According to Sleewaegen & Boon (2001), the multipath performance of GPS L1 C/A signal is comparable with that of the Strobe Correlator: slight improvement for very short delays (i.e., delays less than 20 meters), but rather significant deterioration for medium delays.

In Phelts & Enge (2000a), a fundamentally different approach is adopted to solve the problem of multipath in the context of GNSS. The proposed technique, named as Tracking Error Compensator (TrEC), utilizes the multipath invariant properties of the received correlation function in order to provide significant performance benefits over nEML for narrow-band GPS receivers Phelts & Enge (2000a), Phelts & Enge (2000b).

One of the most promising advanced multipath mitigation techniques is the Multipath Estimating Delay Lock Loop (MEDLL) Nee (1992), Nee et al. (1994), Townsend et al. (1995) implemented by NovAtel for GPS receivers. MEDLL is considered as a significant evolutionary step in the receiver-based attempt to mitigate multipath. It uses many correlators in order to determine accurately the shape of the multipath corrupted correlation function. According to Townsend et al. (1995), MEDLL provides superior long delay multipath mitigation performance compared to nEML at the cost of multi-correlator based tracking structure.

A new technique to mitigate multipath by means of correlator reference waveform was proposed in Weill (1997). This technique, referred to as Second Derivative correlator, generates a signal correlation function which has a much narrower width than a standard correlation function, and is therefore capable of mitigating multipath errors over a much wider range of secondary path delays. The narrowing of the correlation function is accomplished by using a specially designed code reference waveform (i.e. the negative of the second order derivative of correlation function) instead of the ideal code waveform used in almost all existing receivers. However, this new technique reduces the multipath errors at the expense of a moderate decrease in the effective Signal-to-Noise Ratio (SNR) due to the effect of narrowing the correlation function. A similar strategy, named as Slope Differential (SD), is based on the second order derivative of the correlation function Lee et al. (2006). It is shown in Lee et al. (2006) that this technique has better multipath performance than nEML and Strobe Correlator. However, the performance measure was solely based on the theoretical MEE curves, thus its potential benefit in more realistic multipath environment is still an open issue.

A completely different approach to mitigate multipath error is used in NovAtel's recently developed Vision Correlator Fenton & Jones (2005). The Vision Correlator (VC) is based on the concept of Multipath Mitigation Technique (MMT) developed in Weill (2002). It can provide a significant improvement in detecting and removing multipath signals as compared to other standard multipath resistant code tracking algorithms (for example, PAC of NovAtel). However, VC has the shortcoming that it requires a reference function shape to be used to fit the incoming data with the direct path and the secondary path reference signals. The reference function generation has to be accomplished a-priori, and it must incorporate the issues related to Radio Frequency (RF) distortions introduced by the front-end.

Several advanced multipath mitigation techniques were also proposed in Bhuiyan (2011), Granados et al. (2005), Granados & Rubio (2000), Lohan et al. (2006). These techniques, in general, offer better tracking performance than the traditional DLL at a cost of increased complexity. However, the performance of these techniques have not yet been evaluated in more realistic multipath channel model with real GNSS signals.

The rest of this chapter is organized as follows. Multipath propagation phenomena is described first, followed by a description on the influence of signal and receiver parameters on multipath error. The following section provides an elaborate description on different multipath mitigation techniques starting from the conventional state-of-the-art techniques to the relatively complex advanced multipath mitigation techniques. An extensive literature review on the contemporary research on multipath mitigation techniques are also provided. The performance evaluations of some of the multipath mitigation techniques are shown in Section 6 via multipath error envelopes and also via simulations in multipath fading channel model. Finally, some general conclusions are drawn based on the discussions provided in earlier sections.

2. Multipath propagation

Multipath propagation occurs mostly due to reflected GNSS signals from surfaces (such as buildings, metal surfaces etc.) near the receiver, resulting in one or more secondary propagation paths. These secondary path signals, which are superimposed on the desired direct path signal, always have a longer propagation time and can significantly distort the amplitude and phase of the direct path signal. This eventually leads to a deformation in the correlation function as shown in Fig. 1, where a direct LOS signal is added constructively

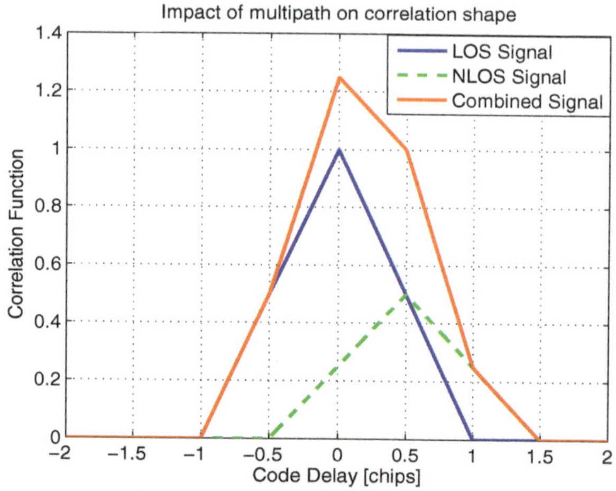

Fig. 1. Received correlation function in two path static channel model, path delays: [0 0.5] chips, path amplitudes: [0 -3] dB, in-phase combination.

with an in-phase (i.e., 0° phase difference), delayed (0.5 chips delayed) and attenuated (-3 dB attenuated) version of it to form a compound signal. The deformed correlation shape introduces an error bias in the pseudorange measurement that resulted in a degraded positioning performance.

In severe multipath environments like those in dense urban areas, it may be possible that the LOS signal is obstructed completely and only the reflected signals are present. These multipath effects on the code phase measurements are most crucial, and the multipath error can reach up to a few tens of meters, or a couple of hundred at most Gleason & Egziabher (2009). Moreover, unlike other error sources, multipath cannot be reduced through differential processing, since it decorrelates spatially very rapidly. All these issues are the main driving factors for the research conducted in the context of this thesis striving for an optimum correlation-based multipath mitigation technique in terms of mitigation performance as well as implementation complexity.

3. Influence of signal and receiver parameters on multipath error

The way multipath affects the tracking and navigation performance of a receiver depends on a number of signal and receiver parameters. Among them, the most influential parameters are:

- Type of signal modulation,

- Front-end filter bandwidth (i.e., precorrelation bandwidth),
- Correlator spacing used in the code tracking,
- Type of discriminator used to run the DLL (i.e., nEML, HRC, etc.),
- Code chipping rate,
- Number of multipath signals,
- Amplitudes, delays and phases of multipath signals with respect to the LOS signal, etc.

The type of signal modulation basically determines the shape of the correlation function. For example, BPSK is used to modulate GPS L1 C/A signal, which has a single significant tracking peak within ±1 chip delay from the correct code delay, whereas CBOC modulations (i.e. CBOC(+) for data channel and CBOC(-) for pilot channel) are used to modulate Galileo E1 signal, each of which has more than one significant tracking peak within ±1 chip delay from the correct code delay. Non-coherent (i.e., absolute value of the correlation function) correlation functions for the above modulations are shown in Fig. 2, where the extra peaks

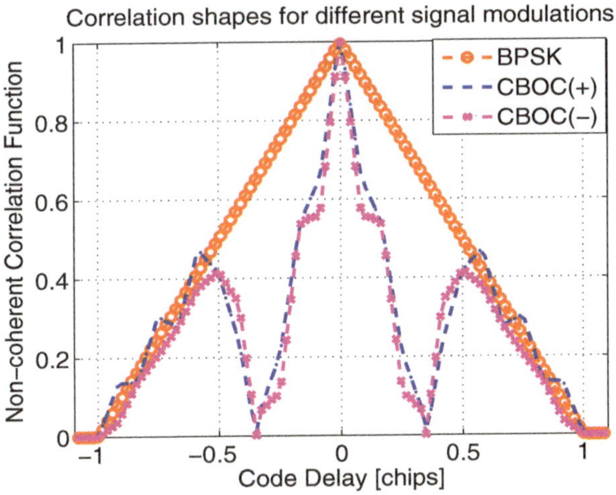

Fig. 2. Non-coherent correlation functions for different signal modulations.

can be clearly observed in case of CBOC modulations. This is the situation in the most ideal single path scenario. The situation would get far worse in the presence of multipath signals, for example, in a typical fading channel model with a two to four path assumption. Fig. 3 shows the distorted correlation shapes of different signal modulations in a two path static channel with path delays [0 0.1] chips and with path powers [0 -6] dB. As seen in Fig 3, the presence of an additional peak (in case of CBOC(+) and CBOC(-)) due to multipath imposes a challenge for the signal acquisition and tracking techniques to lock to the correct peak. If the receiver fails to lock to the correct peak, a multipath error in the order of few tens of meters is of no surprise.

The front-end filter bandwidth used for band-limiting the received signal also has some impact on the correlation shape. The bandwidth, if not chosen sufficiently high, may round off the correlation peak as well as flatten the width of the correlation function, as shown in Fig. 4. For this particular reason, the choice of correlator spacing depends on the receiver's available front-end bandwidth (and off course, on the sampling frequency), that follows the

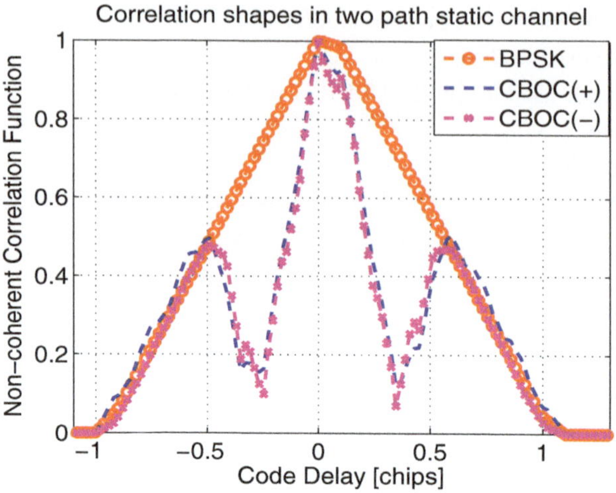

Fig. 3. Non-coherent correlation functions for different signal modulations in two path static channel.

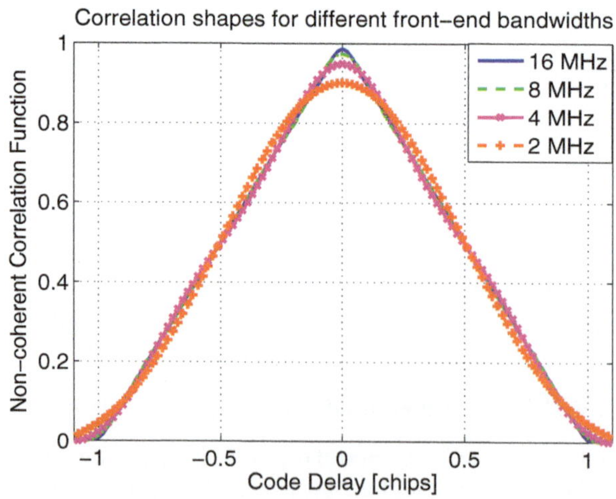

Fig. 4. Non-coherent correlation functions for BPSK modulated GPS L1 C/A signal in different front-end bandwidths.

relation: the more the bandwidth, the smaller the correlator spacing. As mentioned in Betz & Kolodziejski (2000), the early-late spacing Δ_{EL} (i.e., twice the correlator spacing) is related to the front-end bandwidth (double-sided) BW and the code chip rate f_{chip} according to the following equation:

$$\Delta_{EL} \geq \frac{f_{chip}}{BW} \tag{1}$$

The type of the discriminator and the correlator spacing used to form the discriminator function (i.e., nEML, HRC, SC, etc.) determines the behavior of the code discriminator that strongly influences the resulting multipath performance. Generally speaking, a narrower correlator spacing leads to a reduced multipath error and a tracking jitter error, as long as sufficient front-end bandwidth is ensured Dierendonck et al. (1992).

The code chipping rate determines the chip length (T_c), which ultimately decides the resulting ranging error caused by the multipath. This means that a signal with a larger chip length results in a smaller multipath error contribution. That is why, the modernized GPS L5 signal can offer ten times smaller multipath error contribution than the legacy GPS L1 C/A signal, as it has ten times higher chipping rate than that of L1.

The remaining multipath related parameters (i.e., amplitudes, delays, phases and number of multipath signals) depend on the multipath environment, and have direct influence on the tracking performance of the receiver. These parameters are generally used to define a simulation model (for example, multipath fading channel model) in order to analyze the performance of different multipath mitigation techniques.

3.1 Multi-correlator based delay tracking structure

A unified multi-correlator based delay tracking structure is developed to fit all the multipath mitigation techniques in one common tracking structure. In a multi-correlator based delay tracking structure, a bank of correlators is generated, unlike the conventional DLL-based tracking structure, where only few correlators (i.e., in the range of three to seven complex correlators depending on the type of techniques) are used. This large number of correlators are required by the advanced multipath mitigation techniques in order to estimate the channel properties and to take a decision on the correct LOS code delay. As shown in Fig. 5, after

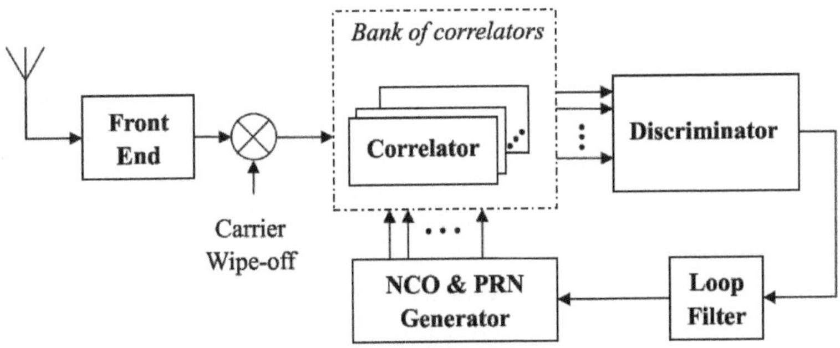

Fig. 5. Block diagram for multi-correlator based DLL implementation.

the necessary front-end processing, and after the carrier has been wiped-off, the received post-processed signal is passed through a bank of correlators. The NCO (Numerically Controlled Oscillator) and PRN generator block produces a bank of early and late versions of replica codes based on the delay of the LOS signal $\hat{\tau}$, the correlator spacing Δ, and the number of correlators M. In case of an EML tracking loop, the corresponding early-late spacing is equal to 2Δ. The received signal is correlated with each replica in the correlator bank, and the output of the correlator bank is a vector of samples in the correlation envelope. Therefore, we obtain the correlation values for the range of $\pm M\Delta$ chips from the prompt correlator, where M is the number of correlators and Δ is the correlator spacing between successive correlators. The various code tracking techniques (named as Discriminator in Fig. 5) utilize the correlation values as input, and generate the estimated LOS delay as output, which is then smoothed by a loop filter. A loop filter is generally used to improve the code delay estimate, reducing the noise present at the output of the discriminator, and to follow the signal dynamics. The order of a loop filter determines the ability of the filter to respond to different types of dynamics, whereas the filter bandwidth ensures that a low bandwidth leads to a good filtering with a high amount of noise filtered, but also requires that the dynamics of the signals are not too high. The loop bandwidth is usually determined by the coefficients of the filter, and can thus be considered as a design parameter for the filter. In accordance with Kaplan & Hegarty (2006), the code loop filter is a 1st order filter, which can be modeled as:

$$\widehat{\tau}(k+1) = \widehat{\tau}(k) + \omega_0 d(k), \tag{2}$$

where ω_0 is calculated based on the loop filter bandwidth, B_n. As shown in Bhuiyan & Lohan (2010), the multi-correlator based tracking structure being used by the advanced multipath mitigation techniques, offers a superior tracking performance to the traditional nEML DLL at the cost of higher number of correlators.

4. State-of-the-art multipath mitigation techniques

The GNSS community has started the correlation-based multipath mitigation studies since early 1990s with the advent of Narrow Correlator (NC) or narrow Early-Minus-Late (nEML) DLL Dierendonck et al. (1992). This section highlights some of the most prominent state-of-the-art techniques, which have gained a lot of interest in the research community by now.

4.1 Early-minus-late delay lock loop

The classical correlation-based code tracking structure used in a GNSS receiver is based on a feedback delay estimator and is implemented via a feedback loop. The most known feedback delay estimator is the Early-Minus-Late (EML) DLL, where two correlators spaced at one chip from each other, are used in the receiver in order to form a discriminator function, whose zero crossings determine the path delays of the received signal Baltersee et al. (2001); Bischoff et al. (2002); Chen & Davisson (1994); Fine & Wilson (1999); Fock et al. (2001); Laxton (1996); Lohan (2003). The classical EML usually fails to cope with multipath propagation Dierendonck et al. (1992). Therefore, several enhanced EML-based techniques have been introduced in the literature for last two decades in order to mitigate the impact of multipath, especially in closely spaced path scenarios. A first approach to reduce the influences of code multipath is based on the idea of narrowing the spacing between the early and late correlators, i.e., nEML or narrow correlator Dierendonck et al. (1992); Fenton (1995); Fenton & Dierendonck (1996). The choice of correlator spacing depends on the receiver's available front-end bandwidth along with the associated sampling frequency Betz & Kolodziejski (2000). Correlator spacings in the range of 0.05 to 0.2 chips are commercially available for nEML based GPS receivers Braasch (2001).

4.2 Double delta (ΔΔ) technique

Another family of discriminator-based DLL variants proposed for GNSS receivers is the so-called Double Delta (ΔΔ) technique, which uses more than three correlators in the tracking loop (typically, five correlators: two early, one in-prompt and two late) Irsigler & Eissfeller

(2003). ΔΔ technique offers better multipath rejection in medium-to-long delay multipath in good C/N_0 Hurskainen et al. (2008); McGraw & Braasch (1999). Couple of well-known particular cases of ΔΔ technique are the High Resolution Correlator (HRC) McGraw & Braasch (1999), the Strobe Correlator (SC) Garin & Rousseau (1997); Irsigler & Eissfeller (2003), the Pulse Aperture Correlator (PAC) Jones et al. (2004) and the modified correlator reference waveform Irsigler & Eissfeller (2003); Weill (2003). One other similar tracking structure is the Multiple Gate Delay (MGD) correlator Bello & Fante (2005); Bhuiyan (2006); Fante (2003; 2004); Jie (2010), where the number of early and late gates and the weighting factors used to combine them in the discriminator are the parameters of the model, and can be optimized according to the multipath profile as illustrated in Hurskainen et al. (2008); Jie (2010). While coping better with the ambiguities of BOC correlation function, the MGD provides slightly better performance than the nEML at the expense of higher complexity and is sensitive to the parameters chosen in the discriminator function (i.e., weights, number of correlators and correlator spacing) Bhuiyan (2006); Hurskainen et al. (2008); Jie (2010). In Hurskainen et al. (2008), it is also shown that ΔΔ technique is a particular case of MGD implementation.

4.3 Early-late-slope

Another feedback tracking structure is the Early-Late-Slope (ELS) Irsigler & Eissfeller (2003), which is also known as Multipath Elimination Technique (MET) Townsend & Fenton (1994). The ELS is based on two correlator pairs at both sides of the correlation function's central peak with parameterized spacing. Once both slopes are known, they can be used to compute a pseudorange correction that can be applied to the pseudorange measurement. However, simulation results performed in Irsigler & Eissfeller (2003) showed that ELS is outperformed by HRC with respect to Multipath Error Envelopes (MEEs), for both BPSK and SinBOC(1,1) modulated signals. An Improved ELS (IELS) technique was proposed by the Author in Bhuiyan et al. (2008), which introduced two enhancements to the basic ELS approach. The first enhancement was the adaptation of random spacing between the early and the late correlator pairs, while the later one was the utilization of feedforward information in order to determine the most appropriate peak on which the IELS technique should be applied. It was shown in Bhuiyan et al. (2008) that IELS performed better than nEML only in good C/N_0 for BPSK and SinBOC(1,1) modulated signals in case of short-delay multipath, but still had poorer performance than HRC.

4.4 A-Posteriori multipath estimation

A new multipath estimation technique, named as A-Posteriori Multipath Estimation (APME), is proposed in Sleewaegen & Boon (2001), which relies on a-posteriori estimation of multipath error. Multipath error is estimated independently in a multipath estimator module on the basis of the correlation values from the prompt and very late correlators. The performance of APME in multipath environment is comparable with that of the Strobe Correlator: a slight improvement for very short delays (i.e., delays less than 20 meters), but rather significant deterioration for medium delays Sleewaegen & Boon (2001).

4.5 Multipath estimating delay lock loop

One of the most promising state-of-art multipath mitigation techniques is the Multipath Estimating Delay Lock Loop (MEDLL) Nee (1992); Nee et al. (1994); Townsend et al. (1995)

implemented by NovAtel for GPS receivers. MEDLL uses several correlators per channel in order to determine accurately the shape of the multipath-corrupted correlation function. Then, a reference correlation function is used in a software module in order to determine the best combination of LOS and NLOS components (i.e., amplitudes, delays, phases and number of multipath). An important aspect of MEDLL is an accurate reference correlation function that can be constructed by averaging the measured correlation functions over a significant amount of total averaging time Nee et al. (1994). However, MEDLL provides superior multipath mitigation performance than nEML at a cost of expensive multi-correlator based delay tracking structure.

4.6 Vision correlator

A completely different approach to mitigate multipath error is used in NovAtel's recently developed Vision Correlator Fenton & Jones (2005). The Vision Correlator (VC) is based on the concept of Multipath Mitigation Technique (MMT) developed in Weill (2002). It can provide a significant improvement in detecting and removing multipath signals as compared to other standard multipath resistant code tracking algorithms (for example, PAC of NovAtel). However, VC has the shortcoming that it requires a reference function shape to be used to fit the incoming data with the direct path and the secondary path reference signals. The reference function generation has to be accomplished a-priori, and it must incorporate the issues related to Radio Frequency (RF) distortions introduced by the front-end.

5. Advanced multipath mitigation techniques

The advanced multipath mitigation techniques generally require a significant number of correlators (i.e., tens of correlators) in order to estimate the channel characteristics, which are then used to mitigate the multipath effect. Some of the most promising advanced multipath mitigation techniques are presented in the following sub-sections. In most occasions, references are made to the original publications in order to avoid too many technical details.

5.1 Non-coherent multipath estimating delay lock loop

MEDLL is considered as a significant evolutionary step in the correlation-based multipath mitigation approach. Moreover, MEDLL has stimulated the design of different maximum likelihood based implementations for multipath mitigation. One such variant is the non-coherent MEDLL, developed by the authors, as described in Bhuiyan et al. (2008). The classical MEDLL is based on a maximum likelihood search, which is computationally extensive. The authors implemented a non-coherent version of MEDLL that reduces the search space by incorporating a phase search unit, based on the statistical distribution of multipath phases. It was shown in Bhuiyan et al. (2008) that the performance of this suggested approach depends on the number of random phases considered; meaning that the larger the number is, the better the performance will be. But this will also increase the processing burden significantly. The results reported in Bhuiyan et al. (2008), show that the non-coherent MEDLL provides very good performance in terms of RMSE, but has a rather poor Mean-Time-to-Lose-Lock (MTLL) as compared to the conventional DLL techniques.

5.2 Second derivative correlator

A new technique to mitigate multipath by means of correlator reference waveform was proposed in Weill (1997). This technique, referred to as Second Derivative correlator, generates a signal correlation function which has a much narrower width than a standard correlation function, and is therefore capable of mitigating multipath errors over a much wider range of secondary path delays. The narrowing of the correlation function is accomplished by using a specially designed code reference waveform (i.e. the negative of the second order derivative of correlation function) instead of the ideal code waveform used in almost all existing receivers. However, this new technique reduces the multipath errors at the expense of a moderate decrease in the effective Signal-to-Noise Ratio (SNR) due to the effect of narrowing the correlation function. A similar strategy, named as Slope Differential (SD), is based on the second order derivative of the correlation function Lee et al. (2006). It is shown in Lee et al. (2006) that this technique has better multipath performance than nEML and Strobe Correlator. However, the performance measure was solely based on the theoretical MEE curves, thus its potential benefit in more realistic multipath environment is still an open issue.

5.3 Peak tracking

Peak Tracking (PT) based techniques, namely PT based on 2nd order Differentiation (PT(Diff2)) and PT based on Teager Kaiser (PT(TK)), were proposed in Bhuiyan et al. (2008). Both the techniques utilize the adaptive thresholds computed from the estimated noise variance of the channel in order to decide on the correct code delay. The adaptive thresholds are computed according to the equations given in Bhuiyan et al. (2008). After that, the advanced techniques generate the competitive peaks which are above the computed adaptive thresholds. The generation of competitive peaks for PT(Diff2) technique is shown in Fig. 6 in two path Nakagami-m fading channel. For each of the competitive peak, a decision variable is formed based on the peak power, the peak position and the delay difference of the peak from the previous delay estimate. Finally, the PT techniques select the peak which has the maximum weight as being the best LOS candidate. It was shown in Bhuiyan et al. (2008) that PT(Diff2) has superior multipath mitigation performance over PT(TK) in two to five path Nakagami-m fading channel.

5.4 Teager Kaiser operator

The Teager Kaiser based delay estimation technique is based on the principle of extracting the signal energy of various channel paths via a nonlinear TK operator Hamila (2002), Hamila et al. (2003). The output $\Psi_{TK}(x(n))$ of TK operator applied to a discrete signal $x(n)$, can be defined as Hamila et al. (2003):

$$\begin{aligned}\Psi_{TK}(x(n)) &= x(n-1)x^*(n-1) \\ &- \frac{1}{2}[x(n-2)x^*(n) + x(n)x^*(n-2)]\end{aligned} \tag{3}$$

If a non-coherent correlation function is used as an input to the nonlinear TK operator, it can then signal the presence of a multipath component more clearly than looking directly at the correlation function. At least three correlation values (in-prompt, early and very early) are required to compute TK operation. But usually, TK based delay estimation utilizes a higher number of correlators (for example, 193 correlators were used in the simulations reported in

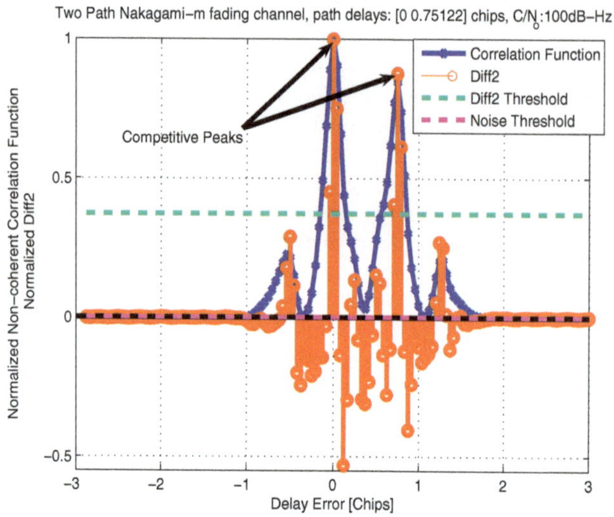

Fig. 6. Generation of competitive peaks for PT(Diff2) technique.

Bhuiyan & Lohan (2010)) and is sensitive to the noise dependent threshold choice. Firstly, it computes the noise variance, which is then used to compute an adaptive threshold. The peaks which are above the adaptive threshold are considered as competitive peaks. Among all the competitive peaks, TK selects the delay associated to that competitive peak which has the closest delay difference from the previous delay estimate.

5.5 Reduced Search Space Maximum Likelihood delay estimator

A Reduced Search Space Maximum Likelihood (RSSML) delay estimator is another good example of maximum likelihood based approach, which is capable of mitigating the multipath effects reasonably well at the expense of increased complexity. The RSSML, proposed by the Author in Bhuiyan et al. (2009) and then further enhanced in Bhuiyan & Lohan (2010), attempts to compensate the multipath error contribution by performing a nonlinear curve fit on the input correlation function which finds a perfect match from a set of ideal reference correlation functions with certain amplitude(s), phase(s) and delay(s) of the multipath signal. Conceptually, a conventional spread spectrum receiver does the same thing, but for only one signal (i.e., the LOS signal). With the presence of multipath signal, RSSML tries to separate the LOS component from the combined signal by estimating all the signal parameters in a maximum likelihood sense, which consequently achieves the best curve fit on the received input correlation function. As mentioned in Bhuiyan & Lohan (2010), it also incorporates a threshold-based peak detection method, which eventually reduces the code delay search space significantly. However, the downfall of RSSML is the memory requirement which it uses to store the reference correlation functions.

In a multi-correlator based structure, the estimated LOS delay, theoretically, can be anywhere within the code delay window range of $\pm\tau_W$ chips, though in practice, it is quite likely to have a delay error around the previous delay estimate. The code delay window range essentially depends on the number of correlators (i.e., M) and the spacing between the correlators (i.e.,

Δ) according to the following equation:

$$\tau_W = \pm \frac{(M-1)}{2}\Delta \qquad (4)$$

For example, in Bhuiyan & Lohan (2010), 193 correlators were used with a correlator spacing of 0.0208 chips, resulting in a code delay window range of ±2 chips with respect to prompt correlator. Therefore, the LOS delay estimate can be anywhere within this ±2 chips window range. The ideal non-coherent reference correlation functions are generated for up to L_{max} paths only for the middle delay index (i.e., $(\frac{M+1}{2})$-th delay index; for $M = 193$, the middle delay index is 97). These ideal correlation functions for the middle delay index are generated off-line and saved in a look-up table in memory. In real-time, RSSML reads the correlation values from the look-up table, translates the ideal reference correlation functions at the middle delay index to the corresponding candidate delay index within the code delay window, and then computes the Minimum Mean Square Error (MMSE) for that specific delay candidate. Instead of considering all possible LOS delays within a predefined code delay window as delay candidates, the search space is first reduced to some competitive peaks which are generated based on the computed noise thresholds as explained in Bhuiyan & Lohan (2010). This will eventually reduce the processing time required to compute the MMSE (i.e., MMSE needs to be computed only for the reduced search space). An example is shown in Fig. 7, where RSSML estimates a best-fitted non-coherent correlation function at a cost of $3.6 \cdot 10^{-4}$ MMSE in a two-path Rayleigh channel with path delays [0 0.35] chips, path powers [0 -2] dB at a C/N_0 of 50 dB-Hz.

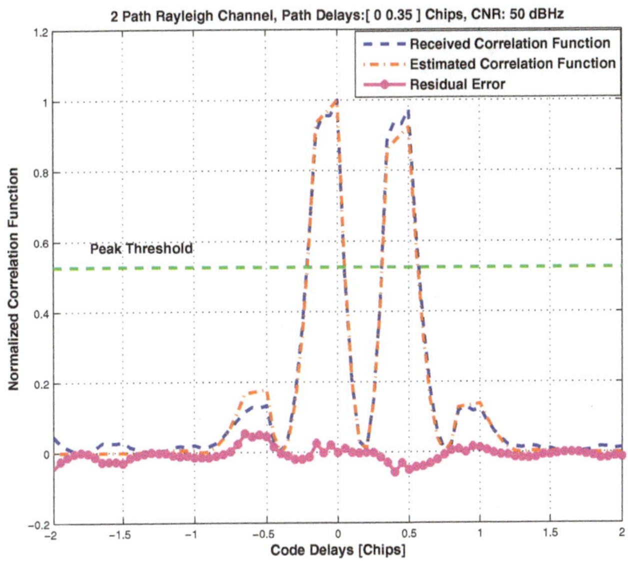

Fig. 7. Estimated and received non-coherent correlation functions in two path Rayleigh channel, path delay: [0 0.35] chips, path power: [0 -2] dB, C/N_0: 50 dB-Hz.

5.6 Slope-based multipath estimator

A simple slope-based multipath mitigation technique, named as Slope-Based Multipath Estimator (SBME), in Bhuiyan, Lohan & Renfors (2010). Unlike the multipath mitigation techniques discussed above, SBME does not require a huge number of correlators, rather it only requires an additional correlator (as compared to a conventional DLL) at the late side of the correlation function. In fact, SBME is used in conjunction with a nEML tracking loop Bhuiyan, Lohan & Renfors (2010). It first derives a multipath estimation equation by utilizing the correlation shape of the ideal normalized correlation function, which is then used to compensate for the multipath bias of a nEML tracking loop. The derivation of the multipath estimation equation for BPSK modulated GPS L1 C/A signal can be found in Bhuiyan, Lohan & Renfors (2010). It is reported in Bhuiyan, Lohan & Renfors (2010) that SBME has superior multipath mitigation performance than nEML in closely spaced two path channel model.

5.7 C/N_0 -based two-stage delay tracker

A C/N_0 based two-stage delay tracker is a combination of two individual tracking techniques, namely nEML and HRC (or MGD). The tracking has been divided into two stages based on the tracking duration and the received signal strength (i.e., C/N_0). At the first stage of tracking (for about 0.1 seconds or so), the two-stage delay tracker always starts with a nEML tracking loop, since it begins to track the signal with a coarsely estimated code delay as obtained from the acquisition stage. And, at the second or final stage of tracking (i.e., when the DLL tracking error is around zero), the two-stage delay tracker switches its DLL discriminator from nEML to HRC (or MGD), since HRC (or MGD) has better multipath mitigation capability as compared to nEML. While doing so, it has to be ensured that the estimated C/N_0 level meets a certain threshold set by the two-stage tracker. This is because of the fact that HRC (or MGD) involves one (or two in case of MGD) more discrimination than NEML, which makes its discriminator output much noisier than nEML. It has been empirically found that a C/N_0 threshold of 35 dB-Hz can be a good choice, as mentioned in Bhuiyan, Zhang & Lohan (2010). Therefore, at this fine tracking stage, the two-stage delay tracker switches from nEML to HRC (or MGD) only when the estimated C/N_0 meets the above criteria (i.e., C/N_0 threshold is greater than 35 dB-Hz).

An example non-coherent S-curve is shown in Fig. 8 for CBOC(-) modulated signal in single path static channel Bhuiyan, Zhang & Lohan (2010). The nearest ambiguous zero crossings for HRC (around ± 0.16 chips) is much closer as compared to that of nEML (around ± 0.54 chips) in this particular case. This indicates the fact that the probability of locking to any of the side peaks is much higher for HRC than that of nEML, especially in the initial stage of tracking when the code delay may not necessarily be near the main peak of the correlation function. This is the main motivation to choose a nEML tracking at the initial stage for a specific time duration (for example, 0.1 seconds or so). This will eventually pull the delay tracking error around zero after the initial stage.

5.8 TK operator combined with a nEML DLL

A combined simplified approach with TK operator and a nEML DLL was implemented in Bhuiyan & Lohan (2010), in order to justify the feasibility of having a nEML discrimination after the TK operation on the non-coherent correlation function. In this combined approach, TK operator is first applied to the non-coherent correlation function, and then nEML

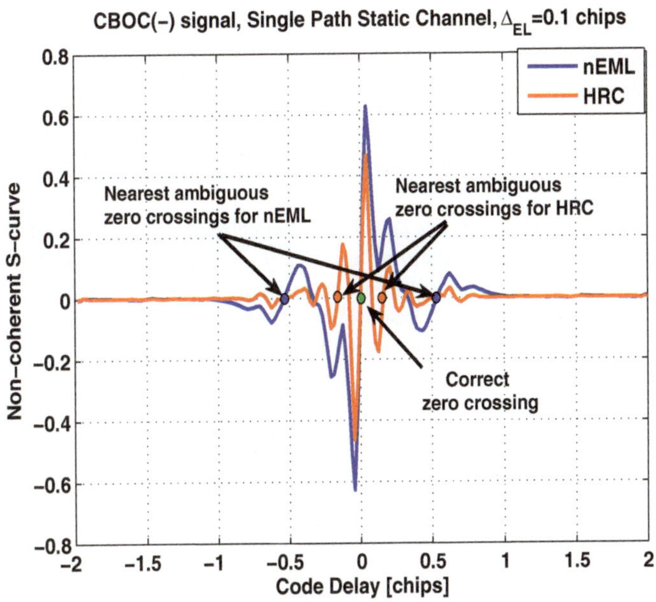

Fig. 8. A non-coherent S-curve for CBOC(-) modulated single path static channel. Bhuiyan, Zhang & Lohan (2010)

discrimination is applied to the TK output. The motivation for this combined approach comes from the fact that, when we apply TK operation to the non-coherent correlation function, it usually makes the main lobe of the non-coherent correlation function (after TK operation) much steeper than before. This eventually reduces the effect of multipath in case of TK based nEML (TK+nEML) as compared to nEML.

6. Performance analysis

The multipath performance of some of the discussed techniques are evaluated in different simulation models, i.e., the semi-analytical simulation in a static channel model and the Matlab-based simulation in a multipath fading channel model. Each simulation model is briefly described first before presenting the results.

6.1 Semi-analytical simulation

The most typical way to evaluate the performance of a multipath mitigation technique is via Multipath Error Envelopes (MEE). Typically, two paths, either in-phase or out-of-phase, are assumed to be present, and the multipath errors are computed for multipath delays up to 1.2 chips at maximum, since the multipath errors become less significant after that. The upper multipath error envelope can be obtained when the paths are in-phase and the lower multipath error envelope when the paths are out-of-phase (i.e., 180° phase difference). In MEE analysis, several simplifying assumptions are usually made in order to distinguish the performance degradation caused by the multipath errors only. Such assumptions include zero Additive-White-Gaussian-Noise (AWGN), ideal infinite-length PRN codes, and zero residual

Doppler. Under these assumptions, the correlation $\mathcal{R}_{rx}(\tau)$ between the reference code of the modulation type MOD (for example, BPSK or CBOC(-)) and the received MOD-modulated signal via an L-path channel can be written as:

$$\mathcal{R}_{rx}(\tau) = \sum_{l=1}^{L} \alpha_l e^{j\theta_l} \mathcal{R}_{\text{MOD}}(\tau - \tau_l) \tag{5}$$

where $\alpha_l, \theta_l, \tau_l$ are the amplitude, phase, and delay, respectively, of the l-th path; and $\mathcal{R}_{\text{MOD}}(\tau)$ is the auto-correlation function of a signal with the modulation type MOD. The analytical expressions for MEEs become complicated in the presence of more than two paths due to the complexity of channel interactions. Therefore, an alternative Monte-Carlo simulation-based approach is presented here, in accordance with Bhuiyan & Lohan (2010), for multipath error analysis in more than one path scenarios (i.e, for $L \geq 2$). First, a sufficient number of random realizations, N_{random} are generated (i.e., in the simulation, we choose N_{random} equals to 1000), and then we look at the absolute mean error for each path delay over N_{random} points. The objective is to analyze the multipath performance of some of the proposed advanced techniques along with some conventional DLLs in the presence of more than two channel paths, which may occur in urban or indoor scenarios.

The following assumptions are made while running the simulation for generating the RAE curves Hein et al. (2006). According to Hein et al. (2006), RAE is computed from the area enclosed within the multipath error and averaged over the range of the multipath delays from zero to the plotted delay values. In the simulation, the channel follows a decaying Power Delay Profile (PDP), which can be expressed by the equation:

$$\alpha_l = \alpha_l \exp^{-\mu(\tau_l - \tau_1)}, \tag{6}$$

where $(\tau_l - \tau_1) \neq 0$ for $l > 1$, μ is the PDP coefficient (assumed to be uniformly distributed in the interval $[0.05; 0.2]$, when the path delays are expressed in samples). The channel path phases θ_l are uniformly distributed in the interval $[0; 2\pi]$ and the number of channel paths L is uniformly distributed between 2 and L_{max}, where L_{max} is set to 5 in the simulation. A constant successive path spacing x_{ct} is chosen in the range $[0; 1.167]$ chips with a step of 0.0417 chips (which defines the multipath delay axis in the RAE curves). It is worth to mention here that the number of paths is reduced to only one LOS path when $x_{ct} = 0$. The successive path delays can be found using the formula $\tau_l = l x_{ct}$ in chips. Therefore, for each channel realization (which is a combination of amplitudes $\vec{\alpha} = \alpha_1, \ldots, \alpha_L$, phases $\vec{\theta} = \theta_1, \ldots, \theta_L$, fixed path spacings, and the number of channel paths L), a certain LOS delay is estimated $\widehat{\tau}_1(\vec{\alpha}, \vec{\theta}, L)$ from the zero crossing of the discriminator function (i.e., $D(\tau) = 0$), when searched in the linear range of $D(\tau)$ in case of conventional DLLs, or directly from the auto-correlation function in case of advanced multi-correlator based techniques. The estimation error due to multipath is $\hat{\tau}_1(\vec{\alpha}, \vec{\theta}, L) - \tau_1$, where τ_1 is the true LOS path delay. The RAE curves are generated in accordance with Hein et al. (2006). RAE is actually computed from the area enclosed within the multipath error and averaged over the range of the multipath delays from zero to the plotted delay values. Therefore, in order to generate the RAE curves, the Absolute Mean Error (AME) is computed for all N_{random} random points via eqn. 7:

$$\text{AME}(x_{ct}) = \text{mean}\left(\left|\widehat{\tau}_1(\vec{\alpha}, \vec{\theta}, L) - \tau_1\right|\right), \tag{7}$$

where $\text{AME}(x_{ct})$ is the mean of the absolute multipath error for the successive path delay x_{ct}. Now, the running average error for each particular delay in the range [0;1.167] chips can be computed as follows:

$$\text{RAE}(x_{ct}) = \frac{\sum_{i=1}^{i} \text{AME}(x_{ct})}{i}, \tag{8}$$

where i is the successive path delay index, and $\text{RAE}(x_{ct})$ is the RAE for the successive path delay x_{ct}.

The RAE curves for CBOC(-) modulated Galileo E1C signal (i.e., pilot channel) is shown in Fig. 9. It is obvious from Fig. 9 that the RSSML and PT(Diff2) show the best performance in terms of RAE as compared to other techniques in this noise-free two to five paths static channel model. Among other techniques, TK+nEML showed very good performance followed by SBME, HRC and nEML. The SBME coefficient and the late slope at very late spacing of 0.0833 chips were determined according to Bhuiyan, Lohan & Renfors (2010) for a 24.552 MHz front-end bandwidth (double-sided). For the above configuration, the SBME coefficient is 0.007 and the late slope is −4.5.

It is worth to mention here that the RAE analysis is quite theoretical from two perspectives: firstly, the delay estimation is a one-shot estimate, and does not really include any tracking loop in the process; and secondly, the analysis is usually carried out with an ideal noise free assumption. These facts probably explain the reason why an algorithm which performs

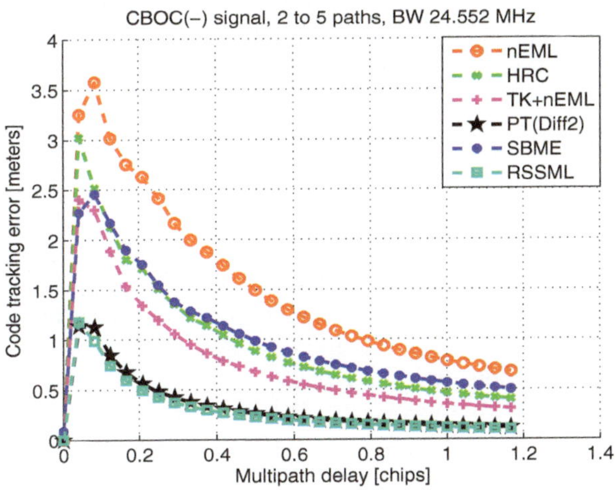

Fig. 9. Running average error curves for CBOC(-) modulated Galileo E1C signal.

very well with respect to RAE may not necessarily provide the same performance in a more realistic closed loop fading channel model, especially in the presence of more than two channel paths. However, MEE or RAE analysis has been widely used by the research community as an important tool for analyzing the multipath performance due to simpler implementation, and also due to the fact that it is hard to isolate multipath from other GNSS error sources in real life.

6.2 Matlab-based simulation

Simulation has been carried out in closely spaced multipath environments for CBOC(-) modulated Galileo E1C (i.e., pilot channel) signal for a 24.552 MHz front-end bandwidth. The simulation profile is summarized in Table 1. Rayleigh fading channel model is used in the simulation, where the number of channel paths follows a uniform distribution between two and five. The successive path separation is random between 0.02 and 0.35 chips. The channel paths are assumed to obey a decaying PDP according to Eqn. 6, where $(\tau_l - \tau_1) \neq 0$ for $l > 1$, and the PDP coefficient $\mu = 0.1$. The received signal is sampled such that there are 48 samples per chip. The received signal duration is 800 milliseconds (ms) or 0.8 seconds for each particular C/N_0 level. The tracking errors are computed after each $N_c * N_{nc}$ ms (in this case, $N_c * N_{nc} = 20$ ms) interval. In the final statistics, the first 600 ms are ignored in order to remove the initial error bias that may come from the delay difference between the received signal and the locally generated reference code. Therefore, for the above configuration (i.e., code loop filter parameters and the first path delay of 0.2 chips), the left-over tracking errors after 600 ms are mostly due to the effect of multipath. The simulation has been carried out for 100 random realizations, which give a total of 10*100=1000 statistical points, for each C/N_0 level. The RMSE of the delay estimates are plotted in meters, by using the relationship:

$$\mathrm{RMSE}_m = \mathrm{RMSE}_{\mathrm{chips}} c T_c \tag{9}$$

where c is the speed of light, T_c is the chip duration, and $\mathrm{RMSE}_{\mathrm{chips}}$ is the RMSE in chips.

Parameter	Value
Channel model	Rayleigh fading channel
Number of paths	between 2 to 5
Path Power	Decaying PDP with $\mu = 0.1$
Path Spacing	Random between 0.02 and 0.35 chips
Path Phase	Random between 0 and 2π
Samples per Chip, N_s	48
E-L Spacing, Δ_{EL}	0.0833 chips
Number of Correlators, M	193[1]
Double-sided Bandwidth, BW	24.552 MHz
Filter Type	Finite Impulse Response (FIR)
Filter Order	6
Coherent Integration, N_c	20 ms
Non-coherent Integration, N_{nc}	1 block
Initial Delay Error	$\pm$ 0.1 chips
First Path Delay	0.2 chips
Code Tracking Loop Bandwidth	2 Hz
Code Tracking Loop Order	1st order

Table 1. Simulation profile description

RMSE vs. C/N_0 plot for the given multipath channel profile is shown in Fig. 10. It can be seen from Fig. 10 that the proposed RSSML clearly achieves the best multipath mitigation

[1] Not all the correlators are used in all the tracking algorithms. For example, nEML only requires 3 correlators.

performance in this two to five paths closely spaced multipath channel. Among other techniques, PT(Diff2) and HRC have better performance only in good C/N_0 (around 40 dB-Hz and onwards). It can also be observed that the proposed SBME and TK+nEML do not bring any advantage in the tracking performance as compared to nEML in this multipath fading channel model. Here also, the SBME coefficient and the late slope were set to 0.007 and −4.5, respectively.

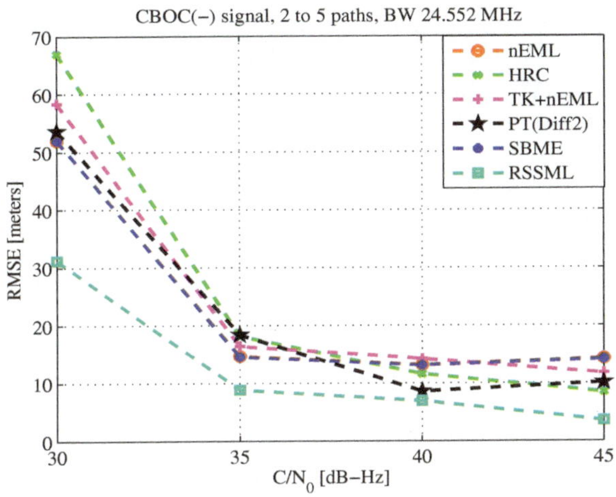

Fig. 10. RMSE vs. C/N_0 plot for CBOC(-) modulated Galileo E1C signal in two to five path Rayleigh fading channel.

7. Conclusion

This chapter addressed the challenges encountered by a GNSS signal due to multipath propagation. A wide range of correlation-based multipath mitigation techniques were discussed and the performance of some of these techniques were evaluated in terms of running average error and root-mean-square error. Among the analyzed multipath mitigation techniques, RSSML, in general, achieved the best multipath mitigation performance in moderate-to-high C/N_0 scenarios (for example, 30 dB-Hz and onwards). The other techniques, such as PT(Diff2) and HRC showed good multipath mitigation performance only in high C/N_0 scenarios (for example, 40 dB-Hz and onwards). The other new technique SBME offered slightly better multipath mitigation performance to the well-known nEML DLL at the cost of an additional correlator. However, as the GNSS research area is fast evolving with many potential applications, it remains a challenging topic for future research to investigate the feasibility of these multipath mitigation techniques with the multitude of signal modulations, spreading codes, and spectrum placements that are (or are to be) proposed.

8. References

Baltersee, J., Fock, G. & Schulz-Rittich, P. (2001). Adaptive code-tracking receiver for direct-sequence Code Division Multiple Access (CDMA) communications over multipath fading channels and method for signal processing in a RAKE receiver, US Patent Application Publication, US 2001/0014114 A1 (Lucent Technologies).

Bello, P. A. & Fante, R. L. (2005). Code tracking performance for novel unambiguous M-code time discriminators, *Proc. of ION NTM*, San Diego, USA.

Betz, J. W. & Kolodziejski, K. R. (2000). Extended theory of early-late code tracking for a bandlimited GPS receiver, *Navigation: Journal of the Institute of Navigation* 47(3): 211–226.

Bhuiyan, M. Z. H. (2006). *Analyzing code tracking algorithms for Galileo Open Service signal*, Master's thesis, Tampere University of Technology.

Bhuiyan, M. Z. H. (2011). *Analysis of Multipath Mitigation Techniques for Satellite-based Positioning Applications*, PhD thesis, Tampere University of Technology.

Bhuiyan, M. Z. H. & Lohan, E. S. (2010). Advanced multipath mitigation techniques for satellite-based positioning applications, *International Journal of Navigation and Observation, DOI:10.1155/2010/412393* .

Bhuiyan, M. Z. H., Lohan, E. S. & Renfors, M. (2008). Code tracking algorithms for mitigating multipath effects in fading channels for satellite-based positioning, *EURASIP Journal on Advances in Signal Processing, DOI: 10.1155/2008/863629* .

Bhuiyan, M. Z. H., Lohan, E. S. & Renfors, M. (2009). A reduced search space maximum likelihood delay estimator for mitigating multipath effects in satellite-based positioning, *Proc. of the 13th International Association of Institute of Navigation*.

Bhuiyan, M. Z. H., Lohan, E. S. & Renfors, M. (2010). A slope-based multipath estimation technique for mitigating short-delay multipath in GNSS receivers, *Proc. of ISCAS in Special Session on Circuits, Systems and Algorithms for Next Generation GNSS*.

Bhuiyan, M. Z. H., Zhang, J. & Lohan, E. S. (2010). Enhanced delay tracking performance of a C/N0-based two-stage tracker for GNSS receivers, *Proc. of the European Navigation Conference on Global Navigation Satellite Systems, ENC GNSS 2010*.

Bischoff, R., Häb-Umbach, R., Schulz, W. & Heinrichs, G. (2002). Employment of a multipath receiver structure in a combined Galileo/UMTS receiver, *Proc. of IEEE VTC Spring*, Vol. 4, pp. 1844–1848.

Braasch, M. S. (2001). Performance comparison of multipath mitigating receiver architectures, *Proc. of IEEE Aerospace Conference*, Big Sky, Mont, USA, pp. 1309–1315.

Chen, K. C. & Davisson, L. D. (1994). Analysis of a SCCL as a PN code tracking loop, *IEEE Transactions on Communications* 42(11): 2942–2946.

Dierendonck, A. J. V. & Braasch, M. S. (1997). Evaluation of GNSS receiver correlation processing techniques for multipath and noise mitigation, *Proc. of ION NTM*, CA, USA, pp. 207–215.

Dierendonck, A. J. V., Fenton, P. C. & T., F. (1992). Theory and performance of narrow correlator spacing in a GPS receiver, *Journal of The Institute of Navigation* 39(3).

Fante, R. (2003). Unambiguous tracker for GPS Binary-Offset Carrier signals, *Proc. of ION NTM*, Albuquerque, USA.

Fante, R. (2004). Unambiguous First-Order Tracking Loop M-Code, MITRE Technical Report. MTR 94B0000040.

Fenton, P. C. (1995). Pseudorandom noise ranging receiver which compensates for multipath distortion by making use of multiple correlator time delay spacing, NovAtel Patent, US 5 414 729.

Fenton, P. C. & Dierendonck, A. J. V. (1996). Pseudorandom noise ranging receiver which compensates for multipath distortion by dynamically adjusting the time delay spacing between early and late correlators, NovAtel Patent, US 5 495 499.

Fenton, P. C. & Jones, J. (2005). The theory and performance of NovAtel Inc.'s Vision Correlator, *Proc. of ION GNSS*, Long Beach, USA, pp. 2178–2186.

Fine, P. & Wilson, W. (1999). Tracking algorithms for GPS offset carrier signals, *Proc. of ION NTM*, San Diego, USA.

Fock, G., Baltersee, J., Schulz-Rittich, P. & Meyr, H. (2001). Channel tracking for RAKE receivers in closely spaced multipath environments, *IEEE Journal on Sel. Areas in Comm.* 19(12): 2420–2431.

Garin, L. & Rousseau, J. M. (1997). Enhanced Strobe Correlator multipath rejection for code and carrier, *Proc. of ION GPS*, Kansas City, USA, pp. 559–568.

Gleason, S. & Egziabher, D. G. (2009). *GNSS Applications and Methods*, Artech House.

Granados, G. S. & Rubio, J. A. F. (2000). Time-delay estimation of the line-of-sight signal in a multipath environment, *Proc. of the 10th European Signal Processing Conference (EUSIPCO)*, Tampere, Finland.

Granados, G. S., Rubio, J. A. F. & Prades, C. F. (2005). ML estimator and hybrid beamformer for multipath and interference mitigation in GNSS receivers, *IEEE Transactions on Signal Processing* 53(3): 1194–1208.

Hamila, R. (2002). *Synchronization and multipath delay estimation algorithms for digital receivers*, PhD thesis, Tampere University of Technology.

Hamila, R., Lohan, E. & Renfors, M. (2003). Subchip multipath delay estimation for downlink WCDMA systems based on Teager-Kaiser operator, *IEEE Communications Letters* 7(1): 125–128.

Hein, G. W., Avila-Rodriguez, J.-A., Wallner, S., Pratt, A. R., Owen, J., Issler, J. L., Betz, J. W., Hegarty, C. J., Lenahan, L. S., Rushanan, J. J., Kraay, A. L. & Stansell, T. A. (2006). MBOC: The new optimized spreading modulation recommended for GALILEO L1 OS and GPS L1C, *Proc. of Position, Location and Navigation Symposium*, pp. 883–892.

Hurskainen, H., Lohan, E. S., Hu, X., Raasakka, J. & Nurmi, J. (2008). Multiple gate delay tracking structures for GNSS signals and their evaluation with Simulink, SystemC, and VHDL, *International Journal of Navigation and Observation, DOI:10.1155/2008/785695* .

Irsigler, M. & Eissfeller, B. (2003). Comparison of multipath mitigation techniques with consideration of future signal structures, *Proc. of ION GNSS*, OR, USA, pp. 2584–2592.

Jie, Z. (2010). *Delay tracker for Galileo CBOC modulated signals and their Simulink-based implementations*, Master's thesis, Tampere University of Technology.

Jones, J., Fenton, P. & Smith, B. (2004). Theory and Performance of the Pulse Aperture Correlator. Tech. Rep., NovAtel, Calgary, Alberta, Canada.

Kaplan, E. D. & Hegarty, C. J. (2006). *Understanding GPS: Principles and Applications*, second edn, Artech House.

Laxton, M. (1996). *Analysis and simulation of a new code tracking loop for GPS multipath mitigation*, Master's thesis, Air Force Institute of Technology.

Lee, C., Yoo, S., Yoon, S. & Kim, S. Y. (2006). A novel multipath mitigation scheme based on slope differential of correlator output for Galileo systems, *Proceedings of 8th International Conference on Advanced Communication Technology*.

Lohan, E. S. (2003). *Multipath delay estimators for fading channels with applications in CDMA receivers and mobile positioning*, PhD thesis, Tampere University of Technology.

Lohan, E. S., Lakhzouri, A. & Renfors, M. (2006). Feedforward delay estimators in adverse multipath propagation for Galileo and modernized GPS signals, *EURASIP Journal on Advances in Signal Processing, Article ID 50971* .

McGraw, G. A. & Braasch, M. S. (1999). GNSS multipath mitigation using Gated and High Resolution Correlator concepts, *Proc. of the National Technical Meeting of the Satellite Division of the Insitute of Navigation*, San Diego, USA.

Nee, R. D. J. V. (1992). The multipath estimating delay lock loop, *Proc. of IEEE Second International Symposium on Spread Spectrum Techniques and Applications*, Yokohama, Japan, pp. 39–42.

Nee, R. D. J. V., Sierveld, J., Fenton, P. C. & Townsend, B. R. (1994). The multipath estimating delay lock loop: Approaching theoretical accuracy limits, *Proc. of IEEE Position Location and Navigation Symposium*, Vol. 1, pp. 246–251.

Phelts, R. E. & Enge, P. (2000a). The multipath invariance approach for code multipath mitigation, *Proc. of ION GPS*, UT, USA, pp. 2376–2384.

Phelts, R. E. & Enge, P. (2000b). Multipath mitigation for narrowband receivers, *Proc. of the IEEE PLANS 2000*, CA, USA, pp. 30–36.

Simon, M. K., Omura, J. K., Scholtz, R. A. & Levitt, B. K. (1994). *Spread Spectrum Communication Handbook*, revised edition edn, McGraw-Hill Inc, New York.

Sleewaegen, J. M. & Boon, F. (2001). Mitigating short-delay multipath: a promising new technique, *Proc. of ION GPS*, UT, USA, pp. 204–213.

Townsend, B. R. & Fenton, P. C. (1994). A practical approach to the reduction of pseudorange multipath errors in a L1 GPS receiver, *Proc. of ION GPS*, UT, USA, pp. 143–148.

Townsend, B. R., Nee, R. D. J. V., Fenton, P. C. & Dierendonck, A. J. V. (1995). Performance evaluation of the multipath estimating delay lock loop, *Proc. of ION NTM*, Anaheim, USA.

Weill, L. R. (1997). A GPS multipath mitigation by means of correlator reference waveform design, *Proc. of the of ION NTM*, CA, USA, pp. 197–206.

Weill, L. R. (2002). Multipath mitigation using modernized GPS signals: How good can it get?, *Proc. of the of ION GPS*, CA, USA, pp. 493–505.

Weill, L. R. (2003). Multipath mitigation - how good can it get with new signals?, *GPS World* 16(6): 106–113.

Permissions

The contributors of this book come from diverse backgrounds, making this book a truly international effort. This book will bring forth new frontiers with its revolutionizing research information and detailed analysis of the nascent developments around the world.

We would like to thank Prof. Shuanggen Jin, for lending his expertise to make the book truly unique. He has played a crucial role in the development of this book. Without his invaluable contribution this book wouldn't have been possible. He has made vital efforts to compile up to date information on the varied aspects of this subject to make this book a valuable addition to the collection of many professionals and students.

This book was conceptualized with the vision of imparting up-to-date information and advanced data in this field. To ensure the same, a matchless editorial board was set up. Every individual on the board went through rigorous rounds of assessment to prove their worth. After which they invested a large part of their time researching and compiling the most relevant data for our readers. Conferences and sessions were held from time to time between the editorial board and the contributing authors to present the data in the most comprehensible form. The editorial team has worked tirelessly to provide valuable and valid information to help people across the globe.

Every chapter published in this book has been scrutinized by our experts. Their significance has been extensively debated. The topics covered herein carry significant findings which will fuel the growth of the discipline. They may even be implemented as practical applications or may be referred to as a beginning point for another development. Chapters in this book were first published by InTech; hereby published with permission under the Creative Commons Attribution License or equivalent.

The editorial board has been involved in producing this book since its inception. They have spent rigorous hours researching and exploring the diverse topics which have resulted in the successful publishing of this book. They have passed on their knowledge of decades through this book. To expedite this challenging task, the publisher supported the team at every step. A small team of assistant editors was also appointed to further simplify the editing procedure and attain best results for the readers.

Our editorial team has been hand-picked from every corner of the world. Their multi-ethnicity adds dynamic inputs to the discussions which result in innovative outcomes. These outcomes are then further discussed with the researchers and contributors who give their valuable feedback and opinion regarding the same. The feedback is then collaborated with the researches and they are edited in a comprehensive manner to aid the understanding of the subject.

Apart from the editorial board, the designing team has also invested a significant amount of their time in understanding the subject and creating the most relevant covers. They scrutinized every image to scout for the most suitable representation of the subject and create an appropriate cover for the book.

The publishing team has been involved in this book since its early stages. They were actively engaged in every process, be it collecting the data, connecting with the contributors or procuring relevant information. The team has been an ardent support to the editorial, designing and production team. Their endless efforts to recruit the best for this project, has resulted in the accomplishment of this book. They are a veteran in the field of academics and their pool of knowledge is as vast as their experience in printing. Their expertise and guidance has proved useful at every step. Their uncompromising quality standards have made this book an exceptional effort. Their encouragement from time to time has been an inspiration for everyone.

The publisher and the editorial board hope that this book will prove to be a valuable piece of knowledge for researchers, students, practitioners and scholars across the globe.

List of Contributors

Fabio Dovis
Politecnico di Torino, Italy

Tung Hai Ta
Hanoi University of Science and Technology, Vietnam

Nagaraj C. Shivaramaiah and Andrew G. Dempster
The University of New South Wales, Australia

Mario Calamia, Giovanni Dore and Alessandro Mori
University of Florence, Italy

Zheng Yao
Tsinghua University, China

Bihter Erol and Serdar Erol
Istanbul Technical University, Civil Engineering Faculty, Department of Geomatics Engineering, Turkey

Marco Pini and Gianluca Falco
Istituto Superiore Mario Boella, Italy

Letizia Lo Presti
Politecnico di Torino, Italy

Paolo Dabove, Mattia De Agostino and Ambrogio Manzino
Politecnico di Torino, Italy

Ahmed El-Mowafy
Curtin University, Australia

Carola A. Blazquez
Universidad Andres Bello, Department of Engineering Science, Chile

Eric North
Canadian Forces Aerospace and Telecommunications Engineering Support Squadron, Canada

Jacques Georgy
Trusted Positioning Inc., Canada

Umar Iqbal
Electrical and Computer Engineering Department, Queen's University, Canada

Mohammed Tarbochi
Electrical and Computer Engineering Department, Royal Military College, Canada

Aboelmagd Noureldin
Electrical and Computer Engineering Department, Royal Military College/Queen's University, Canada

Mohammed Ziaur Rahman
University of Malaya, Malaysia

Masahiko Nagai
Asian Institute of Technology, Thailand

Nabil Drawil and Otman Basir
Department of Electrical and Computer Engineering, University of Waterloo, Canada

Nel Samama
Institut Telecom / Telecom SudParis, France

M. Mainul Hoque and Norbert Jakowski
German Aerospace Center (DLR), Institute of Communications and Navigation, Neustrelitz, Germany

Shuanggen Jin
Shanghai Astronomical Observatory, Chinese Academy of Sciences, China

Mohammad Zahidul H. Bhuiyan and Elena Simona Lohan
Department of Communications Engineering, Tampere University of Technology, Finland